Arthur T. Winfree

The Geometry
of Biological Time

With 290 Illustrations

Springer-Verlag
Berlin Heidelberg NewYork London Paris
Tokyo Hong Kong Barcelona

Arthur T. Winfree
Department of Ecology and Evolutionary Biology
326 BSW, University of Arizona
Tucson AZ 85721, USA

AMS Subject Classification (1980): 92A05

This material is based upon work supported by the National Science Foundation under Grants No. GB 16513, GB 37947, BMS 73-06888A 01, CHE 77-24649, and by National Institutes of Health Research Career Development A ward 5 K04 GM 70660.
Any opinions, findings, and conclusions or recommendations expressed in this publication are those of the author and do not necessarily reflect the views of the National Science Foundation or the National Institutes of Health.

ISBN 3-540-52528-9 Springer-Verlag Berlin Heidelberg NewYork
ISBN 0-387-52528-9 Springer-Verlag NewYork Berlin Heidelberg

Corrected printing of Biomathematics Vol. 8, 1980 with the
ISBN 3-540-09373-7 Springer-Verlag Berlin Heidelberg NewYork
ISBN 0-387-09373-7 Springer-Verlag NewYork Berlin Heidelberg

Library of Congress Cataloging-in-Publication Data
Winfree, Arthur T.
The geometry of biological time / Arthur T. Winfree.
"Springer study edition."
Includes bibliographical references and indexes.
ISBN 3-540-52528-9 (Springer-Verlag Berlin Heidelberg NewYork :
alk. paper). -- ISBN 0-387-52528-9 (Springer-Verlag NewYork
Heidelberg Berlin : alk. paper)
1. Biological rhythms -- Mathematics. I. Title
QH527.W55 1990

The use of registered names, trademarks, etc. in this publiclation does not imply, even in the absence of a specific statement, that such names are exempt from the relevant protective laws and regulations and therefore free for general use.

Offsetprinting: Mercedes-Druck, Berlin; Binding: Lüderitz & Bauer, Berlin
2141/3020-543210 – Printed on acid-free paper

I dedicate this book to my parents, Dorothy and Van, who first gave me tools.

And I dedicate this book to those readers who, expecting wonders to follow so grand a title as it flaunts, may feel cheated by its actual content. I will be delighted if you take this beginning as a serious challenge.

Preface

As I review these pages, the last of them written in Summer 1978, some retrospective thoughts come to mind which put the whole business into better perspective for me and might aid the prospective reader in choosing how to approach this volume.

The most conspicuous thought in my mind at present is the diversity of wholly independent explorations that came upon phase singularities, in one guise or another, during the past decade. My efforts to gather the published literature during the last phases of actually writing a whole book about them were almost equally divided between libraries of Biology, Chemistry, Engineering, Mathematics, Medicine, and Physics.

A lot of what I call "gathering" was done somewhat in anticipation in the form of conjecture, query, and prediction based on analogy between developments in different fields. The consequence throughout 1979 was that our long-suffering publisher repeatedly had to replace such material by citation of unexpected flurries of papers giving substantive demonstration. I trust that the authors of these many excellent reports, and especially of those I only found too late, will forgive the brevity of allusion I felt compelled to observe in these substitutions. A residue of loose ends is largely collected in the index under "QUERIES."

It is clear to me already that the materials I began to gather several years ago represented only the first flickering of what turns out to be a substantial conflagration. According, I took a liberty with the reference list. You will notice that about 30% of its entries are not to be found in the page index of publications cited. That is because they are not explicitly cited. Readers who like to browse will easily find these extra papers: they lie among papers on similar topics by much-cited authors. They lead in the directions of significant expansion.

And what comes next? Well, one never knows; that is half the fun of doing science. But one inevitable development is especially conspicuous by its absence here. In fact, the original 30 chapters came down to 23 in purging it for a later volume. You will find here almost no mention of rhythmic driving of biological dynamics. Plainly that must contain the essence of any practical application, be it in hormonal gating of cell division, in cardiac or gastric pacemaking, or in agricultural photoperiodism. Many surprises await discovery in connection with alternative modes of entrainment, the consequences of synchronization, and evolution in periodic environments. This topic is the natural successor to the present volume on autonomous periodicity. It is now undergoing rapid development, mainly at the hands of neurobiologists, mathematicians, and engineers, and will be riper for harvest a few years hence.

It has been my good fortune to visit lively investigators in many laboratories. I have been stimulated by early exposure to their discoveries (which fill out so much of the following chapters), and their critical attention to my own seminars has refined into presentable form most of what is presented here. But I have never found an opportunity to teach on these subjects, as you can see by the lack of problem sets in this presentation. I suspect that substantial improvements of content and clarity as well as significant new directions would inevitably emerge through contact with students who are eager and ready to study living systems in a mathematical spirit. That is a hard clientele to locate; I could use some help.

I wish you good reading and wish you to send me marginal notations to collect on my copy. Who knows? There *might* even be a second edition.

April 1980   Arthur Winfree

Acknowledgments

I wrote this book but its authors live all over the world. In a broad sense the list of authors is the bibliography. But in a more precise sense this gathering of facts and ideas was shaped by about twenty-five individuals whose conversation and correspondence molded every topic represented here. Many others also will recognize in these pages the distorted reflection of their own imagination and skepticism. Rather than belabor the apologies and disclaimers usual in books of this sort, let me just remark that without the impact of Ralph Abraham, Arthur Brill, Robert DeHaan, Wolfgang Engelmann, Brian Goodwin, Herman Gordon, Joseph Higgins, Stuart Kauffman, Richard Levins, Robert MacArthur, Graeme Mitchison, Jay Mittenthal, George Oster, Theodosios Pavlidis, Colin Pittendrigh, John Platt, Kendall Pye, John Rinzel, Frank Rosenblatt, Otto Rössler, René Thom, John Tyson, and Trisha Woollcott, my explorations into the dynamics of evolved life would have lacked the special richness and color that I here seek to share.

Perseverance in this line of enquiry was made possible by the generous financial support of the National Science Foundation since 1965 and of the National Institutes of Health during 1973−1978. I am especially indebted to my department chairmen, Jack Cowan, Henry Koffler, and Struther Arnott, for safe escort through the three grades of professorship while I remained lost in the dreamworld here described.

Finally, I wish to acknowledge the frequent restoration of my sanity by the turquoise waters, white sands, and blinding sunlight of the Isle of Palms, South Carolina, where most of these pages were first drafted in 1977.

Contents

Introduction   **1**

1. Circular Logic   **4**

A: Spaces   4
B: Mappings   6
C: Phase Singularities of Maps   25
D: Technical Details on Application to Biological Rhythms   30

2. Phase Singularities (Screwy Results of Circular Logic)   **40**

A: Examples   40
B: Counterexamples   70
C: The Word "Singularity"   71

3. The Rules of the Ring   **74**

A: Basic Principles, Paradigms, Language Conventions, Epistemology   74
B: Dynamics on the Ring   77
C: Derivation of Phase-Resetting Curves   82
D: Historical Appendix   91

4. Ring Populations   **95**

A: Collective Rhythmicity in a Population of Independent Simple Clocks   95
B: Communities of Clocks   112
C: Spatially Distributed Independent Simple Clocks   121
D: Ring Devices Interacting Locally   125

5. Getting Off the Ring   **131**

A: Enumerating Dimensions   131
B: Deducing the Topology   132
C: The Simplest Models   134
D: Mathematical Redescription   136
E: Graphical Interpretation   140
F: Summary   144

6. Attracting Cycles and Isochrons   **145**

A: Unperturbed Dynamics   145
B: Perturbing an Attractor Cycle Oscillator   159
C: Unsmooth Kinetics   168

7. Measuring the Trajectories of a Circadian Clock   **176**

A: Introduction   176
B: The Time Machine Experiment   178
C: Unperturbed Dynamics   185
D: The Impact of Light   191
E: Deriving the Pinwheel Experiment   194
F: So What?   198
G: In Conclusion   204

8. Populations of Attractor Cycle Oscillators   **205**

A: Collective Rhythmicity in a Population of Independent Oscillators: How
   Many Oscillators?   205
B: Collective Rhythmicity in a Community of Attractor Cycle Oscillators   207
C: Spatially Distributed Independent Oscillators   212
D: Attractor Cycle Oscillators Interacting Locally in Two-Dimensional
   Space   225

9. Excitable Kinetics and Excitable Media   **231**

A: Excitability   231
B: Rotors   235
C: Three-Dimensional Rotors   250

**10. The Varieties of Phaseless Experience: In Which the Geometrical
Orderliness of Rhythmic Organization Breaks Down in Diverse Ways**   **258**

A: The Physical Nature of Diverse States of Ambiguous Phase   259
B: The Singularities of Unsmooth Cycles   273
C: Transition to Bestiary   275

11. The Firefly Machine   277

A: Mechanics   277
B: Results   280
C: Historical   283

12. Energy Metabolism in Cells   285

A: Oscillators   285
B: The Dynamics of Anaerobic Sugar Metabolism   286
C: The Pasteur Effect   288
D: Goldbeter's PFK Kinetics   289
E: Phase Control of the PFK/ADP Oscillator   291
F: More Phase-Resetting Experiments   291
G: Results: The Time Crystal   292
H: A Repeat Using Divalent Cations   296
I: A Repeat Using Acetaldehyde   296
J: Phase Compromise Experiments   298

13. The Malonic Acid Reagent ("Sodium Geometrate")   300

A: Mechanism of the Reaction   302
B: Wave Phenomena   304
C: Excitation in Non-oscillating Medium   307
D: Wave Pattern in Two- and Three-Dimensional Context   308
E: Pacemakers   312

14. Electrical Rhythmicity and Excitability in Cell Membranes   315

A: Rephasing Schedules of Pacemaker Neurons   317
B: Mutual Synchronization   323
C: Waves in One Dimension   325
D: Rotating Waves in Two Dimensions   329

15. The Aggregation of Slime Mold Amoebae   337

A: The Life Cycle of a Social Amoeba   337
B: Questions of Continuity   339
C: Chemistry in the Single Cell   340
D: Phase Resetting by a cAMP Pulse   342
E: Historical Note   343

16. Growth and Regeneration   345

A: The Clockface Model   345
B: An Alternative Description   349

C: Redescription in Terms of a Map Without Singularities   350
D: Applying the TSS Image Rules   353
E: Experiments Needed   355
F: Summary   358
G: Historical Note   358

17. Arthropod Cuticle   **361**

A: Rules for Development   361
B: Insect Eyes   363
C: Micromechanical Models   365

18. Pattern Formation in the Fungi   **367**

A: Breadmold with a Circadian Clock   368
B: Breadmolds in Two-Dimensional Growth   369
C: Pattern Polymorphism in Bourret's *Nectria*   370
D: Integration of Pattern   373

19. Circadian Rhythms in General   **375**

A: Some Characteristics of Circadian Rhythms   377
B: Clock Evolution   388
C: The Multioscillator View of Circadian Rhythms   394

20. The Circadian Clocks of Insect Eclosion   **401**

A: Basics of Insect Eclosion Clocks   401
B: Phase and Amplitude Resetting in *Drosophila Pseudoobscura*   412
C: Other *Diptera*   424

21. The Flower of *Kalanchoe*   **427**

A: Type 0 Resetting   **427**
B: Resetting Data at Many Stimulus Magnitudes   430
C: A Phase Singularity   433
D: Arrhythmicity Not an Artifact of Populations   433
E: Amplitude Resetting   434

22. The Cell Mitotic Cycle   **435**

A: Three Basic Concepts and Some Models   436
B: Regulation of Mitosis by the Circadian Clock   441
C: Further Developments in the Area of Circadian Rhythms, Applied Back to
   the Cell Cycle   442
D: *Physarum Polycephalum*   444

23. The Female Cycle   **450**

A: Women, Hormones, and Eggs   450
B: Statistics ("Am I Overdue?!")   452
C: Rephasing Schedules   455
D: The Question of Smoothness   457
E: Circadian Control of Ovulation   457

References   **459**

Index of Names   **508**

Index of Subjects   **522**

Imagine that we are living on an intricately patterned carpet. It may or may not extend to infinity in all directions. Some parts of the pattern appear to be random, like an abstract expressionist painting; other parts are rigidly geometrical. A portion of the carpet may seem totally irregular, but when the same portion is viewed in a larger context, it becomes part of a subtle symmetry.

The task of describing the pattern is made difficult by the fact that the carpet is protected by a thick plastic sheet with a translucence that varies from place to place. In certain places we can see through the sheet and perceive the pattern; in others the sheet is opaque. The plastic sheet also varies in hardness. Here and there we can scrape it down so that the pattern is more clearly visible. In other places the sheet resists all efforts to make it less opaque. Light passing through the sheet is often refracted in bizarre ways, so that as more of the sheet is removed the pattern is radically transformed. Everywhere there is a mysterious mixing of order and disorder. Faint lattices with beautiful symmetries appear to cover the entire rug, but how far they extend is anyone's guess. No one knows how thick the plastic sheet is. At no place has anyone scraped deep enough to reach the carpet's surface, if there is one.

Martin Gardner
Sci. Amer.
March 1976, p. 119

Introduction

Ubi materia, ibi geometria

J. Kepler

This is a story about dynamics: about change, flow, and rhythm, mostly in things that are alive. My basic outlook is drawn from physical chemistry, with its state variables and rate laws. But in living things, physical and chemical mechanisms are mostly quite complex and confusing, if known at all. So I'm not going to deal much in mechanisms, nor even in cause and effect. Instead I will adopt the attitude of a naturalist-anatomist, describing morphology. The subject matter being dynamics, we are embarked upon a study of *temporal* morphology, of shapes not in space so much as in time. But by introducing molecular diffusion as a principle of spatial ordering, we do come upon some consequences of temporal morphology for the more plainly visible shapes of things in space.

This is a story about dynamics, but not about all kinds of dynamics. It is mostly about processes that repeat themselves regularly. In living systems, as in much of mankind's energy-handling machinery, rhythmic return through a cycle of change is an ubiquitous principle of organization. So this book of temporal morphology is mostly about circles, in one guise after another. The word *phase* is used (over 1,200 times) to signify position on a circle, on a cycle of states. Phase provides us with a banner around which to rally a welter of diverse rhythmic (temporal) or periodic (spatial) patterns that lie close at hand all around us in the natural world. I will draw your attention in particular to "phase singularities": peculiar states or places where phase is ambiguous but plays some kind of a seminal, organizing role. For example in a chemical solution a phase singularity may become the source of waves that organize reactions in space and time.

This book is intended primarily for research students. Readers who come to it seeking crystallized Truth will go away irritated. I suspect that the most satisfied readers will be those who come with revisionist intentions, seeking the frayed ends of new puzzles and seeking outright errors that might lead to novel perspectives. I am confident that you will find plenty of both, since this project has ramified into more specialty areas than I can keep abreast of, ranging from topology through biochemistry. I've done all I can to eliminate nonsense from earlier drafts, with indispensable help from my very critical friends, especially Herman Gordon and Richard Krasnow. But the material is kaleidoscopic. As long as I work at it, the pieces keep rearranging themselves into tantalizing new patterns. Most of these dissolve under continued scrutiny but more remain than I can pursue. I have chosen to lay them out in the one-dimensional way inevitable to written communication as follows.

This volume has two roughly equal parts. The first half mainly develops a few themes in an order natural to the fundamental concepts involved. The second half is organized more like a "dramatis personae". I call it the Bestiary. It tells about the organisms or other experimental systems from which the conceptual themes arose. In more detail:

Experiments with clocks and maps constitute the principal theme of this book. Phase singularities figure prominently in these experiments. Secondary themes will be played through again and again in different contexts: the progression from dynamics in a single unit (a cell, an organism, a volume element) to collective dynamics in populations of independent units, then to populations of promiscuously interacting units, and to populations arranged in space with interaction restricted to immediate neighbors (as by molecular diffusion). New phenomena emerge at each level.

Apart from those themes, the material gathered here may at first seem to have few unifying features. I have chosen examples from a diversity of living organisms and non-living experimental models. Each recurs in several places, illustrating different points. Our trail through this jungle of exotic flowers intersects itself in several places as these themes surface again and again in new combinations, in new experimental contexts.

The material is handled in three ways:

1.   A single thread of text proceeds through the first half of the book under 10 chapter headings. By the following two devices this part is kept as lean as possible to enable the reader to scoot through for perspective before choosing where to invest more critical thought.

2.   Along the way, frequent allusion is made to enclosed "boxes" of finer print, each elaborating a particular point, raising questions for exercise or research, or offering an anecdote. These stand aside from the main thread of text like scenic overlooks along the toll road. I think they contribute interest and perspective, but you may want to pass over them until you locate the chapters closest to your particular interests.

3.   The second half of the book is a Bestiary of 13 chapters about particular experimental systems. These provide background facts about the organisms or phenomena most frequently alluded to in the first half of the book. These might be the most interesting chapters for readers with little use for theories and models and for readers unfamiliar with the experimental laboratory. I had to put one or the other perspective first and naturally some people think I should have chosen the other way around. I think that only your taste should determine whether you read this Bestiary first, or the preceding abstract notions that inter-relate its contents.

The whole is sandwiched between a table of contents and an extensive bibliography.

In your first glance through these pages you will notice a mathematical flavor about some topics. Though my aim is to avoid mathematical "models" wherever possible, some reasoning with symbols seems inevitable to this subject matter. Mathematics enters in four ways:

1. Simply using numbers to quantify experimental data for presentation as graphs and for comparison with quantified ideas about their meaning ("models").

2. Using digital computer languages to implement data handling and to extract implications from numerical statements of an hypothesis. For example, I might assume that in some useful approximation cells divide when some constantly synthesized substance accumulates to a critical concentration, and digitally seek the implications for a compact tissue in which this substance leaks between adjacent cells.

3. Using standard undergraduate mathematics to extract such implications when conjectures can be formulated in terms tractible to geometry, elementary differential equations, and so on. I've made a big effort to avoid mathematical equations. This is only in part because I seldom manage to get equations right. My deeper motivation is a feeling that numerical exactitude is alien to the diversity of organic evolution, and pretense of exactitude often obscures the qualitative essentials that I find more meaningful. My aim is to get *those* right first. And that purpose seems better served by words and pictures supplemented by occasional numerical simulations and lots of experimental measurements.

4. Employing a little of the language of topologists in an effort to extract from models or from observations what seems to be the essence of their behavior, independent of quantitative variations. Such efforts are fraught both with technical difficulties and with the ever present danger of extracting mathematically trivial tautologies while discarding as "mere quantitative variations" all that is of factual interest about a particular phenomenon. Nonetheless, I believe such efforts constitute an indispensable prelude to explanation: "What is the phenomenon to be explained, as distinct from incidentals to be glossed over? What aspects require explanation in terms of empirical cause and effect, and what aspects are merely mathematical consequences of facts already known?"

It is my belief that the life sciences in particular have much to gain from, and perhaps something to contribute to, mathematical developments in the general area of topology. I wish the reader to consider this. Thus I will dwell on such topological notions as I have found useful in designing experiments and in interpreting their results. From heuristic beginnings, my own efforts seldom go far toward logical rigor, yet I have found much satisfaction in the fruitful dialog between theory and experiment that this approach has fostered.

We turn now to the simplest abstractions about "rhythms," "cycles," and "clocks," with a few examples. Examples are merely *mentioned* here, pending their fuller description in later chapters, where the context is riper.

1. Circular Logic

Philosophy is written in this great book—by which I mean the universe—
which stands always open to our view, but it cannot be understood unless
one first learns how to comprehend the language and interpret the symbols
in which it is written, and its symbols are triangles, circles, and other geometric
figures, without which it is not humanly possible to comprehend even one
word of it; without these one wanders in a dark labyrinth.

Galileo Galilei, 1623

My objective for this chapter is to draw your attention to a few peculiarities
inherent in the logic of periodic functions. I find a visual approach the most fruitful
for thinking about such matters. As the pictures involved consist mainly of mappings between circles and products of circles, I must first say a few words about the
notions of topological spaces and mappings. This chapter thus has four sections:

A. Spaces, with emphasis on rings (i.e., closed loops. To avoid the more exact
connotations of the word *circle* I use *ring*, trusting the reader to not confuse my
meaning with algebraic rings.)

B. Mappings, with emphasis on the winding number of mappings to a ring

C. Phase singularities of maps (Parts I and II), with emphasis on the consequences of a nonzero winding number

D. Technical details on the application of circular logic to biological rhythms

The next chapter goes on to examine experimental examples of mappings to
the ring that contain phase singularities. Discussion specifically focusing on the
physical nature of phase singularities in each case is reserved to Chapter 10.

A: Spaces

As used in this book, a topological space is a collection of points connected in
some way. Spaces are distinguished by the distinctive manner in which these connections are made, i.e., by the rules which associate to any given point an immediate
neighborhood of other points. Six examples follow:

1. The real number axis, denoted by $\mathbb{R}^1$, is a topological space. Each real number lies arbitrarily close to a slightly bigger number and a slightly smaller number. This ordering of points stretches from minus infinity to plus infinity or from zero to infinity, as in rays from an origin. Any endless open line or curve that does not cross itself is topologically equivalent to $\mathbb{R}^1$. The great majority of familiar variables of concern in science and in schools are represented by real numbers: time, chemical concentration, temperature, energy, etc. A finite interval of $\mathbb{R}^1$ is called $\mathbb{I}^1$. Geneticists, for example, derive from breeding experiments a number between 0 and $\frac{1}{2}$ that tells how far apart any two genes are along the chromosome to which both belong. The space of genetic map distances is thus $\mathbb{I}^1$. (For interesting exceptions, see Example 4 below.)

The variables of concern in this volume are of a fundamentally different sort (Example 4).

2. The unlimited two-dimensional space of plane geometry, denoted $\mathbb{R}^2$, is another topological space. Its connectivity is richer in that points can be neighbors without being greater or less than each other. There are two dimensions, a north-south axis in addition to an east-west axis. This is the space of complex numbers or, if you like, pairs of numbers. It is the topological product of two independent $\mathbb{R}^1$'s (e.g., two perpendicular real number axes), thus the notation $\mathbb{R}^1 \times \mathbb{R}^1 = \mathbb{R}^2$. A finite section of a two-dimensional space, a bed sheet or a blackboard, for instance, is topologically equivalent to $\mathbb{I}^1 \times \mathbb{I}^1 = \mathbb{I}^2$, and is also equivalent to the unit disk $\mathbb{D}^2$, whose boundary is the unit circle $\mathbb{S}^1$ to be considered below.

3. In three-dimensional space the neighborhood relationship is richer still: Points can be distinguishable neighbors even when identical in respect to east-west and north-south position. Unbounded three-dimensional space is denoted by $\mathbb{R}^1 \times \mathbb{R}^1 \times \mathbb{R}^1 = \mathbb{R}^3$. A cubical chunk of it is $\mathbb{I}^3$ (the product of three intervals). A solid spherical chunk of it, a ball, is denoted $\mathbb{S}^3$. Topologically, $\mathbb{S}^3$ and $\mathbb{I}^3$ are equivalent (just round off the cube's corners).

4. Returning to one-dimensional spaces, consider the isolated segment $\mathbb{I}^1$. By joining together (or identifying, as topologists would say) the ends of the line segment, one forms the circle or ring. The circle is conventionally denoted $\mathbb{S}^1$ for "one-dimensional sphere," and is of paramount interest in this monograph. All closed one-dimensional manifolds are topologically equivalent to $\mathbb{S}^1$.

Several quantities of familiar practical science take their values not on the real number axis $\mathbb{R}^1$ but on the ring $\mathbb{S}^1$: angles or compass directions, perceived hues of colored objects, orientations of a receptive field of a ganglion cell in the retina, and phases in a cycle, such as the time of day or the season of the year.

Shortly after Lederberg's discovery of sex in bacteria made it possible to apply the standard techniques of genetic mapping to bacteria, geneticists encountered an amusing dilemma that could only be resolved by abandoning the notion that genetic maps are mappings from an observable frequency of certain genetic events to a segment $\mathbb{I}^1$ of the real line. That assumption implied that there is a unique distance between any two genetic loci. The following experiment violated that implication. Up until about 1961 it had always been found that if three genes lie in order ABC (Figure 1) along the genetic map and if a fourth gene D is found to lie

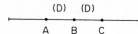

Figure 1.   Three loci on a genetic map assumed topologically
equivalent to $\mathbb{I}^1$.

in order ADC, then it will necessarily lie in segment AB or in segment BC. But this
proved not to be true of bacteria and, later, of viruses! The genetic crosses often
showed that the genes lay in order DAB or BCD, seemingly incompatible with the
view that D is between A and C. More crosses showed that bacterial and viral
genetic maps have the connectivity of a ring. On a ring, D can lie between A and C
along any of *three* arcs (Figure 2). The surprising topological discovery that the
numerically constructed genetic map has the connectivity of a ring of course suggested that the genetic material is physically ring-shaped. This was eventually
proved by direct observation of bacterial (Stahl, 1967) and viral (Thomas, 1967)
chromosomes.

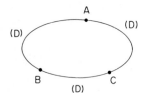

Figure 2.   Three loci on a genetic map assumed topologically
equivalent to $\mathbb{S}^1$.

5.   One two-dimensional analog of a ring is the surface of a doughnut, alias
torus, $\mathbb{S}^1 \times \mathbb{S}^1 = \mathbb{T}^2$. (The one-dimensional torus $\mathbb{T}^1$ is the same as $\mathbb{S}^1$: a ring.)
This surface can be formed by swinging a ring around in a ring perpendicular to its
plane; such a surface is the product of two rings. Each point of $\mathbb{T}^2$ can be identified
by giving its position on its two generating circles as two angles or two phases.
This is the natural space for representing a state given by two independent angles,
e.g., the space of physiological times, compounded of the hour of the diurnal cycle
and the season of the annual cycle.

6.   The alert reader may have noted that the product of $\mathbb{S}^1$ by itself is called
$\mathbb{T}^2$, not $\mathbb{S}^2$. $\mathbb{S}^2$ is reserved for the two-dimensional surface of a sphere. The
connectivity of $\mathbb{S}^2$ is natural for representation of directions in three-dimensional
space, e.g., the orientation of a polar molecule floating in solution, or of a space
shuttle during maneuvers in orbit, or the direction to a star or the position of a
ship at sea.

There are many other finite-dimensional topological spaces which in a natural
way represent the possible values of quantities of concern in experimental science.
Those listed above constitute a sufficient catalog for our present purposes, so we
turn now to mappings.

B: Mappings

A "map" takes points in one space (the source space) to certain points which
the map identifies as the "corresponding points" in another space (the target space).
This could be done in a higgledy-piggledy way or it could be done continuously.
By continuous, I mean that sufficiently nearby points in the source space remain

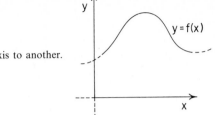

Figure 3.   A map from one real number axis to another.

nearby in the target space.[1] Thus the mapping amounts to a distorted image of source space in the target space. The maps we work with in this volume will always be continuous, even smooth, except at isolated points or along a special locus. Such points or loci constitute the main focus of all that follows. Before proceeding to examine such irregularities, you may want to refresh your familiarity with smooth maps by reading through the following 14 examples. A good *Scientific American* reference is M. Shinbrot, 1966.

1. $\mathbb{R}^1 \to \mathbb{R}^1$: Input a number, output a number. This kind of map, a function of one variable, is most conveniently portrayed as a graph (Figure 3), depicting, for example, the light emission as a function of time in a culture of glowing cells, or land elevation above sea level as a function of distance along U.S. Route 90 from Chicago.

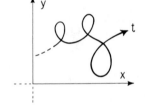

Figure 4.   A map from the real number axis to the plane, geometrically viewed in the product space $\mathbb{R}^3$, seen in planar projection.

2. $\mathbb{R}^1 \to \mathbb{R}^2$: Input a number, output a point on a plane. This kind of map maps the line onto the plane (Figure 4). It could be represented as an image of the source space $\mathbb{R}^1$ in the product space $\mathbb{R}^1 \times \mathbb{R}^2 = \mathbb{R}^3$ of the source space and the target space, as in the first example. Figure 4 is just this three-dimensional plot seen in two-dimensional projection. (Any map can be so represented but it becomes visually inconvenient except in the very simplest cases.) For example, $\mathbb{R}^1$ might be time and $\mathbb{R}^2$ might be the concentration of two substances in a chemical reaction. The map then constitutes a "trajectory" of the changing chemical composition. (If the reaction is exactly cyclic, so that its state in concentration space moves in a circle, then the map in $\mathbb{R}^1$ (time) $\times$ $\mathbb{R}^2$ (concentrations) looks like a corkscrew or helix.) Such trajectories will be useful in Chapter 7.

---

[1]   Continuity defined: $F(X)$ is continuous at $X = A$ means that
   1. F is defined at A.
   2.   The limit$_{X \to A}$ $F(X)$ exists.
   3.   The limit $F(A)$ is the same when X approaches A from any and all directions.
   Such a map is additionally "smooth" if the rate of change of F as X varies is also continuous.

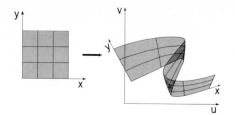

Figure 5.   The *xy* plane is mapped into the *uv* plane (not intended to portray specific features common in retinotecal maps).

3. $\mathbb{R}^2 \to \mathbb{R}^2$ (or $\mathbb{I}^2 \to \mathbb{I}^2$ if finite): Input a point on a plane, output a point on some other plane (or on some two-dimensional surface topologically equivalent to a plane). For example, the source space might be the retina of a frog's eye and the target space might be the frog's optic tectum. The map is then the retinotectal projection of ganglion cell fibers onto their cortical targets. A convenient trick for visualizing such a map (Figure 5) is to rule a coordinate grid on the source space (e.g., retina) and then examine its distorted image in the target space (e.g., tectum), or vice versa, as is in fact the habit of visual electrophysiologists (Gaze, 1970; Jacobson, 1970). In Chapters 8 and 9, the source space will be a two-dimensional disk of reacting fluid and the target space will be the composition space defined by two concentrations. The map is the image of the disk in composition space. In this format the kinetics of reaction interacting with molecular diffusion turns out to be more readily comprehensible than in more familiar formats. We will find many occasions throughout this book to think in terms of a two-dimensional medium's image in some state space.

4. $\mathbb{R}^2 \to \mathbb{R}^1$: Input a point on a plane, output a number. This is a simplified version of the preceeding map, in that instead of associating *two* independent numbers with each point in the source space (the plane), we pay attention to only *one* target coordinate. The contour map is a familiar convenience for representing such a mapping. Each contour line depicts the locus of points on the plane which all map to the same point of the real number axis. In other words, a coordinate grid has been ruled on the target space. This "grid" in one dimension consists merely of equispaced tick marks. We examine its image in the source space. The grid image in the $\mathbb{R}^2 \to \mathbb{R}^2$ maps considered above is just a superposition of two contour maps, one for the east-west coordinate in the target space, and one for the north-south coordinate. In $\mathbb{R}^2 \to \mathbb{R}^1$, we deal with only one of these. As an example consider the temperature distribution in a sheet of encapsulated liquid crystals used for temperature sensing. Each temperature corresponds to a color. The bands of color on the plastic sheet constitute a contour map of the temperature field showing how the two-dimensional sheet maps onto the temperature axis. Another example is the familiar geodetic survey map of any two-dimensional area, showing contours of altitude above mean sea level.

Dimension-reducing maps such as $\mathbb{R}^2 \to \mathbb{R}^1$ correspond in a natural way to the process of measurement: Some aspect of a more richly variable phenomenon is singled out for attention by such a map, typically by reduction to a single number.

5. $\mathbb{S}^2 \to \mathbb{R}^1$: Input a direction in space or a point on a sphere, output a number or a point on a line.

The principle is the same as above. Examples: the familiar weatherman's contour map of barometric pressure around the globe, or the amplitude of radio reception as a function of orientation of an antenna.

The next four kinds of map involve conversions to and from the circle or ring $\mathbb{S}^1$. Topologists call the line $\mathbb{R}^1$ (for one-dimensional real number) and the circle $\mathbb{S}^1$ (for one-dimensional sphere) because these are very different objects. This book exploits the differences.

Few indeed are the instruments of modern science that report back a point on a circle (compass, goniometer, circular counters) rather than a point on the line (thermometers, volt meters, pH meters). The mainstream of experimental science has always flowed along the real number axis $\mathbb{R}^1$ and its product spaces. It wasn't until 1953 that anyone even went to the trouble to formalize an equivalent of Gaussian statistics for observables on $\mathbb{S}^1$ (Gumbel et al., 1953; Greenwood and Durand, 1954). Nonetheless, just as the practical mathematics of engineering and science is mostly about numbers on the real line, so the practical mathematics of things periodic and rhythmic is mostly about points on the circle.

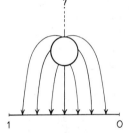

Figure 6.   A map from each point on the ring to its own number between 0 and 1.

6. $\mathbb{S}^1 \rightarrow \mathbb{I}^1$: Input an angle or a point on a ring, output a number or a point on a line segment.

Such a map is inscribed on every compass and protractor and on the face of every clock. Around the circular edge one finds numbers from $0°$ to $360°$, or from 0 to 12 or 24 hours. If each point on the ring maps to a different number, there must inevitably be a discontinuity, e.g., at the point in Figure 6 that maps to 1 or 0.

Note that in this volume I adhere to the navigator's and clock maker's convention when it is necessary to label points on the ring as though they were numbers: We go clockwise from north at the top. Mathematicians and the makers of protractors tend to go anticlockwise from east at the right.

7. $\mathbb{S}^1 \rightarrow \mathbb{R}^1$: Input a point on a circle, output a point on a line.

The tangent function of trigonometry is an example (Figure 7). It produces a number ranging anywhere from $-\infty$ to $+\infty$ depending on the angle given. The

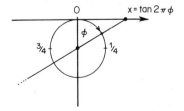

Figure 7.   A two-part map from the ring to the infinite line. There is a discontinuity at $\pm\frac{1}{4}$ cycle.

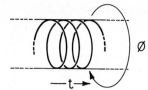

Figure 8.   A map from the real number axis to the ring, geometrically viewed in the product space.

act of measuring a biological rhythm might be considered another example of this
kind of map; given a phase in the organism's daily cycle, there corresponds the
numerical value of some quantity that we find it convenient to measure, such as a
diurnally fluctuating body temperature or a rhythmically varying chemical concentration or a rate of egg laying.

8. $\mathbb{R}^1 \to \mathbb{S}^1$: Input a point on a line, output a point on a circle.

An example of such a map is the visual angle between the moon and the sun as
a function of time. As the year advances this angle varies through one full monthly
cycle and repeats itself during the next month and the next and the next. It can be
thought of geometrically as wrapping the real line (the time axis) around and
around the circle. As noted above, any map can be thought of as an image of the
source space in the product of source and target spaces. In this case the product
space is a cylinder, $\mathbb{R}^1 \times \mathbb{S}^1$, and the source space's image on the cylinder is a
helix (Figure 8).

9. $\mathbb{I}^1 \to \mathbb{S}^1$: Input a number between 0 and 1, output a point on a circle or ring.

An example is the inverse of map 6, which necessarily suffers the same discontinuity though in reverse: While nearby numbers always map to nearby points on
the ring, so do some quite different numbers near the extremes of the interval. This
underlies the necessity of an International Date Line (see Box A) and accounts for
the principal annoying feature of compasses and protractors: We have to remember
that 360° is the same as 0°. Figure 9 depicts this map as an image of $\mathbb{I}^1$ in the
product cylinder $\mathbb{I}^1 \times \mathbb{S}^1$. This picture is equivalent to Figure 6.

Any map to $\mathbb{S}^1$ can also be colorfully depicted as a contour map on its source
space using bands of different hues. It is a curious fact of human vision that we
classify spectral distributions of light in a way that has circular connectivity:
Red is close to orange is close to yellow is close to green is close to blue is close to
indigo is close to violet is close to red is close to orange . . . . So the different kinds
of map from any space to $\mathbb{S}^1$ can be represented by the qualitatively different ways
to paint that space with different hues.

Maps to $\mathbb{S}^1$ are a primary theme of this book.

10. $\mathbb{S}^2 \to \mathbb{S}^1$: Input a point on a sphere, output a point on a circle.

A weather depiction of worldwide wind directions is such a map. Some "homing"
and migrating birds apparently convert the three-dimensional orientation of the
earth's magnetic field (which corresponds to a point on a sphere) to a compass

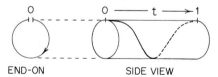

Figure 9.   The map of Figure 6, geometrically
viewed in the product space.

END-ON   SIDE VIEW

Box A: The International Date Line

> The roads by which men arrive at their insights into celestial matters
> seem to me almost as worthy of wonder as those matters themselves.
>
>   J. Kepler, *Astronomia Nova*, 1609

As Example 6 ($\mathbb{S}^1 \to \mathbb{I}^1$) suggests, a difficulty arises in trying to reconcile the notion of cyclic daily time (a point on a clock rim) with the notion of progressive linear time (in a history book). When an astronaut traces the path of Pan American #2 eastward from San Francisco back to San Francisco in $1\frac{1}{2}$ hours, flying 200 miles above the sea, he passes through zones of local time that advance through one full day. So he must arrive back over San Francisco $1\frac{1}{2}$ hours later, but in tomorrow! That is not what an observer on the ground would say. Nor would the astronaut.

The 18 survivors of Magellan's expedition around the world were the first to present this dilemma for the bewilderment of all Europe. After three years westward sailing, they first made contact with European civilization again on Wednesday 9 July, 1522 by ship's log. But in Europe it was already Thursday! Pigafetta writes (translated in *The First Voyage Around The World*, Hakluyt Society, vol. 52, 1874, p. 161):

> In order to see whether we had kept an exact account of the days, we
> charged those who went ashore to ask what day of the week it was, and
> they were told by the Portuguese inhabitants of the island that it was
> Thursday, which was a great cause of wondering to us, since with us it
> was only Wednesday. We could not persuade ourselves that we were
> mistaken; and I was more surprised than the others, since having always
> been in good health, I had every day, without intermission, written down
> the day that was current.

According to S. Zweig (*The Story of Magellan*, Literary Guild of America, New York, 1938), this phenomenon excited the humanists of the sixteenth century with the same sense of wonder we feel about the space-time singularities that our relativity theorists impute to black holes, perhaps more so inasmuch as the missing day was a direct human observation.

So far as I am aware, the young Lewis Carroll (author of *Alice in Wonderland*) was the first to suggest what to do about it, in a whimsical essay in *The Rectory Umbrella* published about 1850 (reprinted by Dover, 1971):

> Half of the world, or nearly so, is always in the light of the sun: as the
> world turns round, this hemisphere of light shifts round too, and passes
> over each part of it in succession.
>
> Supposing on Tuesday it is morning at London; in another hour it
> would be Tuesday morning at the west of England; if the whole world
> were land we might go on tracing[1] Tuesday Morning, Tuesday Morning
> all the way round, till in 24 hours we get to London again. But we *know*
> that at London 24 hours after Tuesday morning it is Wednesday
> morning. Where then, in its passage round the earth, does the day change
> its name? where does it lose its identity?
>
> Practically there is no difficulty in it, because a great part of its
> journey is over water, and what it does out at sea no one can tell: and

---

[1]   The best way is to imagine yourself walking round with the sun and asking the inhabitants as you go "what morning is this?" if you suppose them living all the way round, and all speaking one language, the difficulty is obvious.

besides there are so many different languages that it would be hopeless
to attempt to trace the name of any one day all round. But is the case
inconceivable that the same land and the same language should continue
all round the world? I cannot see that it is: in that case either there would
be no distinction at all between each successive day, and so week, month
&cc so that we should have to say "the Battle of Waterloo happened
to-day, about two million hours ago," or some line would have to be
fixed, where the change should take place, so that the inhabitant of one
house would wake and say "heigh ho! Tuesday morning!" and the
inhabitant of the next, (over the line,) a few miles to the west would wake
a few minutes afterwards and say "heigh ho! Wednesday morning!"
What hopeless confusion the people who happened to live *on* the line
would always be in, it is not for me to say. There would be a quarrel
every morning as to what the name of the day should be. I can imagine
no third case, unless everybody was allowed to choose for themselves,
which state of things would be rather worse than either of the other two.

Thus we have the International Date Line, established 180° from Greenwich, England
in 1878. When an astronaut crosses that line in the mid-Pacific Ocean moving from
west to east, he pops back 24 hours into yesterday, so when he arrives back over San
Francisco it is $1\frac{1}{2}$ hours later *today*, not tomorrow. Each new date starts along this line
as it rolls through midnight. Tomorrow has then begun. The International Date Line
leads an expanding crescent of tomorrow around the east side of the earth into the
dawn light, around through noon and dusk and back to midnight (see figure below).
When the Date Line reaches the midnight point it starts the next day in the same manner.
Thus the citizens of Tafahi in the Tonga islands are among the first to report back to
work after each weekend. Meanwhile, their neighbors on Savia in nearby Western
Samoa, 150 miles to the east, greet the same day as a Sunday.

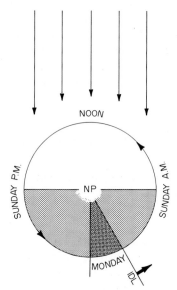

The earth viewed from a station above the North Pole (NP). Monday started when the
International Date Line (IDL) rotated through midnight. A satellite poised above
the midnight point can connect Sunday morning and evening with Monday morning.

By glib reference to this line the modern astronaut excuses himself from what seemed a bewildering conundrum to Pigafetta. Yet there remains something peculiar about this line. What kind of line is it? Evidently it is not a circle since it was transgressed only once in a flight from San Francisco to San Francisco. So it ends somewhere. What if I go to the end and walk around it, cross over, walk around the end again, cross over again, and so on? Can I rachet myself into the remote future or past, a day at a time? Unfortunately not. The International Date Line has its end points at the poles, the two phase singularities implicit in mapping the globe onto an equatorial circle of 24 hours [see Figure 10 (right) and Example 3, next chapter]. It gains back for us the day lost in walking a loop around the pole. Viewed in this way, the Date Line is a necessary consequence of the geometry of a rotating planet; but somehow its inevitability doesn't make it any more intuitive.

heading (a point on a ring). Another instance of such maps occurs in the crystallization of synthetic polypeptides and polynucleotides (artificial analogs of biological proteins and genetic material). Such substances are fibrous, giving to each point on the surface of a crystalline ball ($\mathbb{S}^2$) an orientation. The orientation may be specified as a point on an abstract sphere ($\mathbb{S}^2$) or, given that the fiber lies in the plane of the surface, on an abstract ring ($\mathbb{S}^1$) of compass directions. There is no way to carry out such a map smoothly. There must be at least one point of confused orientation. These points are clearly visible by polarized light microscopy (see Robinson, 1966; Wilkins, 1963; Anderson et al., 1967).

11. $\mathbb{R}^2 \to \mathbb{S}^1$: Input a point on a plane or any curvy two-dimensional surface topologically equivalent to the plane, output a point on a circle.

Consider, for example, a sheet of cells, all of which are regularly carrying out some rhythmic function such as cell division. Given a sufficiently clear snapshot of such a sheet of cells, one could write labels all over it indicating at each point in the plane what stage of the cycle the cells have reached. Such maps will play a central role in our discussion of peculiarities in the rhythmic morphogenesis of slime molds and of fungi, and of the periodic organization of oscillating chemical reactions in space.

Glass (1977) calls such maps "phase maps" and uses them to interpret the peculiar symmetries of regeneration and reduplication in the growth of embryos and in the restoration of amputated limbs in lower animals capable of such tricks (see Chapter 16).

Phase maps will also serve us well in discussing circadian rhythms in flies (Chapter 20). The source space is either a two-dimensional physical space such as a desk top or the more abstract two-dimensional space of different stimuli determined by two quantities at the experimenters' disposal. The target space is the phase of a circadian rhythm reset at that place or by that stimulus. Such a map is drawn point by point by simply measuring the phase in $\mathbb{S}^1$ associated with each point in $\mathbb{R}^2$. It is convenient to run a smooth curve through all points in $\mathbb{R}^2$ that correspond to the same phase. I call such curves "isochrons" ("same-times") on account of the usual association of $\mathbb{S}^1$ with a rhythm in time (see Chapter 6).

12. $\mathbb{S}^2 \to \mathbb{R}^2$: Input a point on a sphere, output a point on a plane.

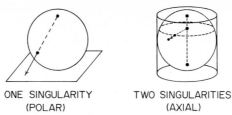

ONE SINGULARITY   TWO SINGULARITIES
(POLAR)   (AXIAL)

Figure 10.   Two ways of mapping the sphere on a flat surface. Mapping onto the unbounded plane on the left requires a singularity (at the North Pole). Mapping onto a cylinder on the right requires two singularities (one at each pole).

These maps include the many projections used by geographers to map the globe, inevitably with some distortion, onto the flat page of a book. As we all realize, perhaps with a residual twinge of perplexity since being intimidated by a fourth grade geography teacher's inadequate explanation, these maps never quite cut the mustard. Unless the earth's surface is printed as at least two disjoint regions, there always arise points or curved paths at which the distortion is *infinite* (Figure 10). These are called singularities. They play a prominent role in this book, starting with the next chapter.

13. $\mathbb{R}^2 \to \mathbb{S}^2$: Input a point on a plane, output a point on a sphere.

In a microtome slice through a fibrous biological material, such as insect cuticle, microscopic fibers at various points on the section plane can be characterized by orientations in three-dimensional space. One geometrically convenient way to visualize the pattern of fiber directions is to think of orientation as a point on the sphere and to map each point of the section onto its corresponding orientation by appropriately stretching an image of the section plane onto a spherical surface. This map may be many-to-one if several regions (or even a one-dimensional locus) of the section plane have the *same* fiber orientation. Just such maps provided a key to a puzzle about the ultrastructure of the insects' corneal lens (Chapter 17).

14. $\mathbb{S}^1 \to \mathbb{S}^1$: Input a point on a ring, output another point on a ring.

Maps from rings to rings resemble the simplest maps treated above (example 1, from the real line to the real line) but with a surprising new feature. The periodicity inherent in a ring results in the kind of quantization which allows us to classify continuous $\mathbb{S}^1 \to \mathbb{S}^1$ maps into distinctly different types. (This is in fact precisely the origin of quantization in the physics of hydrogen atoms around 1920.) Each "type" is associated with an integer-valued "winding number". The winding number is the net number of times the output value runs through a full cycle around a ring, as the input value is varied once forward along its ring. Because these topologically distinct types arise again and again in our experimental observations, I take time here to give eight examples and to dwell at some length on the logic of maps among rings.

Just as the map from $\mathbb{R}^1$ to $\mathbb{R}^1$ is most conveniently displayed as a curve in the plane $\mathbb{R}^2$, the product of $\mathbb{R}^1$ and $\mathbb{R}^1$, so any map from $\mathbb{S}^1$ to $\mathbb{S}^1$ is most conveniently displayed as a curve on the product space $\mathbb{S}^1 \times \mathbb{S}^1 = \mathbb{T}^2$, the surface of a doughnut. If there is a unique output value ϕ' ("phi-prime") corresponding to each

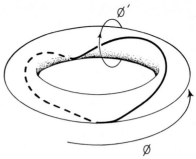

Figure 11. A map from $\mathbb{S}^1$ to $\mathbb{S}^1$, viewed geometrically in the product space.

input value ϕ, then the curve relating ϕ' to ϕ is a closed ring, reaching once around the ϕ axis and necessarily returning to the same ϕ' value after it is followed once around (Figure 11). But notice that such a curve can link any number of times through the hole. It can wind any number of times around the ϕ' axis during each circumnavigation of the ϕ axis. This integer winding number distinguishes qualitatively between the kinds of $\mathbb{S}^1$ to $\mathbb{S}^1$ map.

Unfortunately, toroidal graph paper does not easily fit into a book. We must map the torus onto the plane of the page. In Figure 11 I attempted this one way but it leaves much to be desired since some regions are badly distorted and half of the surface is hidden behind the other half. A convenient alternative is to map $\mathbb{S}^1 \times \mathbb{S}^1 \to \mathbb{R}^1 \times \mathbb{R}^1$ by unrolling the two circular axes along perpendicular linear axes. This is equivalent to cutting the torus open along two perpendicular circles and stretching it into a flat square, then repeating these square unit cells in all directions (Figure 12). It would suffice to present one unit cell since all the rest are identical, except that to do so hides the continuity of the data from the right side of one unit cell to the left side of the next. So I will usually present 2×2 unit cells or one surrounded by fragments of the eight adjoining cells.

Figure 12. A ring on the torus, laid flat for convenient inspection. Repeated as 2×2 unit cells on the left, and on the right as a unit cell flanked by fragments of adjacent reduplications.

In this format, the winding number of a map is revealed by the number of unit cells through which the data climb vertically in the space of one unit cell horizontally. Examples follow:

Example A. Red is close to orange is close to yellow is close to green is close to blue is close to indigo is close to violet is close to red is close to orange: ROYGBIVRO. Thus hue (disregarding saturation and intensity) is a psychophysical quantity with the connectivity of a ring (see Box B). In a photograph taken under a colored light, hues appear altered. The change of hue can be depicted as a map from the color ring to itself: $\mathbb{S}^1 \to \mathbb{S}^1$. Suppose for example that a colorful butterfly is taken indoors to be photographed by the bluish glare of fluorescent lights. Each color on its wings is mapped to a similar, but not identical, color on the photographic print through a combination of the emission spectrum of the

Box B: The Colors of Flickering Lights

It turns out that the purely abstract ring of hues can be put into a practically useful correspondence with the phase ring of a rhythm in real time. If a light-dark alternation is displayed at a frequency on the order of several cycles per second, one sees a flickering gray. Now if a line drawing is briefly illuminated during the non-dark halfcycle, its black ink seems tinged with color, and the color varies according to the phase of its presentation in the light-dark cycle as suggested in Figure a. The eighth of over a dozen published rediscoveries of this phenomenon, by C. E. Benham, was implemented in 1894 as a popular toy in England, a spinning disk decorated with annular patterns of black and white called "Benham's top" (Figure b). The "wave form" of the black-white cycle determines a color. My own encounter with the phenomenon [unrecorded rediscovery #3729 (October 1964)] provides what may be the easiest way to see the effect, though in its least quantifiable form. While studying Moire afterimages in my room near Cornell University, I rotated a celluloid transparency covered with parallel black tapes 1 mm apart. This gives a vigorous but indistinct impression of variegated color.

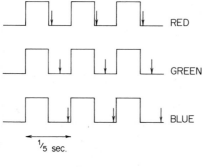

a) Timing of black-and-white picture presentations to elicit various color sensations. High position indicates picture cutoff (dark). The arrows show when to project a black and white image in order to tinge its black areas with the indicated color.

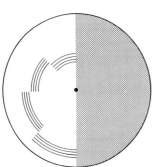

b) When this disk is rotated clockwise at 300 rpm, the innermost rings appear red and the outermost appears blue (or vice-versa if counterclockwise). The middle rings are greenish.

The "explanation" usually preferred is that the three kinds of color receptor in the human retina have different time courses of response to impulse illumination. Thus the mean amplitude of response of each receptor depends on the frequency and wave form of a periodic stimulus. In a flickering pattern the local wave form, and so the local color, varies spatially. So far as I am aware, this matter has not been pursued experimentally far enough to convincingly exclude other interpretations. Can one obtain, e.g., red flicker sensation in monochromatic blue light? Do flicker colors occur in color-blind individuals who have only a single-color receptor? Positive results in such experiments might suggest that color is encoded in the human nervous system by mutual phasing of neural rhythms. This possibility is considered worthy of investigation by contemporary workers in vision research (e.g., see Butterfield, 1968; Sheppard, 1968; Von Campenhausen, 1969; Festinger et al., 1971; Young, 1977).

Despite ignorance of its physiological origins, television engineers managed to parlay Benham's Top into a scheme for broadcasting color pictures over black-and-white transmitters and black-and-white home receivers. The device is called the Butterfield color encoder. Three color filters are arranged on a rapidly spinning wheel to alternate with three consecutive opaque segments in front of the television camera. The red-blocking filter immediately follows the opaque $\frac{1}{2}$-cycle. Then comes a green-blocking filter, then a blue-blocking filter. Since black lines seen early in the non-opaque cycle appear red, and those seen in the middle appear green, and those seen at the end appear blue, this phased presentation of the three complementary color images reconstructs a colored picture on the black-and-white television screen.

Its first public demonstration was a soft-drink advertisement over station KNXT in Los Angeles. Nothing was said about color, to the consternation of thousands of viewers who seemed suddenly to be hallucinating. Its last public demonstration was in Argentina during political advertisements by Madame Peron.

Unfortunately the perceived color proved to vary from individual to individual and depends unpredictably on the colors of its surrounding background. Moreover the effect works best at five to ten cycles per second. The resulting flicker of the color picture was so distasteful to television audiences that they preferred to buy the more costly RCA tri-phosphor system or to forego the color and stick to black-and-white transmissions in which the flicker can be made so fast as to escape perception.

lamp and the absorption spectra of the photographic emulsion and its dyes. If we restrict our interest to hues, ignoring saturation, then the color map is $\mathbb{S}^1$ to $\mathbb{S}^1$. It is continuous because similar colors similarly affect the emulsion. The map might have winding number $W = 1$ if all hues still appear under the bluish light [Figure 13 (right)] or it might have winding number $W = 0$ if the blue is so pure that every color is captured as some shade near blue [Figure 13 (left)].

Example B. A rowboat in a strong current holds a compass heading in direction ϕ. It proves convenient to measure ϕ on a scale 0 to 1: north is 0, east is $\frac{1}{4}$, south is $\frac{1}{2}$, west is $\frac{3}{4}$. The boat's actual track is a straight line in some other direction ϕ'. The map from heading to track is $\mathbb{S}^1$ to $\mathbb{S}^1$. The character of this map is determined by the speed of the boat in the water, relative to the current. Figure 14 depicts the rowboat R at a point in the stream. The stream's velocity is indicated by the vertical arrow. Its speed is v^*. If the rowboat heads in direction RH (compass heading ϕ) at speed v relative to the water, while the water flows downstream at speed v^*, then the rowboat's actual track is RR' relative to terra firma. The

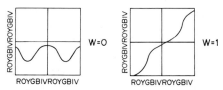

Figure 13. Two qualitatively distinct maps from the color ring to the color ring.

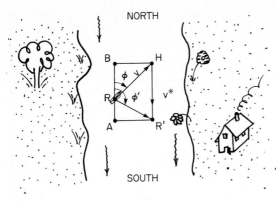

Figure 14. A rowboat heads in direction ϕ at speed v while being carried south in current v^*. Its actual track is in compass direction ϕ'.

rowboat's velocity component toward the shore, BH or AR′, is $v \sin(2\pi\phi)$. Its velocity component parallel to the shore, RA, is $v \cos(2\pi\phi) - v^*$. The ratio of these two components is AR′/RA = $\tan(2\pi\phi')$, so

$$\tan(2\pi\phi') = v \sin(2\pi\phi)/[v \cos(2\pi\phi) - v^*]$$

or in terms of "slowness", $s = 1/v$, additionally adopting the more compact notation sn $\phi = \sin(2\pi\phi)$, and similarly cs for cos and tn for tan:

$$\text{tn } \phi' = \frac{\text{sn } \phi}{\text{cs } \phi - s/s^*}.$$

Figure 15 plots several of these heading vs. track or ϕ' vs. ϕ conversion curves for various rowing speeds v. Note the qualitative transition at $v^*/v = s/s^* = 1$. At this speed the boat cannot track upstream against the current and the map's winding number switches from 1 to 0. We will see this formula again in context of circadian rhythms.

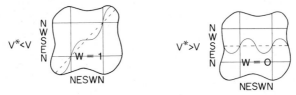

Figure 15.   Four qualitatively distinct maps $\phi' = f(\phi)$; the dashed curves represent the extreme cases $v = 0, v \to \infty$.

Example C. Local time varies with geographic longitude. At any moment, one is a function of the other: a map from $\mathbb{S}^1$ (longitude) to $\mathbb{S}^1$ (clock phase). This map (Figure 16) constituted the whole motivation for the design of mechanical

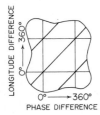

Figure 16.   Clock phase varies one-to-one with longitude.

Box C: Clocks and Maps

In Isaac Newton's day, ships sailed east and west with no practical means for determining longitude. The sextant sufficed for latitude, but without the longitude, life was too frequently tragic for navigators and their crews, pounded to fishmeat on unexpected reefs while the best commercial fruits of Europe littered the sandy bottom. Navigators wrote books on the problem. Outlandish schemes were tried. The best, originally due to Galileo but quietly filed away after certain questions of heresy arose, permitted determination of longitude at all the chief ports of call in both hemispheres, and did so with startling precision. This scheme used a telescopic sighting of Jupiter's four moons together with a table of their anticipated positions as a function of Parisian local time. The Jovian satellites provided a universal clock of unfaltering precision by which to know Paris time. A sextant observation determined local time by determining the moment when the sun or some other star reached its maximum altitude. The difference, a fraction of 24 hours, is a phase point on a circle which can be reinterpreted as the circle of longitudes at $\frac{1}{4}°$ of west longitude per minute of delay ($360°/(24 \times 60$ minutes$)$).

It was seldom possible to observe the celestial chronometer from a rolling ship's deck. Nonetheless, 200 years after Magellan's extravagant experiment of 1513 (see Box A), it was clearly recognized that the longitude could best be acquired as a map from one ring to another: from the face of a 24-hour chronometer to the required coordinate on the circle of 360°. So serious was the need that the British Admiralty offered a prize of 20,000 pounds (in 1714 currency!) for a sufficiently accurate ship chronometer. So taxing was the challenge that for 45 years "the longitude prize" was household metaphor for "the impossible"—somewhat like "the perpetual motion machine" today. But in

John Harrison's first attempt at the Longitude Prize. Its great grandchild (no. 4) claimed the prize. Photograph Courtesy of National Maritime Museum, London.

1759 John Harrison's #4 claimed the prize. On a storm-racked trip to the West Indies it erred by only 1/17,000 of a cycle, corresponding to $1\frac{1}{2}$ miles at equatorial latitudes.

Those who, like myself, are stirred to transports of excitement by this sort of tale will want to savor the marvelous essay entitled "The Longitude" by Brown in vol. 2 of *The World of Mathematics* (J. R. Newman, ed., Simon and Schuster, New York, 1956).

clocks with such accuracy as we take for granted today. Who would have cared to monitor Greenwich Mean Time (GMT) precisely in a world still free of bus, train, and jet schedules? Maybe astronomers. But the practical impetus to develop clocks came from merchants, whose mariners could reckon a ship's longitude *only* by the four-minute-per-degree discrepancy between local noon and the GMT chronometer on board (see Box C). It may be noted that San Francisco time changes as GMT changes. The dependence of San Francisco time on GMT is the same kind of map from $\mathbb{S}^1 \to \mathbb{S}^1$ (Figure 17): Both have winding number $W = 1$.

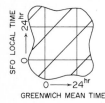

Figure 17.   San Francisco time varies one-to-one with Greenwich time.

(Examples A, B, and C are elaborated as Examples 1, 2, and 3 in the next chapter.)

Example D.   A starling is ready to migrate south, a direction it knows relative to the sun. A physiologist (Hoffmann, 1960) resets the phase of its circadian clock by six hours (one-quarter cycle) and releases it. It flies off in some other direction relative to the sun's azimuth, specifically about 90° away (one-quarter cycle). The map from clock phase to flight direction is $\mathbb{S}^1 \to \mathbb{S}^1$. It has been measured in a few kinds of organism. A type 1 map (winding number $W = 1$) was found in the sunfish (Figure 18). A type 0 map (winding number $W = 0$) was found in the pond skater (insect) (Figure 19).

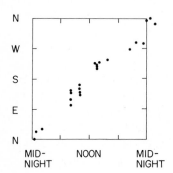

Figure 18.   The direction of sunfish movement vs. time of day (data replotted from Braemer, 1960, Figure 7). Note that the left and right edges are the same (midnight) and the top and bottom edges are the same (north).

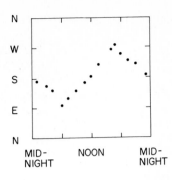

Figure 19. Angular orientation of a water strider to an artificial sun vs. time of day (data replotted from Birukow, 1960, Figure 2). Note that the left and right edges are the same (midnight) and the top and bottom edges are the same (north).

Example E. A regular oscillation goes on in all the growing cells of a fungus. The fungus is a thin sheet, growing only along its border. The border is a ring. The phase of the oscillation varies locally. How does phase vary with position along the ring-like border? The answer can be represented as a map from position (along a ring) to phase (along a ring) (Figure 20). See Chapters 4, Section C; 8, Section C; and 18.

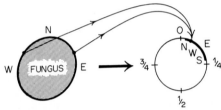

Figure 20. Each point of the frontier ring of a rhythmic fungus maps to the phase of the local rhythm at that point. In this example every point maps within about one-fifth of the cycle. Points at the same north-south latitude are mapped to the same phase.

Example F. Microorganisms sometimes make seawater glow around a swimmer at night. Hastings and Sweeney (1957) made the sensational discovery that, when brought into a dark laboratory, such water glows again at 24-hour intervals. Moreover, if exposed to light at phase ϕ of the rhythm, the suspended animals reschedule their glow as though the phase ϕ had been some other phase ϕ'. The map from old phase ϕ to new phase ϕ' is $\mathbb{S}^1 \to \mathbb{S}^1$. This example is developed further in Section D.

Example G. Consider an optical device, in outward appearance resembling a telescope, in which you view a scene. As you rotate the eyepiece, the scene rotates. The angle of eyepiece rotation ϕ is indicated against a degree scale by a fiduciary black dot. The angle of scene rotation θ can be read against the same scale by projecting the obvious upward direction from the center of the scene (as in Figure 21). Plot θ vs. ϕ must be a smooth curve of some integer winding number. By thinking of simple experiments (such as rotating yourself, rotating the degree scale, rotating the scene), can you show that the winding number is necessarily

IMAGE ROTATOR

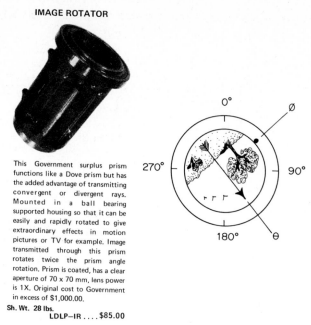

Figure 21.   A catalog ad for a Dove prism and a scene viewed through a Dove prism image rotator.

zero? If so, what do you make of the accompanying mail-order catalog advertisement which makes it appear that one rotation of the eyepiece results in two rotations of the scene: winding number ± 2? I was so perplexed by this paradox that I ordered one of the optical devices advertised (in a less expensive version). When it came, I looked through it at my page of notes purporting to prove that winding number ± 2 is impossible. It was impossible to read the proof! The reason for this and the reason for the irrelevance of the "proof" turns out to be that the optical device presents a mirror image of the scene viewed through it (see Box D).

Example H. Consider a wave circulating around a ring, such as a nerve impulse going around and around a ring of heart muscle or around the circumference of a jellyfish. For our simple purposes, we might think of each little piece of tissue as repeatedly traversing a continuous cycle of states with landmarks conventionally denoted "rest", "excited", "refractory", "relatively refractory", "rest", "excited," and so on. At every instant each point on the physical ring is somewhere in this cycle. At any instant (on a snapshot of the ring) the map from ring to cycle is $\mathbb{S}^1 \rightarrow \mathbb{S}^1$. If a solitary wave revolves clockwise, the mapping has winding number $+1$. It has winding number -1 if the wave circulates anti-clockwise. Two pulses chasing each other around a ring correspond to a map of winding number ± 2, i.e., weaving around a line of slope ± 2, etc. (Figure 22). If the ring beats almost homogeneously, then the map varies about the horizontal, with mean slope 0, and so its winding number is 0.

Box D: The Dove Prism Paradox

Suppose you have a device, something like a telescope in outward appearance, through which you can peer at the standard optical object, an arrow (Fig. a). We will consider rotations about the common axis of your eye and the device. The arrow is mounted perpendicularly on this axis. Suppose first that we rotate the device and the arrow together, as though rigidly connected, through an angle ϕ. This is equivalent to rotating your eye (yourself) through $-\phi$ about the same axis. The image of the arrow on your eye must then rotate to angle ϕ. Now rotate the device alone back to its former position, through angle $-\phi$. The arrow's image has thus been rotated through some additional amount $\theta = f(-\phi)$ characterizing this device, for a total rotation of $\phi + f(-\phi)$. This result is the same as though you and the glass had remained fixed while you watched someone rotate the arrow to ϕ: its image must also rotate to $\phi(\ddagger)$. So we have that $\phi + f(-\phi) = \phi$, which means $f(\phi) = 0$ contrary to the advertised miracle of $f(\phi) = 2\phi$. Yet the device *can* be purchased and the image of the arrow *does* revolve fully twice, uniformly, as the device is turned through a full circle! What's going on?

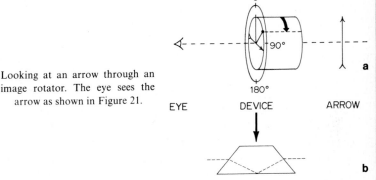

Looking at an arrow through an image rotator. The eye sees the arrow as shown in Figure 21.

When the prism arrived I re-read my proof of its nonexistence, then opened the box. Unwrapping the prism, I eagerly peered through it at the nearest object, the proof. It was unreadable! No, not quite, but just mirror-imaged, besides being rotated. It turns out that the arrow's image rotates once *clockwise* as the arrow turns once *counterclockwise*. So at the ($\ddagger$) above, ϕ must be replaced by $-\phi$, whereupon $f(\phi)$ becomes 2ϕ as advertised.

In fact the wondrous device is nothing more than a mirror with refracting prisms to straighten out the reflected optical axis (Figure b).

15. $\mathbb{S}^1 \times \mathbb{S}^1 \to \mathbb{S}^1$: Input two independent points on a ring (two phases, a phase and an angle, two directions, etc.), output a point on a ring (a phase, a direction, etc.).

These maps, like the $\mathbb{S}^1 \to \mathbb{S}^1$ maps considered above, can be classified by an integer winding number. Maps with different winding numbers have qualitatively different properties. Such maps and their winding numbers play an essential role in much that follows. We begin with four examples:

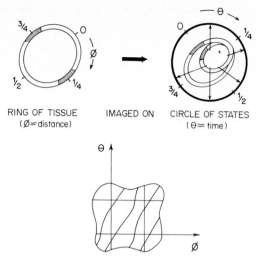

RING OF TISSUE   IMAGED ON   CIRCLE OF STATES
(∅≈distance)   (θ≈ time)

Figure 22.   Upper left: Distance around a ring of tissue is denoted by $0 \leq \phi < 1$. Upper right: Each point of the ring maps to a phase $0 \leq \theta < 1$ in the cycle of local excitation and recovery. Given two circulating pulses, the ring's image wraps twice around the cycle of states. Lower: The same map from ϕ to θ is diagrammed in starker format.

Example A.   Suppose a smooth chemical oscillation, e.g., glycolysis in yeast cells (Chapter 12), is going on in each of two equal volumes. Ghosh et al. (1971) made the intriguing discovery that when such volumes are abruptly combined, the hybrid volume continues to oscillate at some compromise phase somehow determined between the two "parent" phases. This compromise would be most naturally described by a mapping from the phase circle of parent A and the phase circle of parent B to the phase circle of the combined volume: $\mathbb{S}^1 \times \mathbb{S}^1 \to \mathbb{S}^1$.

Example B.   Suppose a motor neuron in a running insect fires at some phase in the rhythmic cycle of locomotion as in the experimental studies of Stein (1974). Let this phase be determined by the timing of two impulses received in each cycle: one from a left leg proprioceptor and one from a right leg proprioceptor or one each from fore and aft segments. The dependence of firing phase on the input phases is most naturally portrayed as a function of two phases: $\phi' = F(\phi_A, \phi_B)$, that is, $F : \mathbb{S}^1 \times \mathbb{S}^1 \to \mathbb{S}^1$.

Example C.   My airplane's compass reads a direction, a phase on the horizon circle. But before it can be useful, installed among the distortions of earth's field caused by the motor and radios, the compass must be adjusted. This is done by moving two little screws, each of which rotates a pair of tiny magnets geared together beneath the compass. With the plane facing in one fixed direction, the compass reading depends on the orientations of the two pairs of magnets: once again a map from $\mathbb{S}^1 \times \mathbb{S}^1 \to \mathbb{S}^1$.

Example D.   At the Kodak Museum in Rochester, New York, there is a device in which two plastic wheels overlap in front of a diffuse light. Each wheel supports around its perimeter a continuum of colored celluloids, running through the whole color circle ROYGBIVROY. By rotating these wheels, you can superimpose in front of the light two filters of any chosen hues and observe the resulting hue. The perceived overlap hue is determined by two hues: Once again this is a map $\mathbb{S}^1 \times \mathbb{S}^1 \to \mathbb{S}^1$.

Many other topological maps naturally organize little areas of experience. Celestial navigation is largely about maps from the product of three circles (time, season, and a sextant angle) to a sphere (position on the globe). The quantum mechanics of electrons and other fermions concerns maps to and from the three-dimensional projective plane $\mathbf{RP}^3$, which is *locally* $\mathbb{S}^2 \times \mathbb{S}^1$. This is also the topological space of fibril orientations in a wide variety of biological polymers (e.g., see Chapter 17). But let's stop with what we have and begin to investigate some of the curious logical implications of maps among circles: what I call "circular logic".

C: Phase Singularities of Maps

A Phase Determined by Two Phases

"But that," said Perion, "is nonsense." "Of course it is," said Horvendile. "That is probably why it happens."

Jurgen (Cabell)

I begin with item 15 above, $\mathbb{S}^1 \times \mathbb{S}^1 \to \mathbb{S}^1$. Call the two input phases or parent phases ϕ_A and ϕ_B, and let the resultant or hybrid phase be called ϕ'. As noted above, the (ϕ_A, ϕ_B) plane goes on and on periodically like wallpaper, in principle to infinity, but we need deal with only one copy of the repeated unit cell since all the rest are the same. In a wide variety of physical and biological situations, the following three axioms hold:

1.   For almost all inputs (ϕ_A, ϕ_B), a small change in ϕ_A or ϕ_B results in at most a small change of ϕ'.

2. ϕ_A and ϕ_B are interchangeable: They play equivalent roles.

3.   When parent phases ϕ_A and ϕ_B are nearly the same, the hybrid phase ϕ' is nearly the same as both.

Referring to Figure 23, consider what happens to ϕ' if ϕ_A is increased through a full cycle while ϕ_B is held fixed (at a value that we may as well call $\phi_B = 0$). If ϕ' changes, it must change in such a way as to return to its first value as ϕ_A completes its scan of the cycle. As in earlier examples, the periodic dependence of ϕ' on ϕ_A has an integer winding number, the number of times ϕ' runs through its cycle as ϕ_A runs forward through its cycle. Whatever it turns out to be, call that winding number W and mark it on path $\alpha\beta$ in Figure 23. Note that $\alpha, \beta, \gamma, \delta$ all represent $\phi_A = \phi_B = \phi' = 0$. Along path $\alpha\delta$ and therefore along equivalent path $\beta\gamma$, ϕ' changes through W cycles again, since this is the same experiment

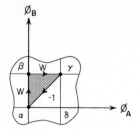

Figure 23.   A phase ϕ varies as a function of position on the (ϕ_A, ϕ_B) torus. Along path $\alpha\beta$ or $\beta\gamma$, ϕ' scans through W cycles. Along path $\gamma\alpha$ it scans through -1 cycle. $\alpha\beta\gamma\delta$ all represent the same point on the torus.

with the A and B labels interchanged. Now returning to α along diagonal $\gamma\alpha$, ϕ' backs up through one cycle according to axiom 3. Thus, from α to β to γ and back to α, ϕ' has changed through some nonzero integer number of cycles $2W - 1$. A contradiction is coming up; watch carefully.

Let us ignore the word "almost" in axiom 1. Then a slight deformation of path $\alpha\beta$, for example, cannot change the behavior of ϕ' along the path enough to alter the fact that it scans through W cycles (rather than $W - 1$ or $W + 1$) from α to β. Since W cannot change except by integer jumps, and jumps are disallowed by axiom 1, therefore W cannot change at all. We can thus deform path $\alpha\beta\gamma\alpha$ as we please. In particular, we can shrink it with impunity to as small a ring as we like. But around that ring, ϕ' still changes through at least one full cycle in response to arbitrarily minute changes of ϕ_A and ϕ_B.

This is called a phase singularity, and it violates the edited version of axiom 1. Without the "almost", in other words, axioms 1, 2, and 3 are incompatible. One *cannot* have a smooth symmetric dependence of phase on two other phases such that ϕ' runs through a full cycle with ϕ_A and ϕ_B when they are equal.

What does this mean in terms of the four examples under item 15 above?

Example A.   Apparently there is at least one critical combination of parent phases ϕ_A and ϕ_B near which the resultant compromise phase is unpredictable and irreproducible. There might also be more than one. For example, there might be a whole line of near-discontinuities along $\phi_A = 0$ and along $\phi_B = 0$ if at that phase the oscillator changes its chemical composition suddenly (Figure 24). The point of Figure 23 is that even if the mechanism is smooth, and is *not* such a so-called relaxation oscillator, then there still must be a phase discontinuity. And if it is not extended along a one-dimensional locus, then it must be the much more violent and condensed kind called a "phase singularity" (Figure 25). In the

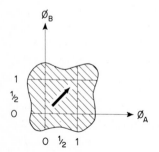

Figure 24.   A relaxation oscillator's contour lines of uniform ϕ' as a function ϕ_A and ϕ_B. Values are not indicated numerically but increase along the arrow. In this case ϕ' changes abruptly by one-half cycle whenever ϕ_A or $\phi_B = 0$.

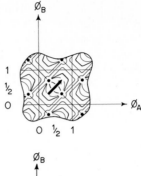

Figure 25.   As in Figure 24, but in this case the inevitable discontinuity is compressed to a phase singularity point in each triangular half of the unit cell.

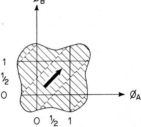

Figure 26.   As in Figures 24 and 25, but the inevitable discontinuity is placed along a 45° line. Here ϕ' jumps by one-half cycle.

simplest kinds of chemical oscillator, the phase singularity is a special combination of parent phases whose intermingling forces the hybrid oscillation to its steady-state (see Chapter 5). But in case of anything more realistic, the phase singularity is not so easy to describe in terms of mechanism. We return to this topic in Chapter 6.

Example B.   In the running insect, it might mean that when ϕ_A and ϕ_B are a half-cycle apart, the third neuron cannot make up its mind whether to fire between A and B or between B and A. This would be the right interpretation if the inevitable discontinuity were a diagonal locus $\phi_B = \phi_A + \frac{1}{2}$ cycle (Figure 26). But if it were an isolated phase singularity, then some more subtle mechanism would seem to be called for.

Example C.   Few of us have ever taken the time to play systematically with the two adjusting screws under a correctible compass. I did once, and subsequently took the thing apart, while preparing an exercise for a course in "The Art of Scientific Investigation". It turns out that the compass follows rules 1, 2, and 3 and that the compass reading becomes extremely sensitive at certain critical combinations of screw angles. This is because at just those angles, the two pairs of little magnets exactly neutralize the earth's magnetic field, or whatever distortion of it reaches the compass through the Plexiglas windshield. The slightest departure from neutrality gives the compass a direction, but it can be any direction, depending minutely on the orientation of each magnet. ϕ' turns out to have a phase singularity (Figure 25) at a critical combination of ϕ_A and ϕ_B.

Example D.   In the case of color perception, it means that on any color wheel *some* combination of two hues must be exactly complementary, resulting in a neutral gray transmission.

These are glib interpretations, not obviously correct. The point I wish to make here is only that there exists a wide variety of mechanistic possibilities underlying the inescapable mathematical fact that there exists no smooth map of the sort required, so we must settle for a map containing at least one isolated phase singularity, or some more extensive kind of discontinuity. That much is mathematics. The *science* comes in locating and making use of the discontinuity, and discovering which of many alternative mechanisms underlie its particular character.

My appreciation of this mathematical fact is the work of Graeme Mitchison, who suggests a more compact statement which I paraphrase as follows:

Theorem.   *The only continuous maps from the disk (e.g., the area inside triangle αβγα) to the circle (the possible values of ϕ') have winding number 0 around the border of the disk. If the winding number cannot be 0 (e.g., because of axiom 3), then the map cannot be continuous no matter what the underlying physical or biochemical mechanism may be.*

For general background and proofs of this theorem in its various guises, see almost any textbook of homotopy theory, e.g., J. Milnor (1965) or E. H. Spanier (1966).

A Phase Determined by a Phase and a Magnitude

With the topological essence of the phase singularity thus extracted and crystallized, it is convenient to turn attention back to the less symmetric case of a phase determined by a single other phase, as in items 14A to H above. For example, in item 14F (page 21) the experimental system determines a phase. But the independent variables at the experimenter's disposal are now a phase ϕ, as above, and a stimulus magnitude M instead of another phase. The map is $\mathbb{S}^1 \times \mathbb{I}^1 \to \mathbb{S}^1$. Thus the resulting phase ϕ' is to be plotted over the ϕ by M plane rather than over the ϕ_A by ϕ_B plane. Let us find a stimulus magnitude which maps ϕ to ϕ' with $W = 0$ and use that as the unit M. Just as we circumnavigated a triangle in the first part of this section, we now circumnavigate a square (Figure 27): Along line PQ with stimulus magnitude 0, $\phi' = \phi$. If Q is placed one cycle beyond P, then along PQ, ϕ' rises through one cycle. Along QR, ϕ' changes by some amount; call it x.

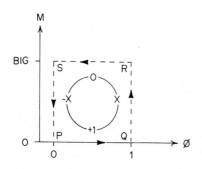

Figure 27.   Analogous to Figure 23 but one axis (M) is a scalar variable rather than a phase. New phase ϕ' is to be evaluated along path PQRSP.

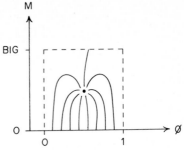

Figure 28.   Contour lines of ϕ' are added to Figure 27 as in Figures 24–26. In this example the contours are allowed to converge to a single point.

The winding number being 0 along RS, ϕ' rises and falls without net progression in either direction. Then from S to P, the change x is undone because PS at phase 0 is identical to QR at phase $1 = 0$. So the winding number of ϕ' around the box is exactly $1 + x + 0 - x = 1$. For the same topological reasons as encapsulated in in the theorem, a peculiar discontinuity must lie within the box. The most localized possibility is an isolated point phase singularity (Figure 28).

One can visualize the necessity of some such irregularity by trying to construct a smooth surface of data points over the square, subject to the constraint that along the boundary PQRSP it must rise up one floor like a parking garage rampway. Think how this is arranged in the garage: It isn't. The concrete apron never closes, but has an inside edge everywhere. It's the same in a helical staircase or sliding board: The central upright is there not just for structural reasons, but also to hide a discontinuity. Architects too, not only the Creator, must design around the facts of topology.

In some cases, the discovery of this screw-like discontinuity constitutes only a trivial insight into the physical mechanisms involved. In some cases, it constitutes a more profound insight, which would be difficult to arrive at in any other way. And in some cases, it has little to do with mechanism but rather points to a fundamental limitation in the process of measurement. As noted above, measurement consists of defining a suitably convenient observable of state and carrying out the map from a system's state space to the space on which that observable is defined. But every experienced scientist knows that we are not free to define observables arbitrarily. One way the use of language in science differs from its everyday uses is that in science one cannot simply define a new term in an intuitively convenient way, obtain consensus on a concise definition, and proceed to use it. For terms refer to concepts, and concepts sometimes prove to harbor latent ambiguities, irremovable ones, e.g., "simultaneity" in relativistic physics. Physicists ran up against this problem, for example, in the 1920s when it was discovered that no one could specify simultaneously the position and velocity of an electron, although position and velocity had served as the basic independent observables of mechanics since the time of Newton. In the same way, if we insist upon forcing nature into a description in terms of phases, we may encounter ambiguities, even paradoxes, implicit in the very definition of that observable because there are situations in which phase is ambiguous or can only be defined by admitting ghastly discontinuities into the description of our observations.

This fact of topology provides us a tool for uncovering phase singularities in a wide variety of contexts. In the next chapter we examine an assortment of such singularities preliminary to entertaining notions in subsequent chapters about their diverse origins in biochemical dynamics. We will then be in a position to inquire into the physical nature of the phase singularity in each example in Chapter 10.

D: Technical Details on Application to Biological Rhythms

This chapter has mostly concerned mappings among circles. As Chapter 2 mainly illustrates phase singularities by empirical data, it will be well to pause before plunging into those examples to make sure that two much-used conventions are clearly grasped. These are:

1.   The correspondence between the ring S^1 and the phases of something periodic in time

2.   The format here used for portraying empirically measured maps from the phase ring to itself

The Phase of a Rhythm

"Phase", like compass direction, is a point on a *circle*, on S^1 We identify directions by a number, on $\mathbb{R}^1$, inscribed on the rim of a compass. We teach our children to think of time schizophrenically both as a thread ($\mathbb{R}^1$) along which we inexorably progress from left to right, from a past lost in the mists of antiquity into a future forever receding before our advances, and also as a circle (S^1) along which we creep like the hour hand that creeps repeatedly around the face of a clock somewhere in Greenwich. According to the Oxford English Dictionary, the original Old English and Old Teutonic stem of our word "time" meant "abstract extension". There was no connotation of recurrence in daily or seasonal cycles such as children now appreciate in asking, "What time is your piano lesson?" The distinctive features of rhythmic systems seem perplexing, even paradoxical, so long as our minds dwell implicitly in $\mathbb{R}^1$. Let us then plant them firmly on S^1 in this chapter about circular logic and phase maps.

By "phase" I mean a point on a ring. But what point and what ring? Let's use the unit circle in the standard (x, y) coordinate plane:

$$x = \sin(2\pi t)$$
$$y = \cos(2\pi t) \qquad (x, y, t \in \mathbb{R}^1)$$

or, in complex notation,

$$\mathbf{z} = x + iy = i\exp(-2\pi it) \qquad (z \in \mathbb{R}^2).$$

As t runs from 0 to 1, the phase point runs clockwise around the circle from north to north (see Figures 6 and 9). As in Example 6 in Section B, our convention for angles is taken from the mariner's compass: clockwise from north, rather than anticlockwise from east as in the mathematicians convention.

In connection with an ongoing rhythm, phase is a function of time: It is the fraction of a period elapsed since the most recent recurrence of whatever event is chosen to mark $\phi = 0$. Thus we write:

$$\phi = f(t) = \frac{(\text{time } t \text{ since marker event})}{(\text{standard period } T)} \text{ modulo 1}$$

which makes $\phi = 1$ equivalent to $\phi = 0$. In an undisturbed, perfectly regular rhythm of period T, this definition implies that ϕ constantly increases:

$$\frac{df(t)}{dt} = 1/T.$$

For notational convenience, we take T as our unit of time henceforth and denote d/dt by an upper dot. Then the above expression becomes

$$\dot{\phi} = 1.$$

Defined this way, "phase" is just a conveniently periodic measure of elapsed time.

To indicate the position of a rhythm on a time axis, we will sometimes speak of its phase as though it were a fixed quantity, not increasing in time. This is only shorthand. What is intended here is $\phi(t_0)$ at some time t_0 which must be specified. For example, the "old phase" of a rhythm is its phase at the moment when a stimulus begins, and "new phase" is its altered phase at the moment when the stimulus ends.

In the preceding paragraphs, we deal in rhythms without giving a thought to the mechanism of rhythmicity. Slipping a little further from pure observation toward interpretation of mechanism, phase is used to denote the state of any process which varies only along a ring or cycle (Chapter 3). Unlike almost any rhythmic process outside the domain of classical mechanics, the only state variable of any consequence in a mechanical clock is the angular position of its meshed gears, as indicated by the position of its hour hand. For many purposes, the same is true in good approximation for circadian clocks. The models of Chapter 3 emphasize this aspect. With such models in mind, one tends to think of the phase $\phi \in \mathbb{S}^1$ of a clock *mechanism*. This is a subtle but weightly shift of meaning from the more innocent use of "phase" above, merely to describe an observed rhythm. It leads to paradoxes in certain kinds of experiments with biological clocks. Thus, when pondering mechanisms we must stick to the purely descriptive definition of phase with fundamentalist zeal; exception is made only in Chapter 3, where the state space consists of no more than a ring, so each state can be unambiguously associated with a phase of a rhythm, and vice versa. But in dealing with real organisms, phase must be defined simply as an observable, a quantity you can wring out of a measured rhythm, without assuming a one-to-one correspondence with the state of the (unknown) mechanism that causes the rhythm. In terms of mechanism, phase is a nontrivial *function* of state and our objective is to infer from experiments the topological features of that function. Such features turn out to have interesting implications.

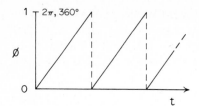

Figure 29.   Phase, considered as a number rather than as a point on a ring, cannot steadily increase. It has to jog back to zero at unit intervals. This notation carries with it the misleading connotation of abrupt change.

I dislike to think of phase as a number from $0°$ to $360°$ or from 0 to 24 hours or from 0 to 2π radians or from 0 to 1 as in Figure 29. To do so gives the impression of a relaxation oscillator, as though something physiologically unique must necessarily occur sometime in the cycle, which point we single out to call $\phi_0 = 0 = 1$. In general, nothing of the sort is so. For instance, no one feels momentary vertigo in crossing the International Date Line (see Box A). The mathematical artifact is the same in both cases: It comes of insisting that the circle should be mapped onto a line. In fourth grade geography we learned not to insist upon mapping the sphere onto a plane but every map in the book nevertheless did it, with apologies. In the same way, we do not always escape it entirely here. Just as we once learned to think in spheres while staring at flat pictures, so in this book the reader must try to think in circles while inevitably handling numbers.

A Format for Plotting Empirical Maps $\mathbb{S}^1 \to \mathbb{S}^1$

We have one more task before going on to Chapter 2. If you are already familiar with phase-resetting experiments, you may find it expeditious to skip ahead, but I urge you at least to scan the following pages on data presentation. For the present book I have committed myself to redundancy in this matter (e.g., see Chapters 14 and 19) in order to be sure that the idea comes across to every reader, but it might not work unless you suffer through some of the redundancy.

Recall Example 14F in which a biological rhythm is rephased from an old (prestimulus) phase to a new (poststimulus) phase. Let's develop this example in greater detail. Experiments in this format will play a major part in chapters to follow. This example focuses attention on the regularities of phase resetting. We experience phase resetting when we cross meridians by jet travel. We inflict it on a wristwatch by pulling out its stem and twisting it to adjust to the time at our landing place. In later chapters (3, 4, 5, and 6) it will be important to deal with the *process* of rephasing. But here we deal only with results: What is the new phase arrived at given the initial phase? The format in which these results were depicted in Example 14F was chosen for compatibility with one of the many formats used to depict experimental results. This chosen format was first used by Pittendrigh and Bruce (1957) and by Hastings and Sweeney (1958) to introduce the notion of a resetting experiment to the field of circadian rhythms. As this is the most direct representation in terms of experimental data and happens also to mesh nicely with convenient theoretical diagrams to appear in later chapters, I adopt it as the standard format for all presentations of resetting experiments in this volume. Data drawn from experimental papers in several different formats (e.g., see Box C of Chapter 4) will all be converted to this format.

Before going into detail about the format, a word is in order about the basic idea of a resetting experiment. The notion is that something rhythmic is going on: Once started by some standard procedure, some event reliably recurs at unit intervals of time. We could just sit and watch it, fascinated by the regularity. But we are soon seized by an impulse to poke at it and see what happens. Very complicated things can happen depending on when in the cycle we poke. We restrict our attention here to just one aspect. In the aftermath of a poke the system eventually reasserts its prior rhythmicity with the event occuring at a time θ (neglecting multiples of 1) after the stimulus. The whole idea of these "hit-and-run" experiments is simply to record the time θ and how it depends on the time in the cycle when the poke is given. Remarkably enough, something can be learned by systematic attention to this simple observation. Moreover, it is an observation of immediate practical relevance whenever one rhythm is to be synchronized by another, or whenever the expected time of an event is to be changed. In order to accomplish this rescheduling we need to know how the time θ depends on the time in the cycle when we do something about it. My objective here is merely to lay out the format of such hit-and-run experiments in a standard way that will lend itself to interpretation of real experiments in subsequent chapters.

Since the late 1950s, experiments of this kind have been published in diverse formats, using rhythms as different as the cell division cycle, pacemaker neurons' electric rhythms, the circadian rhythms of whole-animal activity, and the periodic ovulation of female vertebrates. It seems as though every rationally conceivable format has been employed, and then some. But if comparisons are to be attempted, a convenient standard format must be adopted. This turns out to be one of those intriguing problems for which our habitual reliance on cartesian coordinates does not provide the optimum choice of format. *Toroidal* coordinates prove most natural for visualizing the relationships of idealized rhythms. (See item 14 of Section B above.) But let's work our way around to that inference gradually, starting with the familiar Cartesian graph format.

Figure 30 presents such a format for laying out the results of hit-and-run perturbation of rhythmic processes. In Figure 30, we see a vertical time axis. The origin of time is at the moment when rhythmicity is started by some standard procedure. For example, circadian rhythms can be started by transfering cells from constant light to constant darkness. Thereafter, some conspicuous event recurs at regular intervals. We take that naturally given interval as our unit of time. By choosing the event to call phase 0, we determine the quantity θ_0 between the standard start and the appearance of phase 0. Thus, the standard procedure starts the rhythm, by definition, at phase $1 - \theta_0$. Bear in mind that we have two time scales going now. We have absolute time t measured from the start all the way down the time axis. And we have periodic time ϕ measured modulo 1 from the phase reference events. Further down the axis I have diagrammed a stimulus. The stimulus need not be instantaneous. It can be *any* procedure, for example, turning on the lights and letting them fade gradually back into darkness, or exposing a cell to a series of action potentials 10 milliseconds apart for as long as the stimulus lasts. The essential point is that up to stimulus time, and after the stimulus ends, the rhythmic system is left alone in a standard environment where

was just the single toroidal surface slit open along two perpendicular circles, laid flat, and repeated for the sake of continuity (see Example 14: $\mathbb{S}^1 \to \mathbb{S}^1$). Though redundant, the planar representation is useful and we will see it again in several contexts because real world data are not exactly periodic and the departures from exact periodism are sometimes systematic and meaningful. They would be lost as "noise" in a strictly periodic procrustean bed such as Figure 11. But for now, that is precisely what we want to do: We want to examine the strict implications of exact periodicity in order to generate a rigid skeleton of ideas and expectations on which to hang more flexible realities later.

Let us turn to the torus and inquire how a continuous trail of data points might lie on its surface. If we concern ourselves only with the logic of continuity and of periodicity, then the data trails must rise through an integer number of unit cells in Figure 34 as they advance horizontally through one unit cell. In terms of Figure 11, this means that the data form a closed ring, threading the hole in the doughnut in one direction or the other some number of times. Such curves fall into distinct groups, according to how many times and in which direction they loop through the doughnut's hole. Two limiting cases will serve as examples:

1.   Suppose the stimulus is so brief ($M \to 0$) that it has almost no effect as in Figure 32, so $\phi' = \phi$. The data then lie in a ribbon exactly superimposed on the controls. This curve parallels the main diagonal on our toroidal graph paper, forming a ring that winds once through the hole in what we henceforth take to be the positive direction (as ϕ increases, so does ϕ'). This is called type 1 resetting, according to its winding number. The winding number is the same as the average value of the slope, $d\phi'/d\phi$. Even if ϕ' differed from the controls by a fixed amount independent of ϕ (e.g., if the stimulus does nothing but lasts a long time so $\phi' - \phi = M$ is not negligible), the data would parallel the main diagonal and link once through the hole. Even if ϕ' depended markedly on ϕ in certain ways (e.g., $\phi' = \phi + 100 \sin(2\pi\phi)$), the topological type of resetting would still be the same: The data ring passes many times through the hole, but each time except once it turns around and comes back out again. And although the slope $d\phi'/d\phi$ reaches great extremes, it averages out to exactly 1 on account of the requirement of periodism.

2.   Suppose the stimulus restarts the rhythm at phase ϕ_0 whenever delivered, so that ϕ' is equal to ϕ_0 independent of ϕ. This curve parallels the equator of our toroidal graph paper, maintaining a fixed vertical level as it rounds the ϕ axis. It never loops through the hole. This is called type 0 resetting. Even if ϕ' were $100 \sin(2\pi\phi)$, the topological type of resetting would still be 0, as the winding number of the data through the hole remains 0 despite its passing in and out of the hole (in opposite directions) many times. The mean slope $d\phi'/d\phi$ is also 0. (For example, see Boxes B and C in Chapter 4.)

Curiously, no other resetting types have been reported from experimental systems in which a reasonably smooth curve can be sketched through the data. In principle, *any* integer type is compatible with the requirements of continuity and of periodicity. What further restriction constrains naturally occuring resetting curves to type 0 (ϕ' vs. ϕ wiggling parallel to the stimulus) and type 1 (ϕ' vs. ϕ wiggling parallel to the start)? This question will be more approachable later, when we have passed beyond classifying data into a discussion of kinds of dynamical systems. It turns out (Chapter 6) that some kinds of oscillator mechanism can yield only type 1 resetting, some kinds can yield both type 0 and 1, and other kinds can in principle yield any integer type. At that point, we will also have to deal with the realistic fact that in certain experiments there are systematic deviations from periodicity along the ϕ axis, or along the ϕ' axis, or along both. Thus, Figure 34 cannot really be strictly rolled up onto the doughnut coordinate system of Figure 11. Such irregularities do not prevent our measuring phase and cophase by extrapolating the poststimulus rhythm back to the first moment of free-run after the stimulus. These particulars are dealt with later in connection with particular classes of mechanism, and in Bestiary chapters about particular organisms.

A second big assumption lurks scarcely beneath the surface here: An assumption of continuity, glossed over earlier in stating that resetting curves thread their way smoothly from one unit cell to the next, in Figures 32 through 34. In imagining a smooth curve threading the data points, we *suppose* that a small change in the time at which stimulus is given makes only a small change in the resulting ϕ. In other words, we neglect the possibility of a threshold phase at which the system's physiology changes discontinuously by a jump or a jerk. Counterexamples abound. We take up some of these cases in connection with some kinds of neural pacemaker in Chapter 6, Section C, in connection with the cell cycle in Chapter 22, and in connection with the female cycle in Chapter 23.

In fact, a resolute pursuit of even the continuous cases with winding number $W \neq 1$ leads relentlessly to a discontinuity of the monstrous sort called a "phase singularity" (Chapter 2). But we cannot appreciate the inevitability of a discontinuity without first pursuing the narrow path of idealization, restricting ourselves slightly to consider systems whose phase-resetting data appear to lie on smooth curves.

2. Phase Singularities
(Screwy Results of Circular Logic)

... beware of mathematicians and all those who make empty prophecies. The danger already exists that the mathematicians have made a covenant with the devil to darken the spirit and to confine man in the bonds of Hell.

St. Augustine

A *phase singularity* is a point at which phase is ambiguous and near which phase takes on all values. My purpose in this chapter is to give examples by somewhat idealized description of phase singularities observed in several experimental systems. In some cases, the phase singularity is at this writing only inferred and not yet demonstrated. Some cases of purely hypothetical and trivial nature are also thrown in to help clarify the principles that I take to be involved in the more interesting biological examples. Much is glossed over here that should disturb a thoughtful person acquainted with the physiology of any one of these systems. These details are dealt with in Chapter 10 and in the Bestiary (Chapters 11–23).

The 15 examples I've chosen are organized in this chapter as follows:

First come three that are intuitively more familiar than the rest, to make the geometry familiar. Examples 5–10 all concern biological time measurement: They describe phase singularities of living clocks. Physical space is not essentially involved, except maybe in Example 10, which features clocks in a growing fungus. From there on, physical space plays an indispensible role. The singularity is actually visible as the organizing center of a two-dimensional pattern. Example 15 describes the most sensationally visible example of all, an oscillating chemical reaction in which colorful red and blue striations are organized around phase singularities in three-dimensional space.

A: Examples

Example 1. Color Vision. My first example is at the same time the most abstract and the most compellingly graphic. It concerns an oddity of human color vision cited in the previous chapter as Examples 14A and 15D. Normal

Figure 1. The plane $x_1 + x_2 + x_3 = 1$ in the three-coordinate positive Cartesian depiction of color receptor excitations.

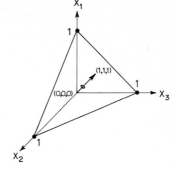

Figure 2. The locus of spectral hues (heavy); its psychophysical closure through purples (dashed); and the uniform-hue contours, all on the triangular plane of fixed intensity from Figure 1.

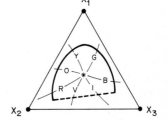

color vision in humans is based on three primary colors. Each color of light falling on the retina excites three distinct receptor processes to different extents. (Other animals may have different numbers of separate color sensors.) The infinite variety of physical colors, each defined by a distribution of energy across the visible part of the spectrum, is thereby reduced in human perception to three numbers, three rates of nerve activity. Each color occupies a place in the three-dimensional space so defined. This is a very simple model, ignoring as it does the effects of surrounding color and texture. However, it is entirely adequate for pure homogeneous color phenomena, the situation for which the colorimetric model was invented by Isaac Newton, James Clerk Maxwell, Hermann Helmholtz, and Thomas Young (see Sheppard, 1968).

If we neglect differences in overall brightness, i.e., if we confine our interest to equally bright stimuli (e.g., by the criterion $x_1 + x_2 + x_3 = 1$ in Figure 1), then we need deal only with a two-dimensional section through this three-space (Figure 2). The boundary of this section is a triangle, topologically equivalent to a ring. Near the triangle's center, we find the least saturated colors, the grays associated with roughly equal excitation of all three receptors. On this triangular section, the spectrally pure colors map to an open U-shaped curve which represents the wavelength axis from about 400 nm to about 700 nm. The corners and edges of the triangle are normally inaccessible because (outside of dreams and clever procedures in the psychophysical laboratory) it is impossible to excite one receptor to the complete exclusion of others, on account of their overlapping absorption spectra.

The peculiar fact about human color vision is that we subjectively perceive mixtures of red and blue as a pure intermediate hue called indigo or purple. These hues complete a ring between the end points of the spectral U locus. They complete it without repeating the spectral hues in reverse order. Thus red is close to orange is close to yellow is close to green is close to blue is close to indigo is close to violet is close to red is close to orange. The essence of any topological space is its connectivity, and we have ring-like connectivity in the case of hues. The closed U is the "color wheel" of grammar school.

Phrased otherwise, the peculiar fact about human color perception is that we subjectively recognize colors not by the Cartesian coordinate system $\mathbb{R}^3$ that our three receptor excitations define, nor by any topologically equivalent coordinate system, but rather by darkness and brightness (topologically equivalent to radial distance from the origin), by saturation (topologically equivalent to distance from the gray diagonal $x_1 = x_2 = x_3$), and by hue (topologically equivalent to an angle about that diagonal). That is, we employ a coordinate system that somehow maps $\mathbb{R}^3 \to \mathbb{S}^1$ so far as hue alone is concerned.

What this means on the color triangle is that the loci of uniform hue (neglecting saturation and brightness) are curves each of which crosses the maximum saturation ring (the color wheel, the closed U) at one point. Inside the color wheel these curves can only converge. The color to which they converge evidently has all hues or no hue: It is gray. It is a phase singularity, in an abstract sense that involves no rhythms in space or in time, but only an irregularity in mapping the triangular disk to a ring. No such map can be carried out continuously unless the winding number is zero along every ring that can be traced on the disk. In this case, the border ring has winding number $W = 1$ so that continuity is precluded. The inevitable discontinuity turns out to be the most localized and violent kind: a phase singularity.

Example 2. Navigation in a Rowboat.   In Example 14B of the previous chapter we dwelt upon the relationship between the compass heading adopted by a boatman and the actual track of his boat as altered by a steady current. We arrived at a trigonometric formula relating track ϕ' to heading ϕ and the relative swiftness of the current. We noted that this relation takes either of two qualitatively different forms, depending on whether the current or the boat is swifter. This peculiarity is associated with a phase singularity implicit in the equation when the boat heads directly upstream ($\phi = 0$) with the same velocity as the current ($v = v^*$). In this situation the boat hangs poised forever motionless in midstream, depending for movement on fluctuations in the heading, the boatman's vigor, or the current. The slightest deviation from equilibrium gives the boat a

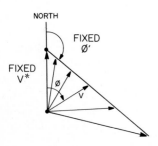

Figure 3.   Taken from Figure 14 of Chapter 1 to indicate the combinations of ϕ and v that are compatible with any fixed ϕ' and v^*

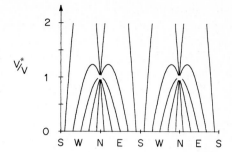

Figure 4. The navigator's mapping: The co-ordinates are heading relative to the current (horizontal) and velocity relative to the current (vertical). The curves are level contours of resultant track direction, relative to the current.

gentle drift in some direction ϕ', and all directions are equally accessible. The combinations of ϕ and v that result in the same ϕ' are indicated geometrically in Figure 3. Figure 4 depicts this dependence as a contour map of track direction ϕ', plotted on the (heading ϕ, velocity ratio v^*/v) plane. The contour lines of each given track direction are identified by their intersections with the heading axis at $v^*/v = 0$ (no current), where heading = track. At $v^*/v = 1$ all contours converge near a north heading (upstream). At greater currents or lesser rowing speeds, all tracks are downstream regardless of heading.

Example 3. What Time is It? What time it is depends on where you are, doesn't it? The clocks behind the newscaster's desk say "Tokyo", "Moscow", "New York", and "Canberra" and each reads a different time. But this madness has its method, as noted in Example 14C of the previous chapter. If we lay out a ring of clocks along a globe-girdling path—let's say the path of Pan American Flight #2—then we have a big circle and along that circle, at any instant, local time advances through a full day. This leads to a difficulty. Our ring of clocks is a big circle lying on the surface of the planet (Figure 5). Let's contract it a little. If we think of it as a necklace of 40 million pocket watches, a meter apart, we can envision taking up a meter of slack, pushing the watches apart, and doing it again repeatedly. As the ring shrinks, we may have to walk a bit north to keep up with it. If we take care that every watch moves north, then each stays in its local time zone and need not be reset as it moves. Eventually, after all but 100 meters of wire have been rolled up, the other side of the ring comes into sight over an ice flow, dragging its millions of pocket watches. Gather up another 100 meters of wire. Now, what time is it? All times, no time, summertime? Look at the sun if the clocks don't agree. When it is at its highest, in the south, call it noon. But the sun is circling at fixed altitude along the horizon and *every* direction is south.

This place is a phase singularity.

Figure 5. A ring of pocket watches, initially on the Equator, during its contraction toward the North Pole.

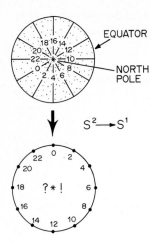

EQUATOR

NORTH POLE

$S^2 \longrightarrow S^1$

Figure 6. A polar view of the North Hemisphere showing boundaries of (idealized) time zones. To assign a time to each place on this hemisphere, the hemisphere must be mapped to the phase ring (below). This cannot be done smoothly.

Let's look at it more abstractly. At every instant the determination of local time as a function of position on the globe is a map from $S^2 \to S^1$. Along the equator the map is quite tidy: local noon advances four minutes for every degree of east longitude. [This simple fact was the whole reason for development of accurate chronometers: They were indispensible for global navigation (see Box B of the previous chapter)]. So the equator is a convenient line along which to cut the sphere open, the better to deal with each of the hemispheres separately as a disk. The two-dimensional disk must be mapped to S^1 in such a way that winding number equals 1 along the disk's border (Figure 6). This cannot be done continuously. The most localized form which the inevitable discontinuity could take is a phase singularity: a point at which all phases converge, a point that the map pulls apart to all phases. This is naturally placed at the pole, being both the earth's rotation axis and a place where almost nobody lives.

Because time and place are symmetrically measured on the earth's two hemispheres, there is a second phase singularity. Both have fascinated navigators since the globe was first charted because both represent inevitable convergences of meridian lines that were initially sketched to separate and to distinguish successive arcs along the equator (see Box A).

Example 4. Timing the Tides.

> It is even said (by footsoldiers returned from Alexander's expedition to the Arabian Sea) that the many ebbings and risings of the sea always come round with the Moon and upon certain fixed times.
>
> Aristotle

Another phase singularity of the same kind is familiar to oceanographers concerned with the prediction of local tides and with understanding the complexities of their geographic variation along the world's shorelines. This example brings out more clearly the involvement of rhythms in time. The tides have many harmonic components based on the earth's period of rotation and on both the

Box A: Living Clocks at a Geographical Phase Singularity

Physiologists as well as navigators have been fascinated by the poles. A garage a few hundred meters from the South Pole became the scene of a unique biological experiment when Karl Hamner arrived by an Air Force transport with hamsters, fruit flies, bean plants, and breadmold (Hamner et al., 1962). Perched on a counter-rotating platform at the end-point of the International Date Line, what time would their biological clocks think it was? Two major possibilities were envisioned:

A. x hours later than it was x hours ago, when they were elsewhere: Expected on hypothesis 1, that the biological rhythms derive from internal biochemical oscillators that take no notice of geography.

B. No definite time because at the pole no clear physiological rhythms will persist by which to evaluate subjective physiological timing: Expected on hypothesis 2, that the rhythm really reflects some local consequence of the earth's rotation, to which the chosen species are more sensitive than are their observers' instruments.

What happened? Outcome A: The organisms continued in their physiological rhythms just as a man's pocketwatch does, not caring that it is carried across the pole. This observation encourages belief in hypothesis 1, that circadian rhythmicity arises from an internal physiological clock.

Unfortunately, the observation of outcome A did not really distinguish rigorously between hypotheses 1 and 2 because organisms *could* have been reading Greenwich Mean Time just as plausibly as they could have been reading other subtle concomitants of the earth's rotation. For example, a Greenwich Mean Time readout is available synchronously around the globe not only from station WWV but also in the vertical gradient of electric potential in the atmosphere. This is caused by a global maximum rate of thunderstorm formation a certain number of hours after the major continental land masses roll into the sunlit hemisphere at 7 P.M. in London (Feynman et. al., 1964, vol. 2, p. 9–3). Because the ionosphere conducts so well, this readout is available unattenuated at the South Pole unless experiments are conducted in a Faraday cage. Palmer (1976) also notes that the *magnetic* South Pole is 1,500 miles away, near New Zealand. Thus organisms on the cosmically stationary platform were subjected to a rotating magnetic field with 24-hour period.

Apart from these quibbles, the idea of the experiment was ingenious: If biological rhythms are geographically caused, then they should vanish when the organism is placed in a geographic phase singularity. They don't. We will see in Examples 5–9 that a similar experiment can be done too: If biological rhythms are physiologically caused, then they should vanish when the organism is placed in a physiological phase singularity. They do.

earth's and the moon's periods of revolution about their primaries, and on local resonances determined by shoreline geography and depth profiles. The excellent approximation provided by linear theory allows us to deal with each component separately. The amplitudes of the components vary locally, but each component is independently subject to the same principles. Consider for example the fundamental harmonic of the lunar semidiurnal tide M_2, which happens to be dominant on the Isle of Palms where I am writing.

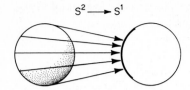

$$S^2 \longrightarrow S^1$$

Figure 7.   The surface of a sphere mapped into a segment of a circle.

In the very first approximation, the tide is a rotating pair of bulges of water above the global mean sea level. Viewed locally, it is a rhythmic rise and fall of the water surface, or a rhythmic advance and retirement of water along the sloping shoreline. The tide cannot be globally synchronous because there is a fixed quantity of water available. So there is a geographic pattern of phase. This is easy to measure along the shores, and it is possible to measure at sea using sonar depth finders. Thus, the tidal pattern on an instantaneous snapshot of the globe could be portrayed as a map of the sphere onto the phase circle ($S^2 \to S^1$). This poses no problem to imagination. A sphere is easily mapped into a circle, e.g., as in Figure 7. But the map I want also has to embody a special fact, namely, that the local rhythm's phase (measured relative to Greenwich Mean Time) scans through two full cycles along any ring girdling the earth near its equator. Since the winding number is 2, not 0, there is no such continuous map. Try to visualize every point on a spherical rubber balloon assigned to a point on the circle while wrapping the equator twice around that circle as required. Figure 8 shows this operation on the Northern Hemisphere. The balloon has to be stretched flat like a pancake, and to finish the job, its surface must be punctured in general at two points in each hemisphere. The edges of each puncture fly out all around the equatorial circle, so each puncture is a phase singularity. This would be the simplest possible geographic pattern of tidal phase.

It turns out that there are quite a *lot* of phase singularities. Some of them perhaps are lost inside the continental land masses, but there are thought to be several in the open sea (Defant, 1961; Platzman, 1972). Oceanographers depict these so-called amphidromic points by drawing "cotidal contours" on the ocean map as in Figure 9 (from Chart 1 of Defant, 1961). A cotidal contour is the geographic locus of synchronous tides, the locus along which the tide crests simultaneously. Thus, each cotidal contour can be assigned a unique clock time, the phase of the 12.4-hour cycle when this locus is at full tide. These contours radiate from the amphidromic point. At the amphidromic point, the tide is full at all

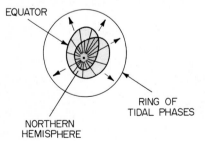

Figure 8.   An attempt to map the North Hemisphere onto a ring with the Equator wrapped twice around. The hemisphere must be punctured at two points.

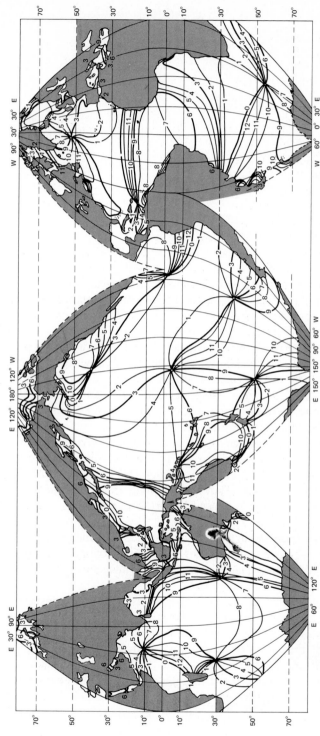

Figure 9. A map of the globe showing the "cotidal" loci along which the tide crests at each given hour. [From Defant (1961) with permission.]

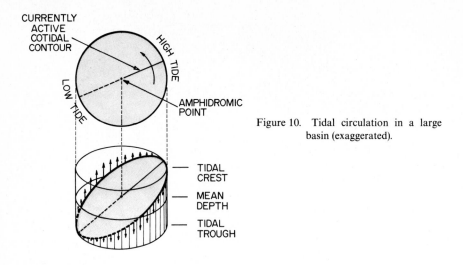

Figure 10. Tidal circulation in a large basin (exaggerated).

times or at no time. That is, there *is no* tide. The water surface pivots about the amphidromic point like a rotating tilted plane, counterclockwise in the Northern Hemisphere (Figure 10).

Example 5. The Fruitfly's Body Clock.   Most living organisms have an innate tendency to repeat their behavior and physiological activities at intervals of about 24 hours. This tendency is ascribed to the working of an internal physiological oscillator called a "circa-dian" clock or pacemaker (see Chapter 19). What little is known of its mechanism in any species at present mostly comes from watching the timing of behavioral rhythms that it indirectly drives. One of the best studied of these is a neurosecretory rhythm that ultimately times the first appearance of butterflies, moths, flies, and wasps after metamorphosis (see Chapter 20). Newly hatched flies appear from a heap of metamorphosing pupae at 24-hour intervals. These bursts of eclosion activity reveal the phase of an internal clock that can be started, stopped, and reset in phase by exposures to light.

The Pinwheel Experiment.   It was in such an experimental system, developed primarily by Colin Pittendrigh, that I was first able to demonstrate a phase singularity in a biological clock. The idea was to conduct the following rather inconvenient experiment:

A lawn of young pupae is spread out on a desk top under constant light (Figure 11). By slowly moving a shadow from east to west like nightfall, oscillations are started in columns of pupae as they fall under the shadow (see Chapter 20). If the shadow takes three days to engulf the desktop, an east-west phase gradient spanning three cycles of phase is established. With all pupae now steadily oscillating in the dark, they are exposed to light, almost simultaneously, by moving the shadow southward, exposing first the north edge of the desk, and eventually the whole desk. Just before it withdraws from the south edge, it is suddenly moved

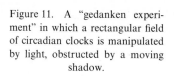

Figure 11. A "gedanken experiment" in which a rectangular field of circadian clocks is manipulated by light, obstructed by a moving shadow.

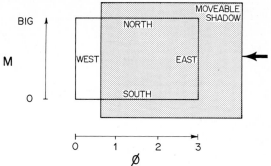

back up to cover the whole desktop. Thus, each row of pupae was exposed for a time proportional to its north-south position. Those at the north edge were exposed for a long time and those at the south edge were not exposed at all. The desk top now contains an orderly array of combinations of phase ($0 \leq \phi \leq 3$, east-west) and duration of exposure ($0 \leq M \leq$ BIG, north-south). We wait to see where pupae simultaneously emerge, and when. This was called the "pinwheel" experiment because the anticipated result was a pinwheel-shaped wave of eclosion or, rather, three identical pinwheels rotating side by side, one in each repeat of the east-west phase gradient. Each should rotate clockwise. This forecast rested on three well-established observations:

1. Of Kalmus (1935) and Bunning (1935): That eclosion follows the light-dark transition by a fixed interval plus multiples of 24 hours thereafter. Thus, the eclosion wave must move along the south edge of the desk from east to west (point D to point A in Figure 12) in each of the three unit cells.

2. Of Pittendrigh and Bruce (1959): That early in the fruitfly's subjective night (point B), a brief but saturating light pulse delays eclosion somewhat whereas early in the subjective morning it (point C) somewhat advances eclosion. It was presumed (and subsequently verified by Engelmann, 1969; Frank and Zimmermann, 1969; Chandrashekaran and Loher, 1969b; Winfree, 1970b and c) that a lesser exposure gives lesser delays or advances.

3. Also of Pittendrigh and Bruce (1959): That with saturating exposures, the transition from the delaying part of the cycle to the advancing part (point B to point C) is made through a region in which the delay inflicted increases by about two hours for each hour into the subjective night. In other words, eclosion follows later after a light pulse given later in the subjective night.

What this adds up to is that if eclosion happens first at point A in Figure 12, then a little while later it is occurring at point B and a little while after that at point C and a little while after that at point D and a little while after that at point A again. The sums of these times necessarily add to 24 hours since everything in Figure 12 has a 24-hour period. In other words, the region of active eclosion circulates clockwise around the borders of region ABCDA in each of the three unit cells of the accompanying figure.

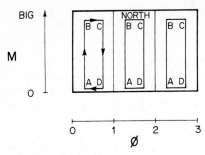

Figure 12.   Figure 11 is repeated after the shadow covers all permanently, indicating the clockwise sequence of activity expected along paths ABCDA.

But if the wave of eclosion rotates, then its pivot point has no definite phase. The pupae in that region cannot be eclosing on a 24-hour schedule. This suggested consequences that might be biologically interesting, so the experiment seemed worthwhile not just to verify what seemed a foregone conclusion but to ask in *what manner* rhythmicity would break down at the pivot of the anticipated rotor. So in 1967 I relocated my graduate studies from the Biophysics Department at John Hopkins University to the Biology Department at Princeton, where Colin Pittendrigh allowed me to carry out this attempt in the constant-temperature darkness of a basement animal room during his tenure as Dean of the Graduate School. The job was finished in a well-insulated basement meat locker in the University of Chicago's Department of Theoretical Biology in 1969 and 1970. Altogether, it required 510 separate experiments in a standard format (Recall Figure 30 of Chapter 1):

1.   The rhythm was initiated in young pupae by transfer from constant light to darkness.

2.   The darkness was punctuated T hours after initiation by an exposure of duration M seconds.

3.   The times of at least three consecutive emergence peaks were observed several days after light pulse.

The 510 separate experiments in this format may be regarded as discrete samples of one big experiment, the pinwheel experiment described above. There is no difference in principle between these two descriptions because (unlike some other insects) *Drosophila* pupae determine their eclosion time on the basis of individually perceived light stimuli, without regard to neighbors' clocks or eclosion time. Thus, it doesn't matter whether they are handled as separate homogeneous populations or as areas in a continuous sheet of differently treated pupae.

The results are idealized in Figure 13 in the form of a computer-plotted contour map. Each contour band indicates the locus in Figures 11 and 12 along which flies are simultaneously active every 24 hours. It indicates the combinations of ϕ and M resulting in emergence peaks after the same delay. Thus after $0 + 24n$ hours, eclosion is along the loci marked by a vertical curve. After $2 + 24n$ hours, eclosion occurs along loci one band clockwise from there, and so on through the 12 bands representing 24 hours at each of the 3 pivot points. As anticipated by the simplistic arguments resorted to above, activity revolves clockwise about pivots disposed at 24-hour intervals in the direction of increasing ϕ (old phase).

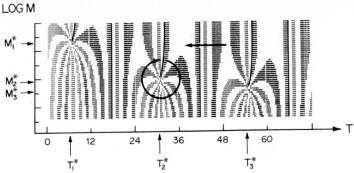

LOG M

Figure 13.   Computer-printed level contours of new phase ϕ' on the plane of exposure time T, to the right, and exposure duration or energy M, upward. [ϕ is $T/24$ modulo 1, offset by a constant; ϕ' is (emergence time θ)/24 modulo 1, offset by a constant. ϕ' approaches ϕ at small M.] Arrows indicate wave movement. This is a contour map of a mathematically-defined surface very similar to the hand sketch in the next figure.

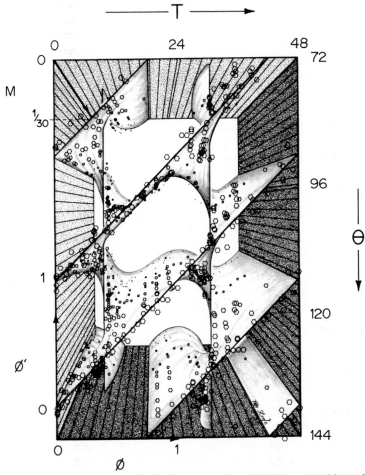

Figure 14.   Six unit cells of the fruitfly clock's time crystal. Data points are plotted larger in the foreground, smaller in background. Stimulus magnitude M increases toward the background from 0 in the foreground. M, T, and θ in hours. ϕ and ϕ' in cycles of 24 hours from the LL/DD transition. See data in stereo without the idealized surface in Figure 5 of Chapter 20.

There is a more artistic and experimentally more direct way to present the data summarized by Figure 13. Figure 13 is a contour map of the surface obtained by plotting activity time (cophase) below the plane coordinated by stimulus time (old phase) and stimulus magnitude. Why not just plot the data and draw the surface in three-dimensional perspective, instead of resorting to a flat contour map? Figure 14 does this. (The cloud of data points through which I sketched the surface is presented in stereo perspective in Chapter 20, Figure 5.) The surface climbs upward as in a spiral staircase winding around poles along the first two singularities.

The essential point of all this is that the activity monitored appears on the stimulus plane as a rotating wave. The position of the wave at any instant is a horizontal cross-section through this screw-shaped surface at the corresponding vertical level. Since the surface is screw-shaped, its cross-sections, at successively lower vertical levels, (successively later times) appear to rotate about the central vertical axis, which serves as a pivot point for the wave (Box B). Actually there are three distinct pivot points spaced 24 hours apart along the east-west phase axis in Figures 11 and 12. These correspond to three repeats of the screw-shaped surface (only two of which are drawn in Figure 14). Since the data are periodic both along the horizontal old phase axis and also along the vertical new phase axis, we have here essentially a crystal lattice of data. Each unit cell of the lattice contains one turn of a screw-shaped surface. I have sometimes referred to this whole structure as a "time crystal" because it resembles a crystal lattice and all three of its coordinate axes represent time. The characteristic screw-like appearance of phase plots near a singularity gives this chapter its subtitle, "Screwy Results", following hard on the heels of our anticipating the phenomenon via "Circular Logic" in the preceding chapter.

This example (the fruitfly clock) was included:

1.   To introduce the equivalence of an abstract space of stimulus parameters to real physical space, as a setting for rotating waves.

2.   To introduce the equivalence of piecemeal resetting experiments to spatial wave experiments, via the time crystal's screw-shaped data surface.

3.   To illustrate again the central role of the winding number in maps of $\mathbb{S}^1 \to \mathbb{S}^1$: The resetting of a rhythm's phase constitutes such a map, and if it has winding number 0 (type 0 resetting), then the winding number cannot be 0 around a certain ring in stimulus space. Therefore that ring must enclose an unpaired singularity (Box B).

4.   To introduce a nontrivial context of application, namely, biological clocks.

The next few examples have the same format and similar results so far as the topological essentials are concerned, each using a different kind of organism. Quantitatively, these experiments do far more than just confirm the generality of the picture first painted by *D. pseudoobscura* because each species has its own surprising idiosyncratic distortions of the basic pattern. But for the sake of keeping our exploration focused on *principles* in this chapter, these details are relegated to Chapter 14 and to the Bestiary.

Box B: Type 0 Resetting Implies a Screw Surface

This seems a good place to draw your attention to a fact more abstractly stated in the previous chapter, i.e., that the helicoidal pattern is very nearly implicit in a much simpler experiment, namely, the measurement of type 0 resetting. You might reasonably ask, "If it is implicit then why bother to carry out the more laborious experiment?" My answer is:

1. Its derivation is not *quite* a tautology. There could be more intricate kinds of singularity, or multiple singularities, which would point to distinct physical interpretations.

2. Behavior in other respects than *phase* resetting (e.g., amplitude resetting) is of critical interest, and such observations come out of the same experiment (Chapter 7).

3. Quantitative aspects of the measurement, not only the qualitative, topological structure of the resetting schedule are also of interest (Chapter 20).

How is it that type 0 resetting betrays the presence of a screw surface within the time crystal? The reason once again derives from the theorem of Chapter 1, Section C, along with some facts and assumptions:

1. Type 0 resetting is obtained with a "big" stimulus.

2. Every rhythmic system exhibits type 1 resetting in response to a sufficiently attenuated stimulus, because in the limit of vanishing stimulus, new phase equals old phase.

3. If the system is really rhythmic, then new phase depends on stimulus magnitude (intensity, duration, or whatever) in the same way at phase ϕ as at phase $\phi + 1$.

4. We suppose, until forced to the contrary, that the dependence is continuous: A small change of stimulus magnitude results in at most a small change of new phase.

5. The winding number must be an integer.

We can reason about the topology of the resetting map on a box diagram in the stimulus plane (Figure a). Let the M axis represent stimulus magnitude from 0 to some "big" stimulus which elicits type 0 resetting. We need to know the winding number of new phase around box ABCDA. From A to B, the new phase increases through one full cycle by item 2. From B to C it changes from 0 to some X. From C to D it goes up and down with no net change according to item 1. From D to A it goes back down from X to 0 by item 3. Thus, the anticlockwise winding number is 1. According to the theorem

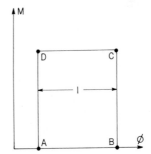

(a) A rectilinear path spanning one cycle of old phase and the range of stimulus magnitude from 0 up to big enough M to elicit type 0 resetting.

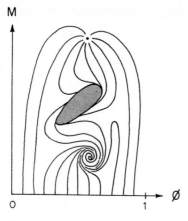

(b) Some of the structures that can appear in a contour map of new phase, superimposed on (a)

of Chapter 1, Section C, there must be an internal discontinuity. This would be a screw axis if the discontinuity is as simple and as violent as an isolated phase singularity. But it could be that boundary ABCDA cannot be shrunk all the way to a point because the inevitable discontinuity is more extensive. For example, it might consist of an isolated singularity and an extra pair of counterrotating singularities or any greater constellation of an odd number of singularities, or even a disk whose boundary constitutes a one-dimensional singular locus inside which phase is undefined (Figure b).

In point of fact, the experimental result for this kind of fruitfly looks about as simple as it could be: a single smooth screw surface with an isolated point singularity. Experiments in which I attempted to improve data resolution near the screw axis (Winfree, 1973a, Appendix) indicated no "inner structure" to the single ostensibly simple singularity seen here at a low resolution.

Another way to look at box ADCBA brings out the connection between type 0 resetting, a singularity, and a screw-like surface. Type 0 resetting with large stimuli implies *nonzero* winding number around the whole box because with small stimuli the resetting is necessarily type 1, and $0 + X + 1 - X = 1$. With winding number $W = 1$ around the box, new phase rises vertically through one full cycle around the box. This establishes the border of the new phase surface within the box. It resembles a helix. Only a screw surface fits within a helical border. A screw surface necessarily has an internal singularity (or some constellation of them).

Example 6. A Mutant Fruitfly's Body Clock (The Singularity Trap Experiment). Box C provides an algorithm for experimentally locating a singularity by checking the winding number of phase along more and more constrictive rings of stimuli on the stimulus plane. Its application was demonstrated using light pulses to rephase a mutant *Drosophila melanogaster*'s eclosion rhythm. In 1972 Konopka discovered type 0 resetting in this mutant whereas only type 1 had been observed in the wild type. So according to Box C, a singularity is implicit in the mutant, though not necessarily in the wild type. A series of rephasing experiments around a box FBCEDGAF in the stimulus plane of Figure 15 produced the eclosion rhythms laid out in Figure 16. Its phase drifts through one full cycle around the box. Thus the winding number around the box is nonzero, betraying the presence of a phase singularity inside that box. Dividing this box by a vertical partition

Box C: Singularity Trap

"The singularity trap" is a procedure for experimentally tightening a noose around the phase singularity in a screw-like resetting surface.

We first draw a coordinate plane with an old phase axis extending through exactly one full cycle and, 90° counterclockwise from it, a stimulus magnitude axis from zero to "big". A big stimulus is one for which the resetting curve (new phase vs. old phase) has winding number $W = 0$, i.e., as old phase is changed through one cycle, new phase rises and falls with no net change. This guarantees that the winding number of new phase around box ABCD in Figure 15(a) is $W = 1$. Accordingly, new phase plotted above the plane of Figure 15 describes a helix along path ABCDA.

How can we most simply fill out the inside of Figure 15 with data points? The objective should be to locate the screw axis (assuming only one) from which the rest of the surface fans out to its helical border. One way is to execute a series of experiments repeatedly bisecting the square ABCD of Figure 15(a). In the first step, the stimulus range (or the cycle) is divided into two equal intervals by a dozen experiments equispaced along line EF in Figure 15(b), all at middle phase with increasing stimulus magnitude. This line divides square ABCD into two rectangular boxes. Depending on how new phase varies along their common border EF, one box still has $W = 1$ and the other has $W = 0$: the boundary of one box (and not of the other) is helical. Take the helical box. Bisect it by a dozen experiments along perpendicular line GH. Of the resulting squares, only one is guaranteed to have a helical border and therefore to contain the screw axis. Take that square and repeat the preceding two-step measurement. In each case retain the same number of experiments, even though they are spaced closer together along the shorter line segments used. In each stage, new phase rises through a full cycle around the box, so that the same number of data points is needed to outline a smooth curve. Three repeats of this procedure locate the screw axis (ϕ^*, M^*) to within one-eighth of a cycle and one-eighth of the original stimulus magnitude range [Figure 15(c)]. This procedure fills the box ABCD with enough data points (automatically concentrated in the steeper parts) to outline the whole surface.

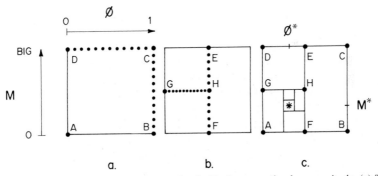

a.   b.   c.

Figure 15. Stimulus parameter plane: one cycle of old phase vs. stimulus magnitude. (a) The first round of measurements is indicated by dots along loci BC and CD. In (b), experiments in the second round of measurements are indicated by dots along loci EF and GH. In (c), after the fourth round, the singularity at old phase = ϕ^* and stimulus magnitude = M^*, is known to be in the innermost square.

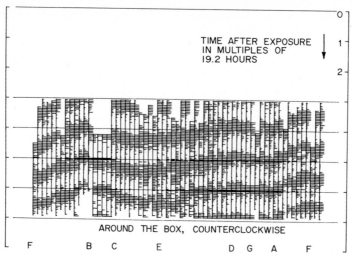

Figure 16.   60 emergence rhythms in *Drosophila melanogaster* populations are plotted vertically downward from stimulus time (exposure to light). Time is measured in periods of 19 hours. The rhythms are stacked up from left to right in order around box FBCEDGAF. The rhythm's phase scans through one cycle in one tour around this box. You can reconstruct the box and the data above it by photocopying this figure and folding it into a cylinder, overlapping the terminally redundant histograms. The data then ascend helically. On boxes not enclosing a singularity, the data form a stack of closed rings instead.

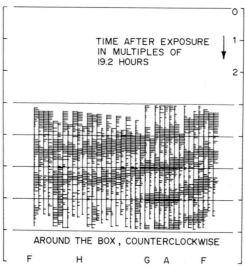

Figure 17.   As in Figure 16 except that path FBCEDG is replaced by shortcut FHG. The rhythm still scans a cycle around this diminished box, so the singularity is not in the omitted part.

(a chain FHE of experiments at fixed phase in Figure 15b), we find that the right half does not contain a singularity and the left half does. Subdividing the left half by a chain GH of experiments at fixed duration, we find that the top half does not contain a singularity. The bottom half does (Figure 17). Thus, by repeated alternations between applications of the theorem and execution of the corresponding experiments, we converge on a singular light pulse of 1,700 erg/cm^2 applied five hours after the light to dark transition (or at 24 hours, or at 43 hours, etc., since this mutant has a 19-hour rhythm). This singular exposure terminates rhythmicity. This is the phase singularity, (ϕ^*, M^*) ("phi-star, M-star"). Putting all the data together in Figure 18, in exactly the format of Figure 14 for D.

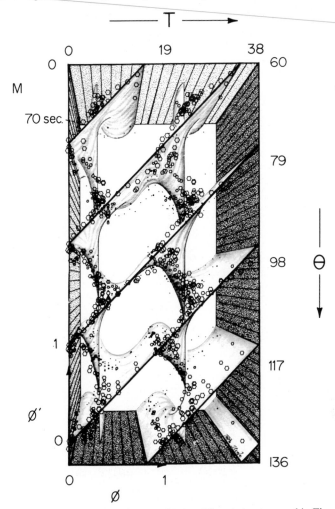

Figure 18. As in Figure 14 but for the mutant of *Drosophila melanogaster* used in Figures 16 and 17. Also only four cycles of new phase are plotted vertically instead of three. All data in the first cycle of old phase are double-plotted, being repeated in the second cycle. See unduplicated data in stereo in Figure 10 of Chapter 20. T and θ in hours. ϕ and ϕ' in cycles of 19 hours measured from an arbitrary origin.

pseudoobscura, once again we behold a single simple screw surface. Figure 10 of Chapter 20 plots the raw data as a stereo pair.

In starker form than any before it, this example illustrates the essential connection between winding number and singularities of maps $\mathbb{R}^2 \to \mathbb{S}^1$. To make the argument more strictly correct, I should note that although winding number 0 around a box is *necessary* for absence of singularities, it does not *guarantee* an absence of singularities within the box. Phase singularities can be clockwise or anticlockwise. Paired, they cancel each other's contributions to a winding number along any ring that encloses them both. The winding number only tells us the net *excess* of clockwise or anticlockwise singularities. If that excess is nonzero, then that many unpaired singularities must lie somewhere within the ring. But if it is zero, there might still be any number of balanced pairs. This fact has a curious implication for regeneration experiments in animals, namely, the occasional production of superfluous left-right paired limbs, as we will see in Example 11.

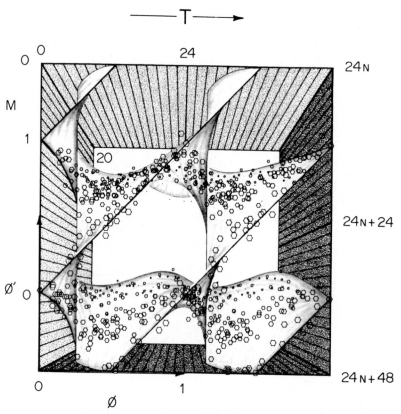

Figure 19. As in Figure 14 but for the fly *Sarcophaga*, according to Saunders's data (1978, with permission). The published data cover only one cycle of old phase and new phase, here repeated twice in each direction. See data in stereo in Figure 11 of Chapter 20. M, T, and θ in hours. ϕ and ϕ' in cycles of 24 hours measured from an arbitrary origin.

Example 7. A Fly that is 1,000 Times Less Sensitive.   David Saunders in Edin-
burgh found type 0 resetting in the pupal eclosion rhythm of the fleshfly *Sarcophaga*
in 1976 (Chapter 20). In this case, as in adult fruitflies (Engelmann, pers. comm.),
the light pulse must continue for several hours at about the same intensity as above:
Stimuli less than four hours long elicit type 1 resetting, and stimuli more than
four hours long elicit type 0 resetting. The several resetting curves of various
durations are assembled to create Figure 19, which suggests once again a screw-like
dependence of new phase on old phase and stimulus magnitude. Figure 11 of
Chapter 20 plots in stereo perspective the data on which this guess is based. The
neighborhood of the screw axis has not yet been explored systematically.

Example 8. Resetting a Flower's Clock.   Turning to circadian rhythms in
plants, Engelmann et al. (1973), working in Tubingen, Germany, carried out

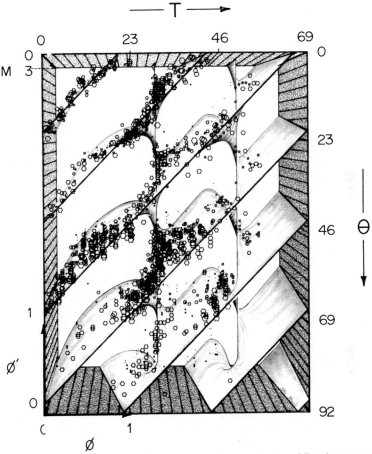

Figure 20.   A time crystal of the plant *Kalanchoe* according to the data of Engelmann et al. (1973),
with permission. Three cycles of old phase are plotted to the right and four cycles of new phase are
plotted vertically. The format is as in Figures 14, 18, and 19 but the first singularity on the left lies outside
the range of these data. See data in stereo in Figure 5 of Chapter 21. T, M, and θ in hours. ϕ and ϕ'
in cycles measured from an arbitrary origin.

measurements in exactly the same format using *Kalanchoë blossfeldiana* (Chapter 21). In this case, a red light pulse of some hours duration resets the rhythmic opening and closing of *Kalanchoë's* little red flower. My idealization of their results is shown in Figure 20. Once again, we have a two-dimensional lattice of screw-like unit cells. This is supposed to fit the data presented as a stereo pair in Figure 5 of Chapter 21. In this data set, pulse durations varied only up to three hours, but the critical stimulus magnitude M^* is more than four hours at the first T^*. Thus the screw axis in this first column of unit cells is outside the field of view.

The topological structure of these results is not dependent on the specific mechanism by which light affects the circadian clock. The same basic structure is obtained by repeating the above experiment, but with the light pulse replaced by a prolonged temperature pulse (Engelmann et al., 1974). Chandrashekaran (1974) obtained similar results using prolonged temperature shocks with the *D. pseudoobscura* eclosion rhythm. However, his measurements don't go quite far enough along the stimulus duration axis: The plot of eclosion time vs. initial phase and exposure duration looks like the first half of Figures 14 and 18, the measurements extending only about as far as the screw axis (or so I interpret it).

Example 9: A Phase Singularity for Beer Drinkers.   Energy metabolism in brewer's yeast proceeds without oxygen by breaking sugar molecules down to alcohol. Due to a pecularity in the feedback regulation of its rate, the passage of sugar through this pathway is commonly pulsatile, and the associated bio-chemistry stably oscillates with a period in the order of a minute. Depending on when in the cycle it is administered, a brief exposure to oxygen induces a phase shift in this oscillation. Reasoning from the topological notions repeatedly used above and from the Pasteur effect (Chapter 12), I proposed in 1971 that the resetting should be type 0. My subsequent measurements verified this, together with the full helicoidal resetting surface and its singularity (Figure 21; raw data in stereo perspective in Figure 4 of Chapter 12). The critical dose is about $\frac{1}{5}$ micromole of oxygen per wet gram of cells, given a few seconds before the NADH maximum. Projecting the screw-like data cloud onto the stimulus plane, we obtain a wave of NADH fluorescence rotating about this point.

This kind of experiment was repeated by Greller (1977) using acetaldehyde pulses instead of oxygen pulses and by Aldridge and Pye (1979) using Ca^{2+} ions. The results were the same in the qualitative essentials here emphasized.

The singularities discovered through type 0 resetting arise from an impossible map from $\mathbb{S}^1 \times \mathbb{R}^1 \to \mathbb{S}^1$ (from a cylinder to a ring, i.e., from the combination of a phase and a stimulus magnitude to a resulting phase) or, as portrayed in Figures 12 and 13, from the disk $\mathbb{I}^2 \to \mathbb{S}^1$. In Chapter 1 we also deal with maps from two phases to a phase $\mathbb{S}^1 \times \mathbb{S}^1 \to \mathbb{S}^1$. We saw (in context of color mixing in Example 1) that regardless of any hypotheses about mechanisms, a certain sym-metric format of experiment induces a map of this sort that cannot possibly be continuous. It might include a phase singularity. This fact has its corresponding expression in the chemistry of yeast cells because oscillating glycolysis in yeast, so easily perturbed by exogenous chemicals, can also be perturbed by contact

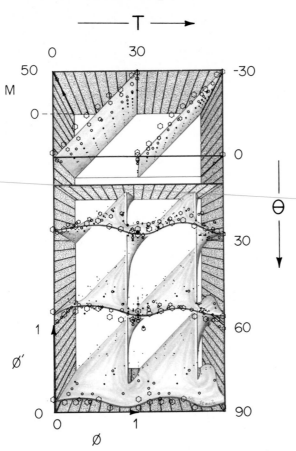

Figure 21.   Similar to Figure 20, but the data concern the 30-second NADH rhythm of yeast cells. Prestimulus data are included at the top, clarifying the relationships of this (and previous perspective drawings) to Figures 30 through 33 of Chapter 1. All data are double-plotted to the right. Also this one plots M into the foreground, so the sense of the screw is reversed. See data in stereo in Figure 4 of Chapter 12. T and θ in seconds, M in units of oxygen dosage. φ and φ' in cycles of 30 seconds from an arbitrary origin.

with differently phased cells. The two methods can be distinguished as follows: A hit-and-run attack with a chemical stimulus (oxygen, calcium, or acetaldehyde) is only one way to probe the reactions of an oscillator throughout its cycle. In some ways, a more natural stimulus is another batch of cells, oscillating on the same cycle, but at a different phase. Thus, instead of testing each part of a cycle with a small to large dose of one of its chemical substrates, we could test each phase by *confrontation* with each other phase: with a variety of stimuli varying not in amount but in a quality that varies along a circle. In other words, two batches of cells could be combined at different phases in the cycle. They interact to establish mutual synchrony. What is the compromise phase?

    This experiment was actually carried out (for other reasons) by Ghosh et al. (1971). I have replotted their data in Figure 22. I divided the cycle into 10 equal-time

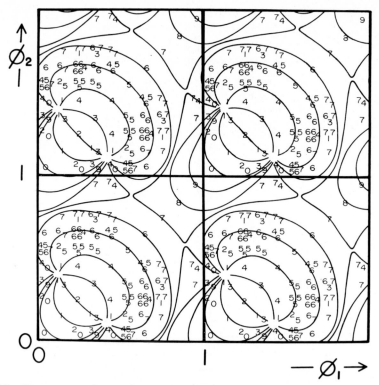

Figure 22.   Phase compromise experiments, plotting hybrid phase as a function of two parent phases:
The digits represent actual hybrid phase in one-tenth cycle intervals from the data of Ghosh et al, 1971.
The digits are placed within one-tenth cycle by one-tenth cycle boxes; their positions within each
box are arbitrary. The contours lines roughly link equivalent data.

parts and ruled a 10×10 square to represent the phases of the two "parent"
volumes of oscillating yeast cells. (The phase was observed as usual by following
the sinusoidal changes of NADH fluorescence; see Chapter 12.) The phase observed
in the mutually synchronized mixture is extrapolated back to the moment of
mixing when the parent phases were as indicated. I tabulated this resultant phase
in the appropriate little square, once again in units of one-tenth of a cycle. The
47 experiments shown do not give a clear impression but they are at least not
incompatible with a continuous map such as I have drawn on an overlay. The
continuous map was designed:

1.   To fit the data as well as any smooth pattern could. (All but four out of
47 data fall within $\pm\frac{1}{6}$ cycle from the contour map.)

2.   To show the resultant mixture phase equal to both parent phases in the
case of control experiments (not actually reported) in which aliquots of equal
phase were recombined.

3.   To reflect, in its symmetry about the main diagonal, the arbitrariness of
calling one aliquot A and the other one B.

4. To confine the necessary discontinuity to as small an area as possible. In this case, we end up with one isolated phase singularity in each triangle, above and below the diagonal. (Using the cell division cycle in *Physarum* (Chapter 22), the experiments in identical format seem to reveal a quite different sort of discontinuity instead of a phase singularity.)

A caveat: These experiments have not been repeated. The uncommonly severe variability of results in the pioneering study of Ghosh et al. leaves room for a radically different interpretation, i.e., that cell mixtures don't compromise at all but always synchronize to one or the other of the parent populations. In all but two out of the 47 trials shown, the "compromise" phase was within one-tenth cycle of one parent's phase. It might be that uncontrolled variables such as the details of mixing determine which parent dominates.

However that may be, the phase singularities discovered by chemical perturbation (oxygen, calcium, acetaldehyde) *are* distinctly visible in those data. I first proposed the experiments shortly after finding the phase singularity in a circadian system. Because of the temptation to believe that the phase singularity might reveal something unique about a universal circadian clock mechanism, it seemed incumbent upon me to demonstrate that the topological argument is really model-independent by applying it to a chemically different clock. Glycolytic oscillations seemed to qualify in view of their short period and acute temperature sensitivity of rate. We know from the pioneering studies of Chance and Pye (see Chapter 12) that yeast oscillations respond to AMP exactly as circadian systems respond to light. Thus experiments could be carried through in identical format using oxygen gas instead of blue light. The results were qualitatively the same.

In the remaining six examples we turn to phase singularities involving physical space in a more substantial way than in the pinwheel experiments previously cited, which were actually carried out piecemeal. The first example is another circadian rhythm which, because it shows type 0 resetting, is a strong candidate for experimental realization of a phase singularity.

Example 10. Morphogenesis in Fungi. The creature I have in mind is the fungus *Neurospora crassa*, the one Hamner took to the South Pole (see Box A). Its rhythm is a banding pattern that marks the loci of the frontier of growth at successive 22-hour intervals. Hyphae formed along the frontier at the same hour of each successive circadian cycle turn out to mature into conidia, the asexual spores by which *Neurospora* reproduces. Hyphae formed at intermediate phases remain forever vegetative, without forming conidia.

Suppose the pinwheel experiment of Examples 5–9 were conducted on a square sheet of *Neurospora* growing on a much larger bed of nutrient agar. According to Figure 13, the winding number of new phase around the square ABCD must be $W = 1$. Suppose phase increases steadily clockwise. That means that as the square continues to grow at its edges, the locus of conidiation moves around the edge, circumnavigating the edge every 24 hours as the frontier moves out. Thus, the conidiation pattern is henceforth a spiral. Another way to visualize this implication can be seen in Figure 23. The continuity of pinwheel experiments

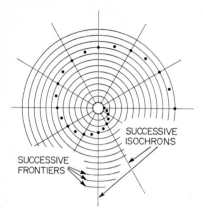

SUCCESSIVE
ISOCHRONS

SUCCESSIVE
FRONTIERS

Figure 23. An idealized colony of *Neurospora* is shown with thirteen successive positions of its frontier. Radial contour lines indicate successive positions of phase 0 of the circadian clock. The locus of ($\phi = 0$ on the frontier) is accordingly an Archimedes' spiral.

using other circadian clocks suggested that in *Neurospora*, too, phase should vary continuously with the timing and intensity of a light pulse, i.e., with position on the mycelium. The usual result in Examples 5–9 is that the contour lines of uniform phase diverge from an internal singularity after the stimulus is given. An hour later, the phase contours are all still there, but the phase each represents is an hour later, as with the cotidal contours of Example 4. Alternatively, one could say that the contour of maximum conidiation-induction moves clockwise from one hour to the next, revolving around the singularity like a searchlight beam as do the eclosion contours of Example 5. As it goes, it induces hyphae along the expanding frontier to differentiate into conidia. Thus the conidia appear along a spiral locus.

This experiment has not yet been tried.

The pinwheel experiment is only one way to arrange a nonzero winding number around the frontier. According to one view of the underlying physiology (Chapter 8), the fungus is liable to fall into a habit of any small integer winding number immediately after germination from a tiny dot of spores. In this way, multiple spirals, both clockwise and anticlockwise, as well as concentric ring patterns should arise spontaneously. This is in fact observed (at least up to winding numbers $+3$ and -3) in *Nectria*, in *Penicillium*, and in *Chaetomium* (see Chapter 18). At present, nothing is known about the phase singularity (or multiple phase singularities) that must lie within spiral or multiple-spiral mycelia. According to one theoretical model (Winfree, 1970a), the rhythmic system of the germinating spore begins in the phase-singular state. In *D. pseudoobscura*, as well, it seems that larvae reared from the egg in conditions of constant temperature and constant darkness remain in the phase-singular state (Zimmerman, 1969; Winfree, 1970b,c, 1971a).

The next three examples are drawn from the recent literature of developmental biology.

Example 11. Regeneration of Limbs.   According to one view of development in higher organisms, each organ emerges from a little two-dimensional patch of tissue that has some kind of organizational coherence. Such patches are called developmental fields. French et al. (1976) and Bryant et al. (1977) have recently gathered an impressive body of evidence under the proposition that (at least in arthropods, and maybe in some vertebrates too) each cell knows its role in the developmental field by retaining within it a quantity defined on the ring— something like a phase, or an angle. At present no one has the slightest idea what this might mean biochemically. It might seem appropriate to remark, as Isaac Newton once did in a similar case, that

> To tell us that every Species of Thing is endowed with an occult specifick
> Quality by which it acts and produces manifest Effects, is to tell us nothing.
>
> Opticks, 1730

However, the complex mechanisms of evolved systems must be unraveled by layers. In this case the deeper layers of molecular mechanism await another generation, while our present concern is with the implications of the putative phase-like quantity. French et al. (1976) argue in effect (and Glass (1977) recasts their argument to bring out this aspect more forcefully) that the center of the field is a phase singularity, and that if experimental manipulations so rearrange blocks of tissue that additional phase singularities are inevitably created, then additional organs of appropriate handedness must emerge at those sites. A number of startling experiments are demonstrated which readily lend themself to this "clockface" interpretation, as it has come to be called. This matter is clearly of central importance in any discussion of the biological role of phase singularities. The biological background and alternative interpretations of the data are elaborated in Chapter 16. The physiological nature of the putative singularity is examined in Chapter 10 together with other kinds of singularity collected in this chapter.

This example and the next are included to play the role of "straw men". Their apparent phase singularities are, in my view, particularly insubstantial. Although these two examples can be described in terms of maps among rings, I believe this language is inappropriate and misleading. I include them because if I am right, the existence of such cases alerts the reader against mechanical application of language: We are trying to do science, not just make mathematical metaphors. I include them also because I might be wrong, in which case they greatly enrich the catalog of biological phase singularities.

Example 12. The Cuticle of the Arthropods.   The lens of the insect eye might be regarded as an organ, somehow assembled by a few underlying cells whose job it is to secrete perfectly transparent cuticle in the right geometric pattern. This pattern is periodic on the scale of wavelengths of light. This periodism was initially ascribed to rhythmic secretion of cuticle (see Chapter 17). Electron micrographs of microtome sections through the lens cuticle consistently exhibit a phase singularity of rhythmic banding in sections transverse to the optic axis. A spiral of more electron dense cuticle is plainly seen winding in toward this center. In sections parallel to the optic axis, a periodic banding is seen. Every vertical line penetrating into the lens parallel to the optic axis exhibits rhythmic alternations of electron density along its length, i.e., along the time-of-deposition axis. In sections perpendicular to the optic axis, the phase of this rhythm is seen to advance through one full cycle as samples are taken at successive positions around any ring concentric to the dark spiral's center. Thus, we have a nonzero winding number of a structural rhythm (originally a temporal rhythm in sequence of deposition) and a phase singularity continuing in depth like a screw dislocation in a crystal. Taken together, these perpendicular sections suggest a screw-like, helicoidal surface of electron density within the block of sectioned cuticle that was lens. Does this have some developmental meaning? Yes, but it is not at all what it appears to be. This mystery is investigated further in Chapters 10 and 17. (My objective for the present chapter is simply to present many cases of phase singularities as phenomena without resolving interpretations; this is reserved for Chapter 10 after a groundwork of theoretical concepts has been laid in intermediate chapters, and after the reader has acquainted himself with more of the pertinent experimental details in Chapters 11–23.)

Example 13. Cellular Slime Molds.   My third example from the lore of developmental biology concerns the marvelous social amoeba *Dictyostelium discoideum*, which has figured so prominently of late in the burgeoning literatures of cAMP and of intercellular communication (Chapter 15).

It has been appreciated for a long time that the preamble to morphogenesis of social amoebae consists of a gathering together of widely dispersed individual cells, mediated by their periodic relaying of a chemotactic signal from a pacemaker cell. This signal guides many thousands of other cells to the pacemaker (Shaffer, 1962). Here they contact one another and assemble a little jelly-like slug which creeps away to find a place suitable for metamorphosis into a fruiting body.

The organizing waves emitted by a pacemaker cell are closed rings concentric to their source, like ripples emanating from a rain drop in a puddle. Thus, some interest attached to Gerisch's observation (1965) of rotating spiral waves in the same situation with the same consequences. The essential differences are that a spiral wave consists of one continuous locus of cell movement rather than a series of disjoint rings of moving cells, and that it requires no special cell at its source. Thus spiral waves commonly arise just as soon as the maturing cells develop a capacity for signal relaying, but before any cells begin to oscillate spontaneously.

How can a wave have no source? It would seem that each wave must first appear somewhere, and that point is its source. But what is manifestly true of concentric ring waves is not necessarily true of rotating spiral waves. If we trace a piece of a spiral wavefront backward in time, we never come to a clear source. Rather, we end up in a vague region of indeterminate periodicity or end up circulating endlessly about a tiny circle bounding such a region, concentric to the pivot (see Section B of Chapter 9). The wavefront at any instant may be taken as a locus of synchronous cells: all cells pulse periodically, at the regular intervals of spiral rotation, and the wavefront is the locus of simultaneous pulsing. A multiple-exposure photograph taken by opening the shutter at intervals of one-tenth cycle would presumably show 10 equispaced parallel spirals, all converging toward the pivot or its bounding circle. If these loci be the contours of uniform phase, then that circle is (or contains) a phase singularity.

Close observations of this interesting region have just begun to appear. In the 3-dimensional slug this singularity is thought to become a 1-dimensional rotation axis. (Clark and Steck, 1979).

Example 14. Pathological Rotating Waves in Heart, Brain, and Eye. The *Dictyostelium* cell's sudden triggerable release of cAMP in many ways resembles the sudden triggerable inrush of sodium ions through a nerve cell's plasma membrane. In both cases, the excited (permeable) state propagates once triggered by a threshold-transgressing stimulus (Chapter 14).

Thus a pinwheel wave similar to *Dictyostelium*'s might be looked for in an extensive planar expanse of nerve-like tissue. The obvious candidate is heart muscle. At least since McWilliam (1887), medical physiologists have postulated some kind of rotating wave, a "circus movement", as an explanation for high-speed rhythmic flutter, and for the much more pernicious mode of asynchrony called "dementia cordis" or fibrillation. Though there are probably other mechanisms too, some kinds of flutter and fibrillation do behave, in many ways, like a sourceless traveling excitation which rotates simply because it got started somehow. But until 1973, the best case for its existence rested on indirect physiological inference and on a well-developed mathematical theory confined almost exclusively to the Mexican and Russian scholarly journals. In 1973, Allessie et al. published the first critical experiment, using a piece of rabbit atrium stretched out flat to accommodate the necessary array of 10 observing electrodes. Using a delicately timed electrical stimulus to initiate the wave, he found instances of persistent, pernicious rotation. Electrical activity circulates as a pinwheel wave, recurrently exciting

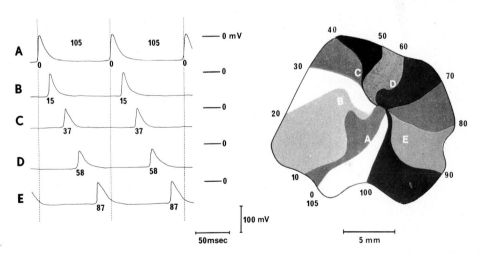

Figure 24. Map of the spread of activation in a piece of isolated left atrial muscle during a period of sustained tachycardia as constructed from time measurements of the action potentials of 94 different fibers. The impulse is continuously rotating in a clockwise direction with a revolution time of 105 ms. At the left are shown the transmembrane potentials of five fibers which lie along the circular pathway (A-E). The moments of depolarization, in msec, are given together with the action potentials and the isochronic lines of the map (Allessie et al., 1977, Figure 1, by permission of The American Heart Association, Inc.).

every patch of muscle to contract (Figure 24). Every point is rhythmically active except the central patch around what would be a pivot of the pinwheel wave. In this central patch, activity is irregular, as befits a phase singularity. The nature of fitful activity in this region poses extremely interesting questions for neuro-physiology and for mathematics both. The question has mathematical interest because the phenomenon is apparently not a consequence of the inevitable inhomogeneities and physiological complexities of real rabbit myocardium. A similar pivoting wave and similar irregularities near its "pivot" were observed by Gulko and Petrov (1972) in a very simple computer simulation based on an idealized notion of excitable membrane, a simplification of the Hodgkin-Huxley equation (see Figure 5 of Chapter 14).

A much more slowly rotating wave of quite different pathological mechanism occurs in the cortex of the brain and in the retina of the eye. Still another kind of rotor, associated with seizures, has been observed in the cortex. See Chapter 14 for detail and references.

Example 15. Self-Organizing Patterns of Chemical Reaction.   I conclude this section with a remarkable example of recent vintage from physical chemistry. It turns out that some kinds of chemical reaction are capable of sustained oscillation and of excitability, very much as in *Dictyostelium* and in nerve membrane. Chapter 13 elaborates in detail about the most convenient of these, a malonic acid reaction in which chemical changes reveal themselves colorfully in water solution at room temperature on a time scale of seconds to minutes.

Consider this reaction in a three-dimensional volume of liquid. Imagine that you could wander through it, monitoring the phase of the red-blue alternation going on everywhere. Let's start at any place P in Figure 25 and follow a closed path through the liquid, ending up back at P. The path PP is a ring. Let's suppose we could monitor phase instantaneously around the ring. The net change of phase along path PP must be an integer number of cycles. Call this integer W, the winding number of phase along ring PP. Unless $W = 0$, the path PP cannot be contracted to a point without encountering a phase discontinuity. Only two possibilities present themselves:

1. W is necessarily always 0. This is simply not true, as the photograph in Figure 26 shows.

2.   Locus PP loses its winding number by unit increments, necessarily abruptly as some special loci are passed during the progressive contraction. These loci would be phase singularities. Eventually there remains only one enclosed within

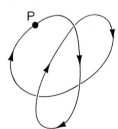

Figure 25.   A closed path in three-dimensional space.

Figure 26. A photograph of rhythmic patterns of chemical activity in a dish of malonic acid reagent. Eight sources are visible. A survey of instantaneous phase along any one of the three closed paths shown leads thus through the indicated integer number of complete cycles of phase. This is the number of left-handed spiral sources minus the number of right-handed spiral sources enclosed.

PP, so PP can be contracted to a tiny ring of winding number ± 1 around that locus, or pulled across and contracted to homogeneous phase. This is, in fact, how it is. Figure 26 gives an example in two dimensions, each singular locus being just a point.

In three dimensions, the phase singularity must be at least a one-dimensional locus, a thread or filament of ambiguous phase in the reacting liquid. This is necessary because each phase singularity must be encountered along *any* path of contraction of PP. Such paths are surfaces bounded by PP. If *every* such surface

Figure 27. Three (out of a continuum of) caps bounded by a closed ring (PP) as in Figure 25. Each cap is a locus along which PP is contracted to a point. Each contains a phaseless point. The locus of such points (**) is thus one-dimensional.

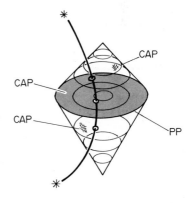

encounters the singular point, then the singularity must be at least a one-dimensional locus (Figure 27).

This thread of ambiguity has been observed visually and I have reconstructed several from chemically fixed serial sections of three-dimensional blocks of malonic acid reagent. These threads typically close in rings for reasons perhaps not evident here but elaborated in Chapters 9 and 13.

B: Counterexamples

That will be enough for examples of the smooth patterns of timing that lead one to anticipate some kind of singularity at the center. A few counterexamples now seem appropriate to forestall any impression that such patterns and their phase singularities are topologically inevitable consequences of using a circular measure, and therefore of no scientific interest. This inference has enough truth to it that I have at times mistaken it for the whole truth. It comes especially close to being the whole truth in connection with the rowboat navigation problem As I elaborate in Chapter 16, much the same might be said of the case of limb regeneration: There I believe the apparition of a phase singularity has more to do with logic and the peculiar but convenient choice of a phase-like measure than it has to do with questions of dynamics, of models, of mechanism.

It is essential in this business to attempt a segregation of issues of logic and measure from issues of fact and mechanism; that comes in later chapters. For now it will suffice to note that a phase singularity of some kind (though not necessarily an isolated point) *is* a topologically generic consequence of continuity and type 0 resetting (Box C). However, many kinds of rhythmic systems violate continuity or do not exhibit type 0 resetting. For example, experiments that might have revealed type 0 resetting or phase singularities in the cycle of cell division still leave those matters in doubt (Chapter 22). Resetting experiments on models of the female cycle, and probably on the real mechanism too, exhibit gross discontinuities, possibly due to threshold processes assumed to underlie hormone release, ovulation, etc. (Chapter 23). Many kinds of nerve, though rhythmic in their firing, either in isolation or in response to specific stimuli, seem to suffer such an abrupt alteration of membrane behavior during the action potential that resetting maps are grossly discontinuous (though curiously, not always *at* action potential time) (Chapter 14).

Turning from particular cases to more abstract categories, there are mechanisms in which rhythmicity is secured through abrupt transition from one state to another, eventually leading back to a prior state. Such discrete state devices, or logical automata as they are called, have been widely employed as metaphors for biological rhythms. But there is no place in such models for resetting curves or surfaces or, for that matter, for considerations of topological continuity.

The class of models taken up in the next chapter are completely continuous and smooth in their behavior but, for reasons to be developed there, can only support type 1 resetting. These too have enjoyed wide popularity as underpinnings for experimental design and inference about biological clocks, but they have no singularity.

Another class of mechanism employs continuous kinetics up to a point at which a fast mechanism, previously inconspicuous, intervenes like a *deus ex machina* to abruptly restart the process at its beginning. These so-called relaxation oscillators have traditionally provided the physiologists' first recourse for explanation of spontaneously rhythmic happenings (see Chapter 6, Section C). But on account of the near-discontinuous jump, their resetting maps are not practically classifiable, winding number goes undetermined, and topological arguments cannot fence in a singularity. If one exists, it cannot be detected by examining the geometry of timing relations elsewhere (but see Chapter 10, Section B.)

Thus, if we in fact observe a phase singularity, then we have some explaining to do in the contingent terms of empirical science, not only by making the graceful gestures of abstract topology.

C: The Word "Singularity"

> The appearance of singularities in a physical theory suggests inconsistency and room for improvement.
>
> Narlikar, 1970

Before departing this chapter, I wish to return to its opening lines, in which the word *singularity* was defined. Let me elaborate on that definition. As used by mathematical physicists and as I use it here, the word *singularity* means a place where slopes become infinite, where the rate of change of one variable with another exceeds all bounds, and where a big change in an observable is caused by an arbitrarily small change in something else. As James Clerk Maxwell observed,

> Every existence above a certain rank has its singular points: the higher the rank the more of them. At these points, influences whose physical magnitude is too small to be taken account of by a finite being, may produce results of the greatest importance . . . .
>
> Cited on p. 972 of Newman, 1956

According to the bible of theoretical physics (Morse and Feshbach, 1953) during the years when those who taught me physics were learning it:

> . . . singularities . . . are usually the most important aspects of scalar and vector fields. The physical peculiarities of the problem at hand are usually closely related to the sort of singularities the field has. Likewise, the mathematical properties of the solutions and differential equations are determined by the sorts of singularities which the equation and its solutions have. Much of our time will be spent in discussing the physical and mathematical properties of singularities in fields.

My involvement with singularities began at a desk in the Eisenhower library in Baltimore, to which I retired each morning in the fall of 1965 with only a pencil and pad to discover some questions natural to an interest in physiological periodism. At length I fumbled to an interest in maps among rings and products of rings. The essence of that subject, according to the quote above, lies in the singularities of those maps. That led to noticing such objects between the lines of current papers from experimental physiologists. A singularity is a kind of discontinuity. It might or might not be interesting. Interest goes with meaningfulness.

Following Fahrenheit and Celsius, I might invent a new temperature scale called Winfree, on which the temperature Winfree is defined as the reciprocal of temperature Fahrenheit. Thus the temperature Winfree would have a singularity somewhat below the freezing point of water, where temperature Fahrenheit goes through zero. But this would have no possible physical significance. This is what is called a "mere coordinate singularity", a mathematical artifact of choosing a peculiar coordinate system.

Other kinds of singularity cannot be removed by any reasonable change of coordinates. In other words, they are defined in terms of variables whose physical import is immediate. Such, for example, are the space-time singularities associated with black holes. One of these is widely believed to be the creation of the universe. That's interesting. On a more modest scale, the singularity of heat capacity that ushered in the quantum mechanics of crystal lattices under the banner of "the ultraviolet catastrophe" was a nonremovable singularity interrelating basic physical quantities. Arguments about winding numbers and the theorem of page 28 have been deployed recently to establish the existence of elementary particles as singularities in quantum-mechanical continua (Rebbi, 1979) and the existence of Bloch points as singularities in magnetic bubbles (Slonczewski and Malozemoff, 1978).

Yet according to one cogent viewpoint, singularities happen only in models. In real life, the singularity is always somehow evaded. Wind velocity falls back to zero in the core of a tornado, color mixtures vanish into grayness, and a rhythm can simply go flat or become very wiggly with no fundamental-frequency component. When a phase singularity results from feeding observed facts into a soundly reasoned model, we know that something has gone wrong in this model and that we are looking at a distillate of the logical contradictions implict in *our notions* of how the real world operates. Experiments that *seem* to imply a singularity signal a contradiction borne of leaving something out. The purpose of dabbling in models and of contriving experiments is to find out what is left out of our thinking, to see what hidden escape hatch a real system takes in evading the demand that it achieve the impossible.

In the case of black holes, there remains a lively debate over the reality of singularities in the space-time "continuum". For example, during the month in which I first wrote up *Drosophila*'s answer to the Pinwheel experiment for publication, Fred Hoyle (Bakerian Lecture, June 1968) accosted the Royal Society with the question, "Do singularities exist in the real world or are they only metaphysical entities?" I suspect that what is ultimately being asked is whether or not our cherished observables (for example, distance, time, and mass) are in fact as fundamental as we imagined before finding that they can be involved in singularities. Maybe they are only observables, poorly chosen in view of their peculiar behavior near singularities, like the Winfree temperature scale. The question then becomes: What are the proper quantities in terms of which the world functions rationally and continuously?

My attempts to answer this question in the cases of biological and biochemical phase singularities require additional familiarization with the experimental systems (Chapters 11–23) and some elaboration of dynamic models (Chapters 3–9).

Only in context of the very simplest dynamic models (next chapter) does the phase singularity acquire an ineluctably paradoxical character. Subsequent chapters provide amendments to the simplest models whereby phase singularities become perfectly tame, though not uninteresting, biological phenomena. We reconvene in Chapter 10 for a look at the physical origins of the phase singularities used as examples above.

Because we must dabble in dynamical systems, I must issue advance warning about a homonym. "Singularity" also means simply "a point where something singular happens". "Something singular" need not necessarily involve a discontinuity. Thus, mathematicians in speaking of the singularities of maps may mean nothing more sensational than the point in deforming a curve at which it first acquires an extra bump. Developmental biologists identify the sites where a hair will emerge on an insect's cuticle as singularities. Engineers use the terms "singular point" or "singularity" interchangeably in speaking of dynamical systems to indicate that unique state at which all rates of change are simultaneously zero, as in a chemical steady-state or the mechanical equilibrium position of a rocking horse. This ambiguity contributes to the terrible confusion in many people's minds about "catastrophe theory", which, in its presently best elaborated manifestation, concerns the zeros of potential flows. This is a subject whose richest domain of clear application seems to lie in engineering physics (Thompson, 1975; Thompson and Hunt, 1977). The zeros of potential flows are the states at which all rates have dwindled to nothing, the equilibrium states, commonly referred to as singularities in the engineering literature. But the singularities of catastrophe theory are in a different space altogether, the space of control parameters. They are the particular combinations of control parameters (e.g., loads on a structure) at which the configuration of equilibria (singularities in the engineer's sense) undergoes a change.

The distinct uses of "singularity" create an opportunity for real confusion in this book because in certain kinds of dynamic systems, described in Chapters 5 and 6, the state which is singular in respect to rates of change, being a state of zero amplitude of oscillation, is *also* a part of the phase singularity, the locus of states where phase does something singular. In fact, in special kinds of dynamical systems (Chapter 5), it is exactly coextensive with the phase singularity. But in less exactingly specialized kinds, e.g., in the chemical rotors described above, the reaction steady-state is in no way related to a phase singularity (Chapters 6, 8, and 9). To help preserve this distinction, I will usually say "phase singularity" rather than just "singularity" in this book, and in referring to the steady-state of reaction, I will say "steady-state", not "singularity".

The arguments explored up to this point have not in any way involved conjectures about dynamic systems or kinetic models. But because some of the most intriguing phase singularities do arise from processes in time, we *would* like to know what kinds of dynamic systems are at least *compatible* with the observed facts and what those kinds would lead us to expect about the nature of the phaseless state. As Sherlock Holmes was wont to enounce, "Singularity is almost invariably a clue." Thus we pass to the next chapter, about dynamics on rings.

Needless to say, we can invent any number of observables. (How many "properties" does an organism have?) But if a system has only D state variables (i.e., its state space, whatever its topology may be, is D-dimensional), then the first D independent observables suffice to determine the state. Any further observables are redundant because given the state, every function of state is determined.

Dynamics

This book focuses on dynamical systems: systems with the interesting property of continually spontaneously changing state. At any instant, each state variable has a rate of change which depends only on the current state of the system.

Here we encounter a linguistic choice. Obviously, any system's behavior depends on environmental conditions: its temperature, the voltage of its power supply, the amount of light falling on it, the position of its knobs. We could consider these as state variables that don't change spontaneously. Then we can say that the rate of change of state depends only on the state. Or we could say that the state variables are those things that change spontaneously, and that things that don't change spontaneously are called *parameters*, not variables. Then the rate of change depends on both. I choose this format. Notationally, then,

$$dX_i/dt = R_i(\mathbf{X}, \mathbf{P})$$

where R_i is the rate function for the ith state variable. Its inputs are all the state variables ($\mathbf{X}$) and all the external parameters ($\mathbf{P}$). Its output is the rate of change of X_i. As there is a rate function for each i, we might denote more compactly:

$$\dot{\mathbf{X}} = \mathbf{R}(\mathbf{X}, \mathbf{P})$$

using also the dot convention to indicate rate of change with time ($\cdot = d/dt$).

Having thus divided our initial collection of state variables into those that change spontaneously and those that don't (parameters), we might back up and note that observables also depend on both parameters and state variables:

$$F_i = F_i(\mathbf{X}, \mathbf{P}).$$

For example, the apparent color of a lizard may depend on its emotional state; but the measurement is affected also by the color of the ambient light.

This frivolous example brings us to a crucial point: Not all observables are numbers, or even combinations of numbers. Some observables, like color or direction or the phase in a cycle, are defined not on the real number line, but on a ring (or on more interesting topological spaces). It is usual to try to salvage the more familiar form by *representing* such a measurement in terms of a number. For example, it is conventional to map the ring onto a line by speaking of "degrees" with the convention that $360°$ is equal to $0°$. A less clumsy device is to map the ring onto a plane (which is the product of two lines), and identify a point on the ring by two numbers, $X = \sin \phi$ and $Y = \cos \phi$. This might be a natural device and it might not be. We will look at cases of both sorts in context of rhythmic patterns in space and time, in living systems, and in chemical reactions.

In this chapter I dwell on cases in which it is *not* natural to think of the ring as embedded in a space of more than one dimension. In these cases, the measurement of phase (an observable) is most naturally regarded as a fairly direct mapping from the system's state space, which is in fact shaped like a ring, onto the abstract ring on which phase is defined ($F_i = X_i$). The topology of rings has some peculiar consequences for the behavior of such systems, and of our measurements.

In Chapters 5 and 6 we dwell on cases in which it *is* natural to think of a phase measurement (defined on a ring) as being embedded in a plane or higher dimensional space. There we take the observable "phase" as a function of state, $F_i = F_i(X_1, X_2)$. It is natural to ask whether any meaning attaches to the inside of the ring. Maybe not. Maybe the very question is nonsense. For example, what is inside or outside of the four-dimensionally spherical three-dimensional continuum of space-time? Only notime and nowhere. Or what is inside the circle of 12 hours on a clock's face? That might make a good Zen koan. But the question does make sense in some cases. For example, what is inside the wheel of saturated colors? Brown and grey. They are real. What's inside the sphere of latitude and longitude coordinates? Gold, coal, and magnetic fields. They are real. Thinking of phase measurement in terms of a continuous map in state space, we see no place for the inside of the ring. Each point in state space has some phase value and the surfaces of constant phase occupy state space completely. If there is a state that has no phase, then it must correspond to some kind of disastrous irregularity in the map: a singularity. The process of measurement being at heart a mapping from · one space to another, it seems natural that the singularities of maps should play a vital role in science. This monograph exploits one aspect of this fact by gathering together experimental systems where phase singularities play a conspicuous role.

B: Dynamics on the Ring

> It will be seen that some cases definitely do not fit the theory. This is not surprising since uncomplicated systems are seldom found by accident. One function of theory is to define the properties of simple systems so that they may be recognized when encountered.
>
> Campbell, 1964

Hourglasses

The world abounds with processes whose states vary in only one way, for most practical purposes, and whose last states are the same as the first. For example:

1. Nerve cells and their analogs in the plant world can be thought of as "resting" until "excited" beyond some "threshold", whereupon a standard sequence of changes ensues by which the cell goes "refractory", then through stages of decreasing "relative refractoriness", and then relaxes back to "rest" (Chapter 14). Diagrammatically, the states of such a system constitute a ring.

2.   The cell cycle is sometimes thought of as an excursion, somehow triggered, from a noncycling state called A or G, through the rest of the growth stage G1, a DNA synthesis stage S, a second interval of growth G2, mitosis M, and back to A (Chapter 22). At this level of abstraction, the human reproductive cycle of ovulation, conception, gestation, delivery, and ovulation has the same form, i.e., a one-dimensional developmental progression that returns to its origin and is thus a ring.

3.   In the female cycle, in "reflex ovulators" such as the rabbit, hormonal events triggered by sexual stimuli precipitate an egg into the uterus and, if it isn't fertilized, they ripen a next egg in anticipation of next coitus (Chapter 23). This cycle of states is a ring.

4.   The circadian cycle in some organisms is thought of by analogy to an hourglass (Chapter 19). It is "turned over" daily by some such event as sunset, then runs through a process of standard duration, terminating with the original pretriggering state.

Adopting the jargon of circadian physiologists, I will refer to such a process—or more exactly to this idealization of such a process—as an "hourglass device". It is an especially simple kind of excitable system (see Chapter 9).

Simple Clocks

But in many systems of interest, that first and last stage is *not* a resting state, so that cycle after cycle ensues without interruption, without need of an external stimulus to initiate each new cycle. For example:

1.   Most kinds of rotating or reciprocating machinery have a fixed cycle of changes through which they repeatedly progress, at whatever rate the engineer chooses. This seems to be how bacteria swim, their helical flagellae being driven by a turbine rotor (Berg, 1974).

2.   The astrophysical cycles that give us the seasons, the tides, and the day repeat without interruption.

3.   In the human menstrual cycle and in the female cycles of other spontaneous ovulators, each hormonal event sets the stage for the next, eventually leading back to another ovulation (unless pregnancy intervenes).

4.   The circadian "clock" of many kinds of animals and plants, according to one useful approximation, is always somewhere in its cycle and advancing under its own power by roughly one hour of subjective circadian time per hour of real time.

5.   The cell mitotic cycle is so often thought of in this way that Campbell (1964) was provoked to define the term "simple clock":

> We shall assume that with respect to the division cycle, the cell behaves like a simple clock ... by simple we mean that there is a single variable on which all the interesting properties of system depend; that any cell can be

assigned a time on the clock, and that the expected behavior of any two clocks which read the same time will be the same regardless of their histories.

Ring Devices

What both categories (the hourglasses and the simple clocks) have in common is their restriction to a one-dimensional ring of states (e.g., the angular position of the meshed gearworks in a mechanical clock). I will henceforth refer to hourglasses (which need to be turned over) and simple clocks (which turn themselves over) as "ring devices", emphasizing the essential fact that they run on a fixed cycle. The notion of a ring of states is naive, and deliberately so. The purpose of such gross approximation is to isolate one essential feature of one class of phenomena. The ring idea constitutes the skeleton, the bare bones, of a realistic analysis of its natural analogs. Chapter 5 hangs flesh on the bones, but to see what is gained by thus complicating the picture, it is worthwhile first to examine carefully what the skeleton alone can do.

I begin with the observation that some systems, and perhaps most, can fall into either category of ring device (hourglass or simple clock) according to external conditions. For example, the cell division cycle in bacteria proceeds apace with regularity in a nourishing medium but is arrested at a certain phase in phosphorus-poor medium (and so it can be gated by "ticks" of injected phosphate; Goodwin, 1969). A nerve cell lingers in its resting state until a stimulus induces it to fire an action potential, but if continually biased with a tiny electrical current or placed in a calcium-deficient solution, it fires spontaneously and rhythmically (Guttman et al. 1979). The circadian clock in the fly *Sarcophaga* cycles spontaneously when warm but arrests at a certain phase when cold (Saunders, 1978).

Rate Equations and Stimuli

Let us put this fact into more general context and make something of it. The point is that a ring device's rate of advance through its cycle is conditioned by an external influence, which we might denote as an influence or intensity parameter I. For example, an alga's cell cycle duration is shorter when it is grown under brighter light (Edmunds, 1974). A sensory neuron fires at shorter intervals when exposed to a stronger smell, hotter surface, brighter light, harder touch, or whatever it is specifically devised to sense.

In general, the effect of I may be different at different phases of the cycle, so that we write:

$$\dot{\Phi} = v(\Phi, I)$$

or, in words, the instantaneous rate of change of phase (the clock's angular velocity v) is jointly determined by its instantaneous state (phase, Φ) and by the external influence parameter. In some standard environment $I = I_0$ in which we initially calibrated the cycle to define "phase", $\dot{\Phi} = 1$ by definition. In other words, $v(\Phi, I_0) = 1$ (Figure 1).

3. The Rules of the Ring

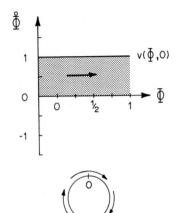

$v(\Phi,0)$

Figure 1. The angular velocity (vertically) as a function of phase (horizontally), using model $\dot\Phi = 1 + I\cos 2\pi\Phi$, with $I = 0$. The circle above depicts by the length of the curved arrow the angular velocity at each phase.

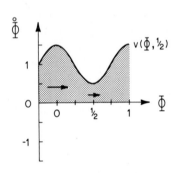

$v(\Phi,\frac{1}{2})$

Figure 2. As in Figure 1 but $I = \frac{1}{2}$.

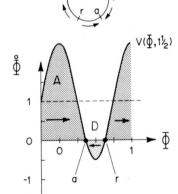

$v(\Phi,1\frac{1}{2})$

Figure 3. As in Figure 1 but $I = 1\frac{1}{2}$, past the bifurcation at $I = 1$. An attractor-repellor pair has opened up from phase $\Phi = \frac{1}{2}$, inverting the angular velocity within that arc.

Box A: The Indecisive Stopwatch

David Saunders drew my attention to the behavior of a standard commercial stopwatch (i.e., standard before the digital revolution). It is a ring device. Its parameter I has only three discretely different values, chosen by punching the wind-up stem: $I = 0$, 1, or 2.

At $I = 0, \dot{\Phi} = 1$: The clock ticks forward uniformly.

At $I = 1, \dot{\Phi} = 0$: The clock freezes where it is.

At $I = 2, \dot{\Phi}$ is very positive in the arc after $180°$ and very negative in the arc before $180°$. As in Figure 3, $\dot{\Phi} = 0$ only at $\Phi_a (0°)$ and at $\Phi_r (180°)$. If Φ is sufficiently close to Φ_r when you change from $I = 1$ to $I = 2$, the poor thing scarcely knows whether to advance or delay and *can* linger, unstably, at Φ_r. This corresponds to a resetting stimulus so strong that extraordinary brevity is required to catch Φ still on the way to Φ_a when the stimulus ends. (I haven't yet succeeded.)

This indecisiveness near Φ_r is not to be confused with the entirely different phenomena associated with a phase singularity (Brady, 1974).

But in any other environment $I \neq I_0$, the ring device's angular velocity (expressed in the phase units of $I = I_0$) generally varies throughout its cycle (Figure 2). Thus, the ring device runs faster or slower, depending on its current phase $\Phi(t)$, so long as exposed to $I \neq I_0$. This modulation of rate, expressed as a sensitivity function (Winfree, 1967a) or as an angular velocity response curve (Swade, 1969; Daan and Pittendrigh, 1976b, pp. 279–280, pp. 287–288) has been invoked to account for several features of circadian rhythms (see Chapter 19).

With large enough I, v may even go negative during part of the cycle (Figure 3). Phase then "sticks" stably at the beginning of this interval, at the attracting stagnation point Φ_a. Unless helped past the repelling stagnation point Φ_r, (e.g., by changing back to I_0 for a while), it won't do the next cycle. Thus, our simple clock has become an hourglass (see Box A). Figures 1, 2, and 3 might have been taken from Figure 4 at levels I_0, I_1, and I_2. The zones of negative angular velocity are indicated as islands in Figure 4. At any fixed I, phase will stick on the upwind shore of any such island.

Figure 4. Arrows depict $\dot{\Phi}$ above the Φ axis at each level of I. The levels of Figures 1 through 3 are indicated on the I axis. Within the shaded region, $\dot{\Phi} < 0$. The left branch of the U is the attracting phase. The right branch is the repelling phase.

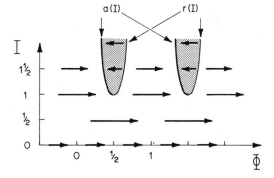

The shape of the function v of course depends on the I_0 chosen for the initial calibration of phase. This arbitrariness in defining phase does not affect the qualitative property essential for what follows, namely, whether v is positive or negative. A different choice of I_0 would only distort the horizontal scale of these diagrams (see Daan and Pittendrigh, 1976b, Figure 10).

Is this Caricature too Simple-Minded?

A pause may be useful to put this ring model in perspective. We are developing the simplest model of a repeatable continuous process. In this simplest case, we have only one variable of state (the phase), only one observable (the same), and only one parameter (I). More complicated models might introduce a second variable of state. For example, in Chapter 5 we abandon the restriction to a unique cycle of fixed amplitude and take amplitude of oscillation as a second variable of state. Alternatively, we might find that our real experimental system responds not so much to the physical parameters of its environment as to their rates of change. For example, such rate-dependent stimulus transduction is important in the light responses of the sporangiophore of *Phycomyces* (Delbruck and Reichardt 1956), in *Dictyostelium*'s response to cAMP (Gerisch and Hess, 1974; and Chapter 15) and in chemotaxis generally. According to our definitions, system behavior depends only on its state and its environment, so that in such cases there must be a second variable of state, a state of adaptation to the external stimulus. This component of the state changes quickly, tracking the stimulus and buffering it out so long as it doesn't change.

One method is to try the simplest interpretation first. To be sure, it will not long remain possible to evade more realistic complexities. But only after pinpointing the simple interpretation's limitations can one recognize the most economical amendment. We adhere in this way to the principle of Occam's razor, but not because of a naive hope that evolved dynamics are simple but rather because the plainest interpretations are the most expeditiously testable. (This is the strategy of looking for an object lost in the dark by working outward from a lamp post while our eyes adapt to the darkness.)

C: Derivation of Phase-Resetting Curves

Figures 1–4 are all drawn using a simple representative model:

$$v(\Phi, I) = 1 + I \cos 2\pi\Phi \tag{1}$$

with $I_0 = 0$. I might represent light intensity for circadian rhythms, or biasing current or pressure for a pacemaker neuron. By setting $I_0 = 0$, we make I the departure from standard conditions. So I might represent deviations from standard temperature or standard ionic composition of the medium surrounding a heart cell. The simple clock range of I is $-1 < I < 1$. Beyond that range, v goes negative at certain phases and we therefore have an "hourglass".

Box B: A Parable of Three Clocks

Suppose, contrary to the principles of simple-clock resetting, that the new-phase vs. old-phase curve smoothly rises and falls back in some region of negative slope. Suppose we take three clocks in this region, $\Phi_1 < \Phi_2 < \Phi_3$. Exposing all three simultaneously to the resetting stimulus, we send them to new phases in the opposite order, $\Phi'_1 > \Phi'_2 > \Phi'_3$ (Figure a). This results in an awkward paradox if we try to describe the mechanism behind the monitored rhythm as a biological clock with some phase that is advanced or delayed at unusual but phase-determined rates during a stimulus—as in resetting a wristwatch. To experience this paradox take a sealed business envelope, cut off the ends with a scissors, and open it into a cylinder. Let its circumference be a phase axis and its length be a "time during the stimulus" axis. We will follow the three clocks' changes of phase during the stimulus as they advance or delay around the cycle. Mark the clockwise direction of the phase axis, and at the left end mark initial phases in order $1 < 2 < 3$. At the right end, mark the three final phases in opposite order $1 > 2 > 3$. Now we draw curves connecting 1 to 1, 2 to 2, and 3 to 3 (Figure b). This may sound rather arbitrary, but it doesn't much matter. The point is that we cannot connect the numbers without some two curves crossing en route. And what's wrong with that? At the moment of crossing (some time during the stimulus) two clocks with identical phases, subject to the same stimulus, proceed to change phase at different rates. That presents a problem if clocks at the same phase are expected to function similarly. They are expected to function similarly in all descriptions of phase control in terms of ring devices (Brinkmann, 1967; Winfree, 1967a,b; Pittendrigh, 1960; Swade, 1969; Kuramoto and Yamada, 1976; Fujii and Sawada, 1978), i.e., that environmental parameters determine, in a phase-dependent way, the rate of change of phase (i.e., phase velocity) of the clock in its cycle. Thus the resetting curves of ring devices, as described in this chapter, cannot have negative slope anywhere.

This observation is actually not so devastating as it appears at first glance because there are two straightforward escapes from this conundrum, both by routes that most physiologists, I think, would accept as plausible in terms of familiar experimental findings:

1.   An organism with a negative slope in its resetting curves may differ from a simple clock in that its biological clock changes amplitude as well as phase during a resetting

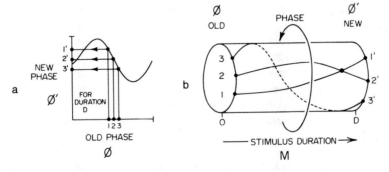

A new-phase vs. old-phase curve with a negative slope region, where the order of phases is inverted. During such an inversion, phase can be followed on a cylinder whose long axis is time during the stimulus. Some pairs of clocks must come to the same phase at the same time in this process.

perturbation. Wever (1963, 1964, 1965a) was the first to argue persistently in the circadian literature against describing every experiment in terms of phase alone, as though that were an unambiguous determination of the state of a circadian clock. Phase *does* tell all there is to know about the state of almost every kind of commercial clock, but it appears that when we kick a *biological* clock, not only do its hands rotate, but they shrink or elongate too: even through length 0 to reappear at positive length 180° away. This trick makes all the difference, as we shall see in Chapter 5.

2. An organism with a negative slope in its resetting curve may differ from a simple clock in that it really consists of many potentially independent clocks. If their aggregate behavior is the basis of our resetting measurements, then the negative slope of the resetting curve could be merely an expression of incoherence. As we will see in Chapter 4, dispersion of phase within a composite pacemaker (a "clockshop") has the same gross consequences as change of amplitude, including the possibility of inversion through zero. There are good reasons to believe that many physiological rhythms reflect the aggregate rhythmicity of many individual circadian clocks, be they simple clocks or not (Chapter 19).

In standard conditions, $d\Phi/dt = v(\Phi, 0) \equiv 1$. If we denote the beginning phase by ϕ, stimulus duration (magnitude) by M, and the final phase by ϕ', then a control stimulus, at normal intensity $I = 0$, gives

$$\phi' = \phi + \int_0^M \frac{d\Phi}{dt}\, dt = \phi + \int_0^M dt = \phi + M. \tag{2}$$

In this case, no phase shift is accumulated during interval M because the clock advances normally.

With $I \neq 0$, the algebra is a little clumsier, but without belaboring details it turns out as follows. During the stimulus, $d\Phi/dt = v(\Phi, I) \neq 1$. Rearranging,

$$\frac{d\Phi}{(1 + I \cos 2\pi\Phi)} = dt \tag{3}$$

$$M = \int_0^M dt = \int_\phi^{\phi'} \frac{d\Phi}{1 + I \cos 2\pi\Phi}. \tag{4}$$

This integral is a bit tricky to solve, but it is a standard lookup. For the case $-1 < I < 1$ (simple clock) it takes one form, and for $|I| > 1$ (hourglass) it takes another (see Box C). In either case, the messy algebraic result is $M = f(\phi', \phi, I)$. This can be used to calculate ϕ' as a function of ϕ, M, and I.

A simple consequence of some utility is that the rate of change of ϕ' with ϕ is necessarily positive: A later final phase is reached at the end of a later starting stimulus of given length. More exactly,

$$\frac{d\phi'}{d\phi} = \frac{v(\phi', I)}{v(\phi, I)}.$$

These two velocities necessarily have the same sign because the sign could only change by going through zero; but when velocity $= 0$, nothing changes. Thus

---

Box C: Integration Formulas For a Simple-Clock Model

Let $\dot{\phi} = 1 + I \cos 2\pi\phi$.
Then for $|I| < 1$,

$$\tan \pi\phi' = A \tan\left(\pi MF + \tan^{-1}\left[\frac{\tan \pi\phi}{A}\right]\right).$$

where

$$A = \sqrt{\left|\frac{1 + I}{1 - I}\right|} \quad \text{and} \quad F = \sqrt{|1 - I^2|}.$$

And for $I = \pm 1$,

$$\tan\left(\frac{\pi}{2} \mp \left(\frac{\pi}{2} - \pi\phi'\right)\right) = M + \tan\left(\frac{\pi}{2} \mp \left(\frac{\pi}{2} - \pi\phi\right)\right).$$

And for $|I| > 1$,

$$\tan \pi\phi' = A\left[\frac{K - 1}{K + 1}\right],$$

where

$$K = \left[\frac{A + \tan \pi\phi}{A - \tan \pi\phi}\right] \exp(2\pi MF).$$

M, like ϕ, is measured as a fraction of the unperturbed cycle duration. The equivalent formulas, rederived in different notation, appear in Kuramoto and Yamada (1976), Aizawa (1976), and Fujii and Sawada (1978), where the same model is applied to wave-like periodism in chemical reactions. (All the figures in this chapter are computed from the above.)

---

resetting curve slope is positive for all mechanisms of perturbation based on modulation of angular velocity, as in simple clocks. This is not true of other classes of oscillating kinetics. This fact is used in Chapter 22, pp. 439–440. It also shows that resetting achieved by modulation of angular velocity cannot produce type 0 resetting curves, which necessarily have a region of negative slope. See Box B.

Graphical Interpretation: How Phase Changes in Time

The curves $\phi'(\phi, M)$ tell what final phase ϕ' is reached by the end of a stimulus of duration M that began at phase ϕ. These dose-response curves are plotted in Figure 5 for the cases $I = 0$, $I = \frac{1}{2}$, and $I = 1\frac{1}{2}$. Phase, Φ, starts at ϕ at $t = 0$ and it increases to ϕ' at $t = M$. In the simple clock case $[|I| < 1$: Figure 5(a, b)] the period τ is indicated as the time elapsed from $\Phi = \phi$ to $\Phi = \phi + 1$. It is the same for any choice of beginning ϕ. The diagrams are sideways (time t plotted upward instead of horizontally as usual) to facilitate comparison with their source, the velocity characteristic $v(\Phi, I)$ plotted above.

A very weak stimulus (I close to 0) has little or no effect even if prolonged. In this case $\phi' = \phi + M$ [Figure 5(a)]. Now, let I differ significantly from 0 so

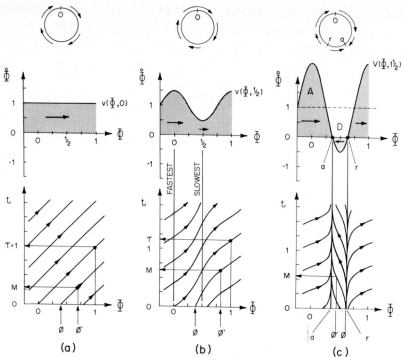

Figure 5.   Parts (a), (b), (c) correspond to Figures 1 through 3. In each, the time dependence of Φ is plotted as a curve, starting from each of various initial phases. The phase axis runs horizontally and the stimulus duration time axis runs vertically. Thus initial and final phases ϕ and ϕ' correspond to times 0 and M on the vertical axis. The time τ from any phase ϕ to $\phi + 1$ is the period $1/\sqrt{1 - I^2}$.

that v is not uniformly 1 as in the previous case, but varies above and below as Φ varies [Figure 5(b)]. So long as $|I| < 1$, the result is that $\Phi(t)$ rises along a sinuous path from each initial ϕ. It rises by one full cycle in time $t = \tau$, which was 1 at $I = 0$ but generally differs from 1 with $I \neq 0$. This is because the ring device is moving around its phase circle at a rate that varies locally. Thus it gets around quicker or takes longer according to the sum of the times taken in each increment of phase. The phase integral of 1/(this altered local velocity) is the period. The mathematical result from Equation (4) (see Box C) is that if $\phi' - \phi = 1$ (i.e., for completion of exactly one full cycle during M at $|I| < 1$), then stimulus duration must be $M = 1/\sqrt{1 - I^2}$. This M is therefore the period τ obtained under chronic exposure to $I \neq 0$. Note that a plot of this M against I looks like a potential well with its bottom at $I = 0$ and vertical walls at $I = \pm 1$. It is flat at the bottom, connoting homeostasis of period near $I = 0$. If v were chosen less symmetric, or if the cycle had been calibrated at some $I \neq 0$, or if v happened to vary less symmetrically about 1, then this well would have its flat part to one side of $I = 0$ (cf. Daan and Pittendrigh, 1976b, p. 278 on the periods of rodents' circadian clocks).

If I differs so much from 0 that the ring device almost stops at a certain phase, then it takes a very long time to complete its cycle. Departures of velocity from 1 are weighted more heavily on the slow side because phase dawdles longer in the slow regions. With any further increase in I we may pass the critical value for transition to hourglass kinetics [Figure 5(c)]. The ring device then lingers forever where $\dot{\Phi} = 0$. The appearance of a range of negative velocities creates a pair of stagnation points, one attracting (Φ_a) and one repelling (Φ_r). The ring device approaches Φ_a and sticks there.

Notice that on each of the preceding diagrams [Figure 5 (a, b, c)], above any point Φ on the horizontal phase axis the curves have the same slope. That is to say, $\dot{\Phi} = v(\Phi, I)$ is given by Equation (1) for that phase. From that phase, the subsequent behavior is the same in every case (supposing constant I), regardless of past history. In other words, the $\Phi(t)$ curves differ only in their displacement along the vertical time axis. This is what it means to say that in any given environment (i.e., exposed to a given stimulus, i.e., given a fixed I), the instantaneous angular velocity v depends only on instantaneous phase. This is the central fact about ring devices, as here defined. It determines their most characteristic properties.

For example, this implies that at any fixed I the paths $\Phi(t, I)$ do not cross each other. If they did, then $\dot{\Phi}$ would not be uniquely determined by Φ and I at that phase. This fact is put to work in Box B.

When the stimulus ends and I is changed back to 0, then Φ carries on from phase ϕ' according to $\dot{\Phi} = v(\Phi, 0) = 1$, i.e., in free run as defined in that standard environment $I = 0$.

Graphical Interpretation: Resetting Curves

Now let's back up and look at this same model from a different perspective. We ask how the new phase of a single ring device depends not on M at fixed beginning phase ϕ, as in the preceding section, but on its beginning phase ϕ, supposing fixed M. What phase shift $\phi' - (\phi + M)$ does the stimulus (I, M) inflict, depending on the phase ϕ when it begins?

Mathematically, the answer is given above where we calculated a formula by which duration M and beginning phase ϕ determine final phase ϕ'. However, it is helpful to summarize in graphical terms. This is done in Figure 6, which plots a whole family of so-called resetting curves. Each resetting curve shows the final phase ϕ' as a function of initial phase ϕ for one fixed stimulus duration M. The family of curves shows how this resetting curve changes as we use stimuli of different durations from 0 up to infinity. The lower row of figures plots not ϕ' but $\phi' - (\phi + M) = \Delta\phi$ against ϕ. These curves have three conspicuous properties:

1.   They are continuous: If the beginning phase ϕ is changed a little bit, then the final phase ϕ' changes by only a corresponding little bit.

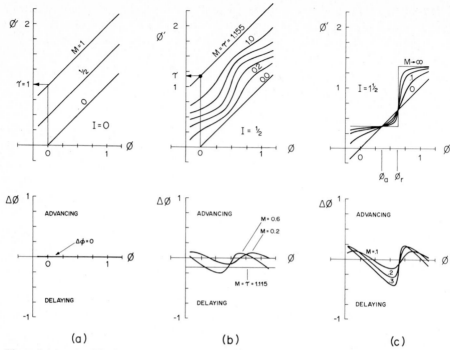

(a)   (b)   (c)

Figure 6. (above): The information portrayed in Figure 5 is redescribed by plotting ϕ' vertically against ϕ horizontally, for each of several stimulus durations M. (below): As above but plotting vertically the phase shift $\phi' - (\phi + M)$, which is the height of the curve above the corresponding curve at $I = 0$. Advances are plotted upward, delays downward. The $M = 0$ curve lies on the $\Delta\phi \equiv 0$ axis.

2. They are periodic along the ϕ axis, since starting a stimulus at phase ϕ in one cycle or in the next makes no difference to the results.

3. They are periodic along the ϕ' axis. The observed result is a rhythm, and saying that its phase is ϕ' is the same as saying that its phase is $\phi' + 1$ or $\phi' - 2$ or ϕ' plus or minus any number of complete cycles.

Thus, our schoolboy habit of plotting things on perpendicular Cartesian coordinates is not the most appropriate or natural in this situation. The Cartesian coordinate in no way represents the fact that the end of the phase axis represents exactly the same state as does its beginning. Actually both phase axes ϕ' and ϕ are rings. This fact would be brought out more clearly if the Cartesian graph paper were rolled up horizontally to close the horizontal ϕ axis in a ring with its endpoint exactly superimposed on its beginning point. Or, if it were rolled in the perpendicular direction to close the vertical ϕ' axis into a ring, bringing its endpoint onto its beginning. If we do both operations, we have rolled the graph paper into a torus, a doughnut-shaped space whose topology exactly captures the logical structure of experiments on a rhythmic system. On this doughnut-shaped graph paper, the resetting curve is a closed ring which links once through the hole in the doughnut. The fact that it is a ring comes from the fact that any ϕ determines a unique ϕ' in this kind of experiment. The fact that it is a *closed* ring comes from the continuity property, i.e., results of nearby experiments are similar. The fact that it links once through the hole comes from the fact that as the beginning phase ϕ is scanned through one full cycle, the final phase ϕ' likewise scans through a full cycle in the same direction. In fact, not only does ϕ' advance through a cycle as ϕ does, but more: ϕ' never decreases as ϕ increases. This is a conspicuous hallmark of ring device kinetics.

Graphical Interpretation: The Resetting Surface

It is now time to parlay the notion of a resetting *curve* into the notion of a resetting *surface*. This extension is necessary in order to incorporate stimulus duration M as a second stimulus parameter, supplementing the time of application ϕ. Take a family of curves in Figure 6, e.g., take 6c at $I = 1\frac{1}{2}$. Set them up side by side on as many vertical coordinate frames, ϕ' vs. M, arranged in order of the beginning phase ϕ. Collectively, they outline a smooth surface $\phi'(\phi, M)$ which tells what final phase ϕ' the ring device achieves (vertical axis) at any ϕ (horizontal axis perpendicular to M). This is the resetting surface (Figure 7). It is a theoretical prototype of the experimentally measured time crystal introduced in Examples 5 through 9 of the previous chapter. This is the time crystal of a simple clock. It differs from those obtained experimentally in that it has no phase singularity linking the (here) separate resetting surfaces stacked periodically along the ϕ' axis.

Think of it as embedded in a cube of clear plastic. The cube could be sliced up into a stack of thin layers perpendicular to any one of its three edges:

1. We have already seen it sliced perpendicular to the ϕ axis. Each such slice is a dose-response curve, showing how ϕ' changes during longer and longer exposures started from a given ϕ (Figure 5).

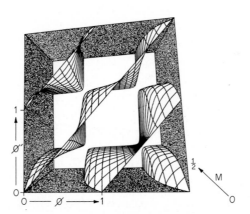

Figure 7.   The preceding figures are gathered into a single three-dimensional plot of ϕ' as a function of ϕ and M at $I = 1\frac{1}{2}$. The corresponding picture at $I < 1$ differs only in that it has no range of phases in which ϕ' decreases with increasing M.

2.   Consider now planes of constant M, i.e., $\phi' \times \phi$ planes, each cutting the resetting surface at one fixed stimulus duration M. The curve along which it cuts the surface is the resetting curve for stimuli of that duration, administered at each beginning phase ϕ [Figure 6(c)]. Such curves are elaborated in experimental context in Chapters 1, 7, 14, 19, and 20.

3.   We can plot what the resetting surface tells us in one more convenient way by sectioning along planes of fixed ϕ'. These planes intersect the surface in a locus of (ϕ, M) combinations that guide the ring device to the same final phase ϕ'. A composite figure such as Figure 8, showing the family of such curves for stepwise increments of ϕ', is a contour map of the resetting surface for one fixed stimulus intensity I. Figure 8, like Figures 5 and 6, shows the qualitative appearance at $I = 0, I = \frac{1}{2}$, and $I = 1\frac{1}{2}$. In each panel, each curve is a level contour of final phase ϕ' on the stimulus plane of beginning phase $\phi \times$ stimulus duration M.

In the pinwheel experiments encountered in the previous chapter the stimulus plane was in real physical space. In that context each level curve of ϕ' may be regarded as the instantaneous position of a wave in space. This wave is marked by appearance of a marker event E, which signals passage through phase $\Phi = 0$. This is because all ring devices at a given final phase ϕ' when the stimulus ends will encounter $\Phi = 0$ simultaneously, at a time $1 - \phi'$ later. Those in the next lower horizontal section plane, at $\phi' - \delta$ when the stimulus ends, will reach $\Phi = 0$ simultaneously at a time δ after that. The contour map representation and the wave representation will both play conspicuous roles in later experiments with yeast cells (Chapter 12), malonic acid reagent (Chapter 13), flies (Chapter 20), and flowers (Chapter 21). For the present it will suffice to note that these contour lines do not converge to a point. It can be shown mathematically that they *cannot* do so, in the case of any resetting process that amounts to modulation of angular velocity. In such a case, a pinwheel experiment cannot possibly evoke a rotating wave.

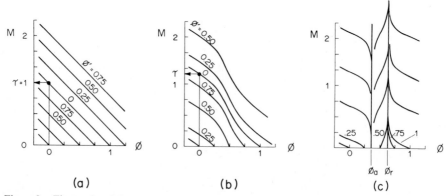

(a)   (b)   (c)

Figure 8.   Figures 5 and 6 represent two perpendicular cross-sections through Figure 7. Here is the third perpendicular cross-section. Initial phase increases to the right as in all preceding graphs, and stimulus duration M increases vertically as in Figure 5. The curves are level contours of uniform ϕ'.

D: Historical Appendix

The circle is the first, the most simple, and the most perfect figure.

Proclus[1]

Lo cerchio è perfettissima figura.

Dante[2]

Applications of the Simple-Clock Idea in Developmental Physiology

It is not unusual for developmental processes to proceed through a sequence of stages, each of which seems to be a necessary precondition for the next. The development of a frog from egg through tadpole to adult to egg, for example, progresses continuously through the Pollister and Moore stages like a train passing stations along its track. Though its speed may vary according to temperature, hormonal supplements, and the availability of food, the stages and their ordering are unchanged.

In the case of temperate zone animals which repeat an annual cycle within their life cycles, Mrosovsky (1970) declared that a system "that could give rise to annual cycles is a sequence of linked stages, each taking a given amount of time to complete and then leading into the next, with the last stage linked to the first again ... as in the case with pregnancy ... it should be possible to alter frequencies by slowing down, speeding up, or reducing certain stages." This is the essence of a "simple clock".

The same may be said of development on the level of a single cell. In this case, there might even be a relatively simple structural analog to a railroad track. Halvorson and Tauro (1971) reviewed experimental indications that the genome (a railroad track) is transcribed in linear sequence, resulting in a sequential ordering of biochemical changes during the cell cycle. The following excerpts from recent research literature point to the fundamental role of simple-clock models in this area:

Edmunds (1975) writes:

> The cell cycle ... comprises a series of relatively discrete morphological and biochemical events, although the specific elements may vary among different systems. These developmental sequences are not necessarily linearly ordered, however, since branching networks (and even "nested do-loops") may provide several alternative pathways, some of which may operate concurrently ....
> In this sense then, the cell cycle is a "clock": the specific events correspond to the numerals on the dial ... under a different set of conditions, the cell cycle consists of the same sequence but ... the relative time spent between such stages is different.

---

[1] Commentary on the first book of Euclid's *Elements*; on Definitions XV and XVI. Cited in Polya (1954), p. 168.

[2] *Convivio II*. XIII. 26. Cited in Polya (1954), p. 168.

Hartwell et al. (1974) write:

> Mitotic cell division in eukaryotes is accomplished through a highly repro
> ducible temporal sequence of events that is common to almost all higher
> organisms . . . . How are (these events) . . . coordinated in the yeast cell cycle
> so that their sequence is fixed? . . . There may be a direct causal connection
> between one event and the next. In this case it would be necessary for the
> earlier event in the cycle to be completed before the later event could occur . . . .
> A second possibility . . . invokes the accumulation of a specific division protein
> and another a temporal sequence of genetic transcriptions . . . . No event in
> this pathway can occur without the prior occurrence of all preceding events.

Tyson and Kauffman (1975) write:

> The periodic event with which we shall be particularly concerned is cell
> division. Mitosis is an event of short duration relative to the cell cycle. The
> exact nature of the system controlling its regular periodicity is not yet known
> for any organism . . . . Some have supposed that the cycle is a cyclic sequence
> of discrete states, each causing the next. Evidence in bacteria and yeast that
> the temporal sequence of genes transcribed during the cycle may be related
> to the sequential linear order of the genes on the chromosomes have prompted
> speculation that the cycle is controlled by sequential transcription around
> circular DNA molecules in eukaryotic chromosomes.

Applications of the Simple-Clock Idea in Circadian Physiology

A tape-reading or railroad track analogy equivalent to those conceived of in
context of the cell division cycle (see above) was explicitly proposed by Pittendrigh
(1966, p. 305) and by Watson (1976, p. 510) as a possible basis for biological timing
and biological clocks in general. Ehret and Trucco (1967) elaborated just such a
molecular model of circadian clocks with the additional proviso that the genes
sequentially activated needn't be physically consecutive along the genome. Their
clock is a sequential machine composed of concentrated lumps of transcribable
DNA. Emphasizing the one-dimensional character of this mechanism, they make
analogies to the one-dimensional clocks used in ancient times:

> The chronon bears an interesting, though somewhat superficial, resemblance
> to the calibrated candle clocks of the medieval period in which the end-to-end
> consumption by a flame of carefully selected lengths of resined rope was a
> standard measure of time.

They suggest that the spontaneous periodism of the clock might even have a
structural basis in the closure of the DNA track in a ring, as in the genomes of
bacteria and of viruses, and more appropriately, of mitochondria.

Even prior to this incarnation of the simple-clock paradigm in a physical
ring, elaborate developments of the idea were propounded by experimental
investigators of circadian rhythmicity. As early as 1940, Kalmus thought of the
circadian clock in fruitflies by analogy to a rotating, repeating gramophone
message or the drum of a music box. (See Chapter 20 for biological background
on fruitfly clocks, to which I will allude frequently in the next few pages.) Brett
(1955) likewise hung his observations of fruitfly clocks on a rotating wheel model.

Bunning (1956a) compared the cycle of activity in plants to the playback of a gramophone record. Enright (1975) continued this theme of tape recorder loop analogies, in the context of the periodic fine structure observed in the activity records of a marine isopod on his beach at La Jolla.

So profoundly did the simple-clock metaphor pervade thinking about circadian rhythms during the 1950s and 1960s that alternative notions seldom seemed to pose a challenge worthy of experimental test. One gains the consistent impression from the literature that, apart from the few writers who do their modelling explicitly by way of equations, most folks do theirs implicitly, in terms of ring devices. The notion thus shapes the choice of descriptive language and the design and interpretation of the experiments. This may be sound procedure: Simpler analogies are easier to test than complex ones and so should be tried first. The clock metaphor also has great appeal due to the fact that time *is* reckoned one-dimensionally both in principle and in almost all human contrivances for time keeping. Introduced by Hoagland (1933), the clock metaphor was adopted in the context of 24-hour rhythms by Johnson (1939). It was first popularized by Brown in the early 1950s in connection with his belief that circadian rhythms are in fact one indicator of external geophysical clocks. Pittendrigh pursued the clock metaphor in the later 1950s in connection with his belief in an internal timepiece whose evolutionary *raison d'être* is to monitor the passage of time.

The idea that a circadian clock goes through a fixed cycle more quickly or slowly according to its phase and environmental factors, such as visible light intensity, was first suggested by Rawson (thesis, 1956) and later by DeCoursey (thesis, 1959). The idea promised an interpretation simultaneously for resetting curves and for the fact of entrainment to a fluctuating light schedule. Palmer's notion of "autophasing" (1959; see Palmer 1976, p. 250) invokes exactly this principle—that the rate of advance depends only on phase and environmental conditions—to rationalize the supposed generation of a circadian rhythm by a geophysical rhythm of different period.

The previously implicit simple-clock paradigm was first made explicit and articulate by Campbell (1964) in the context of mitotic cycles, as noted above. Stimulated by Campbell's analysis and by six months with a population of simple-clock-like oscillators in Pittendrigh's Princeton laboratory (Chapter 11), I contrived an analysis of a simple clock's entrainment by a rhythmically fluctuating stimulus (Winfree, 1967a). Swade (1969) applied similar principles to entrainment of rodent activity cycles in the fluctuating light of the Arctic summer.

The simple clock interpretation of circadian rhythms was most vigorously advanced by Pittendrigh, starting with Pittendrigh and Bruce (1957). Goodwin (1976) reviewed:

> The minimum number of variables assumed to be necessary for an oscillation in biological system is two ... however, there is an even simpler type of oscillation than this, dimensionally speaking, and that is what has been referred to as a generalized relaxation oscillation [see Chapter 6, Box A; read "simple clock"]. This is a periodic function of time which can have different periods, but practically no change of amplitude. The state of such an oscillator can be defined in terms of one variable only, which is its phase; i.e., it is like

> an ordinary clock which can be set to any time and will then run at fixed speed with the only visible indicator of state being the position of the hands . . . from very extensive and ingenious studies of the behavior of biological clocks in response to various types of perturbation, Pittendrigh and Bruce (1957) proposed a [two-oscillator] model in which the chronometer (A) was assumed to be an oscillator of this relaxation type [i.e., a kind of simple clock] whose phase could be instantaneously reset by a light or other signal, but whose amplitude remained unaffected. Any transients in the observable response to the signal were then regarded as occurring in that process (B) which is controlled by the clock (A) and by which its existence is made manifest. This model is widely used and has had very considerable success.

(All bracketed insertions are mine.)

This A-B two-oscillator model has had a profound influence on the development of circadian physiology. Its genesis lies, once again, in the simple-clock paradigm. Pittendrigh and Bruce (1959, p. 491) argue:

> Any model based on a single oscillator is unable to explain the concurrence of the three features that strongly characterize resetting in the fly: (1) ultimate determination of phase by a signal seen three cycles previous to the new steady state, (2) the presence of transients, and (3) the dependence of transient length on the time of the cycle at which the signal fell. These features are, on the other hand, all explained by a model for the system based on two coupled oscillators.

These facts really place no constraint on an oscillator model unless by "oscillator" one implicitly means a simple clock.

The point I am trying to make here is not that the simple clock interpretation is wrong or an inconvenient approximation (although we *will* be more concerned with its deficiencies in later chapters) but that it has been important in the development of thought and experiments in several areas of physiology and biochemistry. It might be added that ring devices lie at the very heart of any industrial society for the simple reason that industry consists largely of automated productive activities, and automation goes farthest quickest in processes that are *repetitious*. In any repetitive fabrication, machines and their operators traverse again and again stereotyped one-dimensional cycles of states. So it is from the internal combustion engine to knitting and weaving machines to the distribution of power synchronized to the rotation of innumerable dynamos.

4. Ring Populations

If thou (dear reader) art wearied with this tiresome method of computation, have pity on me, who had to go through it seventy times at least, with an immense expenditure of time . . .

Johannes Kepler, 1609,
Astronomia Nova, Chapter 16

My intent in this chapter is to direct your attention to several idealizations of rhythmic behavior in *collections* of many similar ring devices. It turns out that some of the peculiar limitations on the behavior of simple clocks do not apply to populations of simple clocks. Here we also encounter our first example in which a phase singularity emerges from an idealized model of the structure and mechanism of a rhythmic system. The chapter is divided into four sections:

A.   Collective rhythmicity in a population without interactions among constituent clocks. This is mainly about phase resetting by a stimulus.

B.   Collective rhythmicity in a population whose individuals are all influenced by the aggregate rhythmicity of the community. This is mainly about mutual synchronization and opposition to it.

C.   Spatially distributed simple clocks without interactions. This is mostly about patterns of phase in space.

D.   Ring devices interacting locally in space. This is mostly about waves.

A: Collective Rhythmicity in a Population of Independent Simple Clocks

Definitions

Before anything else we need a definition of *collective rhythmicity*. If we were pooling sine waves of various phases and amplitudes, all with the same period, the result would always be another sine wave. So there would be no ambiguity about its phase: Whatever is the measure of phase on a single clock, the same measure is used for the aggregate (e.g., phase = fraction of a cycle past a maximum). But what if the output of each clock is a sharp action potential as in the neural oscillators of Chapter 14, or a sawtooth wave as in the neon glow tubes used in Chapter 11? Then the sum of many bears little resemblance to the familiar waveform in terms in which phase was defined. If we are going to ascribe some "phase" to such a collective waveform, then we must have some additional rule. The choice is essentially arbitrary, a matter of convenience, but it must be made definite. I choose the following definition.

Any periodic waveform can be depicted as a superposition of sine waves of suitable amplitudes and phases. This superposition includes one sine of each frequency, from the fundamental on up by integer multiples. We want a number to characterize the phase of a rhythm of unit period. (We take its period as our unit of time.) A number at least as good as any other is the phase of the fundamental, scaled to range from 0 to 1. Whatever the rhythm's shape this measure changes by ε (modulo 1) if the whole waveform is delayed by a time ε. This measure is unaffected by changing the overall amplitude. In nearly synchronous populations it registers the same phase as the phase of each single oscillator. In slightly less synchronous populations, the collective phase is the mean of the phase distribution.

Geometrically this measure is especially convenient. If we represent the cycle of each simple clock as a perfect circle traversed at uniform speed (letting the central angle on this diagram be the fraction of the cycle elapsed), then a population of clocks can be depicted as so many dots moving with constant angular velocity around this circle (Figure 1) The circle represents the fundamental sine wave. The X coordinate of each dot traces in time the fundamental sinusoidal component of that clock's rhythm. The superposition of all sinusoids contributed by the many clocks of a population is traced by the center of mass (∗) of the cloud of dots on this diagram. The phase of that aggregate rhythm is the central angle of the center of mass. We can rule the diagram with radial lines each of which corresponds

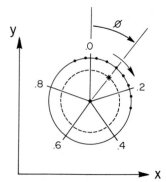

Figure 1.   The outer circle represents unit amplitude of the first harmonic of any rhythm. Dots represent clocks at various phases. The asterisk represents the phase and amplitude of their collective rhythm (averaged). It follows the inner circle so long as the clocks are independent and unperturbed. The radial lines link positions of equal phase.

to a given phase, namely, the phase at which it intercepts the unit circle. Any aggregate rhythm's phase is read off by simply locating the population's center of mass on one of these radii.

This choice of a phase measure dovetails nicely with an analysis of physiological "transients" put forward by Mercer in 1965 and more explicitly by Kaus in 1976. According to this interpretation, the physiological observables that we choose to monitor in the laboratory sometimes do not happen to participate directly in the mechanism of the "clock". Rather, they are only indirectly driven by the clock mechanism. In some cases, it apparently suffices to regard the driven observable as a linear filter, selectively transmitting the fundamental frequency of the driving clock. Its higher harmonics vary too quickly to have much influence on such a sluggish driven system. In such a case (and the fruitfly's circadian rhythm seems to be one of them) the observed rhythm would be determined by a population of clocks in exactly the way suggested above.

Having chosen a collective measure of phase, we now apply it to the case of two, three, and more noninteracting simple clocks.

Two Clocks: Unperturbed Kinetics

The state of a population of N clocks consists of the phases $(\phi_1, \phi_2, \ldots, \phi_N)$ of its N members. The state space thus $\mathbb{S}^1 \times \mathbb{S}^1 \times \cdots \times \mathbb{S}^1 = \mathbb{T}^N$, the N-dimensional hypertorus. In the case of just two clocks, this is $\mathbb{S}^1 \times \mathbb{S}^1 = \mathbb{T}^2$, the familiar surface of a doughnut. We unroll it to form a doubly periodic presentation on flat paper in Figure 2. Let the horizontal phase coordinate be clock 1 and let the vertical phase coordinate be clock 2. We have two things to draw on this surface:

1. Contour lines indicating the phase and amplitude of the pair (considered as a unit) as a function of the two individual phases, and

2. The unperturbed trajectories of the pair of clocks in this state space

(1) Taking the two clocks to have equal periods and equal weight, the phase of the pair is the central angle of a point midway between them on this circle diagram (Figure 3). This can be written trigonometrically, but after a little algebra it reduces to something geometrically obvious: The phase of the aggregate rhythm of two equal clocks lies in the direction midway between the two, at an amplitude less than either by an amount that increases with their phase difference. When the two are one-half cycle apart, amplitude of the fundamental harmonic passes

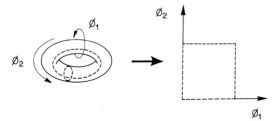

Figure 2.   The state space of a pair of simple clocks, being a torus, can be depicted on flat paper as one unit cell of a sheet of wallpaper.

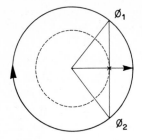

Figure 3.   By the convention adopted in Figure 1, the phase of a pair
of simple clocks lies midway between their individual phases. The average amplitude of the pair is the radius of the dashed circle.

through 0 and phase is therefore ambiguous. If we move either oscillator's phase
a little bit, the aggregate phase switches back and forth by one-half cycle through
amplitude 0. Note that amplitude 0 in this case means only that the *fundamental*
has vanished. All the even harmonics are still there but, because of the twofold
symmetry within each period, the aggregate rhythm's phase is indeterminate,
Note also that (unless the two clocks are exactly one-half cycle apart) advancing
both in phase by any amount advances the aggregate rhythm in phase by the
same amount.

The aggregate phase depends on ϕ_1 and ϕ_2 symmetrically, with a discontinuity
along the 45° degree line where ϕ_1 and ϕ_2 are one-half cycle apart, as indicated
in Figure 4. Pending fuller definition in Chapter 6, let's call any locus of uniform
phase an *isochron* ("same time"). We might also indicate amplitude contours as
$A = \cos 2\pi[(\phi_1 - \phi_2)/2]$. The phase discontinuity is the "amplitude = 0" contour.

(2) Next we need to depict this system's natural motion in its state space. If
both oscillators have the same period, then $\dot\phi_1 = \dot\phi_2 = 1$ (in which an upper dot
denotes the rate of change in time). So the system's path, a trajectory from any
initial (ϕ_1, ϕ_2) combination, is a 45° line. Aggregate phase thus advances uniformly
and aggregate amplitude is steady (Figure 5).

If one oscillator moves a bit faster than the other, then the path's slope T_1/T_2
is no longer 1. It therefore slants at a different angle, slowly crossing amplitude
contours. It passes through the zero amplitude discontinuity at intervals
$1/(1/T_1 - 1/T_2)$. This path on the torus depicts geometrically the familiar beat
note seen or heard by superposing nearby frequencies.

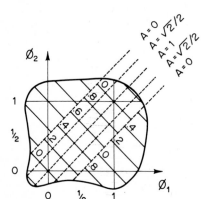

Figure 4.   Contours of ϕ_1 and ϕ_2 along which
the phase (ϕ, solid) or amplitude (A, dashed) of a
pair of simple clocks remain the same.

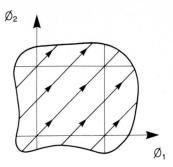

Figure 5. Motion of a pair of identical independent clocks across the (ϕ_1, ϕ_2) torus in the environment $I = 0$ used to calibrate phase in proportion to time. These paths are closed rings of period $\tau = 1$.

Two Clocks: Perturbed

Thus far we have implicitly assumed undisturbed cycling of the simple clocks in whatever standard environment was used to define phase in proportion to elapsed time. In the notation of the previous chapter, this means $I = 0$. Let us now consider $0 < I < 1$. This is the range in which each simple clock continues to cycle, but its positive angular velocity is greater or less at different phases. Each clock lurches and pauses around the common cycle. If two clocks start at different phases, then the phase angle between them constantly increases and decreases back again with the period of each oscillator's altered traversal of the common cycle. Thus the pair's trajectories on the (ϕ_1, ϕ_2) torus are no longer a series of parallel straight lines but now comprise a set of snaky lines (Figure 6).

If $I > 1$ then each clock will come to rest at ϕ_a before it has traversed a full cycle. This ϕ_a is the front edge of the range of phases in which angular velocity has gone negative (Figure 4 of Chapter 3). This behavior is shown simultaneously for both oscillators in Figure 7.

Thus if $I = 0$ for a long time, but then we make $I \neq 0$ for a while, then during that interval of exposure to the stimulus the oscillator pair is moved from its prior 45° trajectory to a new one. When I reverts to 0, the pair continues in motion along the new 45° trajectory. At that time it will generally be on a different phase contour than is an unperturbed control for which I remained 0 during the same interval. The new phase reached by the time the stimulus ends depends on two

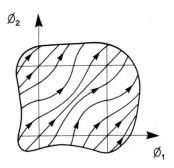

Figure 6. Motion of a pair of identical independent clocks across the (ϕ_1, ϕ_2) torus in an altered environment $I = \frac{1}{2}$, which still permits cycling. These paths are closed rings of period $\tau = 1/\sqrt{1 - I^2}$. This figure and Figures 7–9, 12–14, are sketched from grainy computer printouts and so are not quantitatively exact.

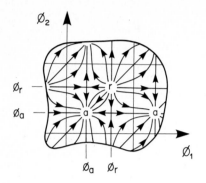

Figure 7. Motion of a pair of identical clocks across the (ϕ_1, ϕ_2) torus with $I = 1\frac{1}{2}$. Each gravitates to an attracting phase ϕ_a.

things:

1. The stimulus (I and its duration)
2. The initial state (ϕ_1, ϕ_2)

Note in (2) that it is not sufficient to specify just the initial phase of the aggregate rhythm. That external observable only locates the state on a certain ϕ contour (isochron) but does not tell where the two-clock system is on that isochron. During exposure to the stimulus, trajectories lead in various directions through the many states on that locus of uniform aggregate phase. We have to expect a different result from disturbing a pair of oscillators at phase ϕ, depending on whether the aggregate phase is ϕ because both are at ϕ or because one is at $\phi + x$ and the other is at $\phi - x$. We need to know the complete internal state of the system. The external observable ϕ is a sufficient identification of a rhythmic system's state only in very special cases. In general, it is one convenient measure of state but not in itself sufficient for design and interpretation of experiments because two systems at the same phase can have quite different subsequent behavior.

To know the complete internal state of a two-clock population, we could know the phase of each oscillator or, equivalently, we could know which contour of aggregate phase the population lies in and where it lies along that amplitude locus in Figure 4. With the aggregate phase measure we have chosen, this means knowing the mean phase of the two clocks and their phase difference. In populations with more simple clocks we would in principle need to know the phase of each clock in the population or equivalently their mean phase and enough higher moments of the distribution to specify the distribution unambiguously. If we could assume that clock phases are unimodally distributed around the mean, then the measures of most importance of our purposes would be the mean phase and the range or variance of phases.

Let us now return to the two-clock case to determine how the mean phase of such a population depends on the timing of the stimulus. It is convenient here and will be convenient again in other contexts to think of the stimulus as a mapping from the plane which describes the stimulus (its duration M and the initial phase ϕ when it starts) into the (ϕ_1, ϕ_2) state plane. This can be done as follows. Choose an initial phase difference between the two clocks. The pair's trajectory is then

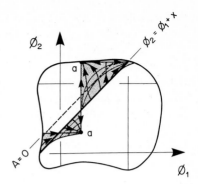

Figure 8. As in Figure 4-7 but a portion is shaded, consisting of all states reachable from the chosen initial unperturbed trajectory $(\phi_2 - \phi_1 = x)$ by increasing durations of exposure to $I = 1\frac{1}{2}$. The curves crossing perturbed trajectories are loci of fixed duration M.

a diagonal that far displaced from the main diagonal, as indicated in Figure 8. Points along this displaced diagonal are, trivially, the states reached by applying a stimulus $I \neq 0$ of duration $M = 0$ at each initial phase ϕ. This diagonal is the ϕ axis of an image of the (ϕ, M) plane in state space. Now expose the population for a duration M at each ϕ, marking off increments of equal duration along the trajectory followed (Figure 8) during stimulation. Through each initial ϕ this establishes an M axis. Collectively these measurements constitute a $\phi \times M$ grid on the state plane. It is an image of the (ϕ, M) plane on which we can immediately read off the *new* phase values arrived at by reference to the overlying contour lines of Figure 4. In imagination, stencil those contour lines onto this distorted image of the (ϕ, M) plane, then pick it up and stretch it out flat and uniform and you have Figure 9. All this can be done by trigonometric equations; in fact, that is how Figures 6–9 were obtained. But the qualitative principles are evident in the graphical methods described above and are independent of quantitative details about the chosen dynamics.

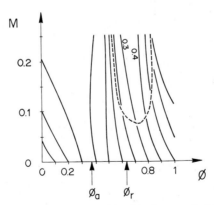

Figure 9. On the (ϕ, M) plane, level contours of ϕ', the phase reached by maintaining $I = 1\frac{1}{2}$ for duration M (in units of unperturbed cycle period), starting at phase ϕ, with $\phi_2 - \phi_1 = 0.3$ cycle initially. The dotted U locus represents amplitude $A = 0$; other amplitude loci are omitted. As $M \to \infty$ contours near ϕ_a diverge to fill the whole space. The U becomes symmetric about ϕ_r and approaches width 0.3 as $M \to \infty$. As M increases across the U, ϕ' jumps one-half cycle.

You will note that so long as the phase difference was initially not 0, there is a phase range within which a stimulus of sufficient duration carries the pair's state across the zero-amplitude locus, where the phase changes discontinuously because the two oscillators are one-half cycle apart. This happens only within a certain range of initial phases. The perturbed trajectory from other initial phases never crosses the discontinuity locus. The phase map of Figure 9 on the stimulus plane shows this piece of discontinuity locus as a U.

We have spent a long time belaboring a trivial example in order to lay the foundations for understanding situations of greater biological interest. The next step in that direction comes by looking at a system of three simple clocks. This is as far as we will need to go with simple clocks because it turns out that three do everything that a larger population can do, so far as our present interests are concerned.

Three Clocks: Unperturbed Kinetics

The principles are the same but the state space is three-dimensional, being $\mathbb{S}^1 \times \mathbb{S}^1 \times \mathbb{S}^1$. That could be thought of as a cube with its opposite faces identified pairwise by analogy to the method of depicting a two-dimensional torus as a square with its opposite faces identified pairwise. The trajectories followed by a state point (ϕ_1, ϕ_2, ϕ_3) within the cube are much the same. The only novel feature is encountered in trying to depict the loci of fixed aggregate phase (the isochrons) in this state space. In the two-clock case they were straight lines transverse to the trajectories, abutting discontinuously along the zero-amplitude locus. In a three-dimensional state space they must be two-dimensional surfaces, once again transverse to all the trajectories because moving all oscillators forward by amount ε must move the aggregate phase forward by the same amount. A theorem often used before can now be invoked to show that something peculiar must happen in the way that these isochron surfaces come together in the three-dimensional state space. Consider Figure 10 and the gedanken experiment indicated by the four-sided closed path ABCDA (heavy arrows). Along path AB the phases of all three clocks increase equally and thus advance through one full cycle of aggregate phase. Along path BC clocks 2 and 3 are held fixed while clock 1 moves through a full cycle. This causes the aggregate phase to change somewhat and change back to its original value *without* scanning through a full cycle, because the single clock affected is in the minority in a three-clock population. This feature of the collective phase measure chosen above is more obvious in a population of thousands of clocks, only one of which is allowed to vary at a time;

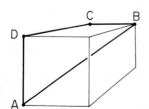

Figure 10.   The $\phi_1 \times \phi_2 \times \phi_3$ hypertorus "unrolled and laid flat" as a cubical unit cell of a three-dimensional crystal. Path ABCDA is followed in a conceptual experiment.

it turns out to be true for any number of identical clocks down to and including three. At point C the phases are all the same as they were at point A and point B. Next clock 2 is varied, holding clocks 1 and 3 fixed. Once again for the same reasons the aggregate phase does not scan·through a complete cycle. The same is true along arc DA (and along any number of additional arcs of the same sort which would be required in describing a population of *many* simple clocks). The upshot is that the aggregate phase has winding number 1 around the closed path ABCDA. Now imagine any two-dimensional surface topologically equivalent to a disk bounded only by path ABCDA. I intend to assign to each point on such a surface some aggregate phase. Thus I propose a map from a two-dimensional disk to a phase circle in such a way that its boundary maps around the circle with winding number 1. Once again (see p. 28), this cannot be done without accepting a discontinuity. Supposing that the aggregate phase function varies only slightly in response to slight changes in the phases of constitutent clocks, then the winding number remains 1 along any distortion of this path ABCDA. By continuing to distort the path by shrinking it, I eventually localize the discontinuity to a unique state. Along a very tiny path around that state, all phases of the aggregate rhythm are realized. Thus this is a state of ambiguous phase. I could have carried out this argument using any surface of two dimensions, bounded by the required closed-ring path. Each such two-dimensional surface contains a point of ambiguous phase. The locus of ambiguous phase consists of all these points, so it is one-dimensional. It is a curve along which all the isochron surfaces converge.

In the three-clock case this is easy to understand in more mechanistic terms. Situations of ambiguous phase will occur only when the three clocks are symmetrically disposed, so that their center of mass lies at the origin, at zero amplitude and ambiguous angle. Figure 11(a) shows the three clocks at the corners of an equilateral triangle. Any rotation of this configuration leaves the collective phase ambiguous. This ring of positions of the triangle corresponds to the one-dimensional locus of phase triplets which correspond to ambiguous phase in the aggregate rhythm. Any slight displacement of ϕ_1 or ϕ_2 or ϕ_3 from such a point can radically change the aggregate rhythm's phase. At that point, not only is phase ambiguous but, additionally, all phases are near at hand. Why? Not just because the aggregate rhythm's amplitude is 0. That was true in the two-clock case, but only a discontinuity, not a phase singularity, was obtained. The underlying difference of mechanism is brought out in Figure 11(b). Figure 11(a) shows the three-clock situation and Figure 11(b) shows the two-clock situation. In both cases symmetry is required to attain the situation of zero amplitude and ambiguous

Figure 11.   Symmetric dispositions of three clocks (a) and of two clocks (b) which may be arbitrarily rotated while keeping aggregate phase ambiguous.

phase. But in the two-clock case, minute adjustment of phases can only move the aggregate phase off zero to the left or right. In the three-clock case, the center of mass can be displaced from zero to *any* angle by minutely adjusting any two phases. This feature, of course, also obtains with any greater number of clocks in the population.

Three Clocks: Perturbed

What does this imply for the response of a three-clock population's aggregate phase to the timing of a stimulus? Precisely as in the previous case, we can visually map the stimulus plane (ϕ, M) into the (ϕ_1, ϕ_2, ϕ_3) state space. Let's assume the same simple clock kinetics as before for $I > 0$. Without going through the details, the image of the stimulus plane turns out to fall across the convergence of all the isochron surfaces. Thus a cross-section through their convergence appears on the (ϕ, M) plane. Figure 12 shows the result of a computer calculation of this map using the same simple-clock dynamics as in Figures 6–9, based on the models of Chapter 3. Figure 13 shows the result with 50 simple clocks. It is essentially the same[1] (see Box A).

Now I wish to draw out four aspects of this result.

1. The Role of Initial Phase Variance.   In all these simulations the clocks' phases were initially uniformly distributed across an arc of 0.3 cycle. During unperturbed operation the trajectory of the population is therefore parallel to the main diagonal of the state space, but displaced from it. With a narrower distribution of phases a population's prestimulus trajectory is closer to the main diagonal $\phi_1 = \phi_2 = \phi_3$. During exposure to the stimulus ($I \neq 0$) the population is driven along trajectories initially diverging from that locus only very slowly. So a little bit of initial variance in phase gives the population a big headstart toward the discontinuity (with two clocks) or singularity (with three or more). With a narrower initial distribution of phase, the phase singularity would occur only after a very prolonged exposure to the stimulus. In the limit of initially perfect synchrony, the population behaves just like a single simple clock: As we see in Chapter 3, it has no phase singularity. In fact it can be shown analytically that the phase singularity lies at critical duration of exposure M^* proportional to the logarithm of the initial variance of phases.

Given a wide initial range of phase scatter in the population, M^* becomes very short. The reason is essentially as follows. To obtain very low amplitude of the

---

[1]   You might expect more fine structure in these maps, especially for small numbers of clocks. It is there. The number of clocks can be counted by counting repeated features in the contour maps. But these features are smaller the greater the number of clocks so they are easily lost in the variability of data. Although they are plainly enough revealed by mathematical analysis, very fine-grained computation is required to map them out numerically. The maps I present in Figures 9 and 12 and in Chapter 8 are deliberately smoothed to emphasize only their gross qualitative structure.

Box A: Screwy Behavior in Schizoid Populations

A phase singularity appears in the collective behavior of a population of identical simple clocks, none of which individually exhibits such peculiar behavior. How can that be? To be rigorous and opaque, it can be shown mathematically by taking ratios of trigonometric functions weighted by the distribution of phases in the population as it changes during a stimulus. But the essence can be appreciated more simply as follows. Figure a once again depicts the fundamental harmonic of each clock's contribution to the collective rhythm. Points around the circle are phases of the simple-clock cycle. The radial line through each phase point connects points of the same phase on sinusoids (circles) of lower amplitude. The *mean* of many unit-amplitude cycles in a population of imperfectly synchronized clocks follows one of those smaller circles while all clocks advance around the unit circle at unit speed.

Now suppose that during a stimulus each clock moves away from ϕ_r toward ϕ_a as indicated by the arrows in Figure b. If the stimulus begins when the population does *not* straddle ϕ_r, then all clocks together advance or delay toward ϕ_a during the stimulus. But if the population does straddle ϕ_r when the stimulus begins, then it splits into an advancing portion ($\phi > \phi_r$) and a delaying portion ($\phi < \phi_r$). If they are about equal in size, then their center of mass drops directly toward ϕ_a, deviating neither clockwise nor counterclockwise. As the population's two halves advance and delay, their aggregate rhythm keeps about the same phase but loses amplitude until, at a critical duration M^*, amplitude reaches a minimum (Figure b). The center of mass of this population is now close to the center of the circle; phase is ambiguous because the sum of oppositely phased fundamentals has very low amplitude.

Now notice that slight adjustments of timing can reposition the center of mass arbitrarily. It can be moved up and down by adjusting the duration of the stimulus. It can be moved left or right by adjusting the proportions of clocks in the advancing and delaying groups, i.e., by starting the stimulus earlier or later, when fewer or more of its clocks have passed ϕ_r. In other words, the population's collective rhythm can be maneuvered anywhere near this convergence of phase lines by slight adjustment of ϕ near $\phi^* = \phi_r$ and of M near M^*. This is a phase singularity in the collective rhythm of the population. Its physical basis is the *splitting* of a population at ϕ_r.

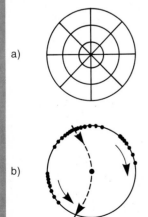

a)

b)

(a) As in Figure 1 but emphasizing the (radial) loci of equal phase connecting orbits of various amplitudes. (b) The cloud of clocks near ϕ_r is split during stimulus $I > 1$ into advancing and delaying subclouds. Their center of mass passes through amplitude zero if the stimulus is suitably timed.

aggregate rhythm requires that clocks be distributed symmetrically around the cycle. This is easily achieved if their phases are already dispersed, but if they are initially close together on the cycle, then they can be scattered only by an exposure that catches them all close to phase ϕ_r. Here slightly precocious clocks are forced still further ahead by the stimulus while slightly retarded ones are drawn backward (Figure 3 of Chapter 3). But in either case, phase velocities near phase 0 are very small. Thus a very long exposure is required.

2. Type 0 Resetting.   Figure 12 and 13 show that by varying through one full cycle the moment at which the stimulus starts, the final phase of the aggregate rhythm is made to vary also through one full cycle, if and only if $M < M^*$. At the end of a stimulus of duration exceeding M^* only *some* final phases are obtained and each is obtained from two different initial phases. In other words, we have type 0 resetting in the aggregate rhythm(see p. 38)even though no clock in the population is capable of type 0 resetting! This phenomenon was first hinted at in the circadian rhythm literature by Johnsson et al. (1973) in connection with populations of *Kalanchoe* flowers, whose petal movement rhythms are customarily assayed

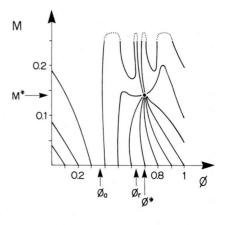

Figure 12.   As in Figure 9 but with a population of three clocks instead of two, spanning 0.3 cycle. The U-shaped discontinuity typical of clock pairs is replaced by the phase singularity typical of all larger populations. The dotted curves show how pieces of certain contours join up at higher M.

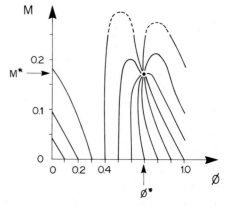

Figure 13.   As in Figure 12 but with a 50-clock population.

Box B: Is There Really Such a Thing as Type 0 Resetting?

It is, of course, a trifle, but there is nothing so important as trifles.

Sherlock Holmes in "The Man With the Twisted Lip"

The existence of type 0 resetting is a live issue in circadian physiology, where the models of Chapter 3 (which allow only type 1 resetting) have been deployed with excellent success, in studies of the cell cycle (Chapter 22), and in neurophysiology, where models involving a real discontinuity of mechanism are widely used as approximations to the less intuitive Hodgkin–Huxley formalism (Chapter 6, last section). What do the data really show?

Even without looking at data, it seems clear that type 1 resetting exists: This is what we get with no stimulus at all, or with one that induces only small advances and/or delays, or induces the same big phase shift whenever administered. Could it be that what appears to be type 0 resetting is really just an extreme version of the familiar type 1 resetting or even type -1 resetting, as in Figure a? Many a data set has been published graphically as though it were (see Box C). It appears to have been tacitly assumed that if a phase shift increases over a certain range of phases, then it must decrease, abruptly if not gradually, during the remainder of the cycle. The fact was not appreciated that what goes up *need not* come down, i.e., that there are alternatives to type 1 resetting.

What are the essential features by which type 0 resetting is recognized in real data? By definition of type 0, new phase ϕ' varies with old phase ϕ in such a way that as ϕ increases the corresponding ϕ' changes and then reverses that change, leaving no net change. Thus each ϕ' is accessible twice, at two different ϕ's (and maybe twice again, in general an even number of times). This situation contrasts with every other resetting map type. Only in type 0 need ϕ' not scan across the full cycle. It *can* (see page 38) but the range of ϕ' accessed can also be very narrow, as in the limiting case of Figure 33 in Chapter 1.

The reality of this pattern is harder to assess when data are presented in terms of phase *shifts*. Phase shift is conventionally measured as the time difference between the nth event in the control rhythm and either the nth event in the perturbed rhythm or the nearest event in the perturbed rhythm. Either way, a discontinuity develops in the curve unless the resetting behavior is type 1. This discontinuity in the presentation looks superficially like a discontinuity in the experimental result. If there were really a big change in the result due to only a small change in stimulus timing, then one might reasonably infer a discontinuity in the underlying mechanism. But in fact the phase shift just before the discontinuity commonly appears to differ by exactly one cycle from the phase shift just to the right: The resulting new phase changes little if any across the putative discontinuity (e.g., see Figure 1 of Chapter 19). Thus I find a different inter-pretation appealing. When the raw data are plotted directly or when the processed (phase shift) data are plotted on graph paper of topology natural to the experiment (the torus), then one is no longer confronted with this topological artifact. In many cases a plausible alternative pattern emerges in which the data are smoothly connected by a type 0 curve (Box C).

This is, of course, an interpretation, not an irrefutable certainty. We're asking about a *curve* but the evidence consists of a set of discrete points. Any kind of curve can be drawn through them. Should we choose the smoothest curve that hits every point? Or the smoothest one that comes reasonably close to most of them? Should smoothness be adopted a priori as a criterion? If not, then the same data can be threaded on curves

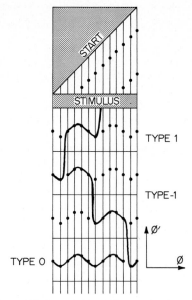

(a) Format as in Figures 30 through 33 of Chapter 2. Phase-resetting data can in principle be threaded by curves of any "type" if there is no strong reason to expect smoothness.

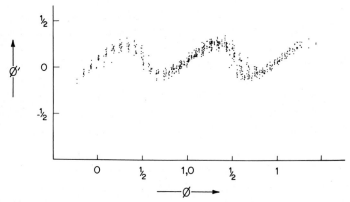

(b) New phase vs. old phase measured by 593 eclosion peaks in about 200 separate experiments with *Drosophila's* eclosion rhythm. The stimuli were saturating light pulses given any time during two consecutive cycles of phase. (Adapted from Figure 12 of Winfree, 1973a.)

of any topological "type" (Figure a). What observations enable us to select the correct type, if such classification is appropriate at all? The question is rhetorical. The answer is to improve data resolution (e.g., Figure b). The curves commonly threaded through plotted data often preserve the appearance of type 1 resetting by means of a very steep segment devoid of data points. If such a curve rightly interprets the data, then points can be placed in this region by suitably close spacing of measurements along the ϕ axis. In Figure b this does *not* happen: The 593 measurements plainly follow the type 0 pattern.

Alternatively one could abandon preoccupation with resetting curves at fixed stimulus magnitude and instead sample the dependence of new phase jointly on old phase and stimulus magnitude. This produces a surface in the time crystal, as described in Chapter 2, Example 5 and the end of Chapter 3. If the resetting curves elicited by strong stimuli are really nearly discontinuous extremes of type 1 resetting, then the data cloud should resemble a stack of separate surfaces, as in Figure 7 of Chapter 3. But if strong stimuli evoke type 0 resetting, then the data cloud should remain a single surface multiply wrapped around a singular axes.

This test was one of the chief purposes of the first pinwheel experiment, done with the cooperation of *Drosophila pseudoobscura*. The result was a helicoid in each unit cell of the time crystal, joining smoothly onto the helicoids in all four adjacent unit cells. Similar observations have since been made in equivalent experiments using *Drosophila melanogaster* (Winfree and Gordon, 1977), *Kalanchoe* (Engelmann et al., 1978), glycolytic oscillations in yeast (Winfree, 1972d; Greller, 1977; Aldridge and Pye, 1979) neural pacemakers in the heart (Jalife, Pers. comm.), and water regulation in oat seedlings (Johnsson, 1976).

My conclusion is, "Yes, type 0 resetting is real." The next question is, "What does it mean?" This chapter presents one possible answer: The clock observed is really a composite of many clocks. Chapter 5 makes another suggestion.

---

collectively in bunches of 16. The analytic basis for this startling phenomenon is described in Winfree (1976).

Given type 0 resetting in response to a sufficiently prolonged stimulus, one can turn the logic around to derive the phase singularity by a technique used repeatedly in earlier chapters (e.g., see Chapter 2, Box B). The empirical basis for belief in type 0 resetting in the real world is elaborated in Boxes B and C.

3. A Rotating Wave and a Screwy Surface.   Given type 0 resetting and a phase singularity (i.e., given a contour map like Figures 12 and 13), one can choose either of two equivalent descriptions of the whole pattern of rephasing of a population's aggregate rhythm:

1.   Figures 12 and 13 have the same format as we found convenient to describe a pinwheel experiment in Examples 5–9 of Chapter 2. Each contour line of new phase (isochron) shows the locus along which $\phi = 0$ recurs simultaneously. This event moves from one isochron to the next, cartwheeling about the phase singularity. This interpretation presupposes a physical layout for the experiment in which a phase gradient at zero stimulus duration lies transverse to a gradient of stimulus duration.

2.   If instead of watching waves circulate on the stimulus plane we plot event time vertically above that plane, then each contour line lies higher than the one before, as in interpreting the contour map of a ski slope. In this case, the surface thus plotted pivots around a singularity as it rises, constructing a helical sliding board. Thus we have a screw of many turns, each turn formed by one day of data.

Box C: Data Sources for Type 0 Resetting

The greatest number of careful measurements come from workers in circadian physiology of protists, fungi, plants, and invertebrate animals. Several neuroelectric rhythms have also shown type 0 rephasing in response to appropriate stimuli. This was expected on principle because the Hodgkin-Huxley equations for periodic nerve firing have an attractor cycle solution (see Chapters 6 and 14). But practically speaking, it came as a surprise that the resetting behavior of pacemaker neurons is sufficiently continuous to *have* a "type". The action potential consists of such a violent and abrupt an excursion of electric potential that resetting behavior immediately after such a discharge would not generally be expected to resemble behavior immediately before the discharge. Yet, in at least the preparations cited below, new phase varies smoothly with old phase right through the action potential. The lower part of the following table cites various biochemical and physiological oscillations that also show type 0 resetting by impulsive stimuli. Examples from the female cycle (Chapter 23) and the cell mitotic cycle (Chapter 22) are conspicuously absent.

Formats for presentation of these data are diverse enough to defy classification. In the third column I indicate my interpretation of the published format, on which basis I replotted ϕ' vs. ϕ to see whether a smooth periodic curve would fit the data and, if so, what is its topological type. Because curves appear in such diverse formats it is essential for clarity to speak of *resetting* type, not a *curve* type.

My notation: ϕ = phase of stimulus beginning, relative to an arbitrary phase in the cycle

ϕ' = phase at end of stimulus or an integral number of cycles later

M = stimulus duration in units of one cycle

K, C = some constants

Format = (upward vertical) vs. (horizontal to the right)

Apart from matters of conceptual convenience, the choice of format makes little difference, as it involves only 45° shearing and choices of direction for axes. The secret to reading data in the many formats of published papers is to find the locus of the unperturbed control experiments then mentally flip and/or shear the whole published diagram as required to align that control locus as desired in the preferred format.

Reference	Figure	Format
Circadian rhythms		
Hastings and Sweeney, 1958	8	$\phi' - \phi - M$ vs. ϕ
Pittendrigh and Bruce, 1959	4	$K - \phi$ vs. $C + \phi - \phi'$
Bruce et al., 1960	1	$K - \phi$ vs. $C + \phi - \phi'$
Pittendrigh, 1960	14($12^h, 4^h$)	$\phi + M - \phi'$ vs. ϕ
Zimmer, 1962	1, 2	$K - \phi$ vs. $C + \phi - \phi'$
Engelmann and Honegger, 1967	3	$K - \phi$ vs. $C + \phi - \phi'$
Honegger, 1967	3	$K - \phi$ vs. $C + \phi - \phi'$
Halaban, 1968	2	$\phi + M - \phi'$ vs. ϕ
Nayar, 1968	5	$K - \phi$ vs. $C + \phi - \phi'$
Sweeney, 1969	3–17	$\phi + M - \phi'$ vs. ϕ
Cumming, 1972	21	$K - \phi$ vs. $C + \phi - \phi'$
King and Cumming, 1972	5	$K - \phi$ vs. $C + \phi - \phi'$

Konopka, 1972	5–2b	$\phi' - \phi - M$ vs. ϕ
Christianson and Sweeney, 1973	III	$\phi' - \phi - M$ vs. ϕ
Engelmann et al., 1973	V	$K - \phi$ vs. $C + \phi - \phi'$
Winfree, 1973a	5 to 17	$K + \phi'$ vs. ϕ
Engelmann et al., 1974	IIId	$\phi + M - \phi'$ vs. ϕ
Karakashian and Schweiger, 1976	5	$\phi' - \phi - M$ vs. $\phi + M/2$
Saunders, 1976a	8	$\phi' - \phi - M$ vs. ϕ
Simon et al., 1976b	3	$\phi - \phi'$ vs. ϕ
Jacklet, 1977	2	$\phi + M - \phi'$ vs. $\phi + M/2$
Winfree and Gordon, 1977	11	$K + \phi'$ vs. ϕ
Wiedenmann, 1977	2	$\phi + M - \phi'$ vs. $\phi + M/2$
Saunders and Thomson, 1977	2	$\phi' - \phi - M$ vs. ϕ
Harris and Wilkins, 1978	2	$\phi - K + M/2$ vs. $C + \phi - \phi'$
Neural rhythms	2	$\phi - K + M/2$ vs. $C + \phi - \phi'$
Perkel et al, 1964	2b	$\phi - \phi'$ vs. ϕ
Walker, 1969	3	$\phi + M - \phi'$ vs. ϕ
Taddei-Feretti and Cordolla, 1976	3	$\phi - \phi'$ vs. ϕ
Pinsker, 1977a	9E	$\phi' + M$ vs. ϕ
Jalife and Moe, 1976	9(5 μA)	$\phi' - \phi$ vs. ϕ
Winfree, 1977	1	ϕ' vs. ϕ
Hartline et al., 1979	1	$K - \phi$ vs. $C + \phi - \phi'$
Hanson, 1978	6	$\phi - \phi'$ vs. ϕ
Yamanishi et. al., 1979	6	ϕ' vs. ϕ
Biochemical rhythms		
Chance et al., 1965a	4B	$\phi' - \phi$ vs. ϕ
Winfree, 1972d	7	$K + \phi'$ vs. ϕ
Greller, 1977	28	ϕ' vs. ϕ
Malchow et al., 1978	5	$\phi' - \phi - M$ vs. ϕ
Aldridge and Pye, 1979a	5a	ϕ' vs. ϕ
Other physiological rhythms		
Karvé and Salanki, 1964	2	$K - \phi$ vs. $C + \phi - \phi'$
Johnsson and Israelsson, 1969	5, 6	$K - \phi$ vs. $C + \phi - \phi'$
Johnsson, 1976	4	ϕ' vs. ϕ

These experiments establish the reality of type 0 resetting. Some may lend themselves to interpretation as population artifacts involving type 1 resetting by individual clocks of an incoherent population. It is essential in such cases to enquire whether the individual clock is by itself capable of type 0 resetting. In the case of circadian rhythms, only since 1977 has type 0 resetting even been documented directly in an individual organism (Saunders and Thomson, 1977; Wiedenmann, 1977; Engelmann and Mack, 1978). In

The eight phase transition measurements in *Acetabularia* (Karakashian and Schweiger, 1976, Figure 5, replotted). A type 0 curve presumably lies somewhere in the shaded area, but alternative constructions are hard to rule out.

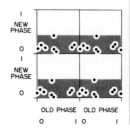

no case has it been plainly exhibited in a single cell. The closest approach to date is the type 1 curve of Karakashian and Schweiger (1976), which lends itself to reinterpretation as type 0; but there are only eight data points (See Figure on p. 111.)

So far as I am aware, *all* other smooth resetting curves ever measured in living systems are type 1. This strikes me as something worth puzzling over, as all but the very simplest models are capable of other integer types of resetting.

4. Amplitude Resetting. The most conspicuous distinguishing feature of the phase singularity deriving from incoherence within a population of independent clocks is the amplitude resetting that goes along with rearrangement of phases. After the perturbation each clock still follows the common cycle at the common period. The distribution of phase does not change, apart from rotating around the cycle. Thus the amplitude of the fundamental is permanently reset (so the waveform is changed) without effect on the period.

It is worth noting that amplitude resetting in this context consists of dividing a population of clocks into two distinct populations, one of which advanced while the other delayed under the stimulus. This splitting of the population provides the clearest experimental test by which to distinguish this mechanism (see Box D).

If the periods are not all exactly equal, then phases can disperse further and the aggregate rhythm eventually runs down. Recovery to the standard amplitude of a single simple clock is obtained only by restoring synchrony within the population. In the absence of an external rhythmic stimulus to synchronize all the individual clocks, this recovery can only be affected by mutually synchronizing interactions among the clocks. This is our next topic.

B: Communities of Clocks

Mutual entrainment is a theme that recurs again and again throughout the physiology of rhythmic systems. I've chosen to place our first encounter with it here, in context of simple clocks. By thus restricting each rhythmic system to a one-dimensional path through its state space, we enormously simplify the analysis, so much so indeed that it becomes tractable. A few typical phenomena emerge that appear to have some physiological interest.

Two Clocks

As in the previous section we begin with two identical simple clocks, but now we let them affect each other: The angular velocity of each now differs from 1 by an amount that depends on its phase and on the phase of its partner.

$$\dot{\phi}_1 = 1 + f(\phi_1, \phi_2)$$
$$\dot{\phi}_2 = 1 + f(\phi_2, \phi_1).$$

Box D: Optical Computation of Phase Scatter

If so much importance must be attached to scatter of phases within a population, then it should be worthwhile to acquire convenient techniques for evaluating the impact of a stimulus on preexisting scatter. The contour map of new phase on the stimulus plane lends itself to this purpose in two ways:

1.   Small scatter is altered in proportion to the derivatives of new phase with old phase and perceived stimulus magnitude:

$$d\phi' = Df = \frac{\partial f}{\partial \phi}\, d\phi + \frac{\partial f}{\partial M}\, dM.$$

This is the *directional* derivative of $\phi' = f(\phi, M)$. It is readily evaluated by shading a finely drawn contour map in alternate black and white bands. Then either place a crystal of Iceland Spar on top of it, producing two images slightly displaced in the direction $(d\phi, dM)$ of the crystal's edge, or hold a Thermofax ®️ transparency over the original displaced in direction $(d\phi, dM)$. The result in each case is a Moiré pattern. The regions of interference form contours of $d\phi'$ on the (ϕ, M) stimulus space. They thus show how any stimulus increases or decreases the population scatter.

Proof: The superimposed replicas form a mesh of parallelograms defined by contour lines, each perpendicular to $Vf(\phi, M)$ or $Vf(\phi + d\phi, M + dM)$. The Moiré runs along the short diagonals of these parallelograms, i.e., along the difference vector perpendicular to $V(f(\phi, M) - f(\phi + d\phi, M + dM))$. A curve perpendicular to a gradient is a level contour (in this case of $Df = d\phi'$).

2.   Large scatter, approaching arrhythmicity, must be handled by finite methods. The simplest is to consider the subject stimulus as a "blob" of width $\Delta\phi$ and height ΔM. This blob is transfixed by a range of isochron curves. Shade them with a yellow Highlighter ®️ pen, following the funnel of contours down to the ϕ axis. That range is $\Delta\phi'$, the scatter after the stimulus. Note that near ϕ^* there is a critical M^* at which an initially compact population is split into two halves about a half-cycle apart.

An experimentally determined contour map, such as Figure 13 in Chapter 2, can be dealt with in this way in order to find out what $(d\phi, dM)$ best accounts for the sharpening or broadening of event timing after a stimulus. In the case of *Drosophila's* eclosion rhythm, this is the change of peak width relative to the minimal peak width in perfectly synchronized populations. These data are tabulated in my computer alongside the mean phase data from which the phase contour map was constructed in the first place. It turns out that most of the scatter of ϕ' is attributable to an initial scatter of ϕ amounting to about one hour. (This was established experimentally in several independent ways, together with similar estimates of the variance of period, photoreceptor sensitivity, and other clock parameters and their roles in experimental quasi-arrhythmicity. This manuscript never reached the publisher due to a secretarial error that resulted in its destruction, with all copies and much of the data, in a Post Office paper shredder during the summer of 1972. Curiously, this disaster didn't arrest the Progress of Science.)

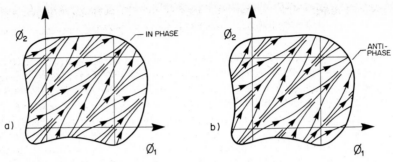

Figure 14.   As in Figure 5 except that the two identical clocks affect each other's rates of change in a phase-dependent way. The result is that one or more trajectories become attracting closed rings at the expense of others. (In this illustration $f(\phi_1, \phi_2)$ is chosen to be a function of $(\phi_1 - \phi_2)$ alone so the antiphase diagonal is straight.)

This situation obviously has a certain symmetry, most readily brought out by plotting the path of the pair of clocks in their toroidal state space as in Figures 5–8. In Figure 14 the unperturbed trajectories curve because each clock's rate is influenced by the other clock. But because clocks 1 and 2 are interchangeable, the whole diagram remains symmetric about the main diagonal. The main diagonal represents the path of perfect synchrony. Because there are mutual influences, the synchronized clocks run faster through some phases and slower through others. They might run faster or slower on the average than either of them would by itself. Trajectories might converge to the main diagonal, conferring on it a local stability as in Figure 14(a). The synchronized solution thereby becomes an attracting cycle (Chapter 6). But the trajectories might alternatively diverge as in Figure 14(b). In this case the synchronized solution is not stable. Due to the diagram's symmetry, there must be another closed path around the torus, midway between repeats of the main diagonal in Figure 14(b). On this path the two simple clocks stay one-half cycle apart on the average, though wobbling about this position as they push and pull each other through the entrained cycle.[2] In Figure 14(b) I drew nearby trajectories diverging from the path of perfect synchrony and converging onto this antiphase path of entrainment without synchrony. (By entrainment I mean a locking together of frequencies, though not necessarily with a stable phase relationship, let alone in synchrony. By synchrony I mean entrainment with, in addition, exact lock-step of phase.) This analysis proceeds without major revision in the more interesting case of unequal periods. The main change is that if the periods are too unequal or the mutual influence is too faint, then the mutual entrainment is lost. For example see the exact analytic solution by Fujii and Sawada (1978) and by Neu (1979b) in connection with chemical oscillators.

---

[2]   This is the case with two identical siphon oscillators such as are commonly assumed to imitate the biochemical regulation of mitosis in blobs of *Physarum* (Scheffey, 1975, pers. comm.; see Chapter 22). In fact it is usual for the more realistically complicated oscillators taken up in Chapters 5 and 6 (e.g., van der Pol oscillators: Linkens, 1976, 1977) to have multiple stable modes of pairwise entrainment. See Box F of this chapter and Box A of Chapter 8.

Box E: An Unexplored Kind of Singularity

A critical point mediates the transition from an incoherent mass of clocks to a synchronously pulsing unit capable of disciplining its members to strict entrainment. This transition resembles in some ways the phase transitions[3] of more familiar materials, e.g., the sudden collective orientation of magnetic dipoles at the Curie temperature or the emergence of long-range order when any liquid crystallizes. In mutual synchronization, the long-range ordering is in time rather than in space, but its emergence is still a critical phenomenon. The nature of such a singularity remains to be explored in connection with the collective behavior of coupled oscillator populations. I think the quantum mechanical applications (e.g., laser optics) are of quite a different nature from what is required in connection with living organisms. This is because the oscillators of physics are all identical harmonic oscillators, whereas in living organisms we confront simple clocks or strongly attracting cycles (Chapter 6). Moreover, the nonzero variance of periods in living clocks plays a dominant role in their collective behavior: With no variance, mutual entrainment can eventually arise even with infinitesimal interaction (Zwanzig, 1976; Grattarola and Torre, 1977; Neu, 1979c); with nonzero variance, a unimodally peaked distribution of native periods is capable of transition to mutual entrainment only above a finite second order critical point (Kuramoto, 1975); a flat-topped distribution has a first order critical point (Winfree, 1967a).

[3]   Beware of a terrible pun. We have dealt with singularities implicit in "phase transitions," in the sense of maps from $\phi \in \mathbb{S}^1$. In this box we use the thermodynamicist's term "phase transitions," meaning a change of collective organization, as from water to ice.

Many Clocks

Now what about populations of many simple clocks, each faintly influencing all the others in the general way considered above? One possibility is that the pooled influence of the many on any one amounts only to faint random noise because the clocks are randomly phased around the cycle. This situation can be stable if the mutual influence is too weak and/or the distribution of native periods is too broad. Another possibility is that with stronger mutual influence or a narrower distribution of native periods, such chaos is unstable. In contrast, mutual synchronization could be stable because in that condition the coherent influence of the many impinges upon each as a strong entraining rhythm, thus keeping them sufficiently synchronous to generate a coherent influence rhythm. Unless all clocks have identical native period, this condition occurs only above a critical point of coupling intensity (see Box E). The first clocks to synchronize are those whose native periods are so close together that their collective rhythm has sufficient amplitude to entrain within that narrow band of periods. With a little increase of coupling intensity, more oscillators are captured; their periods were a little further removed from that of the densest nucleus. This adds to the aggregate rhythm's amplitude, so that the mutually synchronized nucleus expands a little more, capturing a few more oscillators, and so on. This process limits itself when the acquisition of additional oscillators requires a greater increase of collective amplitude than their acquisition provides.

Box F: Mutual Entrainment of Simple Clocks

When a pair of identical simple clocks with identical natural periods lock together, i.e., entrain each other, they do not typically synchronize. Rather there is a phase difference, commonly quite a substantial phase difference. To see why, imagine that in the periodic environment provided by clock B, clock A runs a little bit faster or slower than normal, depending on the phase difference $\delta = \phi_A - \phi_B$. Clock B is subject to the same law at phase difference $-\delta$. Mutual entrainment requires that the period at phase difference δ equals the period at phase difference $-\delta$. To solve this problem, we plot the period as a function of δ and the period as a function of $-\delta$ together on the same graph. One curve (solid) now depicts A's period and the other (dashed) depicts B's period, in both cases as a function of δ, B's phase lag behind A (see figure). These curves necessarily intersect at one or more pairs of points, providing at least two solutions. One is at $\delta = 0$, with the two clocks running synchronously at an altered common period. Another solution is elsewhere, near $\delta = \frac{1}{2}$ in this case. Are these solutions stable? An equilibrium is stable if and only if A's curve lies below B's to the right of the crossing, because then if A lags B more, it's period is shorter than B's and it catches up. So the synchronous equilibrium need not be stable. If it isn't, then mutual synchronization of large populations cannot be stable.

Clock A runs faster or slower, on the average over a cycle, by amount Δf depending on the relative phase δ of clock B. Clock B follows the same rule with respect to clock A in relative phase $1 - \delta$. At two values of δ both clocks run at the same rate.

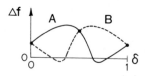

What is required for synchrony is that the mechanism of mutual coupling should so delay or phase shift the influence of one clock on another that a stable equilibrium is moved closer to $\delta = 0$. Lacking this adjustment, a population of interacting clocks not only might fail to synchronize, but they would actually oppose entrainment by any external rhythm; or, if started synchronously, they would disperse around the phase circle much faster than would be expected in view of their range of natural periods. I have seen this phenomenon in computer simulations.

In addition to this phase requirement, coupling must be of sufficient magnitude so that each clock takes up a leading or lagging position as required to match its period to the common period taken up by the others. Contrary to a frequently repeated conjecture, this common period need not be the fastest nor the slowest nor the average in a population of isolated individual clocks. The situation can be visualized as follows. Suppose all clocks are entrained by some external rhythm. They adopt a certain phase lead or lag depending on its frequency. Their influence rhythm is thus synchronous with the entraining rhythm at a certain frequency. It can thus substitute for that rhythm only at that frequency, which is wholly unrelated to the distribution of native periods. If there is *no* such frequency, then mutual entrainment is actively opposed.

All these requirements are formulated more precisely in Winfree (1967a), additionally taking into account the distribution of natural periods in the population. As caricatured in that paper, each individual clock traverses a fixed cycle at a varying rate, as in Chapter 3. This approximation may be adequate for strongly attracting limit cycle oscillators, such as seen in common physiological contexts. In other cases it may not be, as we will see in Chapter 8. In such cases, the amplitude or waveform of the oscillator is altered when it is entrained by rhythmic external influence, e.g., by the aggregate influence emanating from its neighbors. For example, van der Pol oscillators coupled in a straight-forward way are unable to achieve mutual synchronization unless initially entrained by an external rhythm. But when synchronized, their mutual influence suffices to maintain synchrony, making the external entraining rhythm dispensable. If the magnitude of coupling is then reduced sufficiently, the mutual synchronization suddenly comes apart and cannot be restored.

As in the two-oscillator case, treated in the first part of this section and in Box F, a mutually entrained population can run at any period, even faster than the fastest individual in the population would in isolation, or slower than the slowest.

Now we are in a position to appreciate the pertinence of the two-oscillator case in thinking about the pooled influence of the many on any one clock. Consider "the one" to be clock 1 and "the many" to be clock 2. The lone clock 1 might synchronize to the group 2, thus contributing to the aggregate rhythm as above. But it might also entrain at some nonzero phase angle, depending on the detailed shape of the aggregate influence rhythm and of the simple clock's rhythm of sensitivity (Box F). If clock 1 entrains without synchronizing, then it either leads or lags the many. But this analysis pertains to every clock in the population. They can't all lead or lag the rest. One possible outcome is a schism in which about half of the population plays clock 1 and the residue plays clock 2. In this "twinned" mode the aggregate rhythm has two peaks per cycle one-half cycle apart on the average, as in Figure 14(b). This is a common pattern in the activity rhythms of mammals (Chapter 19). Similar behavior was observed in a population of mutually coupled neon glow tube oscillators (Winfree, 1965 and Chapter 11). Pavlidis (1971, 1973, pp. 154–156) offers additional analytic models of such splitting; Daan and Berde (1978) offer still more.

I obtained these phenomena by computer simulations and reduced their quantitative analysis to a description of each simple clock by two properties: its phase-dependent contribution to some aggregate "influence" and its phase-dependent "sensitivity" (in terms of angular velocity) to that aggregate influence (see Box A of Chapter 6). By plotting influence against sensitivity throughout one cycle (Figure 15), one obtains a closed loop whose area and moment of inertia determine the phenomena described above (Winfree, 1967a).

So far as I'm aware no naturally occurring observation of mutually synchronizing or mutually repelling clocks has been analyzed in these terms. Candidate systems include glycolytic oscillations in yeast cell suspensions (Chapter 12) and circadian rhythms in suspensions of single cells (Chapter 19).

The phrase "or mutually repelling" above refers to a surprising situation implicit in the simple-clock model. As noted above, mutual synchrony may be impossible if the influence and sensitivity rhythms are so phased that Figure 15 has negative area. In this case each clock interferes destructively with the entraining rhythm. So not only does spontaneous synchronization fail, but even if it is established as an initial condition, it fails actively. I mentioned above that there are organized alternatives to mutual synchrony, e.g., the twinned mode with two

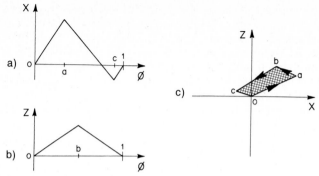

Figure 15.   (a) One cycle in the rhythm of influence $X(\phi)$; (b) one cycle in the rhythm of sensitivity, $Z(\phi)$; (c) X plotted against Z is a closed ring.

groups of clocks. But that doesn't have to be stable either. If each clock's rhythmic influence is very smooth, it is nearly devoid of the harmonics which alone survive superposition of symmetrically phased subpopulations. Then the amplitude of the aggregate rhythm is quite low in two-group or three-group modes of temporal organization. Commonly the amplitude is too low to entrain enough clocks to maintain that organization. In such cases chaos prevails, but not just for want of mutual coupling: it prevails actively, "bucking" any external would-be synchronizer by developing a countervailing rhythm, to the extent that any synchronization at all is imposed (Winfree, 1967a and later unpublished computer simulations).

I know of no definite biological examples, but Ghosh et al. (1971) suggest that acetaldehyde may be an agent of phase-scattering mutual influence among yeast cells. Richter (1965) suggests a need for such mutual influence to maintain steady-state output in the thyroid gland, which may be viewed as a population of rhythmically secreting follicles.

Unless a time delay mediates each oscillator's effect on the others, active scattering does not occur when oscillators interact in the special way typical of chemical reactions coupled by diffusion. Thus no such possibility emerges from Kuramoto's (1975) and Neu's (1979c) reductions of this case to a simple-clock approximation.

Communication by Rhythmic Impulses

In some kinds of natural oscillator populations, mutual influence does not vary smoothly in time but is episodic and pulse-like.

In the case of pulse-like influence, the individual oscillator might be characterized by a resetting map, showing what new phase it recovers to after receipt of an aggregate pulse from the whole population. The analysis of entrainment by Perkel et al. (1964) provides necessary and sufficient conditions on the $\mathbb{S}^1 \to \mathbb{S}^1$ map from old phase before one such stimulus to new phase after the stimulus. This analysis was independently derived in context of circadian rhythms by Pittendrigh and Minis (1964) and Ottesen (1965).

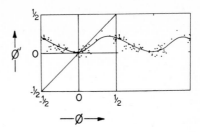

Figure 16. New-phase ϕ' vs. old-phase ϕ, double-plotted from data of Walker (1969) from the chorusing rhythm of crickets perturbed at various phases by the sound of their own call. Phase 0 is the moment of calling.

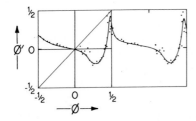

Figure 17. The new phase of rhythmic activity in cardiac pacemaker cells, electrically stimulated at each old phase, replotted from unpublished data of Jalife (1975). The action potential occurs at phase 0.

There arises a dilemma of internal phase compatibility similar to that encountered above. For mutual synchronization the phase shift ($\phi' - \phi$) should be changing from small advances to small delays at the phase of pulse emission. More exactly, the resetting map, ϕ' vs. ϕ, should cross through the $\phi' = \phi$ diagonal at slope between $+1$ and -1 at the phase of pulse emission. Any other resetting map would result in more complicated aggregate behavior than simple synchrony. In four cases the ingredients of such an analysis have been obtained experimentally:

1. The mutual synchronization of cAMP pulsing in suspensions of slime mold cells mentioned above gives a resetting map of the anticipated sort (Malchow et al., 1978; see Figure 3 of Chapter 15.)

2. Mutual synchronization of chorusing in populations of tree crickets does also (Walker, 1969; see Figure 16).

3. Electrical synchrony of cells in the pacemaker of the heart seems to be mediated by a similar resetting curve (Jalife and Moe, 1976, Figure 9; Sano et al. 1978, Figure 5; also see Figure 17). Peskin (1975, pp. 250–278) analyze the mutual synchronization of cells in the heart's pacemaker in terms quite similar to those alluded to above. Exact equations were derived for the two-oscillator case, but the population problem awaits completion.

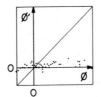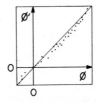

Figure 18. The new phase of the flashing rhythm in fireflies of three species perturbed by the sight of their own flash at various old phases. The flash occurs at phase 0 on this scale. Each box is exactly one cycle by one cycle. Replotted from Hanson (1978, Figure 7).

4.   The same behavior has been obtained from studies of mutual synchronization of flashing in two different kinds of southeast Asian fireflies, one with type 0 resetting and one with type 1 resetting [Hanson, 1978; see Figure 18(a, b)]. Hanson also found that entrainment cannot always be understood quantitatively as repeated resetting à la Perkel, i.e., by iterating the appropriate resetting curve (in his case 7c). This means that simple-clock models fail in at least this one species. The next theoretical effort might entail a study of impulse response and entrainment in attractor-cycle oscillators of the sort familiar to neurophysiologists (see Chapter 14), developing the consequences for a population whose aggregate output is itself the entraining rhythm.

If each clock responds in a way describable neither as a jump along a fixed cycle nor as a modulation of angular velocity along a fixed cycle, then this whole format must be abandoned. We therefore return to this subject in Chapter 8 after developing a fuller appreciation of oscillating kinetics in Chapters 5 and 6.

Collective Enhancement of Precision

A little studied question of some physiological significance concerns the *precision* of rhythmicity in mutually synchronized populations. It is often stated in research papers (it is even alleged that I *proved* this in 1967, which I did not) that mutual synchronization disciplines each oscillator to much improved regularity of oscillation at the common frequency.

One argument might run as follows. Suppose mutual entrainment is stable with deterministic oscillators of native period τ_i, each entrained to the aggregate rhythm at period τ_0. Each locks to that rhythm at some phase $\psi(\tau_i - \tau_0)$. Suppose small $\tau_i - \tau_0$, weak coupling, and a strongly attracting cycle, so that ψ changes slowly and little else changes at all. Then we can write $\dot\psi = 1/\tau - 1/\tau_0 + M(\psi)$ for some suitably-invented function M. Entrainment requires that $\dot\psi = 0$ and $M' < 0$. Now suppose we have not a population of individually precise oscillators of various periods but rather a population of oscillators of identical statistical behavior, individually somewhat irregular of rhythmicity. If they do achieve mutual entrainment, does the aggregate rhythm drone on with a steadiness orders of magnitude greater than any individual's, disciplining each individual to adhere to the collective rhythmicity? If τ_i should slowly change, then $\dot\psi \neq 0$ nudges ψ toward the equilibrium typical of an oscillator with the revised τ_i. Should τ_i fluctuate more rapidly within the range of τ_i's characteristic of this population, then ψ will fluctuate about its stable position for oscillators of middling τ_i. Either way the situation remains qualitatively the same. Thus it would appear that the aggregate rhythm does discipline each individual to enormously enhanced regularity.

Some biologists are inclined to account in this way for the uncanny accuracy of some circadian rhythms, but so far as I am aware, the mathematical essence of such a mechanism has never been revealed. The matter awaits the attention of a master of stochastic dynamics. Substantial beginnings are made in Stratonovich (1967, Chapter 9) and in Kuramoto (1975). A valuable contribution could be made by following them up.

Box G: Active Transport by a Row of Simple Clocks
(The Wave Broom)

I am indebted for this notion to Dr. R. H. Wilhelm, who died before we could develop it further (August 6, 1968). Wilhelm's keenest interest at the time was in rhythmic principles that might underlie active transport in cell membranes (Wilhelm, 1966,1968). After hearing him lecture on parametric pumping, it occurred to me that even a row of simple clocks, properly phased, suggests a mechanism for transport of substances in the direction of wave propagation. Consider a line of cells along which a periodic disturbance is propagating from left to right. Let its period be our unit of time. Consider any substance involved in this disturbance. Let its intracellular concentration vary periodically, for whatever reason: $c(t) = c(t + m)$, m being any integer number of cycles. Given such periodism in cell metabolism it would seem unnatural not to let the permeability of the cell membrane also vary periodically: $p(t) = p(t + m)$. The rate of transport of substance across the cell partition is given by Fick's law of diffusion as:

$$j = (c_1 - c_2)p.$$

Then recalling that c_2 is just a delayed version of c_1, and integrating over a full cycle, the mean flux to left or right, integrated over a cycle, is:

$$j = \oint j\, dt = \oint (c(t) - c(t - \delta))p(t)\, dt \cong \delta \oint p\, dc,$$

where δ is the time delay between adjacent cells. This net flux is just δ times the area of the loop formed by plotting $p(t)$ against $c(t)$. It could be either positive or negative, but it would be exactly zero only by deliberate selection of appropriate shapes of c and/or p. Thus material will in general be transported to left or right as the wave passes. It is tempting to invoke some such "wave broom" to account for maintenance of pacemaker activity, once established, in any potentially rhythmic medium, e.g., the malonic acid reagent (Chapter 13) or the cellular slime mold (Chapter 15). More sophisticated wave-broom-like models have been pursued in various contexts, supposing various mechanisms, by Hejnowicz (1970) and Goodwin (1973, 1974, 1975, 1976).

C: Spatially Distributed Independent Simple Clocks

In this section my intent is only to outline a format of description that receives further development in later chapters. Our subject here is a population of simple clocks distributed in space. We are not here concerned with pooling their outputs nor with their local neighbor interactions. We are concerned with maps again, this time from a physical space in which the clocks are distributed to a ring-like state space on which each clock's phase is described. Two examples will suffice (see also Box G).

The Ascomycete Frontier as a One-Dimensional Population
of Simple Clocks Without Interaction

Some kinds of fungus grow across a food surface at uniform speed of several millimeters per day (see Chapter 18). While the organism propagates in this way

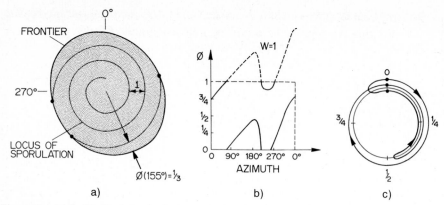

Figure 19. (a) An idealized sketch of a fungus colony on which spore density is rhythmic along each radial growth path. The phase of the radial rhythm drifts through one cycle from north (0°) to east (90°) to south (180°) to west (270°) (b). In (c) each point on the frontier ring is mapped to its momentary phase on the cycle.

at a fixed velocity, its internal metabolic rhythms affect the style of growth: The mycelium becomes locally thick or thin, eventually making spores or not. The moving frontier of the colony thus leaves behind a permanent record of its metabolic rhythm in the form of periodically alternating bands of conspicuously different-looking tissue. Time lapse movies of a disk of such a fungus show that the pattern is laid down at the growing edge, the circular frontier of the colony. As in the construction of rings in deciduous trees, or deposition of the colorful decorations in sea shells, the pattern does not change after the foundations are laid at the growing frontier (Winfree, unpublished movies).

The frontier is a ring, a one-dimensional continuum each point of which may be regarded as harboring a physiological clock. The phase of this clock can be evaluated by measuring back either in time or in space along the radial path of growth to the most recent occurrence of some marker event chosen for convenience to be called phase 0, for example, a maximum of sporulation density [(Figures 19(a), 20(a)]. As it happens, phase almost always varies continuously along the

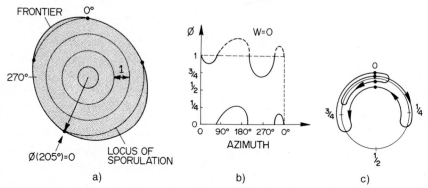

Figure 20. As in Figure 19 but phase meanders back and forth without net effect from north (0°) to east (90°) to south (180°) to west (270°).

frontier ring. (There are exceptions and some physiological interest attaches to these exceptions as opportunities to observe the outcome of a natural experiment. See below.)

It will be convenient to plot the observed dependence of phase on azimuth. Azimuth is the compass direction from some internal origin to a boundary point. A natural choice of origin is the inoculum from which the whole disk has expanded. To help distinguish rings in physical space from the abstract phase ring, we calibrate azimuth in degrees from 0 to 360, while phase is expressed in fractions of a cycle from 0 to 1.

Figure 19 idealizes the phase vs. azimuth plot from the frontier ring of a mycelium on which a single spiral intersects the boundary in three places. Figure 20(b) idealizes the phase vs. azimuth plot from the frontier of a mycelium on which concentric rings intersect the boundary in four places. (See Chapter 18, Figures 5 and 6 for real measurements.)

It would be more natural to conduct a measurement by mapping the ring-shaped frontier onto the clock cycle, idealized as a ring as in Figure 20 of Chapter 1. Choosing a point on the cycle to represent phase 0 (e.g., the phase at which sporulation is fated in newly created tissue), we plot each frontier point on the ring according to its radial distance (its time) from phase 0. This is done in Figure 19(c) and 20(c). In Figure 19 phase increases more than it decreases as we traverse the frontier clockwise. It increases exactly one full cycle more, resulting in winding number $W = 1$. In contrast, in Figure 20 the image of the frontier has winding number $W = 0$ around the phase circle. Phase increases and decreases through more than a full cycle around the frontier, but it decreases back again exactly as much as it increases. A completely synchronous mycelium has $W = 0$. Its frontier maps to a point circulating around the phase ring. In this case sporulation patterns consist of concentric rings. A map with winding number $W = \pm 1$ corresponds to clockwise or anticlockwise spiral morphogenesis. In the case of a two-armed spiral, the map turns out to wind twice around the phase circle ($W = \pm 2$) in a direction ($+$ or $-$) determined by the handedness of the twin spiral.

The phase circle might portray the states of a ring device functioning in each little patch of tissue, if each frontier point harbors a simple clock. Since phase advances (by definition) uniformly in time, this map is to be thought of as rotating rigidly once in each period of the ring device. This is 16 hours in the case of the *Nectria* fungus, which provided the examples described above.

In all the foregoing, we have paid no attention to the likely possibility that neighboring frontier cells interact to some degree. This seems particularly noteworthy in cases where phase changes rapidly over a short arc of frontier, i.e., where there is a radial edge dislocation in the periodic banding of the fungal disk. One might reasonably expect that physically nearby cells (or hyphae, to be more exact) would compromise in phase, smoothing over the steep spatial phase gradient. Smoothing of near-discontinuities of phase does occur, but slowly. A discontinuity typically lasts through many cycles before it is really smoothed out. So in this first approximation we ignore interactions.

Note that this example deals only with the colony's growing frontier. Suppose we had certain proof (as Dharmananda and Feldman, 1979 might for the similar

ascomycete *Neurospora*) that interior points (old frontiers) also retain the same
rhythmic activity as the present frontier more plainly exhibits. Then it would be
appropriate to map the whole disk onto the phase ring. As we have often seen
before, this cannot be done smoothly unless the frontier's winding number is 0.
In the other cases, a violent phase discontinuity is implicit somewhere inside the
disk. There the phase gradient is too steep to permit neglect of interactions. In fact
it is too steep for any finite accounting. This will force us from simple-clock models
to a biochemically more realistic view in which the mapping is not a ring but a
two-dimensional state space (Chapters 5 and 6). The mapping to a two-dimensional
space is smooth regardless of winding number and it permits neglect of interactions
in a first approximation even in situations with $W \neq 0$. But for the present we
must confine our attentions to the frontier.

A Liquid Chemical Oscillator Viewed as a Population of Simple Clocks Without Interaction

Imagine a fluid in which each tiny volume element periodically executes a cycle
of changes, returning to an initial state at regular intervals of time. If nearby
volume elements are in nearly the same state, then (apart from concern about the
stability of this situation) we can neglect coupling through molecular diffusion
because there are no significant concentration gradients to drive enough flux of
any substance from one volume element to the next.

Despite the abruptness of one stage in its reaction cycle, the best studied example
at present is the oscillating Belousov-Zhabotinsky reagent (Chapter 13). Using
one-dimensional columns of this reagent with sufficiently shallow phase gradients,
Beck and Varadi (1971, 1972), Thoenes (1973), Kopell and Howard (1973a), Varadi
and Beck (1975), and Beck, Varadi and Hauck (1976) studied wave-like patterns
of chemical activity in terms of spatial patterns of phase. As in the *Nectria* fungus,
disks of this reagent typically show concentric ring patterns and spiral patterns of
chemical activity, indicated by color. The color varies between red and blue. In a
ring pattern, all points along any ring map to the same point on the phase circle.
Adjacent rings map to adjacent phase points. The geometry is suggested by
Figure 21: The whole disk maps exactly as does each radial wedge of the disk.

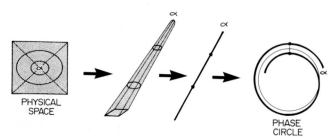

Figure 21. A two-dimensional oscillating chemical medium momentarily has uniform phase along
closed rings concentric to point α. Each ring is shrunk to a point in imagination, identifying points of
equal phase. This collapsed image of the plane is mapped onto the ring of phases.

If there are many cycles of phase along each radius, then the map winds many times around the phase ring.

As indicated in the case of *Nectria*, a two-dimensional spiral wave cannot be mapped continuously onto the ring. Apart from the theorem about winding numbers, the problem can be visualized as follows. Consider the border of the disk of reacting fluid. Having winding number $W = 1$, it maps once around the phase ring. So does the ring of fluid just interior to this border, and so on by concentric rings inward toward the pivot. An arbitrarily small ring of liquid around the pivot also maps once around the phase ring. Inside that, there is either a discontinuity in the state of the fluid or the tiny disk at the very center covers the whole inside of the phase ring. But this "inside" is not part of the state space and has no conceivable interpretation in terms of a simple-clock mechanism. To avoid this paradox, we will have to abandon interpretation of spirals in terms of phase in a cycle, turning to a state space of two dimensions in order to support the required map: a space which gives insides to the phase ring. This is reserved to Section C of Chapter 8.

In the one-dimensional case, each point in a column of oscillating fluid maps to a point of the phase ring $\mathbb{I}^1 \to \mathbb{S}^1$. The image of the column rotates steadily around the ring. This picture was given more interest by contriving a temperature gradient (Kopell and Howard, 1973a) or an acidity gradient (Beck and Varadi, 1972; Thoenes, 1973) so that one end of the column cycles faster than the other. In this situation the image of the column continually stretches out along the phase ring as the leading (hot or acid) end wraps through more and more turns than the lagging (cold or less acid) end. There are as many cycles of wave-like activity along the column as there are windings of the image. Eventually there are so many, and the waves are packed so close together, and the image is stretched so taut, that local interactions can no longer be left out of the picture. At this point a new phenomenon first appears: real propagating waves of chemical activity.

D: Ring Devices Interacting Locally

Linear Coupling to Neighbors

Consider a simple clock whose motion along the phase ring can be described (vacuously) as $\dot{\phi} = f(\phi)$. Suppose as in example 14H of Chapter 1 that we have a physical ring of such clocks in which the angular velocity of each is affected by its fore and aft neighbors. To construct the simplest case, suppose there are so many clocks on a ring that neighbors are at very nearly the same phase. Let the influence of each clock on its neighbor's rate then be simply proportional to the phase difference between them:

$$\frac{d\phi_i}{dt} = f(\phi_i) + k(\phi_{i-1} - \phi_i) + k(\phi_{i+1} - \phi_i) \text{ for each cell,} \qquad i = 1 \text{ to } N. \quad (1)$$

In words, this rate equation says that each clock speeds through its cycle as it would if alone but, additionally, it is hurried along or retarded by its two neighbors, depending on whether they lead or lag. Each clock then tends to take up a phase midway between its neighbors. This set of N ordinary differential equations can be iterated in a computer to follow the changes of each ϕ_i. I did this in 1973, collaborating with R. Casten and J. Mittenthal over those aspects of the problem that could be approached analytically. Before I tell what we saw, it should be noted that problems of this sort have a tradition, and mathematicians know lots of solutions, though apparently not for exactly the situations that caught our attention then. Equation (1) could be written:

$$\frac{d\phi_i}{dt} = f(\phi_i) + D \frac{\left(\dfrac{\phi_{i-1} - \phi_i}{h}\right) - \left(\dfrac{\phi_i - \phi_{i+1}}{h}\right)}{h} \tag{2}$$

in which h is the physical spacing between cells, D is a diffusion coefficient, and k is written D/h^2. Letting $h \to 0$, the interaction term is seen to be the second derivative of ϕ along the chain of cells. So dropping subscripts, we think of ϕ as a continuous function of time and position along the line of cells, and write (2) as a parabolic partial differential equation:

$$\frac{\partial \phi}{\partial t} = f(\phi) + D \frac{\partial^2 \phi}{\partial \theta^2}, \tag{3}$$

where θ is distance along the line. This is a familiar challenge. There exists a big and rapidly growing literature of mathematical solutions for special forms of $f(\phi)$ (e.g., see Murray, 1977 and Fife, 1979).

Surprisingly, however, not much attention has been lavished on the case in which ϕ is a point on a ring rather than a point on the line (a *phase* rather than, let us say, a voltage or a concentration) and θ is also a point on a ring rather than on a line (i.e., the physical medium [in which $\dot{\phi} = f(\phi)$ is going on everywhere] is a closed loop rather than an open thread). Nonetheless the qualitative features introduced by incorporating spatial interactions are the same in any context. The main feature introduced is wave propagation.

Waves

Two kinds of waves invite distinction as extreme cases. In the first case suppose $f(\phi) > 0$ for all ϕ so that we have a simple-clock medium, and suppose D is so small or that ϕ changes so slowly with θ that $Dd^2\phi/d\theta^2$ can be ignored in Equation (3). (In the limit, this case would belong in Section C: noninteracting clocks). The medium can still give the *appearance* of conducting waves because wherever ϕ varies with θ, $\phi = 0$ is reached first in one place, then a little further along, and so on. These "pseudo" waves differ from real propagating disturbances (see Box C of Chapter 13) in that their wave shape and velocity are determined only by $\phi(\theta)$. In fact that apparent velocity, being $1/(d\phi/d\theta)$, varies locally and can even be *infinite*. I speak of this case only to provide contrast with the real waves that I call "trigger" waves. In a trigger wave a local displacement of ϕ tends to travel along the θ axis, eventually shaping itself into a certain waveform moving at a fixed speed.

Both simple clocks and hourglasses behave similarly in this situation: Either kind of ring device will conduct a trigger wave. Waves of this sort are quite unlike those of classical physics. Unlike sound waves, light waves, and waves on water, they vanish at the ends of the line instead of reflecting. They annihilate each other in head-on collisions instead of passing through each other. In these respects they resemble the waves of electrical activity typical of nerve, heart, and brain tissue (Chapter 14) and their chemical analogs (Chapter 13).

Quite different mathematical approaches to waves in ring device continua are detailed in the appendix of Goodwin and Cohen (1969), in Ortoleva and Ross (1973), and in Neu (1979a).

Mappings and Their Winding Numbers (Again)

I now return to the special case in which Equation (3) represents a continuum of identical ring devices arranged in a physical ring. As in the previous section it is helpful to visualize this situation by mapping the physical ring onto the ring of states, $\phi(\theta): \mathbb{S}^1 \to \mathbb{S}^1$. The image of the physical ring necessarily winds some integer number W of times around the phase ring. Once established, there is no way that winding number can change without cutting the image. Cutting the image is forbidden so long as $\phi(\theta)$ is continuous: Arbitrarily nearby points on the ring cannot have finitely different phases so long as molecular exchange continues, guaranteeing local uniformity. The diffusion term confers coherence on the ring's image. Equation (1) shows that diffusion is analogous to *elasticity* of the image of the physical ring: Each element of length moves (on the phase ring) toward its neighbors at a speed proportional to their separation on the phase ring (see Chapter 8, Box D). The physical ring's image then behaves much like a massless rubber band, moving in a viscous medium which itself moves according to $\dot\phi = f(\phi)$. Each little piece of the ring follows this local rule of angular velocity but additionally stretches to relieve tension in locally more taut places (where phase ϕ in state space changes more quickly with distance θ in real space). Each little piece, traversing the same cycle as its fore and aft neighbors, is continually adjusting its velocity to stay poised midway between them. That fact and the conservation of winding number constitute the whole story of qualitative behavior of ring devices with smooth kinetics interacting locally by such simple rules as Equation (1) expresses.

Computational Experiments

Consider for example an hourglass ring device described by:

$$f(\phi) = \phi(1 - \phi)(\phi - a) \tag{4a}$$

or

$$f(\phi) = 1 + I \cos 2\pi\phi. \tag{4b}$$

There is a repelling equilibrium ϕ_r [at $\phi = 1 = 0$ in (4a); elsewhere depending on $I > 1$ in (4b)] and an attracting equilibrium ϕ_a [at $\phi = a$ in (4a); at $\phi = -\phi_r$ in (4b)].

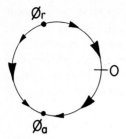

Figure 22.   Kinetics on the phase ring during strong perturbation, analogous to Figure 3 of Chapter 3.

Without spatial coupling, all volume elements of a ring of such a medium would follow the $\dot{\phi}$ arrow of Figure 22 to ϕ_a. But that would cause a rupture of continuity at ϕ_r. Volume elements initially just to the left and just to the right of ϕ_r separate increasingly as they approach ϕ_a anticlockwise and clockwise, respectively. If local interaction is allowed, that cannot happen. What *does* happen now depends on the winding number of the image. If $W = 0$ as in Figure 23(a), then the whole image slides around to ϕ_a. If $W \neq 0$, that can't happen. If $W = 1$ as in Figure 23(b), then the image becomes tauter near ϕ_r and denser near ϕ_a, but the counterclockwise circulation of $f(\phi)$ prevails. The whole image then continuously rotates. Thus even though the local kinetics during perturbation describes an hourglass, incapable of oscillating by itself, yet the whole system still continuously oscillates just as it did at $I = 0$. Every volume element along the physical ring is pulled around the cycle each time a wave of activity circumnavigates the physical ring. If $W = -3$, there are three such waves, rotating clockwise, and they never stop.

We observed these behaviors for diverse choices of $f(\phi)$, using the computer to iterate through Equation (1). It doesn't much matter whether $f(\phi)$ describes

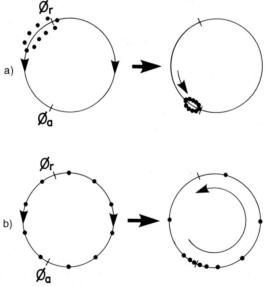

Figure 23.   A ring of clocks (dots) mapped onto the phase ring. If $W = 0$ as in (a), then the ring can homogeneously contract to ϕ_a. If $W \neq 0$ as in (b), then it cannot, but instead rotates forever.

a simple clock, an hourglass, or something else. The velocities of wave propagation and the wave shapes agreed well with analytic solutions for a single wave on the open line when we made the physical ring large. For example, in Equation (4a) the velocity is close to $V = \sqrt{2}(1/2 - a)$ and the wave shape $\phi(\theta - vt) = 1/(1 + e^{1/2(vt-\theta)})$ as Huxley found analytically (McKean, 1970) and as Nitzan et al., (1974) derived in more general context. More generally, covering cases which cannot be approximated analytically, we found that as ring circumference is increased, velocity increases toward an upper limit equal to the speed of a solitary pulse on the open line. As ring circumference is decreased toward 0, wave speed decreases toward 0. Without being able to make the mathematical proof completely rigorous, Casten conjectured that the period of circulation (circumference/velocity) should also increase with the circumference. This was found numerically in every case. The reciprocal period appears to approach the mean $\dot{\phi}$ as ring circumference approaches 0:

$$\frac{1}{T} \to \int f(\phi)\, d\phi.$$

In this limit the wave appears to flatten out to $\phi(\theta, t) = (t/T) + \theta \times$ (any integer), as might be expected with molecular diffusion averaging reaction rates throughout the whole of a very small ring. Using the above expression for T, it appears that the stable solutions to Equation (3) as $D \to \infty$ on a closed ring are:

$$\phi(\theta, t) = W\theta + t \oint f(\phi)\, d\phi$$

for any nonzero integer W.

I present these fragmentary results because their apparently great generality seems to invite a more elegant mathematical resolution.

From One Dimension to Two

Cases with $W \neq 0$ are important for biological applications such as the ring-shaped frontier of a fungus engaged in spiral morphogenesis in Section C. But the really interesting cases come when we try to treat two-dimensional media. This could be attempted by expanding the physical ring to an annulus and filling in its central hole until it is closed and we have a disk. If ϕ has nonzero winding number around the edge of the disk, then there has to be a ring of discontinuity inside the disk. In the simplest case this ring is as small as can be, namely, a point phase singularity. This is unavoidable but it is an embarrassment to models in which the local state can vary only along a one-dimensional state space because then all these different phases which converge at the phase singularity represent discretely different states. The phase singularity then represents a confrontation of distinct states in an arbitrarily small space. This is physiologically and bio-chemically unrealistic. Using Equation (4b), Kuramoto and Yamada (1976) tried to account for spiral waves in chemical media (Chapter 13) but were frustrated by the central phase singularity (their Sections 3.D and 4). Thoenes (1973) had previously attempted a less mathematical argument of the same sort, overlooking the contradiction implicit at the pivot of a spiral wave in a medium whose state varies only in respect of phase. Wiener and Rosenblueth (1946) had concluded that rotating waves could *not* occur in excitable media with only a one-dimensional state space unless by circulating around a central *hole*. They didn't give reasons in the form of a proof, but I presume that prominent among them was the un-palatability of a phase singularity.

In the next chapter we begin to remedy these contradictions by widening our view to encompass models whose internal state can vary independently in two ways.

5. Getting off the Ring

To bring a quality within the grasp of exact science, we must conceive it as
depending on the values of one or more variable quantities, and the first step
in our scientific progress is to determine the number of these variables which
are necessary and sufficient to determine the quality.

James Clerk Maxwell

My purpose in this chapter is to start "putting flesh on the bones" of the simple
clock metaphor. Up to this point, I've tried to hold your attention on "phase"
and its rate of change by confining discussion to the simplest metaphor of smooth
cyclic dynamics, namely, the ring device. I have studiously avoided allusion to
other degrees of freedom of the "state" of any biological clock. To make the transition to a broader perspective in an orderly way, I now wish to introduce just one
additional notion, i.e., that a rhythmic process might be adjustable not only in
phase, our exclusive preoccupation in previous chapters, but also in some measure
of its vigor, amplitude, range, or degree of variation during the cycle.

I do this by elaborating on a few strictly idealized examples, drawing out the
main features in which they differ from simple clocks. These new features are met
again in more realistic models in Chapter 6 and in several sets of measurements
on biological systems recently examined for such features in the laboratory.

A: Enumerating Dimensions

Just as people have asked how many components there are in the sensation of
color (three) or of taste (four) so it also seems basic to ask how many components
there are to the sense of *time*. Here that question is restricted to the particular sense
of circadian, rhythmic time. A commonly accepted answer is "one", by analogy
to the unidirectional character of linear time. The inadequacies of linear descriptions of time have been met by recognizing that the one-dimensional time axis, in
cases of rhythmic time measurement, describes a circle. No significance is attached
to the insides of the circle, just as we attach no significance to places off the one-
dimensional historical time axis. In preceding chapters we adhere to this viewpoint

and test its limits by mathematical inference and comparison with observations. The observations of negative slope in resetting curves, of type 0 resetting, and of phase singularities present paradoxes for this paradigm. The result is that we are forced in this chapter to contemplate the possibility that the sensation of time has two or more components.

There are serious conceptual problems about enumerating the degrees of freedom of a real system. It is not obvious that there is any unique number of this sort because every real system involves myriads of variables, some of dominant importance, some of negligible interest. But neither is it usual for people to deal with real systems. We deal instead with certain *aspects* of their behavior that we consider "relevant". In our heads, we deal with adequate approximations, caricatures, metaphors, models which acknowledge only the dominant variables of real systems. For a model system I use the term "degrees of freedom" to indicate the integer number of independently varying quantities which jointly determine its state, in the sense that if and only if all are simultaneously known (together with any relevant environmental parameters), then their immediate rates of change are all unambiguously determined. This is the sense in which "degrees of freedom" is used in thermodynamics and statistical mechanics. It is the dimension of the state space. It is the number of oscillators only if each oscillator is a simple clock with a single state variable (the phase). In certain other literatures usage differs. For example, in mechanics the state variables always come in conjugate (position, momentum) pairs, so "degrees of freedom" has come to mean the number of pairs, i.e., half the dimension of the state space (Feynman et al., I 49-6). This is the number of oscillators if each is a simple harmonic oscillator with two independent variables of state. In electrical engineering "degrees of freedom" means the number of changing quantities in an equation, not counting their rates of change as distinct degrees of freedom. This can be a much smaller number than the dimension of the state space (for example, see Pavlidis, 1973).

In this chapter we entertain the notion of exactly two components, but the qualitative inferences emphasized here are also valid for any greater number of components.

B: Deducing the Topology

So long as we envisioned only a one-dimensional set of states with the connectivity of a ring, there was no latitude of choice in selecting a state space. All one-dimensional closed manifolds are topologically equivalent to a ring.

But if we are driven to recognize different flavors of each phase, then we must choose one of many possible state spaces. The issues at stake are dimension and connectivity. Consider some of the options for introducing just one more variable, bringing the dimension of state space to two. We might consider involving a second clock for the second degree of freedom. The state space would be the product of two rings: ($S^1 \times S^1$), the two-dimensional torus. Without belaboring detail, this turns out to entail some discontinuities not observed in practice. But a system of three or more simple clocks would serve (Chapter 4, Section A). If our "system" is really a composite of many similar simple clocks, the required additional degrees of freedom might lurk in the distribution of phase about the mean phase. In many respects the narrowness of the distribution would determine an amplitude, intensity, or vigor of aggregate rhythmicity.

Alternatively, we could supplement the "phase" description of a simple clock with a second quantity, an "amplitude" of oscillation. Diagrammatically this could be done by picturing the circle of phases as girdling a cylinder. The cylinder's long axis distinguishes oscillations of different amplitudes. Dynamics in systems with cylindrical state spaces has been systematized by Minorsky (1962, Chapter 8) and by Andronov (1966, Chapter 7) among others. Without belaboring detail, I also reject the cylindrical state space because its qualitative features don't correspond to the qualitative behavior of the biological systems all this is supposed to be about.

Our state space might alternatively be a surface of a cone or its topological equivalent, a disk. The cycles of different amplitude would be rings concentric to an axis or point of zero amplitude. The singularity of zero amplitude might then be regarded as a degenerate ring in which all phases are indistinguishable.

In principle there are lots of alternatives, each characterized by a topology and corresponding idiosyncracies of behavior. If we want a state space of two dimensions, the state might be described by two real numbers or by two positive real numbers (like chemical concentrations) or by a phase and a real number (the cylindrical state space) or by two phases (a torus or a Klein bottle) or in ways that lend themselves less readily to verbal description (as on the surface of a sphere with 17 handles, six of which are knotted together). And if more than two dimensions are countenanced, the possibilities exceed the capacity of human imagination. How is one to choose the state space natural to a given phenomenon? If I may quote Ambrose Bierce (on "sleep", from *The Devil's Dictionary*):

> It is hardly a burning question: it is not even a problem that presses for solution. Nevertheless, to minds not incurious as to the future, it has a mild, pleasing interest, like that of the faintly heard beating of the bells of distant cows that will come in and demand attention later.

For present purposes I evade the whole issue by choosing $\mathbb{R}^D$, in fact only its positive part, as though the process with which we will be concerned is determined simply by D chemical concentrations (D stands for "dimension"). By this choice I conjecture that for the systems we will be concerned with there is a continuum of realizable states between any two states and that any closed ring of states can be contracted through realizable states to a point. (This is *not* true, for example, of $\mathbb{S}^1$ nor of $\mathbb{S}^1 \times \mathbb{S}^1$, nor of $\mathbb{S}^1 \times \mathbb{R}^1$ contemplated above.) Beyond that conjecture I make no more. For example we have little reason to suppose that $D = 2$, as though *only* two quantities suffice to adequately determine the state of a biological clock. This *is* commonly assumed for reasons of convenience in constructing mathematical models, but rarely is more substantial justification given. To assume $D = 2$ is to postulate the nonexistence of a number of pecularities that require a third degree of freedom, as we will see in Chapter 6. The models in this chapter all suppose $D = 2$ for simplicity but I will deal only in those properties of the models that generalize straightforwardly to $D > 2$.

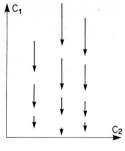

Figure 3.   Figure 2 is altered to depict revised kinetics during a perturbation which quickly destroys substance 1.

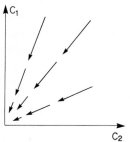

Figure 4.   Kinetics during removal of both species at a rate proportional to their abundances.

harvestor that selectively removes species 1 (reduces C_1, Figure 3) or unselectively takes species 1 and 2 in proportion to their abundances (reduces C_1 and C_2 together, Figure 4). The timing of rhythmic fluctuations in species abundance is governed by the same geometric relations as in the previous cases.

D: Mathematical Redescription

The purpose of this section is to articulate more exactly the very simplest models of phase resetting in an oscillator characterized by an amplitude as well as by a phase. Modulation of angular velocity, the mechanism explored in Chapter 3, is inadequate to describe or analyse the phase control of such an oscillator. We will see why and how that is by doing as we did in Chapter 3: We will write equations for a few general styles of perturbation (such as Figures 3 and 4) and graphically examine their implications. The four models to be presented here all resemble a pendulum in that under unperturbed conditions ($I = 0$) trajectories are concentric circles traversed clockwise at uniform angular velocity, the same on every circle. The radius of the circle is the oscillation's "amplitude", and angular position along the circle (measured clockwise from the positive Y axis) is the oscillation's instantaneous phase (Figure 1).

The four models to be presented here differ in the ways they react to a stimulus $I \neq 0$. During an interval when $I \neq 0$, the oscillator does not behave as it does when $I = 0$. Its state changes in a different way. Our analysis proceeds by following the state along a trajectory of unperturbed dynamics ($I = 0$) up to a certain phase, then following it from there along trajectories of perturbed dynamics ($I \neq 0$) during the stimulus, then reverting to unperturbed dynamics ($I = 0$) after the stimulus. This is essentially what we did in Chapter 3 for simple clocks except that there the trajectories were geometrically identical, all being limited to a common one-dimensional path. They differed only in speed along that path. Now we admit perturbations of speed in two directions. The perturbed path in state space thus differs *geometrically*. To keep the story simple we consider only one nonzero value of I without describing how the dynamic depends continuously on I.

This method of alternating flows has been used in diverse contexts that involve perturbed oscillators; by Danziger and Elmergreen (1956) in connection with an endocrine cycle and its control by hormonal stimuli; by FitzHugh (1960, 1961) to analyze the effects of electrical stimuli on nerve membrane; by Kalmus and Wigglesworth (1960) to describe the impact of daylight on a circadian clock; by Campbell (1964) to analyze mitotic synchrony; by Strahm (1964), by Moshkov et al. (1966), and by Pavlidis (1967) to rationalize resetting experiments in the fruitfly's circadian clock.

Example 1. Let us suppose that the stimulus removes substance C_1 at a constant rate k. If this rate is large and the stimulus is brief, then we can ignore other processes occurring during the stimulus and write:

$$\frac{dC_1}{dt} = -k, \quad \text{so} \quad C_1' = C_1 - kt$$

where the prime indicates the new, reset value of C_1 after a stimulus of duration t. Let's call kt the "magnitude" M of the stimulus. (Note that usage is a little different in Chapters 3 and 4, where stimulus magnitude is interpreted strictly as duration at fixed I.)

In the case of a pendulum, C_1 might be the momentum coordinate, the stimulus being a shove; the magnitude of the stimulus is then the mechanical impulse, the momentum change. In ecological context C_1 might be deer population density, the stimulus being the hunting season. Continuing to use unprimed symbols to represent the state at the beginning of an interlude of perturbed dynamics and primed symbols to represent the state at the end of that interlude, i.e., at the beginning of unperturbed dynamics after the stimulus, we have (see Figures 2 and 5):

$$C_2' = C_2 = C_2^* + A' \sin 2\pi\phi' = C_2^* + A \sin 2\pi\phi.$$

Figure 5.   An impulse changes the momentum of a pendulum or the density of a species by amount M. The vector M from the initial state to the final state determines the new phase and amplitude trigonometrically. Essentially the same diagram can be found in Mercer (1965, Fig. 12).

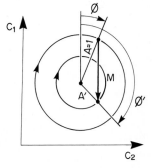

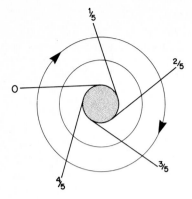

Figure 10.   The $\phi = 0$ locus of Figure 9 is rotated to consecutive positions one-fifth cycle apart to construct isochrons (supposing the angular velocity of trajectories on this diagram is everywhere the same).

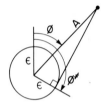

Figure 11.   Given threshold level ϵ and amplitude A of oscillation, the two alternative definitions of phase ϕ and $\phi^{\ddagger}$ are geometrically related.

so

$$\phi^{\ddagger} = \phi + \text{the correction term } \text{cs}^{-1}\frac{\epsilon}{A},$$

where A is the amplitude expressed in terms of ϕ and M.

E: Graphical Interpretation

Now what does this little game get for us? It gets us equations by which to instruct a computer to draw the anticipated consequences of experiments in which a single harmonic oscillator is perturbed for any duration at various times in its cycle. The behavior of such an oscillator presents a marked contrast to that of the simple clocks examined in equal detail in Chapter 3. This new behavior is intuitively accessible in terms of pushing on a pendulum, just as the simple clock's behavior was intuitively accessible in terms of modulating the angular velocity of a rotating machine. In certain qualitative essentials the new behavior found here resembles the behavior typical of a much wider and more realistic variety of biochemical and physiological oscillators. The reason for this resemblance will come out in the next chapter. This chapter finishes with a display of some of the computer plots alluded to. These plots have three main qualitative features:

1. $\phi' = \phi$ in the limit of very brief stimuli, $M \to 0$.

2.   Near phase $\phi = 1$ cycle, near stimulus magnitude $M = 1$, the new phase ϕ' has a singularity. Writing ϕ as $1 + x$ and M as $1 - y$, we have in good approximation for small departures x and y,

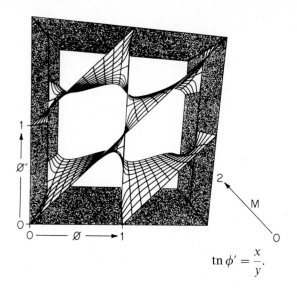

Figure 12.   As in Figure 7 of Chapter 3 except that new phase is determined by old phase and stimulus magnitude according to the equation of Example 1 rather than according to simple-clock kinetics.

$$\operatorname{tn}\phi' = \frac{x}{y}.$$

In other words, the new phase of the oscillator is entirely determined by the ratio of the tiny departures from $\phi = 1$ and $M = 1$. The slightest error in approximating that stimulus timing can therefore evoke any phase, but at a very small amplitude (in models 1, 2, 3; in model 4 phase is undefined at amplitudes less than threshold).

3.   At very large M, $\operatorname{tn}\phi' \to 0$, i.e., a sufficiently prolonged stimulus essentially resets the oscillator to a unique phase regardless of the initial conditions.

The easiest way to see how ϕ' jointly depends on ϕ and on M is by plotting ϕ' in three dimensions above the (ϕ, M) plane. It is a screw surface repeated in each repeat of ϕ' along the vertical axis and in each repeat of ϕ along the horizontal axis. It all consists of one single surface wound around vertical poles at $\phi =$ any integer, $M = 1$ (Figure 12 and Figure 4 of Chapter 2). The corresponding plot of new, reset amplitude A' of oscillation would show a deep pit descending to zero at $\phi =$ any integer and $M = 1$, rising to unity (by definition) along the front wall at $M = 0$ and taking other values elsewhere according to the particular dynamics adopted.

It is convenient to portray these surfaces as contour maps on the stimulus plane as we did for the new reset phase of a simple clock in Chapter 3. These contours depict the (ϕ, M) combinations along which ϕ', and therefore $\operatorname{tn}\phi'$, is constant. Thus their shapes are given by $\alpha \operatorname{sn}\phi' = \operatorname{cs}\phi' - M$. These are arcs of sinusoids, all converging to $M = 1$, $\phi = 1$ and all touching the $M = 0$ line at $\phi' = \phi$. Figure 13 shows the contour map of new reset phase for Examples 1 through 3.

The new phase ϕ' at the moment the stimulus ends is the complement of the event latency θ, the time until phase $\phi' = 0$ recurs after the stimulus ends. I call this latency the *cophase* because phase plus latency (cophase) is necessarily one cycle. Thus a plot of event latency downward from the stimulus plane is identical to a plot of new phase upward.

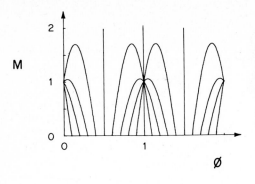

Figure 13.   As in Figure 4 of Chapter 2 the (ϕ, M) plane is depicted with the contour lines of fixed new phase, ϕ', at intervals of 1/10 cycle. In the simple clock case (Figure 8 of Chapter 3) they were parallel displacements of a common curve whereas in this case they are various pieces of sine curves converging to a singular point.

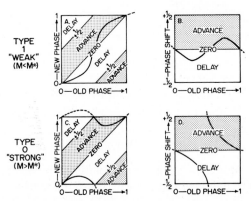

Figure 14.   The top two panels (a, b) show type 1 resetting; on the left (a) as a resetting curve, plotting new phase upward (or cophase downward) against old phase to the right, and on the right (b) as a phase response curve (PRC), plotting phase advances upward and delays downward, also against old phase to the right. The bottom two panels (c, d) do the same for type 0 resetting.

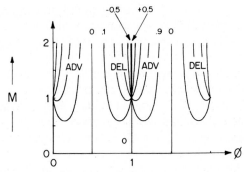

Figure 15.   As in Figure 13 except that the contours represent (ϕ, M) combinations resulting in the same phase shift, $\Delta\phi$, rather than the same new phase. The phase shift is the new phase minus the old phase minus the stimulus duration. Stimulus duration is taken to be negligible in this figure. Contoured at intervals of one-tenth cycle as in Figure 13, using equation $M = \operatorname{tn} \Delta\phi / (\operatorname{sn} \phi + \operatorname{cs} \phi \operatorname{tn} \Delta\phi)$ from Example 1.

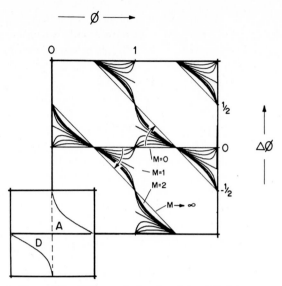

Figure 16.   Similar to Figure 6(c) of Chapter 3, showing phase shift (advances A, upward) vs. phase for several stimulus magnitudes ranging from $M = 0$ to ∞. Curves of this sort are more commonly depicted in various experimental literatures as indicated in the lower left. The dashed segment should not be interpreted as part of the curve, but is commonly drawn or even drawn solid.

In some fields it has been convenient to think instead in terms of phase shifts. This is a convenience, for example, in thinking about one model of entrainment, according to which an oscillator of native period τ keeps up with a periodic driving stimulus of period T by doing a little phase shift $\Delta\phi = T - \tau$. In terms of old phase and new phase, the phase shift $\Delta\phi$ is $\phi' - (\phi + \text{stimulus duration})$. Thus a plot of θ or ϕ' is readily converted to a plot of $\Delta\phi$ by shearing it 45° along the ϕ axis (Figure 14). A contour map of $\Delta\phi$ on the stimulus plane is sketched in Figure 15 for Examples 1–3. Figure 16 plots the same $\Delta\phi$ vs. ϕ for several stimulus magnitudes ranging from $M = 0$ to $M = 2$. Note the abrupt change of topology at $M = 1$, where the phase singularity is encountered at $\phi = 1$. This family of phase response curves (PRCs) should be contrasted with the family of phase response curves following from simple clock models, in which phase shifts accumulate by modulation of angular velocity on the ring [Figure 6(c), Chapter 3]. In that family there is no topological change. The simple clock's strong stimulus PRC is simply an extreme version of its weak stimulus PRC. It includes a very steep narrow upslope region. Nothing of the sort is seen in the models of this chapter. Instead, all the steepness is packed into that one critical stimulus duration $M = 1$ at the phase singularity.

The reason for this qualitative difference of behavior lies in the qualitatively different mechanisms of rephasing. The simple clock achieves its big advance or big delay to the opposite side of the cycle by advancing or delaying through half the cycle. Because the choice between advancing and delaying depends on whether the initial phase was just before $\phi = 1$ or just after $\phi = 1$, large stimuli applied sufficiently close to this phase can have extremely variable results. In contrast, the pendulum-like oscillator achieves both big advances and delays by a process that entails a reduction of amplitude toward zero, and through zero, and a rebuilding of amplitude on the opposite side of the cycle. The only sensitive stage is the passage through zero amplitude, where phase is ambiguous.

F: Summary

In Chapter 3 we thought of a single simple clock as behaving differently along its cycle according to the intensity of some environmental parameter such as light intensity. Here we see the same principle in a two-dimensional state space. In the presence of some external modifier of the system's dynamics, its state changes along different paths in state space. In Chapter 3 we calculated the consequences in terms of phase-dependent phase resetting. The result was a "time crystal": a periodic lattice of unit cells composed of old phase and new phase repeats along two axes, with a third perpendicular axis depicting some measure of stimulus magnitude. Within each three-dimensional unit cell lies a two-dimensional surface describing new phase as a function of stimulus magnitude and the old phase at which that stimulus was given. This "surface" proved to be a stack of *separate* surfaces repeating along both phase axes.

We redo the calculation in this chapter to see whether a topologically different kind of dynamic model implies a qualitative difference in resetting behavior. If so then resetting behavior might provide a useful criterion for rejecting models that are inappropriate to given experimental system. It transpires that there is one conspicuous qualitative difference: Resetting in this two-dimensional state space consists of a resetting of phase *and* a conjugate and equally stable resetting of amplitude. There is a peculiarity of the phase response when the amplitude is reset to zero. A specific, exacting combination of stimulus timing and magnitude is required to guide the system to this steady state and leave it there. At this special old phase and magnitude, new phase is arbitrarily sensitive to details of the stimulus. The new phase surface tilts vertically and winds around a screw axis at that point. This phase singularity links all repeats along ϕ and ϕ' directions into one single surface quite unlike the stack of separate surfaces that characterize a simple clock. We have seen this pattern in experimental data. Chapter 4 gives us one interpretation for it in terms of internal incoherence among many clocks. Here we have another in terms of the dynamic steady-state of a single oscillator.

In coming chapters we ask which features of Figures 1 through 4 are essential for this result. It turns out that the behavior exhibited in Figures 12 through 15 is typical of a much broader class of less idealized biochemical dynamics and has little to do with the equilibrium state or with stable resetting of amplitude. But this simplest example is convenient for introducing the result, preliminary to perceiving it in less contrived models. (It also turns out that the fruitfly's circadian clock behaves remarkably like this simplest caricature: compare Figures 13 of this chapter and Chapter 2.)

6. Attracting Cycles and Isochrons

In Section A of this chapter we associate a phase with each state of a limit cycle oscillator during dynamics in the absence of any perturbing influence. In Section B a stimulus smoothly alters the trajectories so that phase changes in peculiar ways, even discontinuously. This analysis is intended to apply to smooth dynamics. Accordingly, in Section C references are compiled to models which violate this precondition and thus do not fall within the purview of this chapter.

A: Unperturbed Dynamics

Introduction

By referring everything to the purely geometrical idea of the motion of an imaginary fluid, I hope to attain generality and precision, and to avoid the dangers arising from a premature theory professing to explain the cause of phenomena. If the results of mere speculation which I have collected are found to be of any use to experimental philosophers, in arranging and interpreting their results, they will have served their purpose, and a mature theory, in which physical facts will be physically explained, will be formed by those who by interrogating Nature herself, can obtain the only true solution of the questions which the mathematical theory suggests.

James Clerk Maxwell, 1856, "On Faraday's Lines of Force"

We saw in Chapter 4 that the collective amplitude of a population of simple clocks can be arbitrarily small. It is determined by the distribution of phase within the population. But this distribution is stable only in the limiting case of complete independence among identical simple clocks contributing to an aggregate rhythm. If any interaction is allowed and/or if the simple clocks' periods differ at all, then the phase distribution does change in time and with it the shape of the collective rhythm changes. The collective rhythm might damp out altogether as phases gradually randomize. Or it might approach an attracting cycle as the many simple clocks entrain one another to a common frequency in a self-stabilizing phase distribution. Under various assumptions about the manner of interactions there might be several different attracting cycles.

Chapter 5 is built around the idea that a single oscillator also can have an amplitude as well as a phase. But only in physical situations characterized by symmetry principles and conservation laws is it usual that dynamic systems can oscillate persistently at arbitrary amplitude, as in the simple idealizations exploited in Chapter 5.

Attracting Cycles

We pass from Chapter 5 to this Chapter through the idea that an oscillator can have a *preferred* amplitude from which it can be perturbed and to which it regulates back again. This type of behavior is typical of oscillators that tend to a stable periodic behavior. Such an oscillator's degrees of freedom, be they few or many, affect each other's rates of change in such a way that all eventually settle into a regularly repeated cycle. Poincaré called this a "limit cycle" because it is asymptotically approached in the limit of infinite time. For application to experimental science we are concerned only with attracting (as opposed to repelling) limit cycles. Because we seldom await the limit of long time, and for brevity, I drop the word *limit*. Thus we embark upon a study of attracting cycles. Geometrically, an attracting cycle is a closed ring in state space. It may be quite irregular, even knotted, in a space of many dimensions, but the essential feature for our purposes is that it is a closed ring path and that all nearby paths gradually funnel onto it. The region occupied by paths leading onto an attracting cycle is called the cycle's *attractor basin*. Any states not in the attractor basin comprise that cycle's *phaseless set*.

A geometrically simple example is shown in Figure 1. In polar coordinates the system is described by instantaneous phase ϕ and instantaneous amplitude R. The angular velocity $\dot{\phi}$ depends on the instantaneous amplitude and so does the rate of change of amplitude $\dot{R}$:

$$\dot{\phi} = A(R)$$
$$\dot{R} = B(R).$$

If $\dot{R} = KR(1 - R)$ then the amplitude R regulates to the unit circle $R = 1$. It does so quickly if K is big, or spirals more gradually toward the unit circle if K is small [Figure 1(a); see also Box A]. If $K = 0$, then $\dot{R} \equiv 0$, so that any amplitude is stable forever as in Figure 1 of Chapter 5.

If $\dot{\phi}$ is constant, then the period is the same at any amplitude, as in a pendulum or a population of simple clocks. If $\dot{\phi}$ is not constant, then the period depends on the amplitude.

The attractor basin is the whole plane minus the origin. The phaseless set consists only of the steady-state at the origin, $R = 0$.

Latent Phase

I draw attention to this geometric caricature of an oscillator because it helps introduce, in an exact way, a generalization on the notion of phase which proves crucial for much that follows. In Chapter 5 we defined phase as the fraction of a

Box A: Simple Clocks as Caricatures of Attracting Cycles

Before leaving this section, note that for some purposes the simple clock of Chapters 3 and 4 is a suitable approximation to either:

1.   A system whose state space really is $\mathbb{S}^1$ for structural reasons (e.g., the state is the position of a polymerase on a ring of DNA) or

2.   A dynamical system of arbitrary complexity in which trajectories flow swiftly toward an attracting cycle under all experimental conditions considered.

In terms of the models of this section, if $\dot{R} = 100000 \times (1 - R)R$ then the oscillator's amplitude regulates very swiftly to $R = 1$, so that for all practical purposes the system always runs on that unique ring, except for a brief moment after violent perturbation. With $I \neq 0$, a simple clock runs on the same cycle but at phase-dependent angular velocity. If I enters only into the expression $\dot{\phi} = A(\ )$ for angular velocity and not into the expression $\dot{R} = B(\ )$ for amplitude regulation then the same is true of this attractor cycle model.

It is also true in good approximation if the stimulus $(I \neq 0)$ adds anything to the oscillator's kinetics that is too weak to substantially oppose the violently amplitude regulating $\dot{R} = B(R)$. Then R is still confined close to the unperturbed cycle and we need deal only with alterations of its angular velocity. Thus its dynamics can be summarized in a sensitivity function $Z(\phi)$, representing the fractional deviation of angular velocity from its unperturbed unit rate, per unit stimulus, that is per unit deviation from standard environmental conditions. This $Z(\phi)$ (Winfree, 1967a) is the velocity response curve of Swade (1969), Daan and Pittendrigh (1976a,b), and Pittendrigh and Daan (1976a,b,c). In the notation of Chapter 3, the angular velocity of a simple clock is denoted $v(\phi, I)$. Performing a Taylor series expansion in I gives $\dot{\phi} = v(\phi, I) = 1 + (I - I_0)Z(\phi)$ + neglected terms in higher powers of $(I - I_0)$. This extreme version of an attracting cycle has been called a generalized relaxation oscillator (Winfree, 1967a; Swade, 1969; Kramm, 1973; Goodwin, 1976). The term originated in the idea that relaxation oscillators (see Section C) adhere very closely to an attracting cycle that includes a jump. The generalization consists in embracing smooth cycles as well as nearly discontinuous cycles. See also Box A of Chapter 9.

Kuramoto (1975) takes this limit explicitly, reducing the $\lambda - \omega$ oscillator to a simple clock in order to simplify the calculation of mutual entrainment in populations. In place of my single influence and sensitivity, he invokes an orthogonal *pair* of conjugate influence-sensitivity pairs, coupling oscillators by

$$\dot{\phi} = 1 + \sin\phi_j \cos\phi_i - \cos\phi_j \sin\phi_i = 1 + \sin(\phi_j - \phi_i).$$

The symmetry of this coupling facilitates analysis.

Neu (1979a,b,c) shows mathematically just how a D-dimensional oscillator with a strongly attracting cycle can be reduced to the simple-clock approximation, as assumed in Winfree (1967a) with merely handwaving arguments.

cycle elapsed since some phase marker event such as the maximum of C_1 or the upward crossing of a threshold. That definition supposed that behavior was strictly periodic following any initial conditions. But in attractor cycle dynamics, behavior is not initially periodic. It is strictly periodic only in the undisturbed system and only in the *limit* of long time. So a more practical definition is called for. A notion that naturally recommends itself is the *latent* phase: One waits until the perturbed system reverts to regular periodism, and then extrapolates that rhythm back to the time in question to discover its latent phase at that past moment. Equivalently, one might start a "test" system at time 0 in some state whose latent phase we wish to know and simultaneously start another on the cycle at phase zero. After a long time both are on the attracting cycle but the test system is some fraction Φ ahead of the control. Then Φ is the latent phase of that initial condition.[1]

This is in fact how physiological experiments are conducted. For example, suppose a periodically firing nerve cell is electrically stimulated. During that stimulus the cell's rhythmicity is disrupted. When the stimulus is removed the membrane is left in some peculiar state not necessarily on the cycle. From that initial condition, its normal rhythmicity recovers during the subsequent unperturbed free run. The phase of that recovered rhythm is recorded. By comparison with the control rhythm or by extrapolation back to the moment when the stimulus ended, the latent phase reached at the end of that stimulus is measured (Chapter 14, Section A). The same format has been used to explore the dynamics of circadian clocks (Chapters 19, 20, 21), of biochemical oscillations (in the slime molds of Chapter 15; and in the yeast cells of Chapter 12), and of the cell cycle (Chapter 22). In these and many other cases, attractor cycle models have been proposed to link the observed rhythmicity to plausible biochemical kinetics.

It is useful to develop a little bit of abstract geometry about the relationship between latent phase and an attractor cycle oscillator's state. The pair of equations above gives us a way to start thinking about it. Now that we have quit thinking of the state space as a ring, phase and state are no longer synonymous. Now phase is some function of state. What kind of function? The next section defines it analytically for the special case of polar symmetric two-variable kinetics represented by the pair of equations above. [Guckenheimer (1975) gives a far more general, precise, and rigorous treatment for state spaces of arbitrary dimension and topology. He also treats examples of the sort to which we now turn. Also see Kawato and Suzuki (1978).]

The Isochrons of Symmetric Attracting Cycles

We resort once again to useful jargon: Any set of states having the same latent phase Φ is called an *isochron*, specifically "isochron Φ" (see Box B). Suppose as

---

[1]   Existence proofs for ordinary differential equations are to be found in Coddington and Levinson (1955, Theorem 2.2, p. 323); Hale (1963, Theorem 10.1, p. 94); Hale (1969, Theorem 2.1, p. 217); in Fenichel (1974, 1977) and, for functional differential equations, Hale (1977, Theorem 3.1, p. 242) where it is called "asymptotic phase."

Box B: On Definitions of Isochrons

I invented isochrons in 1967 to help me think about timing relations in oscillators perturbed off their attracting cycles. This is important in physiological applications because biological oscillators are almost never on their attracting cycles. The term *isochron* has since been used in a number of senses, some only subtly different and some radically different. This Box provides a brief overview.

As used in the series of papers begun in 1967, an isochron is a certain set of phase-equivalent initial conditions in the state space of any process that immediately or eventually runs at a fixed period that we take as a unit of time. These initial conditions are equivalent in the sense that they lead to rhythms which are synchronous in whatever operational sense we have adopted. Thus an isochron is the inverse image of a point in $\mathbb{S}^1$ for a map $\Phi : X \to \mathbb{S}^1$ from the state space X of an oscillator to the circle of phase $\mathbb{S}^1$. In the simple clocks of Chapter 3, the only admissible initial conditions are all on the cycle: $X = \mathbb{S}^1$. Thus no two initial conditions $x_1, x_2 \in X$ are equivalent unless they are identical. If they are not identical, the rhythms begun from such initial conditions differ by a phase shift $\Phi(x_2) - \Phi(x_1)$. The notion of an isochron first acquires nontrivial interest in Chapter 5, where we consider models which run stably on cycles of various amplitudes. In this situation, rhythms of different amplitudes might be considered equivalent if they go through their maxima at the same time or if they cross a threshold at the same time. If, as in the present chapter, we are talking about a limit cycle process and initial conditions in the attractor basin of that cycle, then initial conditions are considered equivalent if oscillators started at those initial conditions funnel synchronously onto the cycle. In this case trajectories from any initial conditions on the same isochron eventually become indistinguishable, and trajectories from any other isochron eventually differ only by a time translation equal to the latent phase difference $\Phi(x_2) - \Phi(x_1)$ or the cophase difference $(1 - \Phi(x_1)) - (1 - \Phi(x_2))$ between those two isochrons.

It is helpful to visualize isochrons by thinking in terms of a stroboscopic presentation of dynamics in the oscillator's state space, as follows. Taking the cycle period as 1, we start an oscillator at any initial condition at time 0 and look again to see what state it

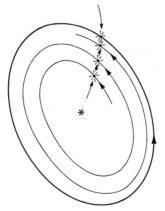

Points on a trajectory spiraling toward the attracting cycle are made to emit a flash of light at unit intervals of time. The sequence of such points converges to a fixed point on the cycle, outlining an isochron.

has reached at unit intervals thereafter. If we happened to start exactly on a closed ring trajectory (the only choice for oscillators of the sort considered in Chapters 3, 4, and 5), then at each later observation it will be back to exactly the same state. The same will be true of an attractor cycle system only if we start the oscillator exactly on the attracting cycle. But more generally, at each successive look we find the oscillator's state closer to the attracting cycle. The sequence of states reached approaches a cluster point on the cycle. All these points lie on the same isochron (see Figure a).

I have used the term *isochron* sometimes to mean a locus in state space as above, but sometimes also to mean a set of stimuli, i.e., a locus in the (ϕ, M) plane or cylinder. These are the stimuli that evoke synchronous rhythms by pushing oscillators onto a given isochron set in state space. Thinking of the stimulus space as a map into state space as in Figures 5–11, we see that a given isochron appears in stimulus space where the image of stimulus space intersects that isochron in state space.

In terms of cophase, an isochron is the locus of constant cophase θ but the cophase intended can refer to a function of state, or to a function of stimulus parameters that achieve the required state.

Guckenheimer (1975) applies the powerful mathematical methods of topological analysis to prove the existence and main properties of isochrons in the attractor basin of a limit cycle. He inquires into the conditions on state space topology and "genericity" of dynamics that hedge in my conjectures (Winfree, 1974c) about isochrons and their phaseless set. Guckenheimer's paper and mine are reprinted together in Levin (1978).

Kauffman (1974) refers latent phase not to the attracting cycle but instead to the next threshold crossing as in Example IV of Chapter 5. This may be viewed as an experimentalist's approximation to the measurement of latent phase in real physiological systems which take an inconveniently long time to approach their attracting cycles. Or it might be viewed as a different definition of latent phase, in which the isochrons have somewhat different topological properties than those considered above. Pavlidis (1973), Pinsker (1977), and Kawato and Suzuki (1978) also use this definition of latent phase, focusing on only the very next event and ignoring the regular rhythmicity presumed to persist a long time after the stimulus (after any transient irregularities have died out).

The word isochron has also been used in quite different senses not to be confused with any of the above by

1.   Bouligand (1972a), to mean the locus in physical space along which cuticle is simultaneously deposited (Chapter 17)

2.   Allessie et al. (1973), to mean the locus of simultaneous excitation during propagation of an electrical wave in heart muscle (Chapter 14);

3.   Pinsker (1977) ("isochronous") to mean "having the same period" as in Christian Huygens's *Horologium Oscillatorium* (1673): "Pendula isochrona vocentur, quorum oscillationes, per arcus similes, aequalibus temporibus peraguntur." [etc.].

above that

$$\dot{\phi} = A(R) \text{ with the unit of time chosen so that } A(R_0) = 1$$

and

$$\dot{R} = B(R) \text{ with } B = 0 \text{ and } dB/dR < 0 \text{ at some positive } R_0.$$

This is the simplest attractor cycle oscillator. If there is only one such R_0 then R_0 is a unique attracting cycle and we may as well scale R to make $R_0 = 1$.

Since the dynamical flow has polar symmetry, the isochrons must also have polar symmetry:

$$\Phi = g(\phi, R) = \phi - f(R).$$

Now the latent phase Φ necessarily increases at unit angular velocity as the oscillator follows its kinetic equation. Thus we write

$$\dot{\Phi} \equiv 1 = \dot{\phi} - \frac{df(R)}{dR}\dot{R}$$

so

$$\frac{df}{dR} = \frac{\dot{\phi} - 1}{\dot{R}}$$

This is a differential equation for $f(R)$. We can integrate it and then obtain $g(\phi, R)$.

Example 1. Suppose

$$\dot{R} = 5(1 - R)R \quad \text{Figure 1(a)}$$
$$\dot{\phi} \equiv 1 \qquad\qquad \text{Figure 1(b)}$$

then

$$\frac{df}{dR} \equiv 0$$

so

$$f(R) = \text{constant, chosen so } \Phi = 0 \text{ at } R = 1 \text{ and } \phi = 0$$

so

$$\Phi = g(\phi, R) = \phi \quad \text{Figure 1(c)}$$

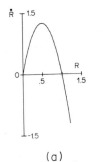
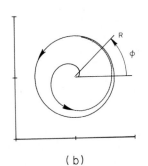
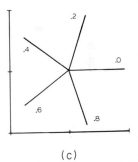

(a) (b) (c)

Figure 1.   (a) $\dot{R} = 5R(1 - R)$. (b) A single trajectory of the dynamical scheme described in the text. The unit circle is an attracting limit cycle. (c) Isochrons of the same dynamic, marking off intervals of one-fifth cycle along all trajectories in the attractor basin of the cycle. Trajectories begun on any isochron reappear on it at unit intervals of time, closer to the point $R = 1$.

In this case, the latent phase corresponds perfectly with the simplest local instantaneous definition of phase, viz the fraction of a cycle elapsed between extrema of x or y. This is exceptional. Isochrons are more commonly complex curves. They usually differ from such local definitions of phase as one might be tempted to invent. This is illustrated in the next example.

Example 2.   Suppose

$$\dot{R} = 5(1 - R)R^2 \quad \text{Figure 2(a)}$$
$$\dot{\phi} = R \qquad\qquad \text{Figure 2(b)}$$

Here the trajectories have the same geometry but the state traverses these trajectories at speeds R times faster than in Example 1. In this case the angular velocity $\dot{\phi}$ is no longer independent of amplitude but increases in proportion to R. Thus

$$\frac{df}{dR} = \frac{R - 1}{5R(1 - R)R} = -\frac{1}{5R^2}$$

so

$$f(R) = \frac{1}{5R} + \text{constant}$$

so

$$\Phi = g(\phi, R) = \phi - \frac{1}{5R} + 0.2 \quad \text{Figure 2(c)}.$$

Thus the isochron structure close to the repelling focus of this two-variable kinetics consists of tight spirals. In contrast to the straight-in radial isochrons of simple clocks (Chapter 3), clockshops (Chapter 4), and harmonic oscillators (Chapter 5), those associated with attracting cycles typically exhibit ornate structure near their convergence (see Box C). Only in perfectly symmetric models with only two variables is this structure as simple as the bundle of spirals derived

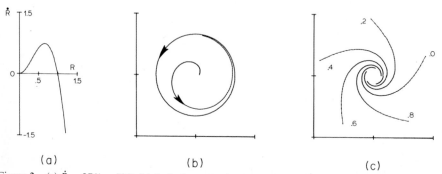

(a)   (b)   (c)

Figure 2.   (a) $\dot{R} = 5R(1 - R)R$. (b) As in 1a except both components of velocity are multiplied by a scalar field, leaving directions unaltered. (c) The changed isochrons reflect the radial dependence of angular velocity.

Box C: Isochrons in Nonsymmetric Kinetics

The example on display here is taken from FitzHugh (1961). FitzHugh sought a useful simplification of the Hodgkin-Huxley equation. He found it in a second order cubic polynomial equation constructed by hybridizing Bonhoeffer's (1948) model of iron wire excitability (see also Box B in Chapter 14) with van der Pol's (1926) model of the heart-beat. This "BVP" model mimics a pacemaker neuron with a repelling steady-state and an attracting cycle. I computed its isochrons as follows:

1. The attracting cycle was found by integrating the rate equations numerically. A point from the cycle is chosen arbitrarily to be phase 0. The cycle period is taken as unit time.

2. An arbitrary initial state was chosen and once again the rate equations were integrated by iterating the difference equation in steps Δt until the state comes within a small tolerance of phase 0 on the cycle.

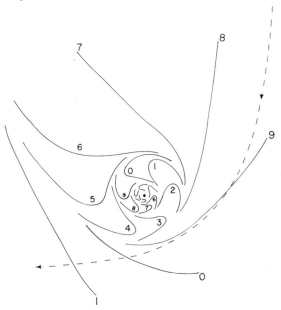

(a) Similar in principle to part (b) of Figures 1 through 4. The isochrons are calculated in the linear region near the steady state of FitzHugh's BVP model of pacemaker activity in nerve membrane:

$$\dot{x} = c\left(y + x - \frac{x^3}{3} + z \right)$$

$$\dot{y} = -(x - a + by)/c$$

$$a = 0.7, \qquad b = 0.8, c = 3.0, \qquad z = -0.4.$$

The isochrons are interrupted where the calculation becomes too delicate, along the spiral separator. Coordinates are as in Figure 6 but the region near steady-state is here greatly enlarged. All trajectories (not shown) are clockwise exponential spirals out of the steady-state, cutting isochrons at unit rate. The dashed curve is the lower right-hand arc of the attracting cycle shown complete in part (b) of this figure.

3. Backing up along that trajectory, the latent phase was noted at each step as $\Phi = -N\Delta t$ modulo 1, N being the step number.

4. This was done for many different initial states.

5. The isochrons were sketched through states of equal latent phase at intervals of one-tenth cycle. From states on one isochron, the time to phase 0 is the same, modulo 1. All trajectories pass from one isochron to the next at intervals of one-tenth cycle.

The results near the steady-state are shown in a figure. For purposes of convenience a linear transformation of variables was effected in plotting the trajectories and isochrons. This was done in order to display trajectories near the steady-state as exponential spirals and to give the whole diagram polar symmetry. The isochrons approach the steady-state in a distinctly complicated way.

Further out, the trajectories approach their attracting cycle and the isochrons behave as one might expect intuitively:

1. They are packed close together near the separator (Figure b): A slight displacement of state where the system is in one of these states elicits a big change of phase in its eventual rhythm.

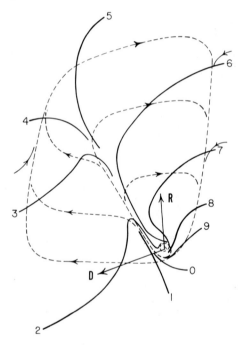

(b) Similar in principle to parts (b and c) of Figures 1 through 4. Isochrons are solid curves. Trajectories (dashed) wind out of the repelling steady-state and diverge along the spiral separatrix. The picture is somewhat skewed from the conventional representation in FitzHugh's Figure 5 [see depolarizing (D) and recovery (R) variables] by the linear transformation required to symmetrize the spirals near steady-state. Figure (a) enlarges that region.

2. A small depolarizing (excitatory) pulse advances phase during the depolarizing phase of the action potential and delays it during the relatively refractory recovery interval. Hyperpolarizing (inhibitory) pulses have the opposite effect.

Box D: A Measurement Problem

Does the ornate structure of isochrons near their convergence point provide an experimental distinction between such oscillators as we considered earlier and the attractor cycle dynamics under consideration in this chapter? It might, but only in principle.

In practice, isochron structure is poorly resolved near the isochrons' convergence. The models of Chapters 4 and 5 have unusually low-amplitude rhythmicity in this region. That implies unusual difficulty in measuring phase with adequate precision near the region of convergence. It also means that the system is particularly susceptible to the random perturbations that chronically afflict any real experiment, especially biological experiments. Maybe in dealing with an attractor cycle mechanism, as in this chapter, one might escape that difficulty because the amplitude will regenerate. Why not wait for normal amplitude to recover, *then* measure phase and extrapolate back to that earlier moment when amplitude was too low? This works fine if regeneration is swift but the longer the wait, the greater the imprecision of phase measurements. Extrapolation is not accurate across many cycles during which random perturbations are always creeping into the experimental system. In every case, the mere fact that the isochrons do converge means that high resolution is essential. Slight variability in stimulus timing, or in the sensitivities of individual biological organisms, has a disproportionate impact on phase measurements near a convergence of isochrons, e.g., see measurements of this variability in Winfree, 1974, Appendix, for the *Drosophila* circadian system. In short, data variance is typically greatest in the very region where highest resolution is needed to distinguish the isochrons. Because measurement is so imprecise near a singularity, there is little point in deploying refined models. The screw surfaces and pinwheel contour maps I have fitted to data have all had the simplest, plainly radial isochron structure near the phase singularity.

---

above. And in such cases, the spirals degenerate to radial lines as in Example 1 only if the angular velocity A is independent of amplitude (see Box D).

Example 3.   Suppose

$$\dot{R} = (1 - R)R] \text{ Figure 3(a)}$$
$$\dot{\phi} = 1 + \varepsilon(1 - R)] \text{ Figure 3(b)}$$

so

$$\frac{df}{dR} = \frac{\varepsilon}{R}$$

so

$$f = \varepsilon \ln R + \text{constant}$$

so

$$\Phi = \phi - \varepsilon \ln R] \text{ Figure 3(c).}$$

This example shows that the isochron spiral can turn either way relative to trajectories, depending on the sign of ε.

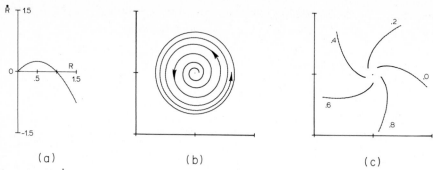

Figure 3. (a) $\dot{R} = R(1 - R)$. (b) As in 1a but radially five times slower and azimuthally somewhat dependent on radius. (c) The isochrons make exponential spirals into the origin, in contrast with the hyperbolic spirals of Figure 2c.

Example 4.   Suppose

$$\dot{R} = 5(1 - R)(R - \tfrac{1}{2})R] \quad \text{Figure 4(a)}$$
$$\dot{\phi} = 1 + \varepsilon(1 - R)] \quad\quad \text{Figure 4(b)}$$

The point here is to examine a self-sustaining oscillator that is not also self-exciting (examples: the circadian clock models of Kalmus and Wigglesworth (1960) and of Pavlidis and Kauzmann (1969); Teorell's (1971) mathematical model of pressure sensitive neural pacemakers; Best's (1976, 1979) computation of isochrons for pacemaker nerve and the subsequent measurement by Guttman et al. (1979); Aldridge and Pye's (1979) theory and measurements of oscillating glycolysis in yeast; Johnsson's (1976) and Johnsson et al.'s (1979) on oscillatory transpiration):

$$\frac{df}{dR} = \frac{\varepsilon}{5R(R - \tfrac{1}{2})} = \left(\frac{1}{R - \tfrac{1}{2}} - \frac{1}{R} \right) * \frac{2\varepsilon}{5}.$$

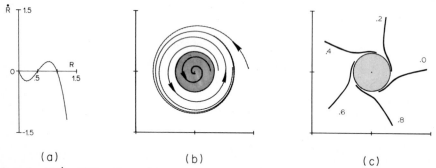

Figure 4. (a) $\dot{R} = 5R(1 - R)(R - \tfrac{1}{2})$. (b) The azimuthal equation is as in Figure 3a, while the radial equation is modified to create a repelling cycle at $R = \tfrac{1}{2}$. Trajectories inside that radius wind into the attracting equilibrium. (c) The phaseless manifold is the disk $R = \tfrac{1}{2}$. Along its border all isochrons converge (as they did in Figures 1c, 2c, and 3c but with $R = 0$).

For $R < \frac{1}{2}$, Φ will go undefined. For $R > \frac{1}{2}$

$$ f = \frac{2\varepsilon}{5} \ln \frac{R - \frac{1}{2}}{R} + \text{constant} $$

so

$$ \Phi = \phi - \frac{2\varepsilon}{5} \ln \left(2 - \frac{1}{R} \right) \Bigg] \; \text{Figure 4(c).} $$

The phase singularity in this case is not a point but the whole circular border of a black hole. (This provides an alternative mechanistic basis for Example IV of the preceding chapter).

Phaselessness vs. Timelessness

In all the above examples, the phaseless set is an equilibrium or its attractor basin. Ambiguous phase is achieved by timelessness in these cases: the system ceases to change state. Do not be deceived by simple examples to imagine that this is necessary or even usual. Even in kinetics involving only two variables as above, the phaseless set could consist in part of quite lively states, e.g., it could be the attractor basin of a different cycle. A phase singularity is not necessarily timeless, but only indeterminate in respect to the phase of a rhythm of specified period.

To assume that any given physiological oscillator has exactly two important degrees of freedom is of course a rather special assumption. It is a common assumption because two is the *minimum* number needed to describe an attractor basin around a cycle. Two is also the *maximum* number of dimensions compatible with a number of widely used analytic and graphic techniques required for exact mathematical solution of models. However, the features that I would bring out do not rely on this restrictive ad hoc assumption. They do not rely on the geometry of the plane, and they do allow us to treat the more complex models which are becoming common in the recent literature of physiological and biochemical control systems (Box E).

Oscillators of Greater Complexity

The main principle involved in generalizing to an arbitrary number of dimensions has been encountered before (p. 28). It is that a disk cannot be mapped continuously onto the phase ring unless the winding number of phase along the disk's rim is zero. This principle applies as follows in context of latent phase in a D-dimensional state space. Suspend from the attracting cycle any simply connected two-dimensional surface, i.e., a cap whose boundary is the attracting cycle. Now ask what is the latent phase Φ of each point in this cap? Around the boundary $\Phi = \phi$: Its winding number is 1. Thus Φ cannot be assigned in a continuous way to all points in the cap. There must be a discontinuity. This is true of *every* cap that can be hung on the cycle (as on pp. 28 and 69). Thus the necessary *point* of

---

Box E: Dynamics Involving Three or More Interacting Variables

By conceptually simplifying a system to a single-variable kinetic model, we lose track of phenomena that are conspicuous in the laboratory. The same is often true of simplification to only two variables. In matters of physiology and biochemistry it is no mere affectation to abjure reliance on mathematical methods that fail when a third or fourth variable comes on stage. Any action involving two or more oscillators, for example, transpires in a space of at least four dimensions (except in the most emasculated approximation, that makes one or both into a simple clock). Any action transpiring in space, in a continuum of coupled oscillators, is properly described by a partial differential equation. This may be written in approximation as a set of many repeats of the ordinary differential equations of kinetics, one for each volume element, thus multiplying the number of dimensions by the number of volume elements considered. Kinetics portrayed in terms of a time delay likewise require a partial differential equation or approximation by a high-dimensional ordinary differential equation. You will also find in perusing the bibliography of Chapters 12–14 that realistic chemical kinetics seldom involve as few as two major interacting variables. Three or four are typically involved in the models employed to mimic oscillating glycolysis in yeast, cAMP pulsation in slime molds, alternating oxidation and reduction in the malonic acid reaction, and the permeability interactions in nerve membrane.

It is not unusual for smooth dynamics involving three or more variables to exhibit chaotic behavior. Oscillators and coupled sets of oscillators seem especially susceptible to such departures from periodicity. See 1976–79 papers of Rössler.

---

discontinuity in each cap is that cap's intersection with a phaseless locus of at least D-2 dimensions. That locus threads the cycle, in the sense that it penetrates every cap.

How does this peculiarity of $\Phi(\mathbf{x})$ relate to anything observable in the phase relations of perturbed oscillators? This is the subject of the next section, but let's first tidy up this section by summarizing the essential and generalizable features of latent phase seen in the foregoing examples:

1.   Within the region of state space from which all trajectories lead to the cycle (its attractor basin), every state has a unique latent phase Φ.

2.   States having the same latent phase lie in a single connected continuum of dimension one less than the dimension of state space. I call these continua isochrons, meaning same-time loci. The isochrons fill up the attractor basin of the cycle.

3.   The isochrons never intersect except on the basin's boundary.

4.   Each of these isochrons cuts the cycle at one and only one point, in order around the cycle.

5. Φ changes at unit rate along all trajectories in the attractor basin. Therefore, the isochrons are transverse to trajectories everywhere inside the attractor basin.

6.   Initial conditions (states) outside the cycle's attractor basin have no latent phase. This set of states includes all steady-states of the dynamics as well as the trajectories leading to such steady-states (their "attracting manifolds").

7.   Along the boundary of the attractor basin all isochrons typically converge in such a way that they all become arbitrarily close together. This boundary has either one or two fewer dimensions than the state space. A piece of it also threads the cycle.

8.   The phase singularity has nothing necessarily to do with timelessness. It only indicates failure to return to the standard mode of rhythmicity.

In the above models we used unperturbed dynamic equations ($I = 0$ implicitly) to determine the latent phase function of state $\Phi = g(\mathbf{x})\colon \mathbb{R}^2 \to \mathbb{S}^1$. Then we plotted its contour map in state space by plotting the locus of states (the isochron) corresponding to each Φ. We did not inquire into behavior during perturbation ($I \neq 0$). The next section does this and in so doing shows how to transfer the isochrons onto a space of experimental stimuli, which is our only window into the state space of a system whose mechanism and state variables are still unknown.

B: Perturbing an Attractor Cycle Oscillator

Introduction

In most of the experimental situations gathered into this book we deal with systems that are eventually regularly rhythmic. Because we don't know what complexities of physiological dynamics mediate between the oscillator we wish to study and our more peripheral rhythmic observable, we cannot easily interpret any observations preceding establishment of that standard rhythmic state. But once that condition is achieved, a phase displacement of the observable rhythm may be taken to reflect an equal phase displacement of the underlying oscillator. In other words, in the last analysis our only reliable observable is the oscillator's latent phase. This function of its state was deduced from sample dynamic equations in Examples 1–4. The states of equal latent phase were there plotted as isochrons. Having drawn the isochrons in state space, we can now forget about trajectories and their equations.

But all this pertains only to the standard environmental conditions, the unperturbed condition we denote by $I = 0$.

Perturbed Trajectories

In the presence of a stimulus (by definition), the oscillator's dynamic is altered. Its state no longer moves from one isochron to the next at unit rate. How does it move? We must draw trajectories again as we did in Chapter 5, Figures 3, 4, 6, 8. Nothing of those diagrams is changed in principle in this chapter about attractor cycles. The only new feature introduced in this chapter is that the isochrons are generally wiggly curves instead of straight lines. The topological outcome is the same. Whether we consider cophase θ(the event latency after stimulus) or latent phase Φ(a function of the state reached at the end of the stimulus) or new phase ϕ' (of the poststimulus rhythm, extrapolated back to the end of the stimulus), the graphs above the stimulus plane (ϕ, M) all resemble a lattice of screw surfaces when plotted in the three-dimensional time crystal. Recall also that we may conveniently speak of that three-dimensional graph in terms of its contour map on the stimulus plane. The contours being loci of equal *time* of an event, latent phase resembles a pinwheel *wave* when plotted as a contour map on the stimulus plane.

The purpose of this section is to put those observations in a more general context. This seems to be called for because the qualitative features of resetting behavior seem remarkably independent of many quantitative details of model equations and because in real physiology we seldom have equations anyway, nor do we know how to define the state space in which to draw the trajectories and isochrons that such equations would describe. Can we set aside these hypothetical constructs to deal only with what we can observe? The following argument may provide a step in that direction, to be pursued experimentally in Chapter 7.

We suppose a dynamical system whose state may be represented by a point in $\mathbb{R}^D$. We suppose the system tends to an attracting cycle which is some closed ring in $\mathbb{R}^D$, not necessarily circular, possibly even knotted. The essential feature is only that it has the topological connectivity of $\mathbb{S}^1$. We suppose also that every point $\mathbf{x}$ in that cycle's attractor basin in principle has a latent phase $\Phi = g(\mathbf{x})$, defined on the circle $\mathbb{S}^1$.

Let us suppose that when a stimulus begins at time $t = 0$ the oscillator is somewhere on the attracting cycle. In general it departs from that cycle during the stimulus. After a time Δt, oscillators initially on the cycle have moved to other states as schematically indicated in Figure 5. Assuming continuity and smoothness of the dynamics, those new states at time $t = \Delta t$ also lie on a ring: a distorted version of the ring of initial conditions along the attracting cycle. If the stimulus continues (and so the perturbed trajectories extend to $t = 2\Delta t$) then the set of states arrived at is another ring, further displaced from the attracting cycle. We are defining a two-dimensional surface in the state space $\mathbb{R}^D$. This surface is fibered by the perturbed trajectories, each from a different initial phase ϕ, and,

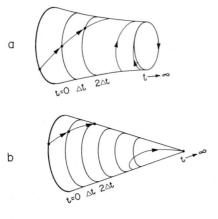

a

b

Figure 5. (a) Points initially on the attracting cycle at time $t = 0$ are "blown" off it during exposure to a stimulus. The trajectory from one initial point is shown with arrows. The ring of all initial points progressively deforms and moves to a new attracting cycle as $t \to \infty$. (b) As above, but the cycle degenerates under continued exposure: The stimulus forces the system to a steady-state regardless of its initial phase. For examples in context of circadian rhythms, see Peterson and Jones (1979).

criss-crossing that ruling, by the rings, each for a different duration t of perturbation. The surface is an image of the stimulus plane (ϕ, t) mapped into state space. Topologically it is a cylinder like a windsock blowing from the attracting cycle. Where it intersects the isochron surfaces in state space we have the isochrons of stimulus space, the loci of stimulus combinations (ϕ, t) that are equivalent in that they evoke synchronously reset rhythms.

Suppose some kind of rhythmicity persists under indefinite prolongation of the stimulus, as is the case for circadian rhythms exposed to dim enough continuous light and for pacemaker neurons biased by a small enough electrical current. Then the rings approach a limiting ring along which the state continues to circulate at a new period. This is the system's attracting cycle under conditions of continual exposure to the stimulus. The set of states reached by such stimuli thus constitutes a cylinder, as in Figure 5(a).

But suppose the perturbing stimulus is such that the system is eventually brought to rest at a unique equilibrium. This is the case for some circadian rhythms exposed to prolonged artificial daylight, and for nerve in a voltage-clamp device and for oscillating glycolysis suppressed by oxygen (the Pasteur effect). In geometric terms all trajectories converge to that rest state, the standard starting state from which trajectories approach the attracting cycle again in permissive conditions. So the two-dimensional cylindrical surface converges to a point, like a cone [Figure 5(b)]. Incorporating that point, the surface is closed. It is a two-dimensional cap bounded only by the attracting cycle (attracting under unperturbed conditions, in the absence of the stimulus). This cap is topologically equivalent to a disk whose boundary $(\phi, 0)$ is that cycle.

Topology to the Rescue

We know a theorem about mapping such a disk to $\mathbb{S}^1$, as we implicitly do in assigning a latent phase $\Phi \in \mathbb{S}^1$ to each stimulus (ϕ, t): the map cannot be continuous unless the winding number of Φ around the boundary $(\phi, 0)$ is 0. But it isn't. By definition, $\Phi = \phi$ around the cycle: The winding number is 1. So Φ cannot vary with stimulus coordinates (ϕ, t) in a continuous way.

What kind of discontinuity might be expected? Since there is no discontinuity along the boundary, it must be a closed locus inside the cap. That locus could be as small as a point at which the isochrons converge. This is our minimum phase singularity. It does not commonly lie at the apex of the cap in Figure 5(b): If it did, the rhythm would start at no definite phase upon reversion to permissive conditions. So in general, the phase singularity can be reached by following some perturbed trajectory from a particular phase, ϕ^* for a particular finite stimulus duration t^*. In other words, regardless of the complexity with which any number of variables may be interacting in a rhythmic system, some point of the phaseless set can be reached by varying stimulus parameters in only two ways.

What does a point in the phaseless set correspond to physically? It is most likely not an equilibrium of the *un*perturbed dynamics. This is because for $D > 2$, any two-dimensional cap of trajectories altogether misses most points (e.g., the equilibria) in $\mathbb{R}^D$. There seems little to be said about it apart from discussion

Box F: Phase Compromise in Chemical Oscillators

One way to construct a cap across an attracting cycle in state space is to perturb the system from the cycle toward a fixed point. We have seen that the set of states so reached necessarily encounters all isochrons and encounters their convergence point in the phaseless set.

There is another way, too. By mixing the chemical oscillators one creates an oscillator of hybrid composition. Suppose the two "parents" are identical but for timing, and both are on their common attracting cycle, and that all their state variables are substance concentrations that quickly reach an intermediate compromise. Then the hybrids initially occupy states on straight lines joining the two parent phases in state space. By fixing one parent's phase and varying the other as well as the volume ratio of the two parents, one can explore a two-dimensional set of states bounded only by the attracting cycle: a cap (Fig. a). This cap necessarily intersects the convergence of isochrons, the boundary phaseless set.

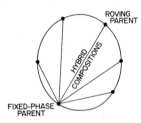

By mixing two copies of a chemical oscillator, of different volumes, at different phases, one achieves chemical compositions intermediate between those mixed. The set of all such hybrids constitutes a cap bounded only by the attracting cycle.

Note that this can be done for any choice of the one parent's fixed phase. Thus in an experiment varying both parent's phases at fixed one-to-one volume ratio the singularity is encountered twice: once as above and once under an exchange of the two parents' phases and relative volumes. (See also Box A of Chapter 22.)

An exception arises in the special case of a plane circular cycle with a repelling steady-state at its center. Then by combining equal volumes of any two geometrically opposite phases, one creates the state at the convergence of isochrons, viz the central steady-state. This phase discontinuity thus appears not as two isolated points but as a one-dimensional curve in the (ϕ_1, ϕ_2) plane. A fuller presentation of the relevant geometry is given in Winfree (1974c). It turns out that the phase singularity occurs along a one-dimensional locus in the three-space of ϕ_1, ϕ_2 and the volume ratio. In the exceptional case mentioned, this locus is a straight line falling exactly in the equal volume plane.

of a specific dynamic system. It is just a point in the attractor basin's boundary through which the unperturbed trajectory, wherever it may lead, does not return to the attracting cycle.

The take-home lesson seems to be that observing a phase singularity in the stimulus plane (observing a helicoid in each unit cell of the time crystal or observing a rotating wave in a pinwheel experiment) in itself suggests nothing very specific about the underlying mechanism. It may be interesting that more complicated phenomena are *not* observed, but the observation of only this much scarcely indicates more than: (a) that the physiological dynamics tends to a unique cycle

Box G: The Dilemma of "The Man Without a Metric"

Figures 1 through 4 assume an underlying state space homeomorphic to the surface on which they are printed. If, as must usually be the case, the actual state space has more than the two dimensions of this paper's smooth surface, we can still get by. We just say that the picture shows only the two-dimensional set of states encountered along trajectories departing under the standard stimulus from the ring of different possible initial conditions. The isochrons are then one dimensional *intersections* of this surface with the higher dimensional isochron surfaces or hypersurfaces of dimension $D - 1$, as in Figure 11. Whatever the interpretation, we are concerned only with the relative geometry of the perturbed trajectories and the unperturbed trajectories. That is what determines how the clock changes phase under diverse combination of stimulus timing.

So someone could come along and deform the underlying state space arbitrarily, stretching it here, twisting it there, compressing it elsewhere. It doesn't matter so long as the deformations are merely quantitative and do not violate the topological connectivity of the space. (In other words, this is rubber sheet geometry, in which scissors, pins, and razor blades are disallowed.) So long as the perturbed and unperturbed trajectories (and therefore the isochrons) are deformed together, the resetting behavior is unaffected. Thus the resetting behavior of any simple smooth kind of oscillator can be obtained from much more complicated oscillators, even from oscillators in which things change abruptly and trajectories shoot across great distances at lightning speed, as in relaxation oscillators. To argue that a pattern of resetting behavior indicates a quantitative feature of the underlying oscillator, it is necessary to know a priori how that oscillator works either in the presence or the absence of the chosen stimulus.[2] Resetting behavior reveals only a comparison between perturbed and unperturbed dynamics.

This simple fact seems to me to invalidate any inference from dynamics observed experimentally to the many kinds of dynamic model or metaphor that populate the *Journal of Theoretical Biology*, the *Bulletin of Mathematical Biophysics*, *Biological Cybernetics*, the *Journal of Mathematical Biology*, etc. Quantitative inference works only in the other direction, i.e., from an assumed mechanism to the expected resetting behavior. Departures from such behavior, if large enough, might be taken to exclude the underlying hypothesis about mechanisms. But concurrence with observation proves nothing.

On the other side of the coin, certain kinds of resetting behavior cannot occur unless the state space has the right connectivity and enough dimensions to allow trajectories the required elbow room in which to maneuver. Thus certain qualitative patterns of resetting behavior (e.g., negative slope in a resetting curve) suffice to exclude certain topological categories of mechanism (in the same example, mechanisms based on modulation of angular velocity as the only state variable). And some conspicuous structures that must occur in $\mathbb{R}^D$, such as a phaseless set threading the attracting cycle, need not occur in spaces of other topology, such as the cylindrical state spaces of rotating machinery [Andronov, et al., Chapter 7 (1966)].

---

[2]   Conditions are more favorable in phase compromise experiments with chemical oscillators. When two volumes of reaction at distinct phases are mixed, each perturbs the other in a known and simple way, viz taking a weighted average of concentrations.

**Box H:	Resetting Curves For Multiple Stimuli**

Campbell (1964) initiated the practice of analyzing entrainment of biological rhythms in terms of regular alternation between two environments, and thus between two dynamics. (He had in mind sychronization of the mitotic cycle by rhythmic alternation between two growth temperatures.) This technique is useful not only for his simple clock, but also for any oscillator that quickly recovers to an attracting cycle in one of the two environments (e.g., during absence of the stimulus). Phase control of such an oscillator by a rhythmic stimulus can be analyzed by calculating the impact of a single (repeated) exposure to that stimulus. This was first spelled out by Perkel et al. (1964) in context of neural pacemakers driven by a volley of action potentials and by Pittendrigh and Minis (1965) in context of circadian clocks driven by a daily light pulse.

We are thus equipped to ask, "What is the winding number W_n of the phase resetting curve induced by a sequence of n regularly spaced stimuli?" This resetting curve tells how well an arbitrarily phased oscillator is synchronized to the entraining rhythm after the nth pulse. So far as the topological type of the curve is concerned the answer is conveniently simple: $W_n = W_1^n$. To see why, reason as follows. As we vary ϕ_0, the phase at which the first stimulus encounters the oscillator, we change the phase ϕ_0' at the end of that first stimulus. As ϕ_0 varies through one cycle, ϕ_0' varies through W_1 cycles. The next stimulus arrives after a fixed interval T at phase $\phi_1' = \phi_0' + T$. For each cycle through which ϕ_1 is varied, ϕ_1' and therefore ϕ_2 varies through W_1 cycles. Thus ϕ_2 varies through $W_1 \times W_1$ cycles per cycle of ϕ_0. And so on.

Thus if W_1 is 2 or -3, let us say, then W_n can be arbitrarily large. In other words the average slope of the resetting curve (to a volley of repeated stimuli) can be arbitrarily large.

On the other hand, if $W_1 = 0$ or 1, then $W_n = 0$ or 1. This doesn't guarantee small resetting curve slopes, but a small average slope is a necessary precondition for small slopes. Interestingly, these are the only two resetting types thus far observed in nature, so far as I am aware.

---

when unperturbed; (b) that it tends to a steady-state under the stimulus used; and (c) that both these processes behave in reasonably smooth ways.

In fact it turns out that convergence to a steady-state is by no means necessary: Displacement toward a new attracting cycle as in Figure 5(a) can serve the same purpose. The same result also emerges from an entirely different style of perturbation (see Box F). Nor is smoothness of the dynamics really necessary (see Section C and Box G). In fact, the same behavior has been found *even* in oscillators which are not describable by finite-dimensional state spaces (Johnsson and Karlsson, 1971). It should also be noted that our interpretation of the t coordinate as stimulus *duration* is more restrictive than necessary. First of all, the stimulus need not be a fixed parameter change. If the stimulus program is a concatenation of simpler stimuli as in Box H or a wildly time-varying schedule of influences, it still drives the system's state along some path. The only geometric change induced by this generalization is that such paths can now intersect each other and themselves. That only means that the (ϕ, t) surface may be folded and may pass through itself, but this is of no concern to our topological results. t may also therefore be interpreted as dosage of a substance or magnitude of a perturbation delivered all at once, which only gradually dribbles into the dynamic system. As in Chapter 5, M here refers to stimulus magnitude, thus generalizing the usage in prior chapters where a stimulus was described as lasting for *duration M*, from $t = 0$ to $t = M$.

Oscillators of Greater Complexity

Given at least three state variables to play with, one can obtain rather complex patterns of phase response. Depicted as isochrons on the stimulus plane, these patterns typically amount to arrangements of singularities smoothly connected together as in Figure 6. Note that in the M ranges indicated, the dependence of ϕ' on ϕ can be type $-1, 0, 1, 2$, etc. In the simpler models considered up to now we have seen only types 0 and 1. Let me explain briefly how these things can happen.

In all the following, we will assume as a stimulus some fixed change in the oscillator's dynamics. This is applied for a while then removed. During such a stimulus the rate of change of state is the same at any given state no matter how or when that state was reached. (This is the *definition* of state, not an assumption.) The trajectories in state space therefore cannot cross through each other. We saw in Chapter 3 that simple clocks respond to such stimuli with type 1 resetting. In Chapters 4 and 5 we saw that by introducing a second variable of state we can also obtain type 0 resetting. The "type" is the winding number of latent phase around the ring along which all the perturbed trajectories end, starting from the attracting cycle. This ring, being a distortion of the cycle under nonintersecting flow in two dimensions, either encircled the phaseless set with winding number 1 (as at stimulus duration 0) or was pushed beyond the phaseless set to obtain winding

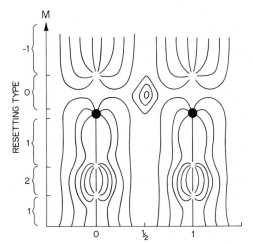

Figure 6. Isochrons as they might appear in the (ϕ, M) plane of initial phases and stimulus magnitudes. The trajectory followed as M increases from initial phase 0 passes through a succession of encounters with phaseless sets. At these critical M's, the resetting "type" changes by ± 1. See also Chapter 2, Box B.

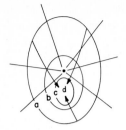

Figure 7.   Rings a, b, c depict states reached during flow toward point d from the attracting cycle of an oscillator with only two independent degrees of freedom (the two-dimensional analog of Figure 5(b)). The radial strokes indicate isochrons. Rings a and b wind once through the cycle of isochrons. Ring c has winding number 0. These are the only two choices for non-self-intersecting flow in the plane.

number 0 (as in the limit of approach to equilibrium under prolonged exposure to the stimulus) (see Figure 7).

What if there are more degrees of freedom than 2? Can other resetting types be obtained? First, note that invoking additional degrees of freedom might be the same as invoking a more complex pattern of perturbation: A constant stimulus such as considered above might appear to an oscillator as a time-varying pattern of interference in its operation if there are internal degrees of freedom that slowly move in response to the fixed stimulus and, as they change, affect the working of the oscillator. In physiological systems it would be quite astonishing to find only two state variables and the corresponding elementary resetting behavior. Up to now only such simple resetting behavior has been found in the several systems examined in detail. But the observations are all crude and are less than a decade old, so it seems to me timely to indicate in broad strokes some other behavior that undoubtedly awaits discovery.

To begin, recall that in a two-dimensional state space the phaseless set is at least an attracting or repelling steady-state point and that in a three-dimensional state space the phaseless set is at least the extension of that point into a one-dimensional curve. This curve consists of the two trajectories leading toward or away from the steady-state.[3] This phaseless set is the convergence point for all the isochrons, arranged in order around the cycle. It threads the cycle. Latent phase has winding number 1 around the phaseless set along any closed path near the attracting cycle. We will now proceed by considering how oscillators initially on that cycle might be moved to new states under a stimulus. We thus consider distortions of a ring of initial conditions assumed to coincide with the cycle at stimulus magnitude $M = 0$. Figure 8 suggests how a stimulus can change this winding number from 1 to 0. Viewed in projection along the phaseless locus, Figure 8 would look a lot like Figure 7. Figure 9 suggests how a somewhat different geometry of change during the perturbation could induce a winding number of 2, i.e., type 2 resetting. Figure 10 suggests another alteration of the flow during perturbation which would result in winding number 0 and then, with a further prolongation of the stimulus, would result in winding number -1. In fact any of the integer winding numbers, all of which are allowed so long as we consider only the abstract logic of phase resetting, can also be achieved by appropriately contrived smooth dynamical flow involving no more than three state variables.

---

[3]   Which must exist, supposing the trajectories point inward from infinity, as they must in all real chemical systems (Wei, 1962).

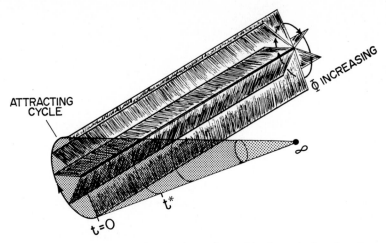

Figure 8. Figure 5(b) is elaborated to include pieces of several isochrons to show how they converge along a curve which threads the attracting cycle from which the cone hangs like a wind sock. Isochrons acquire their phase labels in the order of intersection with states along that cycle at $t = 0$. At any $t > 0$ the ring of states reached either does or does not encircle this locus. The resetting type is the winding number of this ring around the phaseless set.

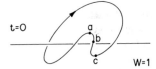

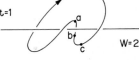

Figure 9. During perturbed dynamics, oscillators initially on the attracting cycle at $t = 0$ are displaced only slightly (mostly in region abc) to achieve $W = 2$. The horizontal line is supposed to be the convergence locus of isochrons, i.e., the phaseless set.

Figure 10. Suppose the dynamical flow during the stimulus is a clockwise rotation about an axis indicated by the dot, perpendicular to the page. Then a ring of states initially with winding number $W = 1$ around the phaseless locus next converts to $W = 0$ and then to $W = -1$.

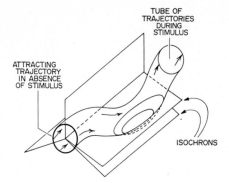

Figure 11.   Similar to Figure 8, but the tube of displaced rings of initial conditions (the (ϕ, M) surface of Figure 6) intersects some isochrons in closed rings (the ring isochrons of Figure 6). This can only happen in a state space of at least 3 degrees of freedom.

Figure 8 shows not just distortion of the initial ring of states achieved at the end of the stimulus, but the whole continuum of progressive distortion, comprising the cylinder of states that can be reached by stimuli of increasing magnitude beginning at any phase on the cycle. This can be made to intersect the phaseless set any number of times in a great variety of ways as in Figures 9 and 10. The two-dimensional surface in Figure 8 is a mapping of the stimulus plane into state space. So isochrons on the stimulus plane (ϕ, M) can in principle converge to any number of phase singularities of either clockwise or anticlockwise orientation. No one has yet observed such behavior experimentally[4] but I feel confident that it only awaits the investment of labor. R. Krasnow (unpubl. ms) argues that almost any circadian oscillator should exhibit multiple phase singularities in the stimulus plane for very dim light at durations comparable to the circadian period.

In Figure 11 we switch emphasis from the phaseless set to examine how a single isochron surface in state space might intersect the (ϕ, M) cylinder. The geometry of three dimensions allows intersections along a closed-ring locus with the consequence that isochrons in the stimulus plane can present the appearance of a bull's eye, a nest of concentric rings as in Figure 6. This is impossible in the case of only 2 degrees of freedom.

C: Unsmooth Kinetics

In deriving the existence of phase singularities from simple observations in Chapters 1 and 2, we relied on topological arguments. We have now seen that phase singularities are also implicit in a variety of dynamical models. In both cases smoothness (in space, in time, in state space) was an indispensible ingredient of the argument. But in some kinds of data and in some kinds of dynamical model, things do not change so smoothly; things sometimes change quick as lightning. The theory of discrete state automata, catastrophe theory, and the like provide an abundance of intoxicating perspectives on processes involving abrupt change,

---

[4]   Note added in proofs: Jalifé et al. (1980) have demonstrated a sequence of resetting types $1 \rightarrow 0 \rightarrow 1$ with increasing stimuli in the sinus node of the kitten's heart. I thank Stuart Hastings for the observation that no third variable is required for this: a concave arc in the planar limit cycle is sufficient.

but my purpose here is only to examine the discontinuities implicit even in smoothness. Accordingly, the remainder of this chapter provides no more than a brief survey of the role played by quick changes of pace in the experimental systems most involved in this volume.

The main points I would submit to your attention in these examples are:

1.   Systems exhibiting abrupt changes have commonly been described by discontinuous models. Most commonly, these models involve only a single internal variable of state. They are essentially simple clocks with a discontinuity in their rate functions. Such models have no place for phase singularities.

2.   Nevertheless, the experimental systems in question commonly turn out to exhibit one or more phase singularities.

3.   A big improvement of realism in modeling is achieved by recognizing the necessary existence of at least one other internal variable of state. This at least makes it possible to consistently describe a dynamical mechanism without invoking magic at the discontinuity. For examples in biophysical contexts see van der Pol and van der Mark (1928), Fitzhugh (1960, 1961), Katchalsky and Spangler (1968), Franck (1968), Lavenda et al. (1971), and Linkens et al. (1976). In more general context, see Chapter 8 of Andronov et al. (1966) and Chapters 26 and 31 of Minorsky (1962).

4.   Such models are topologically equivalent to their smoother relatives and their phase behavior is qualitatively the same as that of smoother models except that a lot of it is compressed into the very brief interval during which lightning changes transpire.

5.   Such models often exhibit phase singularities similar to those considered in prior chapters. But sometimes no singularity is evident because it is squeezed into the brief interval of rapid change, where data resolution is particularly poor.

Let us now turn to some examples in order of their appearance in the Bestiary.

Relaxation Oscillators

The neon glow tube used in the coupled oscillator population of Chapter 11 provides an example of a relaxation oscillator. It works by accumulating a voltage on a capacitor. The voltage rises toward a limit. If the voltage exceeds a threshold, then the tube discharges before reaching the limit, and the cycle restarts. If the threshold exceeds the limit, then the tube waits to fire until stimulated and *then* restarts its accumulation. This caricature invokes arbitrarily rapid change in one state variable at a certain phase in the cycle, in every cycle. In ring devices it occurs only in returning toward the cycle after a disturbance (Box A). These are two different limiting cases of a smooth attractor cycle, both of which eliminate all but one state variable (corresponding to phase) by turning the others into a seldom invoked "*deus ex machina*". Such approximations usually do violence to the model's accuracy in situations other than routine circulation along the attracting cycle.

The Malonic Acid Reaction

Though the kinetics of any reaction are necesarily continuous, certain steps can go very much faster than others. In the malonic acid reaction (Chapter 13) the transition from red to blue, with all that it entails, is so abrupt that we may as well think of it as a discontinuous change at a threshold, or so it would seem. But cutting corners in this way leads to contradictions when we come to consider reaction geometries in which phase has a nonzero winding number around some ring. This is because such a ring necessarily contains a phase discontinuity, as we have often seen. If phase is identified with some chemical concentration, that means a concentration discontinuity. But that is impossible in a physically continuous medium. So it is necessary to retain the more complex description involving rapid but continuous changes in two or more mutually regulating reactions.

The malonic acid reaction can be thought of as a relaxation oscillator whose primary variable of state is bromide concentration. Bromide concentration falls exponentially as bromide is consumed. At a threshold, it is recreated during a momentary explosion of oxidative activity. But within this small fraction of a second, a second variable of state $HBrO_2$ increases 100,000-fold (according to the carefully fabricated kinetic scheme of Field et al. (1972)). Hastings and Murray (1975) showed an attracting cycle implicit in this scheme. Vavilin et al. (1968) also showed that the system can be phase-shifted without permanently altering the reaction's parameters, using a pulse of ultraviolet light. Thus the malonic acid reaction presents us with all we need in order to find a phaseless manifold: an enterprise that makes no sense in terms of single-variable relaxation oscillator models. By locating the singular combination of dosage and timing that puts the system in its phaseless manifold, one might discover whether the steady-state is locally an attractor or a repellor. However, such matters have not yet been pursued in the laboratory so we don't really know how much of the topological logic of resetting actually will prove useful in an oscillator with such stiff kinetics.

Electric Rhythms in Cell Membranes

Relaxation oscillators are widely used to depict nerve kinetics. In fact it was in context of neural rhythms that A. V. Hill (1933) initially declared that all biological oscillations are of this sort, a remark echoed in contemporary papers even as sophisticated as Nelsen and Becker (1968) and Linkens et al. (1976). In the latter paper the relaxation oscillator is really continuous, following a tradition instigated by van der Pol (1926), in which a periodic jump can be made as swift as desired by choosing a parameter in a differential equation. But another tradition makes no compromises with continuity. According to this interpretation of pacemaker activity in neurons, the membrane voltage slowly falls as sodium leaks in. Meanwhile the threshold for sudden increase of sodium permeability gradually rises, recovering from a previous action potential. When the two meet, an action potential occurs and both are discontinuously reset to widely separated values, resuming their approach to equality. For example, see models in Perkel et al., 1964; Rescigno,

1970; Knight, 1972; Fohlmeister et al., 1974; Peskin, 1975; Hartline, 1976; Pinsker, 1977; Allessie et al., 1976, 1977; and Glass and Mackey, 1979. This kind of model of a neural oscillator resembles a leaky bucket in which a single quantity accumulates until the bottom springs open momentarily, after which accumulation resumes. In such models there is no place for type 0 resetting nor for rotation about a pivotal phase singularity.

Nerve kinetics has also been caricatured in terms of the other kind of single-variable discontinuous kinetic scheme, the ring device. The first well-developed model came from Wiener and Rosenblueth in 1946. Wiener envisioned the heart cell as having an excited state followed by a prolonged refractory period, which terminates in a resting state. With the resting state and the excited state identified as adjacent, and the refractory state portrayed as an interval along which state advances at a steady rate, the model resembles a ring device. Krinskii (1968) stretched the excited state out to a finite duration to recapture some phenomena lost in Wiener's more ascetic simplification, thus anticipating experimental results which later induced a similar amendment of the Hodgkin-Huxley equations for the heartbeat (McAllister, Noble, and Tsien, 1975). Those two early papers led into substantial mathematical developments about the behavior of ring devices in sheets, spatially coupled to admit the possibility of wave propagation as in the heartbeat, in epileptic seizure, and in spreading depression (see Chapter 14). One upshot is that there cannot be a stably rotating wave of excitation.

The rotating wave and type 0 resetting are both admitted only when a second variable is admitted. It is *not* necessary to exclude discontinuous kinetics though. For example, Zeeman's (1972) "cusp catastrophe" model of nerve membrane and heart muscle, whatever its other defects, does support a rotating wave, given suitable quantitative adjustments (Winfree, 1973c).

The fast kinetics of the threshold process is described in continuous detail by the Hodgkin-Huxley equations using *four* variables of state. Along most of the cycle two variables are changing slowly and the other two faster variables behave almost like instantaneous functions of the slow pair. They exhibit interesting behavior only during the moment of the nerve's firing, which is reduced to instantaneous discontinuity in the more compact approximations.

Best in his dissertation (1976) inquired whether type 0 resetting and the phase singularity are implicit in the Hodgkin-Huxley equations. Best computed resetting curves for a periodically firing squid axon using the Hodgkin-Huxley equation with parameters as measured on squid. He used a current bias to induce oscillation, described by an attracting cycle in the four-dimensional state space. One of the state variables is membrane voltage. At various times in the cycle, Best increased or decreased this voltage by some fixed amount by charging or discharging the membrane capacitance. He then continued the computation until the membrane had returned close enough to its attracting cycle so that its phase was increasing very nearly uniformly in time. Extrapolating back to the moment of the voltage impulse, he plotted the new phase just after the stimulus as a function of the old phase at which the stimulus was given. For stimuli of middling magnitude, either positive or negative, the resetting curves looked discontinuous. But by enormously increasing the time resolution in the ostensibly discontinuous region, Best was able to show that the curves are in fact continuous, and that they change topology from type 1 to type 0 at a critical pulse magnitude.[5]

According to the Hodgkin-Huxley equation the singular depolarizing (excitatory) pulse and its actual discontinuity of phase are found in midcycle, between action potentials. At nearby stimulus magnitudes the resetting curve has a region too steep to distinguish experimentally from a discontinuity. For hyperpolarizing (inhibitory) pulses, this lies near the end of the cycle, just before the action potential. Suggestively similar experimental results appear in Moore et al. (1963, Fig. 14), Perkel et al. (1964, Fig. 2A), Schulman (1969), Jalifé and Moe (1976, Figs. 5 and 9), Pinsker (1977, Fig. 5E-1 Sec.), Ayers and Selverston (1979, Figs. 5, 6, and 11), Jalifé and Antzelevitch (1979), Jalifé and Moe (1979, Fig. 14) and Scott (1979). Discontinuities are not commonly associated with the action potential itself. Scott (1979) argues that they represent encounters with the separator emanating from the singularity (see Box C, pp. 153–154).

Best's result also illustrates the importance of using a continuous causal model for real dynamic systems, however quick some stages of their dynamics may be. When we settle for a threshold relaxation caricature of membrane kinetics, we invoke magic at the threshold to bring about the necessary changes. All questions of process and of causation are obviated by collapsing that critical interval to a single structureless instant. After resorting to such an approximation, it seems difficult to understand the type 0 resetting observed in pacemaker neurons, in which the Hodgkin-Huxley cycle is evidently very much smoother than it is in squid.

Biochemical Excitability in Slime Molds

A sheet of aggregating social amoebae behaves as an excitable medium capable of relaying a pulse of cAMP (see Chapter 15). The triggering of cAMP synthesis and release is abrupt. Nonetheless, pacemaker cells exhibit type 0 resetting. And sheets of cells support rotating waves, pivoting around a phase singularity that later becomes the assembled animal (Clark and Steck, 1979).

Circadian Clocks

Relaxation oscillators have been invoked as models for the mechanisms of circadian clocks since the earliest days (e.g., Pittendrigh and Bruce, 1957, p. 84; Bunning, 1960 and 1964; Roberts, 1962; and Pavlidis, 1967a). A recent addition to

---

[5] Because Best chose parameters such that the steady-state is locally an attractor competing with the attracting cycle, the geometry was not quite so simple in detail. Resetting curves too close to the critical pulse magnitude are not either type, because each has a zone of old phase in which new phase is undefined. This happens because the system falls into the locally attracting steady-state after any such stimulus. Only along the boundary of this attractor basin does new phase behave in an honestly singular way. A slight adjustment of the biasing current would presumably eliminate this peculiarity without affecting much else.

this literature by Rössler (1975) ties the relaxation oscillator notion to a chemical mechanism for temperature independence. But the reasons for adopting discontinuous models seem to me less than compelling. They are:

1.   Tension discharge models are especially easy to visualize and to test experimentally.

2.   Resetting curves, when plotted in terms of phase shift, commonly suggest a discontinuous change from one-half cycle advance to one-half cycle delay. This "phase jump" was thought to be the "discharge" of a physiological accumulator.

Because smooth attractor cycle kinetics lead to the smooth screw surface actually observed in the time crystal, I rejected the relaxation oscillator idea (Winfree, 1970b). That was wrong too. Only the single-variable discontinuous caricature is excluded by this observation. Nonsmooth kinetics can readily result in smooth resetting behavior (see Box G), just as smooth kinetics can result in nonsmooth resetting behavior (e.g., Figure 9 of Chapter 4).

The Cell Cycle

Extensive resetting measurements have been conducted using the 10-hour cycle of nuclear division in the acellular slime mold (Chapter 22). These and other experiments have led many to believe that cell division is governed by a cyto-chemical oscillator which once per cycle triggers the sequence of processes required for DNA synthesis and nuclear fission. Is the oscillator a relaxation oscillator, constrained to a fixed cycle with an abrupt jump in it? Several mechanisms of this sort have been proposed (e.g., Rasmussen and Zeuthen 1962; Rusch et al., 1966; Fantes et al., 1975). Or might it be a feedback process that functions like a smooth attracting cycle? Several models of this sort have also been proposed (Selkov, 1970; Burton and Canham, 1973; Gilbert, 1974; Kauffman, 1974; Tyson and Kauffman, 1975). These two classes can be distinguished by examining their resetting behaviors even prior to learning more about the physical mechanism. Tyson and Kauffman (1975) and Kauffman and Wille (1975) have tried to put *Physarum* in a position to reveal the smoothness of its cycle using such measurements. Unfortunately, *Physarum* doesn't cooperate. Most experiments up to now show a sharp discontinuity in resetting behavior less than a hour before mitosis (Sachsenmaier et al., 1972; Tyson and Sachsenmaier, 1978:) The particular simplicity of the discontinuity observed in *Physarum* strongly recommends a model in which the concentration of a mitotic inhibitor or activator abruptly doubles or halves when receptor sites on the DNA are doubled during replication.

More fundamentally discontinuous models have been elaborated. Kauffman (1969), for example, envisions the cell cycle as a sequential machine not unlike a digital computer, in which each configuration of gene activity induces a unique next state until the whole dance has worked through to a new beginning. Nothing like a phase singularity would be expected in such a case, either. In fact there seems to be no compelling evidence to date that the cell cycle need be described by anything more than the one dimension of phase.

Box I: Alternatives to the Limit Cycle Interpretation
of Biological Clocks

Kalmus and Wigglesworth (1960) provided one of the first applications of the qualitative theory of ordinary differential equations to understanding the behavior of circadian rhythms. This model invoked a pair of limit cycles, one attracting and one repelling. Its architects were aware that they were instigating a ccnceptually important departure from the then widespread implicit assumption of simple-clock dynamics: "We hope that they may serve us better than the alarm clock which our children hear ringing from the inside of Peter Pan's crocodile, and which has misled many of us in the past." This hope was well placed, as we shall continue to see in subsequent chapters. Nonetheless we shall also see that the limit cycle metaphor fails in several important ways.

First, in the real world one never does get to the limit. What if recovery to the attracting cycle takes much longer than the typical interval between successive stimuli? Then the existence of the limit has no more than academic pertinence. It is therefore important to measure the rate of recovery. This is undertaken for *Drosophila*'s circadian clock in the next chapter.

Second, it may not be usual to find a single isolated physiological oscillator. More commonly they come in pairs or in larger populations, interacting at least weakly. Two mutually synchronized oscillators have at least one attracting cycle, and generally several alternative attracting cycles in a space of twice the single oscillator's number of dimensions. In principle this case may still be regarded in terms of dynamics around an attracting cycle, but that may not be too helpful in situations involving significant departures from the cycle because then the complexities of flow in high-dimensional space can become quite unmanageable.

Third, it is not obvious that real physiological systems commonly have a limit cycle at all, though they may nonetheless be attracted into a ring-like volume of states. It is entirely possible for a system involving three or more variables to meander aperiodically without ever converging either to a stationary state or to a unique attracting cycle. Laboratory examples of such irregularly periodic dynamics are just beginning to appear in the research literature of 1977. Many of the theoretical examples, notably due to Rössler, suggest that the chaotic dynamics of so-called strange attractors readily arises in connection with slightly perturbed oscillators and pairs or populations of mutually interacting oscillators.

It seems particularly ironic that most of this chapter was developed in an effort to account for the reaction of *Drosophila*'s circadian clock to a single phase-resetting stimulus. But more recent experiments gave scant support to the notion that *Drosophila*'s eclosion clock has an attracting cycle; if it does, it attracts much more slowly than anyone had imagined. More recent theoretical models of this circadian clock tend to emphasize its resemblances to a population of oscillators whose collective behavior is not dominated by an attracting cycle. These matters are pursued in Chapters 7 and 8.

Rhythmic Ovulation

The female cycle (Chapter 23) has usually been described in terms of sudden discrete changes of state (e.g., ovulation, hormone levels transgressing thresholds, etc.) which eventually recreate a prior state. The early mathematical model of Danziger and Elmergreen (1957) is typical in this respect. In such a case the cycle's rephasing by exogenous hormonal stimuli would be expected to reflect these discontinuities. This is the case in experiments on Bogumil's numerical summary of the endocrine kinetics thought to underlie the human menstrual cycle. Estrogen infusions of various sizes applied at various phases do reveal stark discontinuities in the model's response (Bogumil, pers. comm.). If there are any phase singularities, they are well hidden in these discontinuities.

The only complete resetting curves presently available from living mammals were obtained with such potent synchronizing stimuli that little can be inferred about the smoothness or discontinuity of the cyclic mechanism (see Chapter 23).

Conclusion

In respect to phase resetting and patterns of phase in space, there seems to be little qualitative difference between attractor cycle oscillators with smooth kinetics and others which change state abruptly during the normal cycle. Resetting behavior is really discontinuous only if the state space is one-dimensional or if the attracting cycle has certain ungeneric exact symmetries (e.g., Box F). However a phase singularity can be caught up in an interval of sudden change, where the time axes are, as it were, very compressed. This can result in such severe distortion of the bundle of converging isochrons that one sees only stark discontinuity. This can happen, but need not, and apparently does not in some of the experimental systems referred to above.

These last remarks are not just theoretical, but conjectural as well. With a parting reference to Box I it now seems appropriate to depart from the extremes of theoretical abstraction approached in this chapter. We will next pursue operational definitions and experimental methods for testing the utility of isochrons and allied notions in connection with a particular physiological system. In the next chapter I present such an attempt using *Drosophila pseudoobscura*'s circadian clock.

7. Measuring the Trajectories of a Circadian Clock

First get your facts; and then you can distort them at your leisure.

Mark Twain

A: Introduction

If, in context of real laboratory experiments, we wish to seriously contemplate models with more than one degree of freedom, then we must find two or more independent empirical measures corresponding to the movements of the system in its state space. We must seek to plot a trajectory in a space of two or more measureable quantities. If we can find a way to do this, then we can distinguish the quickly attracting cycle of Chapter 6 from the orbitally stable kinetic schemes of Chapters 4 and 5.

In mechanical systems such as the pendulum of Chapter 5, position in space and its rate of change (the velocity or momentum) comprise a natural choice of coordinates for the state space. In chemical systems, this kind of plot is much less natural because reaction rates depend only on concentrations, not on their rates of change. Ideally we would monitor two or more concentrations or functions of the concentrations and plot these against each other to measure a trajectory. Observations in this style have been collected by Ghosh and Chance (1964), Betz and Chance (1965b), Betz (1966), Degn (1969), Betz and Becker (1975a), Blandamer and Roberts (1978), and Wegmann and Rössler (1978).

Lacking such complete information, people have resorted to plotting any single accessible observable against its rate of change. The hope is that both quantities—the observable and its rate of change—are reasonably smooth, independent functions of the state variables (concentrations). If so, then a plot of the one function of state against the other might constitute a recognizable, though distorted, image of a two-dimensional projection from the state space.

The least risky attempt of this sort is a plot of concentration against its rate of change in a system believed to be well understood and to involve only two important variables of state, one of which is the measured concentration.

A more adventuresome attempt with the same procedure might use a system believed to importantly involve interactions among *several* quantities. Trajectories of this kind are familiar to neurophysiologists using membrane voltage (one of the four coordinates in Hodgkin's and Huxley's state space, essentially a ratio of inside/outside ion concentrations) as the only direct observable (e.g., Jenerick 1963; Gola, 1976; Pinsker, 1979; and Guttman et al. 1979. See Chapter 14 for background).

More precarious is the attempt to use a conveniently observable function of state without the support of any well-tested idea of what the basic variables of state might be or what kind of function of state the observable might be. In this category we might consider any circadian rhythm in which some quantity can be monitored continuously. For example, the openness of *Kalanchoe's* flower is such a quantity. A plot of openness against its rate of change winds 'round and 'round a center, with variations of amplitude, at reasonably uniform angular velocity. Such plots provided the first direct measurement of "isochrons" (see Chapter 21).

Still more perilous is the effort undertaken in this chapter to use an observable that is not even defined on the same kind of space as are the presumed variables of state. The observable most natural to our interest in things rhythmic is the phase of a circadian rhythm. Of course that makes a rather dull plot since phase, being defined in proportion to elapsed time, has rate of change identically 1. But persevering in this preoccupation with phase, we might instead take the reset phase of a rhythm as our observable and plot this value against its rate of change as we vary the time at which a stimulus is given. Those two observations presumably constitute two independent measures of the unperturbed clock's state. To put it another way, we do as in the previous paragraph, but our rhythmic observable is "the phase to which the rhythm would be reset if perturbed right now in some standard way." We plot that against its rate of change. This is simply the familiar rhythm of sensitivity, the resetting map, plotted against its slope.

The labor involved in such an effort is nearly prohibitive. Why go to the trouble? What will be gained by it?

The result will be some kind of distorted look into a circadian oscillator's dynamical space: the first ever obtained. In principle it might reveal any kind of un-anticipated dynamical complexity, any unforeseen pattern of organization among attracting and repelling steady-states, attracting and repelling limit cycles, thresholds, and separatrices, etc.: No one has ever before examined the dynamics of a circadian clock for its essential, qualitative features. In fact, as you will see, what emerges is startlingly simple and disconcertingly unlike the relaxation oscillator model and quickly attracting limit cycle models previously entertained.

Such an experiment requires thorough automation. The remainder of this chapter mostly presents the results of one such experiment, using *Drosophila's* circadian rhythm of pupal eclosion and my University of Chicago "time machine". (See Winfree, 1973a and Chapter 20 for background.) The analysis sketched here was presented verbally at the Biophysical Society meeting in Philadelphia, February 1975; the graphs were distributed at the Dahlem Conference on the Biophysical Basis of Circadian Rhythms in Berlin in November 1975; both were presented together at the Circadian Clock Symposium in La Jolla in January 1976; but they have not previously appeared in print.

B: The Time Machine Experiment

It is a capital mistake to theorize before one has data. Insensibly one begins to twist facts to suit theories, instead of theories to suit facts.

Sherlock Holmes,
"A Scandal in Bohemia"

Background

To put the forthcoming trajectories in context of our earlier encounters with *Drosophila*, let me remind you of the pinwheel experiment and its resulting time crystal (Chapter 2, Chapter 20). The pinwheel experiment was contrived to distinguish the clock's states by keeping it for a while in one environment (constant dark or red light up to a time T or phase ϕ) and then for a while in another (constant blue light for a duration M). Eclosion timing then provided one measure of the clock's state at the end of that treatment. The measured dependence of the resulting cophase, θ, on T and M was depicted in two equivalent ways:

1. As a lattice of screw surfaces in which a screw axis is the only conspicuous irregularity, and

2. As a contour map or a wave rotating on the stimulus description plane, in which the pivot corresponds to the screw axis at $T = T^*$, $M = M^*$).

This θ provides *one* measure of the clock's state at the end of the stimulus. However, as we saw in Chapter 3, single-variable models proved inadequate to account for all the facts of phase-resetting. So we need to discriminate some additional way the clock's state can change. We thus need to figure out how to assay the state of the clock as it follows some path possibly off its "normal" cycle. We can then ask whether the path observed is or is not the same old usual cycle. If not, how quickly does it return to the standard cycle? Or does it noticeably approach the standard cycle at all? In performing this exercise I will abjure theory and stick consistently to facts taken from a single circadian system, viz *Drosophila pseudoobscura* eclosion.

It seems appropriate to start by defining observables in strict operational terms. How can we monitor the state of a circadian clock without knowing what a circadian clock *is*? We don't know the clock's biochemical state variables, but we do have some observables that may qualify as *functions* of state (Figure 1):

A.   For a first measure take the time from t hours after the initializing stimulus (T, M) until eclosion in continuous red light or darkness. So long as eclosion occurs in reasonably narrow unimodal peaks, the center of mass of the peak (the daily average eclosion time) will adequately characterize the interval from t to t_{eclosion}. That measure of the clock's state $\mathbf{x}$ at time t will be called $\theta(\mathbf{x}(t)) = t_{\text{eclosion}} - t$,

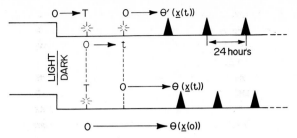

Figure 1.   Defining θ and θ' (θ'' is $d\theta'/dt$). The horizontal time axis starts with transfer of pupae to darkness (downstroke), then leads through a stimulus after T hours and the assay pulse (or its absence) after t more hours, then ends in a sequence of eclosion peaks about 24 hours apart.

referring to eclosion in the singly pulsed pupae. Because this is merely proportional to time the after the initializing stimulus, it is not an especially informative measure of clock behavior after that stimulus.

B.   For a second measure, do as above but also give the organisms a standard exposure of blue light at time t, prior to leaving them undisturbed in red light. Because of the blue light exposure, which we call an "assay" pulse, the time to eclosion turns out to be different. Call that revised time $\theta'(\mathbf{x}(t)) = t_{\text{eclosion}} - t$, referring to eclosion in the doubly pulsed pupae. Two details:

1.   The assay pulse should be strong enough so that variations in its duration or intensity do not cause the resulting eclosion times to vary much. In other words the assay pulse should saturate the phase response. This will prove to be an important technical convenience below.

2.   The blue exposure maps the state $\mathbf{x}(t)$ to a new state $\mathbf{x}'(t)$, so $\theta'((\mathbf{x})t)$ is the same as $\theta(\mathbf{x}'(t))$. This relation will be useful because through it we will be able to measure the mapping $\mathbf{x} \to \mathbf{x}'$ imposed by blue light.

C.   For a third measure evaluate the rate of change with time of the second measure:

$$\dot{\theta}' \equiv \frac{d\theta'(\mathbf{x}(t))}{dt} \equiv \frac{d\theta(\mathbf{x}'(t))}{dt}.$$

(Note that I consistently use the upper dot for the time derivative. The prime has no such connotation in this book.) Let me be clear about the actual operational meaning of the time derivative. Because the measurement has to be done with discrete aliquots of the population of flies, θ and θ' cannot be measured continuously. θ is measured once and θ' is measured in replicate populations at two-hour intervals of t. With θ' plotted against t, we put a smooth curve through the data points by hand, digitize it at intervals of one-fifth of an hour (i.e., at 10 points between actual measurements) and define $\dot{\theta}'$ as five times the difference between successive θ' values.

There are obviously other possible measures. For example, we could take the time to eclosion if we sprinkle cigarette ashes on the pupae at the same time as we

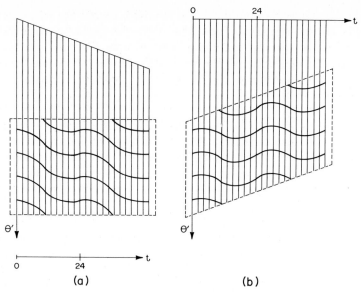

Figure 3.   (a) Observation window from Figure 2, rotated 90°; (b) part (a) with the zeroes of the 24 time axes aligned horizontally.

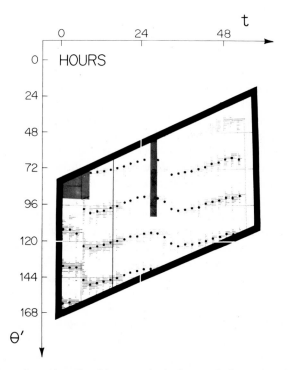

Figure 4.   Actual experimental results with $M = 0$, in the format of Figure 3(b). The shaded regions represent missing data (mostly mechanical failures).

θ' is plotted downward instead of upward so that the "new phase" axis extends vertically upward on the same diagram.

In Figure 3(b), Figure 3(a) is sheared by one hour vertically per hour horizontally in order to bring the zeroes of all 24 θ' axes into horizontal alignment. Thus θ' is measured downward from a common zero. This shearing puts the data into the format of Figures 31–33 of Chapter 1, in which type 0 resetting has average slope 0 and type 1 resetting has average slope 1.

Figure 4 shows the results of an actual measurement in this format, using *Drosophila's* eclosion rhythm. In this particular case the prior $M = 0$, therefore T is arbitrary. It shows that the eclosion peaks are of uniform narrow width. (I believe this is because the 120-second duration of the saturating assay pulse evokes rhythmicity of standard high amplitude in the circadian clock. I disallowed myself this approximation in the more rigorous exposition of Winfree (1973a) but for our present purposes it works well enough). Thus we can use the centroid of each peak (the average eclosion time) as a sufficient measure of eclosion behavior.

In Figure 5 for contrast, the assay pulse is omitted. The 24 eclosion records are handled exactly as in Figure 4, i.e., the time when the assay pulse *should* have been given is made time zero and all the zeros are aligned horizontally. The qualitative change in the data layout shows that the assay pulse *is* doing something in

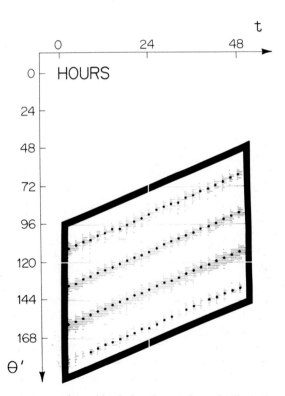

Figure 5.   For contrast, an experiment identical to the one shown in Figure 4, except that the assay pulse was omitted.

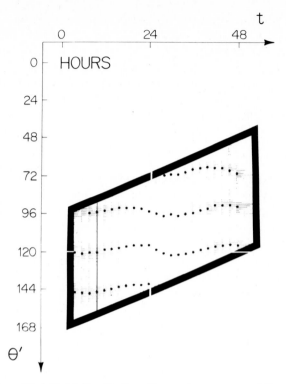

Figure 6.   As in Figure 4 but $M \neq 0$. Specifically, a 20-second exposure was delivered 6.2 hours after oscillation began, then θ' was measured by assay pulses at two-hour intervals for the next 48 hours, as in Figure 4.

Figure 4. Without it in Figure 5 the 24 aliquots are indistinguishable, except for the systematic displacement imposed on their time axes.

In Figure 6, $T = 6.2$ and $M = 20$ seconds. The point here is to show that the prior pulse (T, M) does set up different initial conditions by showing the different subsequent configurations of data. The θ' measurements differ systematically from those encountered in Figure 4. Peaks are still of standard width, so the centroid remains a sufficient numerical measure of the subsequent rhythmicity.

Figure 7 (right) simplifies the data more, focusing interest only on the centroids. Because the centroids recur periodically along the θ' axis at intervals close to 24 hours, it will suffice to reduce them to a single collective measurement, viz θ' modulo 24 hours. Figure 7 (right) shows the data replotted in this way. At each t there are now two or three dots representing the two or three successive eclosion peaks, all reduced to nearly equal θ' values by removing multiples of 24 hours. These θ' points lie in a reasonably well-defined corridor through which I have presumed to draw a smooth curve by hand. [In my first approach to these data I tried a "hands off" method, automatically fitting a ninth degree polynomial to the digitized data points. But these polynomials proved to fit the data no better

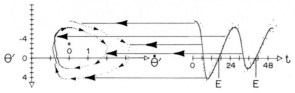

Figure 7. (right) As in Figure 4 but only the centroid of each peak is retained, and all centroid times θ' are reduced modulo 24 hours. The θ' scale is given on the left to minimize clutter. (left) The vertical downward axis is θ' in units of 1 hour. Each dot projected from the right is located horizontally by its $\dot\theta'$ (the local slope along the curve on the right). Eclosion time in the unassayed control is marked E on the right.

than the hand-sketched curve (standard error $= \pm 0.6$ hour) and they often had unnecessary bumps and wiggles in them.] This is Figure 4 with $M = 0$. Thus θ, θ', and the slope $\dot\theta'$ are being measured along the standard cycle initiated by a light-to-dark or blue-to-red transition. Note that the curve is not strictly periodic along the t axis. This is the reason why I use rectilinear measures T and t rather than circular measures such as ϕ_1 and ϕ_2 for the times of the first and second pulses: The flies' behavior deviates systematically from perfect periodicity and so it is inappropriate to implicitly map our observations onto the phase circle. This can be done later as an explicit approximation if that seems appropriate after we have looked at the raw data without first coloring them by theory. Note the arrows marked E for eclosion: These mark the times at which $\theta(t) = 0$ along the t axis. This is how the θ measure is brought into this diagram. Now all 3 observables (θ, θ', and $\dot\theta'$) are presented in a single plot.

C: Unperturbed Dynamics

As a matter of convenience we could eliminate the t axis, plotting θ' against $\dot\theta'$ in Figure 7 (left) (redrawn in more convenient format as Figure 8). This kind of plot necessarily gives us a clockwise rotation since θ' is increasing (downward) when its rate of change is positive (to the right). (If it is a rotation, does it rotate about some *point*? If so, that point must be on the $\dot\theta' = 0$ line. I have drawn an asterisk at $\dot\theta' = 0$ at $\theta' = -3$ hours: watch this spot in the subsequent diagrams.) This clockwise rotation provides some kind of picture of the standard cycle. It might be called a trajectory in $(\theta', \dot\theta')$ space. If $\theta'(\mathbf{x}(t))$ and $\dot\theta'(\mathbf{x}(t))$ are functions of the state $\mathbf{x}(t)$ then this diagram must be some two-dimensional projection out of $\mathbf{x}$ space. What kind of projection? There is no way to know in advance. Maybe θ' and $\dot\theta'$ are such peculiar functions of $\mathbf{x}$ that the projection will be twisted and folded so that trajectories appear to criss-cross. Fortunately, this turns out not to be so, as we will see by examining trajectories started from 30 different initial conditions.

Note that no theory or models have been involved so far nor will they be in what comes next.

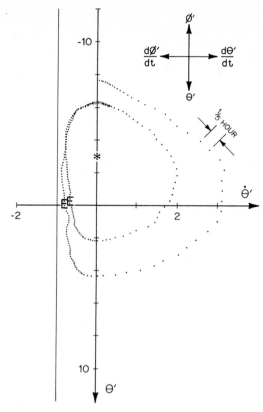

Figure 8.   As in Figure 7 but rescaled. Letters E are placed on the trajectory where $\theta(t) = 0$ by observation in the unassayed control. At states along locus $\dot{\theta}' = -1$, $\theta' + t = $ constant along trajectories, i.e., the assay pulse inflicts the same phase shift whenever given. Note that the θ' axis is really a circle: -12 hours is the same as $+12$ hours, modulo 24 hours. The inset coordinate arrows at the upper right show how the $(\theta', \dot{\theta}')$ directions are related to "new phase" ϕ'.

Many Trajectories

The next job is to compare trajectories from other initial conditions, that is $(T, M \neq 0)$. These initial conditions might or might not prove to be on the standard cycle. Let's see. Figure 9 shows about half of the 30 experiments, in the format of Figure 7, starting from diverse initial states by using diverse initial (T, M) exposures. Even though we cannot define a metric on $(\theta', \dot{\theta}')$ space, it seems clear that $\mathbf{x}(t)$, as assayed by $(\theta'(t), \dot{\theta}'(t))$, does not follow a unique cycle. The changing state of the clock cannot be regarded as a mere time displacement along one preferred trajectory.

A big transient is seen in a few of these trajectories. In all cases it lies entirely within the early parts $T + t < \frac{2}{3}$ cycle after pupae are removed from the prior continuous light in which they were reared. This feature does not repeat at 24 hour

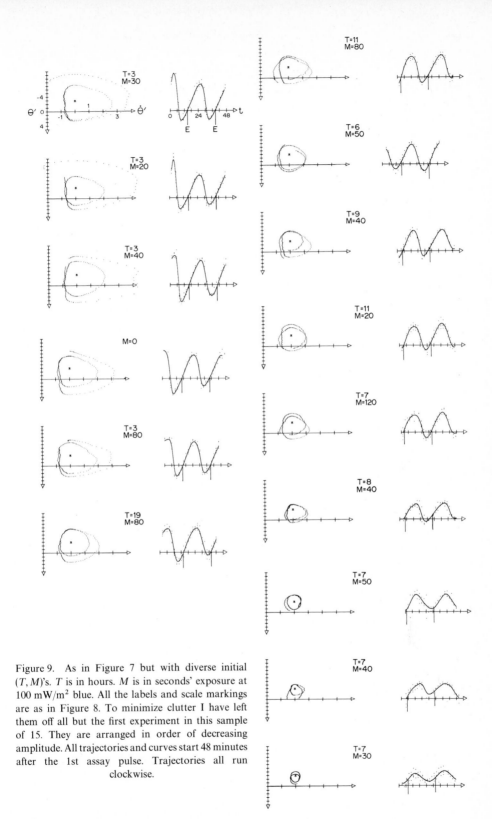

Figure 9. As in Figure 7 but with diverse initial (T, M)'s. T is in hours. M is in seconds' exposure at 100 mW/m^2 blue. All the labels and scale markings are as in Figure 8. To minimize clutter I have left them off all but the first experiment in this sample of 15. They are arranged in order of decreasing amplitude. All trajectories and curves start 48 minutes after the 1st assay pulse. Trajectories all run clockwise.

intervals. It is a one-time transient. A number of other circadian rhythms and other biological clocks have since revealed a similar transient upon release from prior inhibition (e.g., in *Kalanchoe*, Chapter 21; see also Johnsson and Israelsson, 1969). The clock may be winding down from the initial state in which it was held by whatever prolonged stimulus was used to arrest clock motion (in this case continuous light). In *Drosophila* this feature is not due to the photoreceptor's dark adaptation, which also follows prolonged exposure to light; nor is it an indication that the assay pulse is less than wholly saturating: A similar anomaly is observed when the assay pulse exposure is easily 10^5 times greater (unpublished experiments, and Pittendrigh, pers. comm. 1972).

It seems noteworthy that this 16-hour transient is all one would ever observe in the wild, where clocks seldom run as long as 16 hours before daylight returns.

At first sight this big transient looks like a winding down to a unique attracting cycle, as in Chapter 6. Such behavior was anticipated in the mathematical metaphors of Wever (1964, 1965a) and Pavlidis (1967). However, there is no corresponding trajectory winding outward toward the same cycle. Rather there seems to be a continuum of parallel concentric orbits of smaller sizes that go right down to zero. After the first 16 hours there is no conspicuous tendency inward or outward, neither to a preferred cycle, nor to a steady-state. Thus the

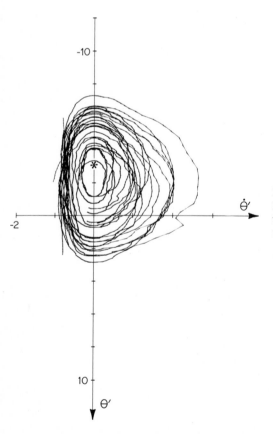

Figure 10. The left-hand parts of Figure 9 are superimposed, omitting the inward transient characteristic of the first 16 hours after release from constant-light inhibition.

effect of the light pulse (T, M) is not just displacement of the rhythm in time but is also something like a resetting of amplitude.

The tendency of each trajectory to repeat its previous path after the initial transient is more conspicuous when all the initial parts of trajectories with $T + t < \frac{2}{3}$ cycle are omitted (Figure 10). Here we see that the trajectories sampled in Figure 9 close more or less concentrically about the point $\theta' = -3$ hours at $\dot{\theta}' = 0$ (the asterisk).

It is also convenient to examine this behavior by constructing a Poincaré return map as follows. Points along the first 24 hours of each trajectory are connected by arrows each to its corresponding point along the same trajectory 24 hours later (Figure 11). In case of perfect periodicity these two points should be identical. In case of perfect periodicity obscured by noise the arrow connecting the two points should have some nonzero length but random direction. Figure 11 shows a distinct inward tendency of the arrows from high amplitude. This transient in the first 16 hours is followed by essentially random variation around periodicity at all subsequent points along the trajectories. There seems to be no systematic dependence of the period on the amplitude.

Orbits are concentric to the asterisk. Viewing this $(\theta', \dot{\theta}')$ plot as a projection from the dynamic space of the clocks' mechanism, we might take this as evidence

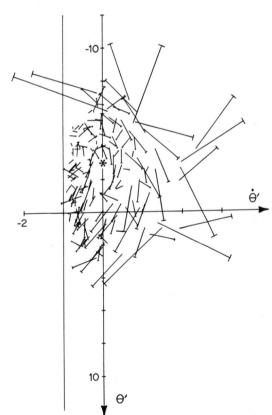

Figure 11. In the format of Figures 8 and 10, the trajectories of the left-hand parts of Figure 9 are used to construct a 24-hour return map. Each "nail" proceeds from a $(\theta', \dot{\theta}')$ point (at the blunt end) to the point reached 24 hours later along the same trajectory (at the sharp end). The initial transient is distinctly inward but it is hard to see any other tendency.

that there is nothing like an alternative attracting equilibrium or a saddle point or a repelling cycle between attracting cycles or a homoclinic point anywhere within the accessible portions of the clock's state space. In other words, its dynamics seem very smooth and simple.

Isochrons

The next question we can approach through these data is, "At what states is the clock found 24 N hours before eclosion occurs?" We used the θ measurement to put an E on the t axis wherever $\theta(t) = 0$ in Figure 7 (right). In Figure 8 these were transferred to the trajectory plot as two E's, one on each of the two cycles during which the trajectory was measured by assay pulses. Collecting together the 30 trajectories, each of two cycles duration starting from a different initial condition, we obtain a cloud of 60 E's, shown as dots in Figure 12. These constitute a ray projecting downward to the left from the singularity at the asterisk. This suggests that eclosion does not occur at any fixed θ' value nor upon crossing any definite $\dot{\theta}'$ threshold, but rather it occurs along a radius cutting trajectories of all amplitudes.

This might suggest that eclosion timing is determined by an extracted first harmonic of the oscillator's movement, as proposed by Kaus (1976).

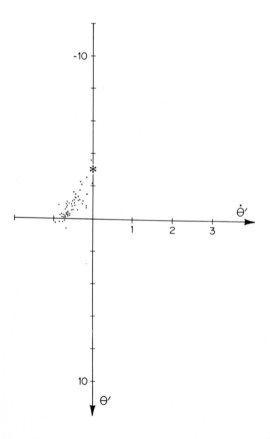

Figure 12.   All 30 trajectories are superposed in the format of Figure 10, but only the E's are retained (as dots). All other points are suppressed. This ray of dots marks the states of the clock that precede eclosion by multiples of 24 hours.

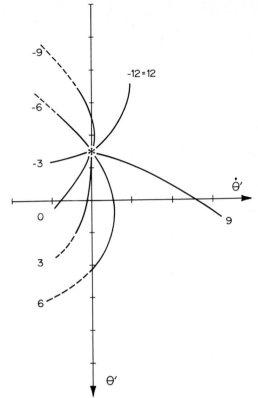

Figure 13.   Figure 12 is extended to show eight isochrons by proceeding for the appropriate number of hours along each trajectory from E before placing a dot. Contours are eyeballed through the dots at three-hour intervals. The dashed parts are extrapolated beyond the presently available data.

More generally one can ask, "Where was the $(\theta', \dot\theta')$ state *any* number of hours before eclosion, or the complementary number of hours after one of the pre-anniversaries of eclosion?" This is answered by marching along any trajectory to that number of hours beyond the ray of E's and placing a dot, and then another one 24 hours further along. After we've done this for all 30 trajectories, each of these new clouds of dots contains about 60 points (Figure 13). All extend radially from the asterisk near the center of rotation of trajectories. Though uncertain now whether the attractor cycle theory of Chapter 6 applies here, I call these sets of states "isochrons": equivalent time loci, so far as eventual eclosion time is concerned. This is the second direct measurement of the isochrons of a circadian clock. (In the first I used a simpler experimental technique with data from the plant *Kalanchoe*; see Chapter 21.)

D: The Impact of Light

All the above was about the trajectories followed in darkness, the nominally "unperturbed" trajectories. Now how does the individual trajectory's shape depend on the timing and duration of the stimulus (T, M) which was used to establish its beginning state? One can ask about the stable amplitude after the initial

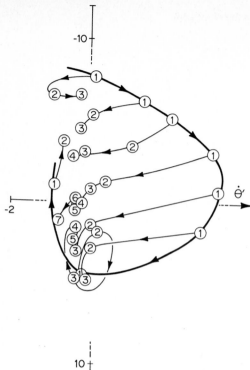

-10

-2

$\dot{\theta}'$

10

θ'

Figure 14. As in Figure 8 but only the first 18 hours of the $M = 0$ trajectory is shown, and the axes are suppressed. The circles mark the approximate starting places of 30 trajectories. These starting places are successive points along perturbed (blue light) trajectories, here connected and numbered in order of increasing M at fixed T.

inward transient: Which preliminary stimuli lead to big cycles and which lead to smaller cycles? In more detail one can ask, "How does exposure to light move the state of the clock from the standard trajectory?" The way to ask the question is to find out at what $(\theta', \dot{\theta}')$ state each trajectory starts, immediately after the initial conditions are set up by exposure for M seconds to a light that started T hours after the oscillation began. At each T we have several durations of M, from $M = 0$ up to the duration (120 s) of the assay pulse. The sequence of states reached by longer and longer M is the clock's trajectory during the blue light exposure.

Figure 14 connects the initial points of 30 trajectories in order of duration in blue light, at each initial T. The initial point of each red light trajectory was obtained by extrapolating the trajectory backward from the first datum (typically at $t = 1$ to 2 hours) to isochron $\theta(0)$. In five cases the first few points along the trajectory were too far apart to permit confident extrapolation: Those five are omitted from this plot. Note that all the remaining blue light trajectories start along the $M = 0$ trajectory (which is here given six "starting points", since $M = 0$ can be assigned to any T.) The trajectories generally shoot leftward to lower $\dot{\theta}'$ in blue light, remaining at roughly constant θ'. This is as expected: the initial θ' is the time to eclosion after a saturating assay pulse; if instead we give a stimulus of duration M followed without interruption by a saturating pulse, that combination pulse is also saturating and should therefore result in the same eclosion time θ'. In fact what is peculiar in these data is that θ' is *not quite* constant as M increases.

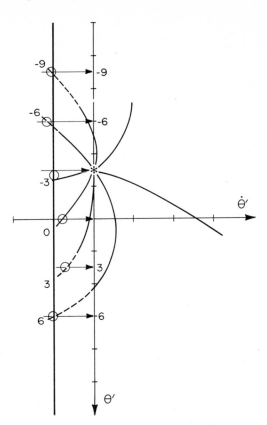

Figure 15.   Intersections $\theta = \theta'$ are esti-mated using the isochrons measured in Figure 13. At such points (marked by circles) the assay pulse has no effect on the clock's phase. These points lie roughly along the vertical line at $\dot{\theta}' = -1$.

I do not know why it is not. Maybe the assay pulse is not really as thoroughly "saturating" as I believed from earlier measurements. In any case the main point to notice is that the blue light trajectories are not along the cycle, and cannot be described as backing up (phase delaying) or hurrying ahead (phase advancing) along the cycle.

The predominantly leftward trajectories must end up along a special locus which can be constructed as follows. Saturating exposure by definition leaves the clock in a state such that its phase is not affected by further immediate exposure, so $\theta = \theta'$. The loci of fixed θ are the isochrons in Figure 13. The loci of fixed θ' are horizontal lines. The intersection between a given isochron and the corre-sponding θ' stratum occurs along the heavy arc in Figure 15. This has to be where all the long exposures end up. In point of fact they don't end up exactly along this locus but Figure 14 suggests that they tend toward it as exposure duration increases.

So much for the direction of blue light trajectories. Now what about marking off time along them? The 30 exposures given to initiate the 30 trajectories all had durations 0, 20, 30, 40, 50, 60, 80, or 120 seconds. So time marks can be roughly placed along the blue light trajectories by drawing links (Figure 16) between the $M = 20-30$ second experiments (row 1), another row of links between the 40 second experiments (row 2), another row connecting the $M = 50-60$ second experiments (row 3), and another connecting the $M = 80$ second experiments

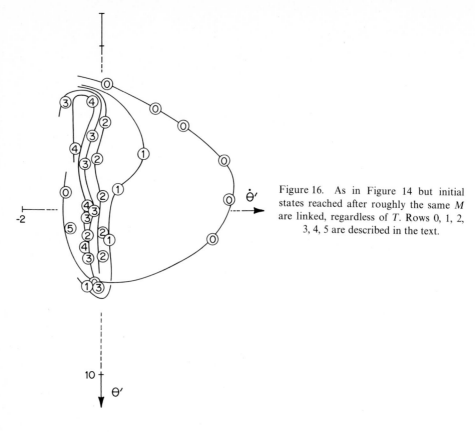

Figure 16. As in Figure 14 but initial states reached after roughly the same M are linked, regardless of T. Rows 0, 1, 2, 3, 4, 5 are described in the text.

(row 4). We already have links connecting the $M = 0$ second experiments (row 0) because this is the unperturbed red light trajectory. These rows of links get closer and closer together as M approaches saturation.

E: Deriving the Pinwheel Experiment

Now, combining the trajectories at fixed T and their cross-links at fixed M we have a (T, M) grid. This superposition of Figures 13, 14, and 16 tells what isochron we get to by starting an oscillation with a light-dark transition, waiting T hours, and then giving M seconds of blue. This experimentally determined diagram (Figure 17) turns out to be essentially identical to the one I presented as a theoretical inference in summer of 1968 at the Federation of European Biochemical Societies's meeting in Prague (Figure 11 of Winfree, 1968). It can be regarded as a contour map of θ above the (T, M) plane. This is made a bit clearer by straightening out the distorted (T, M) grid of Figure 17 to create Figure 18. We have seen this picture before as Figure 13 of Chapters 2, 4, and 5. It is the contour map of one unit cell of the time crystal portrayed in Figure 14 of Chapter 2. The time crystal came from the pinwheel experiment, which I was

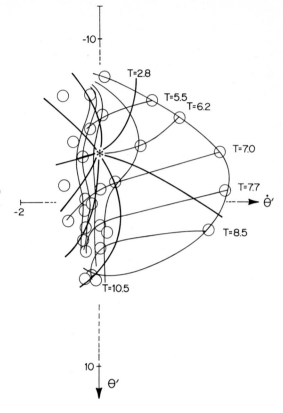

Figure 17. The isochrons on a (T, M) grid constructed by superposing Figures 13, 14, and 16.

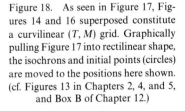

Figure 18. As seen in Figure 17, Figures 14 and 16 superposed constitute a curvilinear (T, M) grid. Graphically pulling Figure 17 into rectilinear shape, the isochrons and initial points (circles) are moved to the positions here shown. (cf. Figures 13 in Chapters 2, 4, and 5, and Box B of Chapter 12.)

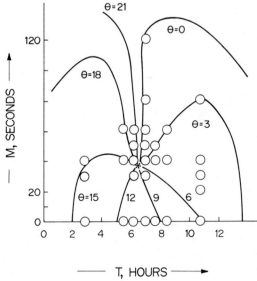

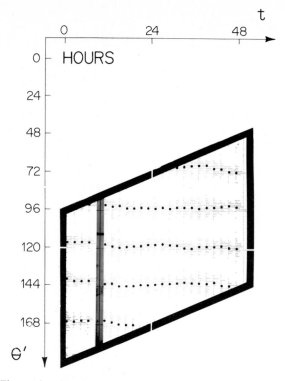

Figure 19.   As in Figure 6 but the prior stimulus was 40 seconds exposure at 6.2 hours after initiating oscillations. This near-singular pulse resulted in almost arrhythmic eclosion. There is very little time dependence of θ' in rhythmic eclosion shown here after a second stimulus (e.g., the assay pulse). The measured trajectory winds close around * (see Figure 20).

conducting while building the time machine to undertake this more elaborate measurement of trajectories. The results of that earlier experiment (fortunately) turn out to be implicit in the more comprehensive results. The phase singularity is the point * about which all trajectories revolve in the absence of a perturbing stimulus.

Note the trajectory starting (at $M = 0$) at $T = 6.2$ in Figure 17. It passes very nearly across the singularity. If blue light (of the standard intensity used in all these experiments) is terminated after only about 40 seconds, the clock should be left in a state of uncertain phase. In fact the eclosion "rhythm" becomes virtually aperiodic after such a stimulus (Figure 19) and the trajectory measured after such a stimulus has exceptionally low amplitude (Figure 20). This is as much as to say that *whenever* light again strikes the clock, it restores normal rhythmicity at practically the same phase, specifically, the phase of the isochron which roughly parallels blue light trajectories proceeding leftward from the environs of the singularity. This is also the phase of rhythm initiation in "naive" flies, suggesting that prior to some stimulus to mark "time zero", the newly created clock in a young fruitfly larva lingers near the * in its state space.

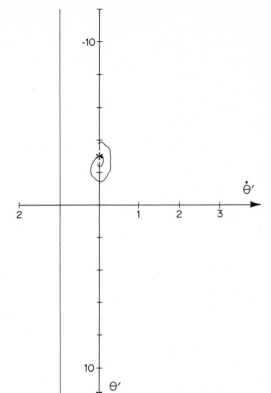

Figure 20.   As in Figure 8, replotting the
data of Figure 19.

What about trajectories under perturbing light of a different color or intensity? If photons get into the clock through a single chromatophore (i.e., if the clock lacks color vision), then color and intensity should be interchangeable. There seems every reason to believe the same trajectories are followed under higher intensity illumination, but faster: The net effect of brief (<1 hour) illumination seems determined by energy. But what of much *dimmer* lights? As intensity approaches zero the trajectories must approach the concentric rotation measured in utter darkness. This part of the story has not yet been filled in. However from Winfree (1974a) we know that *Drosophila*'s clock is arrested under prolonged illumination brighter than $\frac{1}{30}$ mW/m^2, and continues to oscillate under illumination 10 × dimmer. From Chandrashekaran and Engelmann (1976) we know that T^* remains the same under illumination as dim as $\frac{1}{10}$ mW/m^2, even though M^* has increased to 14 *hours* in such dim light. Thus it would appear that the transition from unperturbed to perturbed geometry occurs in the range between $\frac{1}{300}$ and $\frac{1}{10}$ mW/m^2 of blue light.

As noted in Chapter 2, several other circadian systems and at least one biochemical oscillator all gave qualitatively similar results in a pinwheel experiment. Do they also exhibit closed concentric trajectories when probed with a second pulse as here described? Engelmann et al. (1973) repeated this two-pulse measurement using the plant *Kalanchoe* and obtained answers similar to *Drosophila*'s, but more detail was needed for an exact comparison. *Kalanchoe*, unlike *Drosophila*, provides a continuous readout of its clock's activity (as opposed to only one datum per cycle, at eclosion time). So it should be possible to construct complete trajectories by the methods of Chapter 21, using only the existing experimental records from which the published phase measurements were extracted. This remains to be attempted in detail, but Engelmann et al. (1978) do report that the *Kalanchoe* clock's amplitude, once reduced by a critically timed stimulus, usually stays reduced for at least several days.

Turning now from circadian clocks to biochemical rhythms of shorter period, we find measurements in the same format by Greller (1977) using the glycolytic oscillation in yeast. Up to now the published data are too few to assemble a convincing trajectory, but just as noted above in regard to *Kalanchoe*, the unpublished fluorimeter traces could presumably be used to plot trajectories more directly by the method of Chapter 21 (in this case, plotting NADH fluorescence vs. its rate of change).

Similar remarks apply to Johnsson's (1973, 1976) and Johnsson et al.'s (1979) recordings of oscillatory water transport in seedlings, perturbed for various durations at various phases. This case is especially interesting because both the steady-state and the limit cycle locally attract trajectories. Thus a second, and repelling, cycle might be revealed when detailed trajectories are plotted from the existing data.

Malchow et al. (1978) inflict single-pulse perturbations on the cAMP clock of *Dictyostelium* (Chapter 15) while continuously recording optical properties of cells in liquid suspension. They too note that recovery to an attracting cycle is not swift: Effects on amplitude are conspicuous.

F: So What?

This is the end of my data presentation for this experiment. What good are these results? First of all they fulfill the objective I proposed at the beginning, namely, to elicit a picture in model-free form, a picture in which nature could reveal herself directly prior to any attempt on our part to impose theoretical interpretations. Any model that can draw this picture or any consistent distortion of it gets by unexcluded. But in so saying, it should be noted that the $(\theta', \dot{\theta}')$ plane is an arbitrary projection from the assumed state space $\mathbf{x}$. This projection could be arbitrarily distorted. The data require only a relationship between the isochron grid, the red light trajectories, and the blue light trajectories. Any distortion of the space on which they are all drawn leaves those relationships intact and therefore is as valid a picture of the dynamical flows as is any other distortion. The particular choice of coordinates employed here happily presents an especially simple and pleasing picture of smooth round concentric red light trajectories, parallel blue light trajectories, and nearly radial isochrons. Though it is pretty, remember it is not unique.

To my mind the most noteworthy and previously unsuspected features of the circadian clock which are brought out by these experiments are:

1. The time crystal with its helicoidal unit cell

2. The phase singularity (and its equivalence to the initial state of the clock in organisms which have never been exposed to an environmental stimulus)

3. The smoothness of the concentric closed trajectories

4. The initial transient following prolonged arrest in a nonpermissive environment (e.g., under continuous light)

5. Lability of the apparent amplitude of oscillation, with very nearly the same period at all amplitudes

6. An effect of perturbing stimuli which to a first approximation resembles parallel displacements of state along one state variable, as though one particularly labile substance were destroyed by the stimulus.

Here are three quite different classes of mechanism which typically exhibit such features.

A Single Non-simple Oscillator

There are abstract dynamical models asserting (by analogy to mechanical, electrical, or chemical kinetic equations; see Box A) that the clock has two or more state variables that affect each others' rates of change in such a way that an oscillation is engendered. So far as *Drosophila* is concerned, the most successful of these models to date emerges from the tradition begun by Danziger and Elmergreen (1956), developed by Strahm (1964) as a masters thesis project specifically to match Pittendrigh's *Drosophila pseudoobscura* data, and further refined by Pavlidis (1967, 1973). Several specific assumptions of this model have been found awkward in the decade since its first publications, but the basic ideas (feedback oscillations around an equilibrium state, lateral displacement of states during perturbation) remain acceptable (see Boxes A and B). Recovery or decay of amplitude in *Drosophila* (as assayed by the resetting curve) proved to be so slow that two whole days can go by without noticeable ($\pm 10\%$) change. This can be accomodated by an ad hoc adjustment of descriptive parameters in such a model to make its lower amplitude trajectories more nearly closed rings.

But it is not really necessary to abandon the now widespread assumption that circadian clocks are typically *rapidly* recovering attractor cycle oscillators. Such mechanisms may still be compatible with observations 1–6 above if we admit the possibility of many relatively independent sources of circadian rhythmicity, for example all the cells of some rhythmic tissue. Phase dispersion may at first seem an outlandish possibility, scarcely worth checking. But the fact is that we have precious little reason to assume that the circadian clocks in most organisms necessarily oscillate as a functional unit, except under conditions of entrainment by an external light-dark cycle. There exists considerable evidence to the contrary,

Box A: Traditions in Modeling Drosophila's Circadian Clock
by Mathematical Metaphors
(see also Chapter 20, Section B)

There was a time when biologists had not heard the words "limit cycle", which have become so popular since about 1965. The first significant incursion of those words into the world of circadian rhythmology was through Kalmus and Wigglesworths' paper presented at the first symposium on biological clocks at Cold Spring Harbor in 1960. That paper expounded the doctrine of attracting cycles and separated it from the notion of repelling equilibria but neglected to delineate in any explicit way the dynamical interpretation of a stimulus (e.g., an exposure to light). This might have been picked up from FitzHugh (1960, 1961) on neural oscillators, but its first explicit application to circadian experiments (in fact, to Pittendrigh's *Drosophila* data) was in Strahm's (1964) unpublished masters thesis at MIT. Borrowing from Danziger and Elmergreens' (1956) interpretation of periodic disease, Strahm outlined in considerable detail essentially the view of clock dynamics that prevails in today's literature, after refinement and quantification by Pavlidis (1967 et seq.).

The Strahm-Pavlidis model involves spiral trajectories winding out from a repelling equilibrium, but limited in amplitude by a "wall" beyond which no further decrease of one variable is allowed (perhaps because this is concentration 0). This nonlinear cut-off thus becomes one arc of an attracting limit cycle. Trajectories do not just approach the attracting cycle, but actually get *onto* it (by hitting the wall) within a day or two from most initial conditions. This seems to be as natural a first model of rhythmic dynamics as the van der Pol oscillator or the λ-w oscillator. After Danziger and Elmergreens' model of catatonic schizophrenia and the *Drosophila* clock model, it apparently arose again independently in Hunter's (1974) analysis of the sleep-wake cycle and in Nicolis and Prigogines' (1977, p. 394) model of lac operon dynamics in *E. coli*.

The Strahm-Pavlidis model postulates that the limited quantity is photolabile: During exposure to light it swiftly approaches the wall. Strahm even introduced the notion of circadian modulation of a threshold of developmental readiness as the basis for the quantization of populations into discrete eclosion peaks (Chapter 20, Box A).

Pavlidis's very influential contribution to this story lay largely in directing attention quantitatively to the implications of brief perturbation. He was the first to calculate resetting curves quantitatively, as a function of stimulus magnitude. The first mention of a *singular* combination of stimulus phase and duration is in Pavlidis (1967), where the further conclusion is drawn from nonobservation of such a disaster that trajectories must diverge with extraordinary haste from equilibrium toward the attracting cycle. Note that except for this, Figures 10 and 14 look remarkably like Pavlidis's diagrams describing the model.

However in the case of *Drosophila*'s eclosion rhythm, the circadian rhythm most studied by perturbations in the 1960s, there was never any evidence specifically indicating either instability of the steady-state or return toward an attracting cycle. I think my own experiments were the first to look for such indication, and none were found. In fact an adjustable amplitude model more like a frictionless pendulum than a limit cycle fits the data splendidly. This Winfree (1972e) model may have seemed retrogressive from the by-then-fashionable assumption of an attracting limit cycle. However, the adjustable amplitude model served all purposes of phase resetting equally well and additionally accounted for the newly observed plasticity of the resetting curves (the subject of this chapter) and the arrhymicity of dark-reared animals. It also successfully predicted suppression of circadian rhythmicity by incredibly dim light. However, it too proved deficient in still further tests. See Box B.

Box B: Are Phase and Amplitude Enough?

The adjustable-amplitude model of Winfree (1972e, Appendix) was formulated in response to 3 discoveries. First of all Winfree (1968) and Zimmerman (1969) had found, contrary to previous report, that the arrhythmicity of "naive" *Drosophila* populations could not be attributed to random phasing of individuals: rather, each animal's clock lay stably at equilibrium. Second, this initial state seemed the same as the "singular, phaseless" state reached by critical perturbation of previously rhythmic *Drosophila*. Third, it had been found (Winfree, 1972b,e) that the *Drosophila* clock's relative sensitivity to light during the first day after constant-light inhibition resembled dark adaptation in a clock-independent photoreceptor.

The principal conjecture embodied in this model is that the clock is strictly periodic but has a second variable of state, an unchanging "amplitude" conjugate to its phase, and that amplitude and phase together comprise a complete identification of its state. Amplitude zero is the phaseless or naive state. The clock can oscillate along any of a continuum of closed concentric trajectories of constant amplitude. As shown in Chapter 5, this leads to the observed resetting behavior (two "types", a singularity, and a lattice of screw surfaces) and to an amplitude dependence of the resetting curves such as we observe in this chapter. Formulated a little differently, the behavior of this model is entirely summarized in a family of phase-resetting curves *and* a family of amplitude-resetting curves. In these families the shape of each curve depends on stimulus magnitude or initial amplitude. This result was first foreseen by Wever (1962, 1963, 1964, 1965a).

The last paragraph of Winfree (1973a) explicitly conjectures that this expansion from one to two variables of state is complete and sufficient. But this has not yet been checked experimentally and there are now plenty of reasons to doubt it. These are

1.   Significant shortcomings of the adjustable amplitude model, conceived of as a single oscillator:

a.   Under sustained light its amplitude should decrease at a rate proportional to the light intensity. It seems not to (Winfree, 1974a).

b.   It provides no interpretation of the transient observed in the first 16 hours of darkness.

c.   As P. Kaus was the first to recognize (pers. commun., 1975), it should blow up under entrainment by light pulses 26 hours apart. (It does not.)

d.   It does not automatically account for the apparent constancy of the period at all amplitudes.

2.   The availability of a different class of model, namely, a population of relatively independent attractor cycle oscillators. Such models typically accomodate the facts of phase resetting and also accomodate points a–d above.

If models of type 2 are appropriate, then the Winfree "last paragraph" conjecture mentioned above should be demonstrably wrong. The reason is that there are many ways for a population to achieve any given amplitude of oscillation, namely, by different distributions of phase. Such different populations should phase-shift differently in reaction to any subsequent stimulus, although they may have the same phase and amplitude according to the experimental criteria used up to the present.

especially in the vertebrates and in plants (see Chapter 19). One need not necessarily expect strong mutually synchronizing interactions among circadian oscillators in small, nearly transparent, non-temperature-regulating organisms, since all of their circadian clocks would normally be independently entrained by the external cycle of light and dark, hot and cold. In no experimental organism have we yet resolved even the simple question whether type 0 resetting betrays scattering of phases within the organism as in Chapter 4 or, as in Chapters 5 and 6, it reflects a single oscillator's dynamics.

Thus we come to the second of three classes of mechanism:

Bilateral Symmetry in the Fruitfly

Within the practical limits of resolution in biological phase-resetting experiments, a smooth screw-shaped resetting surface and smooth concentric trajectories would be expected of a population of three or more simple clocks (Chapter 4). How many independent attractor cycle oscillators would be required to give this appearance in the individual organism? Under the rule used in Chapter 4 to pool the outputs of many oscillators into a single aggregate observation, the answer is "just two". Pair-wise redundancy of the circadian mechanism has long been suspected in bilaterally symmetric organisms and is especially clearly brought out in the recent experiments of Page et al. (1977) using the cockroach and of Koehler and Fleissner (1978) using a beetle, and of many workers intrigued by the splitting of activity rhythms in various vertebrates (see Chapter 19). The left- and right-brain clocks of Koehler and Fleissners' beetle apparently function independently. Similarly, let *Drosophila*'s two oscillators be loosely enough coupled so that they show no measureable tendency to synchronize during the two days throughout which their composite trajectory was monitored. It turns out that an excellent caricature of the data is given by combining the outputs of only two independent oscillators, separately reacting to a light pulse in the ways characteristic of the rapidly attracting oscillators explored in Chapter 6. In fact, I have quantitatively fitted all my *Drosophila* data, both for the one-pulse perturbations used in the pinwheel experiment and for the two-pulse perturbations used in this trajectory experiment, to a very simple model of this sort. The root-mean-square fit is quite as good as for any other models thus far contrived, and almost as good as the reproducibility of phase measurements allows.

Lots of Attractor Cycle Oscillators

Chapter 4 presented the third class of models which mimic the trajectories of *Drosophila*'s clock. In these models a population of very weakly interacting or completely independent oscillators of almost any kind individually pursue a common cycle, be that an attracting cycle or the unique cycle of a ring device. In such a population, phase corresponds to the mean phase of the population and amplitude corresponds to the dispersion of phase within the population. The blue light trajectories reflect the changing phase and amplitude of the fundamental harmonic of such a population's aggregate rhythm during the independent changes of phase of all of its many constituent oscillators.

In the first 16 hours after release from constant light, the clock's trajectory winds swiftly inward before repeating along a cycle of whatever amplitude it finally adopts. This inward transient is hard to account for in terms of simple clocks. But it could represent the collective behavior of many attractor cycle oscillators individually winding inward to a standard amplitude. The final closed cycle could be the collective behavior of those oscillators all moving on the common cycle, but not exactly synchronously.

Demonstration that the *period* of the whole organism's rhythm depends on its amplitude would suffice to exclude this notion of superposition of independent cellular clocks.

Besides the initial transient, there is another way in which the resetting behavior of a population of attractor cycle oscillators differs from that of a population of simple clocks. In the simple-clock populations, the singular stimulus duration goes to infinity as the initial dispersion of phases goes to zero. In contrast, the singular stimulus duration in an attractor cycle population goes to a finite limit, namely to the singular stimulus duration of the single oscillator.

This suggests an experiment, not yet undertaken so far as I know, in which an organism would be subjected to a particularly severe entraining rhythm prior to the phase-shifting experiment. The idea here is to impose the strictest attainable synchrony on whatever oscillators might comprise the individual's clock. If this treatment fails to increase the singular duration then it might be supposed that dispersion of circadian phases within the individual organism plays a negligible role in determining the phase singularity. This would indicate at least that the individual circadian oscillator does have a helicoidal resetting surface, unlike a simple clock. It would remain to be determined whether this clock's trajectories approach an attracting cycle slowly enough to explain (without invoking phase dispersion) the nearly concentric trajectories found above. A more direct resolution of the question might employ a histological assay in which the pertinent part of the fly's brain would be microscopically examined for synchrony of its cellular secretory rhythms. Ostensibly arrhythmic flies might show distinct rhythmicity in their randomly phased cells or they might show arrhythmicitiy all the way down to the cellular level (Figure 21). Suppose internal homogeneity and coherence of circadian function or nonfunction could be established by some such assay. Then phase-resetting experiments on whole organisms could be interpreted at face value,

Figure 21.   Contrasting two microscopic causes of macroscopic arrhythmicity: in (a) individual cells remain normally rhythmic but are randomly phased; in (b) each cell is arrhythmic. (Experiment suggested by Bunning, 1959, p. 522.)

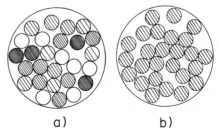

a)   b)

providing much stronger constraints on interpretations of the clock mechanism than they do at present. But suppose it were found that the cells or tissues of a single organism can persevere in circadian rhythmicity at arbitrary relative phases. Then phase singularities and less drastic changes of amplitude would not tell us much about the individual clock in multicellular organisms or in populations of single-celled organisms until a way is found to guarantee and monitor coherence among their many clocks.

At least in the case of certain free-swimming unicellulars, the question lends itself to resolution by a tidy experiment which seems more immediately feasible than the above. Owing to the continual sedimentation of cells suspended in water, it is not unusual for circadian rhythms of motility, density, adhesiveness, etc., to manifest themselves as circadian variations in the vertical density distribution of cells. Arrhythmicity induced by diversifying phases within the population could then be distinguished from arrhythmicity of every cell by removing samples of liquid from various depths into new bottles. This either will or will not isolate observably more coherent subpopulations. The same kind of experiment could be used to determine whether the type 0 resetting observed in such organisms (e.g., *Gonyaulax*) represents type 0 resetting of each cell or only a splitting into advanced and delayed subpopulations.

All such experiments presuppose that the cells' clocks do not influence each other. Such interactions have been looked for and none have been found (see Chapter 19).

G: In Conclusion

No one expected the phenomena demonstrated in *Drosophila* in this chapter, least of all myself. Yet in retrospect they can be derived with almost no specific assumptions on oscillator mechanisms, without even invoking interdependence among the individually competent clocks. Going still further, Aldridge and Pavlidis (1976) and Aldridge and Pye (1979a,b) point out that incoherence can last a long time even *with* strong coupling among multitudes of clocks.

The moral seems to be that it is of the greatest importance to check for coherence (e.g., among cells) in the circadian rhythms of multicellular organisms. This remains the crux of the dilemma stressed by Wilkins (1965):

> Unfortunately, nothing is known about the amplitude of the basic oscillating system or how it is related to the amplitude of the rhythms in the physiological or biochemical process used to monitor the behavior of the basic system.

8. Populations of Attractor Cycle Oscillators

Chapter 4 provides a preliminary look at the phenomena to which we now turn: the phenomena typical of aggregates of oscillators. Just as the oscillator populations of physics comprise a very special case with very special properties (associated with linearity, energy conservation, etc.), so did the simple clocks of Chapter 4 comprise another very special case with very special properties (associated with the one-dimensionality of their state space). My objective in this chapter is to organize under the same four headings as in Chapter 4 some discussions and examples of what I take to be the characteristic behavior of attractor cycle oscillators in populations and communities. Such oscillators can have any number (≥ 2) of variables mutually determining their rates of change in nonlinear ways. Linear oscillators, conservative oscillators, and simple clocks are special limiting cases of the attractor cycle oscillators considered in this chapter.

The chapter is organized in four sections:

A. Collective rhythmicity in a population without interactions among constituent oscillators. Mainly about phase resetting by a stimulus.

B. Collective rhythmicity in a community with completely promiscuous interactions (all individuals influenced by the aggregate rhythmicity of the population). Mainly about mutual entrainment and mutual repulsion.

C. Spatially distributed oscillators without interactions. Mainly about patterns of phase in space.

D. Oscillators interacting locally in space. Mainly about *smoothing* of concentration gradients that would become discontinuities were neighbors not coupled.

A: Collective Rhythmicity in a Population of Independent Oscillators: How Many Oscillators?

As in the case of simple-clock populations, we can ask of a population of attractor cycle oscillators whether there are characteristic features of the collective phase response by which to distinguish the type and number of independent

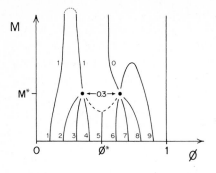

Figure 1.   Ten equispaced contours of new phase on the (old phase, stimulus magnitude) plane, calculated for a pair of independent attractor cycle oscillators similar to Example 1 of Chapter 6. The two are initially 0.3 cycle apart. At the dashed line, new phase shifts by one-half cycle.

oscillators involved. For this purpose I adopt the same measure of collective rhythmicity as before: the phase and amplitude of the *fundamental* harmonic of the aggregate rhythm. This aggregate rhythm is simply the sum of the rhythms $f(\phi_i)$ of N identical oscillators at phases ϕ_i on the attracting cycle. The scalar function f of phase is arbitrary but might be one of the chemical concentrations involved in the oscillation.

To recapitulate, we found in Chapter 3 that in a single simple clock new phase depends on old phase and stimulus magnitude in an especially simple way: All the contour lines of uniform new phase are parallel translations of a single monotone curve. The behavior of a pair of simple clocks is characterized by a U-shaped locus along which new phase jumps one-half cycle. A population of three or more simple clocks dissembles as a single pendulum-like oscillator in that its phase-resetting pattern is organized around a singularity and its amplitude is as labile as its phase.

A single attractor-cycle oscillator also typically exhibits one or more phase singularities, but its amplitude recovers promptly to normal. Now a *pair* of such oscillators, like a pair of simple clocks, turns out to exhibit a U-shaped locus of (old phase, stimulus magnitude) combinations along which new phase shifts abruptly (if the two clocks contribute exactly equally) by one-half cycle. But in this case the locus does not go to infinity along the magnitude axis: It terminates at two half-singularities as in Figure 1. The reason for this odd behavior is understandable. At any time when the two oscillators straddle ϕ^*, there is a stimulus magnitude, roughly M^*, that sets them to opposite phases (Figure 2). Here collective ampli-

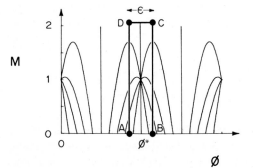

Figure 2.   Figure 13 of Chapter 5 is repeated with emphasis on resetting from two old phases ε apart, straddling ϕ^*. New phase has winding number $W = 1$ around path ABCDA. The phase difference along BA at $M = 0$ is ε. At sufficiently large M, along CBAD, it is close to 1. If there are no discontinuities along this path then the phase difference must pass through $\frac{1}{2}$ at some smaller M.

tude is 0 and a slight change of stimulus magnitude makes it positive or negative (i.e., positive at either of two opposite phases). At either extremity of this range, one oscillator can be placed on its singularity by $M = M^*$. A suitable slight change of ϕ or M will change that oscillator's phase to any desired phase, as we saw in Chapter 6, while negligibly altering the phase response of the other oscillator. Now suppose that both oscillators recover to their attracting cycles, and suppose that we continue to employ the collective phase measure introduced in Chapter 4. Then the phase of the pair ranges through half a cycle in the near neighborhood of each of these two points in the stimulus plane.

A more stringent test of the two-clock interpretation at the end of Chapter 7 might make use of Figure 1, in which the split singularity presents the signature of a pair of independent oscillators. To test for this experimentally, the first thing to do is to ensure that the putative two oscillators are well separated in phase. A way to arrange this reliably in many replicate organisms remains to be contrived.

Three or more independent attractor cycle oscillators behave pretty much like one, except that unlike the single oscillator, a population's collective amplitude does not recover from adjustment of the phase distribution. Also its apparent M^* is less than the single oscillator's by an amount that increases with the initial variance of phases.

As noted on page 104, the finer details of resetting behavior, if they can be resolved experimentally, do reveal the number of independent oscillators collectively observed. However, if $N > 3$, such details are easily obscured, especially in data collected from populations of individual animals or plants.

B: Collective Rhythmicity in a Community of Attractor Cycle Oscillators

From populations of independent oscillators we now turn to consider communities of (interacting) oscillators. Boxes A and B provide a compact literature survey of recent physiology in which mutual entrainment between two coupled attractor cycle oscillators is prominent.

We now turn directly to mutual entrainment among many clocks. This part of the story is not much altered by exchanging the community of simple clocks considered in Chapter 4 for the community of attractor cycle oscillators considered here. However, owing to the additional degrees of freedom of the attractor cycle oscillator, there do exist additional instabilities through which such a population can escape the mutually synchronized state to other modes of organization in time. So far as I am aware these have not been systematically explored.

Oscillators of Identical Period

A particularly tidy mathematical study of this sort was carried through by Zwanzig (1976) using an idealization of a mechanical clock: a piece-wise linear attractor cycle oscillator (Minorsky, 1962, Figure 26.2). He specified the conditions

Box A: Bibliography of Experimental Systems Involving Pairs of Coupled Similar Oscillators

Mutual entrainment between two identical oscillators plays a conspicuous role in a wide variety of physiological processes. (See Box B for dissimilar oscillators.) Analysis has proceeded independently on a dozen fronts up to now. A decade of experimental and theoretical work is contained in the following references and their bibliographies; see corresponding Bestiary chapters for context.

Chapter 12. Differently phased populations of yeast cells synchronize their rhythms of energy metabolism (Ghosh et al., 1971; Winfree, 1974c, 1978b).

Chapter 13. Chemical oscillators in continuously stirred tank reactors sometimes synchronize (Marek and Stuchl, 1975). It should be noted that there are quite a lot of other possibilities, e.g., a non-uniform steady-state or asynchronous oscillation. Some of these are described in Ashkenazi and Othmer (1978). Fujii and Sawada (1978) and Neu (1979a) describe some of the almost periodic alternatives in the course of an analysis of synchronization. Rössler (1976c) notes that coupling between two identical oscillators can also lead to chaos (Box B of Chapter 13). In fact, the first observation of chaotic solutions to ordinary differential equations resulted from Poincare's coupling of two pendulum oscillators (see Arnold and Avez, 1968). A recent electronic experiment by Gollub et al. (1978) is of interest in this regard.

Chapter 14. Individual cells of the pacemaker node in the heart normally synchronize each other (Berkinblit et al., 1975, using Noble's adaptation of the Hodgkin-Huxley equation; Peskin, 1975, pp. 250–278, using the "integrate-and-fire" model of membrane kinetics; Linkens and Datardina, 1977, using Hodgkin-Huxley equations; Torre, 1976 and Grattarola and Torre, 1977, using the van der Pol model). Spontaneously rhythmic muscle units of mammal intestine synchronize each other electrically (Nelson and Becker, 1968; and Linkens, 1976, 1977, again using van der Pol's model). Linkens (op. cit.), Boon and Strackee (1975), Zwanzig (1976), Ashkenazi and Othmer (1978), Pavlidis (1978b) and Kawato et al. (1979) note that it is just as common for two oscillators to lock together *out of step* as to synchronize. (See also Box F of Chapter 4.) Krinskii et al. (1972) make this case especially pertinent to irregularities of the heartbeat.

Chapter 19. The left and right clocks of the insect brain function independently in the one species of beetle examined (Koehler and Fleissner, 1978), but in the cockroach they synchronize one another unless surgically decoupled (Page et al., 1977; Page, 1978). As noted on page 202, a model invoking two independent oscillators suffices to account for resetting of amplitude as well as phase in *Drosophila* (Winfree, 1976), but Pavlidis (1976) showed at the same meeting that even if the two oscillators are strongly coupled, it can take them so long to synchronize on the attracting cycle that the required phenomena are still obtained.

Chapter 22. Two blobs of rhythmically mitosing *Physarum* quickly establish synchrony at a compromise phase when mixed together (Tyson and Sachsenmaier, 1978). Kauffman and Scheffey (1975, pers. comm.) found that entrainment between two of the single-variable "siphon" oscillator models traditionally associated with this mechanism does not necessarily entail synchrony: Out-of-step entrainment is common. Tyson and Kauffman (1975), using a two-variable attractor cycle model more recently advanced as a model of mitotic control, find the same phenomenon. In fact, this is quite ordinary in diffusion coupled oscillatory reactions as noted under chapter 14 just above.

Box B: Bibliography of Experimental Systems Involving Pairs of Coupled Dissimilar Oscillators

In at least five separate areas of physiology it has proven useful to abstract the dynamics of a rhythmic process in terms of a pair of unequal oscillators. In some cases one drives the other asymmetrically; in other cases the two interact reciprocally.

Chapter 14.   So far as I know, the first instance of this sort was van der Pol and van der Marks' (1928) analysis of the heartbeat, in which the S-A pacemaker drives the otherwise autonomous A-V pacemaker. The most significant direct descendents of this innovation include Grant (1956), Nadeau and Roberge (1969), Ličko and Landahl (1971), Bhéreur et al. (1971), Katholi et al. (1977) and Plant (1979).

Chapter 19.   The next phenomenon to benefit from this outlook was the 24-h circadian rhythm, beginning with the introduction of a light-sensitive master clock A driving a more passive and physiologically more temperature-dependent oscillator B (Pittendrigh and Bruce, 1957). The AB model was developed mainly in context of *Drosophila's* eclosion rhythm (Chapter 20) and was refined mathematically by Pittendrigh et al. (1958) and Kaus (1976). A more interactive version of the two-oscillator model has since been introduced by Pittendrigh (1974), elaborated by Daan et al. (1975), and refined by Daan and Berde (1978) under the labels M and N or E and M. This model supposes two closely coupled circadian oscillators with conspicuously different responses to visible light, together determining the activity rhythms of rodents. A variety of other circadian rhythms have been variously interpreted as summations of two rhythms, one initiated by dawn and one by dusk. This tradition began independently with Takimoto and Hamner (1964) and Tyshchenko (1966, in Russian) was applied to insect rhythms by Chandrashekaran et al. (1973) and Saunders (1974) and was nicely reviewed by Hamner and Hoshizaki (1974).

Chapter 23.   Of more recent vintage is the realization that the estrous cycle in laboratory hamsters, rats, mice, etc., is an oscillator whose period is constrained to exact multiples of the circadian clock's period. It appears that the circadian oscillator provides a daily neurohumoral input to the female's longer period endocrine oscillator, and this unilateral coupling mediates entrainment. (See Chapter 23 for references.)

Chapter 22.   There seems increasing reason to think of the cell mitotic cycle as a dynamic interplay between two separately competent cycles, one involving DNA replication and one involving cytoplasmic growth and cleavage. Explicit two-oscillator modeling here began with Goodwin's 1966 model, which is essentially isomorphic to Pittendrigh and Bruce's (1957) AB model of circadian rhythms. For more recent work see Mitchison (1971) and Chapter 22.

Chapter 12.   Oscillations in anaerobic sugar metabolism have been subjected to intensive experiments and dynamic modeling since about 1964. It appears that this pathway contains two sites of oscillation coupled through a common metabolic pool (Chance et al., 1967; Dynnik and Selkov, 1973, 1975a,b; Dynnik et al., 1977).

under which mutual synchronization is stable, at least for the case of identical oscillators. Grattarola and Torre (1977) achieved the same for populations of identical van der Pol oscillators, suitably coupled. Neu (1979c) did it for chemical oscillators with a circular limit cycle.

I think the restriction to populations of *identical* oscillators is probably a serious one. At least in simple clock communities (Winfree, 1967a), mutual entrainment is a threshold phenomenon requiring contributions from a number of oscillators proportional to the range of native periods they span. But in a population of identical noise-free oscillators, that range is zero.

A Finite Range of Native Periods

Envision a two-variable dynamic flow with polar symmetry. As in Section A of Chapter 6, suppose that phase increases at a uniform rate dependent only on the amplitude and that amplitude also changes at a rate dependent only on the amplitude. This is the λ-ω model exploited by Kopell and Howard (1973b), and the A-B model of Ortoleva and Ross (1974). In the limit of very rapid regulation of amplitude to the attracting cycle, this is almost a simple clock (see Box A of Chapter 6). If the rate of change of phase is now made to depend instant by instant upon the aggregate of all the oscillators' outputs, then we have a close approximation to the situation encountered in Chapter 4. The main result there was that, in a population of somewhat dissimilar native periods, interaction might encourage *or discourage* synchrony, depending on the magnitudes and phase relations of an influence and a sensitivity to that influence within the model oscillator. In the case of encouragement, synchrony arises by a collective process, above a critical density. I expect these features to carry over to populations of attractor cycle oscillators. Kuramoto (1975) successfully dealt with this model, in the special case of interactions similar to molecular diffusion between chemical oscillators. More recently Grasman and Jansen (1979) generalized the results presented in Chapter 4 for the case of mutually-coupled relaxation oscillators with at least two degrees of freedom.

Systematic Reorganization of Amplitudes

Unlike simple clocks, any attractor cycle oscillator is able to deviate from the common cycle during rhythmic perturbation. In the simplest cases each oscillator adjusts not only its phase but also its amplitude, which depends on the relative phase while entrained. This not only alters each oscillator's contribution to the aggregate rhythm but also alters its mode of response to further rhythmic influence and, typically, its period too. One consequence is that mutual synchronization may be stable, but only after it is initiated: If mutual coupling fails briefly, mutual synchronization is lost and cannot again arise spontaneously from the disorganized state (Winfree, 1967a, and unpublished computer simulation 1978–1979 circulated privately). Another consequence of labile amplitude is that chemical oscillators, mutually synchronized by free exchange of reactants through molecular diffusion, are susceptible to instabilities of a wave-like nature in space (Nicolis and Prigogine, 1977; Ashkenazi and Othmer, 1978). Through such diffusive instabilities, spatial gradients of phase or even discontinuities akin to shock waves can arise (Howard and Kopell, 1974, 1977; Neu, 1979a).

Prodded by instructive conversations with J. Mittenthal in 1969, I ran extensive computer simulations of a population of van der Pol oscillators in order to explore the range of validity of the simple clock approximation. Except in the limit of very strong amplitude regulation (large μ), new modes of behavior were readily demonstrable, all of which involve systematic dependence of the individual oscillator's entrained amplitude on its native period. These phenomena included splitting of the population into two oppositely phased parts and inability of the initially randomly phased population to approach mutual synchrony even though synchrony was stable once achieved (by entraining the whole population to an external rhythm). Informal circulation of these results, and discussions with Mittenthal and Pavlidis, led to much mathematical analysis but no broad generalizations. You may wish to consult more recent analyses of special cases by Pavlidis (1969, 1973), Linkens (1974), Aizawa (1976), and Aldridge (1976, p. 77). Aldridge and Pye (1979a,b) point out that if the oscillators' limit cycle and steady-state *both* attract, then perturbations may have lasting effects on collective amplitude even when oscillators are strongly coupled: by bumping some of them into the steady-state's attractor basin.

Hysteresis

Van der Pol oscillators are typical of attractor cycle devices in their well-known habit of entraining to rhythmic input in more than one stable way. These alternatives give rise to hysteresis if the driving frequency is varied up and down. They also suggest that mutual synchronization in a population may admit almost as many waveforms as there are oscillators to choose between alternative modes of entrainment. Viewed in the aggregate, a population of many such oscillators would then appear to have a continuum of neutrally stable waveforms and amplitudes, rather like a Hamiltonian oscillator. We saw such behavior in populations of coupled simple clocks in Chapter 4. And we saw it in the fruitfly's circadian rhythm, which might reasonably be suspected of a composite origin in many imperfectly synchronized cells.

The Chemistry of Coupling

At a less abstract level of analysis, one would like to know what mediates the interactions between biological oscillators of each kind.

The *Physarum* plasmodium, in which myriad nuclei synchronously fission, seems to represent a case of very strong interaction (see Chapter 22, Section B). The chemical nature of the coupling factor remains to be discovered. Nor does anyone know what substances synchronize mitotic timing in yeast cells dividing in suspension culture (Halvorson et al., 1971) or cultured mammalian cells (Dewey et al., 1973).

Turning to shorter period regulatory oscillations, Gooch and Packer (1971) impute to the ATP-ADP ratio a dominant role in mutual synchronization of respiratory oscillators in suspensions of mitochrondria. Aldridge and Pavlidis (1976) and Aldridge (1976) consider certain oddities in the reaction of a well-stirred suspension of oscillating yeast cells to chemical perturbation, from which they reason that the cells are metabolically coupled in a phase-dependent way. The troublesome challenge is to guess what molecular messenger couples the cellular oscillators. None of the presently envisioned candidates satisfy all the presently envisioned requirements (see Chapter 12). The adenyl cyclase oscillation in social amoebae presents a well-studied example of attractor cycle oscillators synchronizing through chemical coupling. The chemical is cAMP. Gerisch et al. (1975) and Malchow et al. (1978) contrived ways to perturb this oscillation in suspensions of single cells, with much promise for combined experimental and theoretical investigation of mutual synchronization.

In the case of circadian rhythms, not much is known about the channels of communication through which the various organs stay synchronized during prolonged residence in an arrhythmic environment (Moore-Ede et al., 1976, Sulzman et al., 1978). It might be noted that sometimes they *don't* stay synchronized.

C: Spatially Distributed Independent Oscillators

Oscillators Arranged in One Dimension Without Interactions

As long as they don't interact and we leave to them to their autonomous kinetics, attractor cycle oscillators run near a fixed cycle almost like simple clocks. Only their reaction to perturbation reveals their lack of constraint to the one-dimensional world of simple clocks. This section examines the spatial consequences of a rephasing perturbation, building on the examples used in Section C of Chapter 4, and leading to a reconsideration of wave-like patterns in space.

Reversing a Pseudowave

The first principle at work here is, again, the principle of a "pseudowave". You saw one of these the last time you noticed the chain of flashing strobe lamps at the end of an airport runway. The lamps flash in quick succession, giving the appearance of a ball of fire moving toward the runway threshold. This is not a wave in the usual sense: It is not a propagated disturbance. It is just a spatial gradient in the timing of strictly local oscillators. Its speed is simply $-1/(d\phi/dz)\tau$, τ being the period. This ball of fire can "move" at any speed, even faster than the speed of light if the phase gradient is shallow enough. Moreover it takes no notice of whatever ostensible barrier might be erected across its path. That is why in my first publication on chemical oscillators and their wave behavior, I distinguished

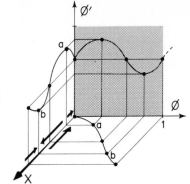

Figure 3. A three-dimensional diagram shows how the ϕ' vs. ϕ resetting curve revises a phase gradient in space (x). The resetting stimulus reverses the gradient between points $x = a$ and $x = b$.

this case as a *pseudowave* (Winfree, 1972c; see also Beck and Varadi, 1971; Thoenes, 1973; Varadi and Beck, 1975). In the slightly more general context of spatially graded period (from which emerges a *changing* gradient of phase), Kopell and Howard (1973a) called it a "kinematic" wave. They demonstrated the irrelevance of an impermeable barrier inserted after the gradient was established.[1]

Now what would happen if a one-dimensional array of independent oscillators, spatially graded in phase, were exposed to a phase-shifting disturbance, spatially graded in magnitude? Consider, for example, exposing a dish of oscillating malonic acid reagent to a flash of ultraviolet light. Vavilin et al. (1968) and Busse and Hess (1973) showed that a phase shift results, though they did not systematically measure its dependence on phase and the energy of the flash. But let us suppose behavior typical of an attractor cycle oscillator. A sufficiently brief exposure would induce only slight phase shifts and would therefore leave the geometry of preexisting pseudowaves qualitatively unaltered. Local phase gradients would be changed in proportion to the ratio $d\phi'/d\phi$ (from $d\phi/dx$ to $d\phi'/dx$) resulting in a proportional change of pseudowave speed in that region. The direction of propagation would remain unaffected.

The story would end here were we still thinking in terms of ring devices, whose response to a temporary disturbance consists entirely of a phase-dependent change of rate along a fixed cycle. But in this chapter we deal in attractor cycle oscillators. A somewhat larger perturbation can therefore elicit a resetting curve with a region of negative slope $d\phi'/d\phi < 0$. This inverts the local phase gradient in space and so turns pseudowave velocity backward (Figure 3). (Note also that at the boundaries of this region, where $d\phi'/d\phi = 0$, there $d\phi'/dx = 0$ so pseudowave speed is infinite, not 0). Figure 3 shows the old and new phases at time $t = 0$ at each point x. We transfer the new $\phi'(x)$ curve from Figure 3 to Figure 4 now (dropping the prime) and watch as ϕ continues to increase everywhere. The pseudowave can be visualized by considering how phase changes everywhere as t increases from 0: Every

---

[1] Their excellent paper in *Science* was delayed in publication by the referee's (my) insistence on that experiment; meanwhile Thoenes's (1973) almost identical paper appeared in *Nature*, but without the critical experiment. The same experiment, in this case proving that electrical wave trains in the brain are *not* pseudowaves, was performed by Petsche et al. (1970) and Petsche and Rappelsberger (1970).

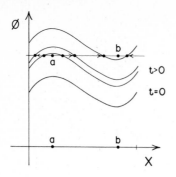

Figure 4.   The revised phase pattern $\phi(x)$ in Figure 3 changes in time as all phases increase together. Pseudowaves appear to emerge from point $x = a$ and vanish at $x = b$.

$\phi(x)$ increases in register. As the $\phi(x)$ curve moves up the ϕ axis its points of intersection with any line parallel to the x axis are found to move. The points at which $\phi = 0$ might be taken as wave front markers. At unit intervals of time two pseudowaves now emerge from point a, and at a later moment they annihilate each other at point b.

Preliminary to incorporating nearest neighbor interactions into this picture, we should note that in the preceding figure we made use of a map from the line segment $\mathbb{I}^1$ to the phase circle $\mathbb{S}^1$ (rolled out along linear coordinate axes). It will be convenient to draw pictures of the line segment mapped into the state space of the oscillator. Let's now put that map into context by changing the target space to $\mathbb{R}^D$, the D-dimensional space of chemical concentrations (or whatever other variables might define the oscillators' state). Such a map can be visualized as an embedding of $\mathbb{I}^1$ in $\mathbb{R}^D$. Specifically $\mathbb{I}^1$ maps along the attracting cycle in $\mathbb{R}^D$ in Figure 5 as specified by $\phi(x)$. Figure 6 shows this map transformed to $\phi'(x)$ by a stimulus. The ostensible source point a and collision point b are indicated.

Rhythmic Fungi Again

For more practice in thinking geometrically, reconsider the case of *Nectria*, the fungus that makes spores only in periodic patterns of ring or spiral topology. When we first considered these patterns it sufficed to think of phase maps from the frontier ring to the ring of phases of a simple clock. The moving frontier left rhythmic zonations in its wake. But how are we to think of the *origin* of pattern in a germinating spore or very tiny mycelium? This question will be examined more elaborately below in context of two-dimensional media, but we can make a beginning here while thinking only about the one-dimensional outer border, the frontier of the mycelium. Because it has the connectivity of a ring, its image in composition space is also a ring. Recognizing that the ungerminated spore is dormant, we might reasonably suppose that the oscillation encountered in the large, mature mycelium began from some uniform state in the germinating spore or in the very small mycelium. Thus the frontier's image is initially very small, lying quite close to the pregermination state. *Suppose* that state is the repelling equilibrium state of whatever reaction engenders rhythmicity later on. The equilibrium lies within the convergence of isochron surfaces (not shown) in composition space. In fact the equilibrium state *is* the unique convergence point of the isochrons if we suppose for a moment that the oscillation is engendered by interaction between only two variables, so that the composition space is only two-dimensional. In this simplification it is easier to see that the frontier ring necessarily has some integer winding number around the ring of phases represented by the converging isochrons. Each isochron indicates the latent phase of the oscillation prior to realization of its destiny on the attracting cycle. So the winding number realized here and now foretells exactly the winding number and therefore the pattern type of the mature mycelium. In this two-dimensional simplification the unperturbed trajectories of Figure 5 diverge from the repelling equilibrium, so that every oscillator on the frontier is independently carried out toward the attracting cycle. [In D-dimensional space, they would diverge from the entire $(D-2)$-dimensional phaseless set.] The frontier image expands onto the attracting cycle without ever changing its winding number (Figure 7). The winding number could only increase or decrease through a stage in which some part of the frontier's image crosses over the convergence of isochrons. But the local reaction flow is *away* from that state.

The crucial point of this story is that because the initial conditions are arbitrarily close to equilibrium, their winding number is determined by initially arbitrarily minute fluctuations about that equilibrium composition (Figure 8). In that respect this diagram rationalizes the observed polymorphism of pattern in mycelia grown under seemingly identical conditions from genetically identical single spores (Winfree, 1970a and 1973d; and Chapter 18).

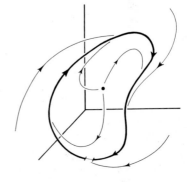

Figure 5. The x axis of Figure 3 before perturbation is mapped onto the attracting cycle of the local oscillation.

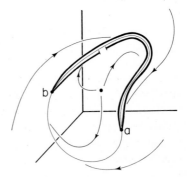

Figure 6. Figure 5 is revised in correspondence with the resetting curve diagrammed in Figure 3. Note that after perturbation point $x = a$ leads the image around the cycle, while point $x = b$ is the last to experience each event.

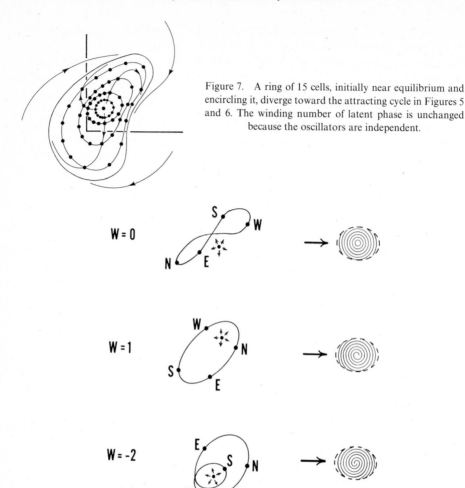

Figure 7.   A ring of 15 cells, initially near equilibrium and encircling it, diverge toward the attracting cycle in Figures 5 and 6. The winding number of latent phase is unchanged because the oscillators are independent.

Figure 8.   Situations analogous to Figure 7 are depicted at an early moment when all cells are near to equilibrium. A ring of cells (with north, east, south, and west points indicated) has winding number $W = 0$, 1, or -2 about the phaseless manifold (here shown simply as an equilibrium point). At the right we see the rhythmic pattern that will form on an expanding mycelium whose frontier has each winding number shown.

Oscillators Distributed Across Two-Dimensional Space Without Interaction

As in the one-dimensional case, the new features introduced by substituting attractor cycle kinetics for simple-clock kinetics are (1) the attractor cycle oscillator's different pattern of rephasing by a stimulus and (2) the existence of a phaseless locus in the attractor cycle oscillator's state space, where latent phase (a function of state) can suffer a discontinuity while the state itself still changes continuously.

The new feature introduced by substituting two dimensions of physical space for the single dimension to which we previously restricted our attentions is the possibility of applying a stimulus gradient transverse to a preexisting phase gradient. This is the format of the pinwheel experiment, by which the time crystal's helicoid and singularity were discovered in circadian clocks, in oscillating glycolysis, and in transpiration in plants.

My purposes for this section are to describe the pinwheel experiment in terms of a map from two-dimensional physical space into state space and to exhibit a similar mapping derived from our consideration of *Nectria's* growth rhythm.

Pinwheel Experiments: Two-Dimensional Arrays of Independent Attractor Cycle Oscillators

Suppose an extended medium such as a two-dimensional rectangle in which every area element is oscillating on a common attracting cycle. For example, imagine a dish of malonic acid oscillator, as in Busse and Hess' (1973) demonstration of local phase shifting by a focused beam of ultraviolet light. Let there be an east-west phase gradient such as might have been established by the earlier passage of a solitary wave of excitation from east to west. As a geometric convenience, roll up the rectangle to superimpose any repeats of the full cycle of phase along the east-west gradient. We now have a cylindrical piece of oscillating medium with a cycle of phase around its circumference as shown in Figure 9. This is only a pseudowave. It sweeps around the cylinder as indicated by the arrow. Now suppose we apply a stimulus graded in magnitude M along the vertical axis, transverse to the circular phase axis. At the bottom of the cylinder let $M = 0$. At the top let M be "big" by the criterion that type 0 resetting is evoked. Consider the fate of a ring-shaped element R of the cylinder, shown in state space in Figure 10. All points on R receive the same stimulus, but these points are at all of the old phases ϕ on the attracting cycle. R initially has winding number $W = 1$ around the circle of latent phases because the attracting cycle does and R maps directly onto the attracting cycle. If the stimulus acts in a way qualitatively resembling the models of Chapter 6, then R is moved off the cycle during the stimulus. By the time the stimulus ends, R's image in state space may or may not still have winding number $W = 1$ around

Figure 9.   Phase has winding number $W = 1$ around a cylinder of oscillating medium. It is exposed to a stimulus for a longer time at higher altitudes, resulting in winding number $W = 0$ along ring R.

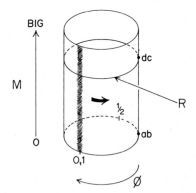

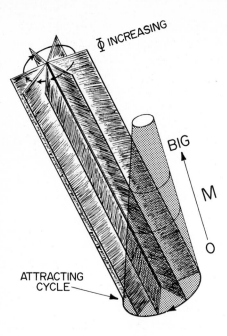

Figure 10.   After the stimulus of Figure 9 this cylinder is mapped into a three-variable state space. Isochrons of the dynamical flow are schematically illustrated by the "paddlewheel". The attracting cycle is the heavier ring at the bottom. Before the stimulus the whole cylinder was on the attracting cycle (similar to Figure 8 of Chapter 6).

the convergence of isochrons. For big enough M, it doesn't. It can have any other interger winding number in general, or none at all in rare cases of discontinuity or degeneracy. But let's suppose $W = 0$ at big M.

Figure 10 shows the terminal states reached by ring R for bigger and bigger M. It is a mapping of the (ϕ, M) cylinder into state space. The convergence of the isochrons necessarily lies within the stippled area. We know this because between the ring $M = 0$ and the ring $M = $ big, W has changed from 1 to 0. More formally, if both boundaries are traversed in one sweep along path ABCDA in Figure 11(a), then the winding number of new phase along the composite border of the stippled area is 1. That means that a full cycle of isochrons enter the stippled area but do not emerge from it. It also means that the stippled area cannot be mapped continuously onto the phase circle. There must be an internal discontinuity. That is where the missing isochrons vanish. The tidiest way to manage it would be a single point where they all converge: a phase singularity. The models of Chapter 6 do it that way. Figure 11(b) shows the isochrons in the shaded area, converging to an internal phase singularity in agreement with Figure 10. The shaded area being an image of the cylinder of Figure 9, these isochrons can be drawn on the cylinder. Slitting the cylinder along vertical element and laying it flat, we have the stimulus plane of the pinwheel experiment [Figure 11(c)]. The isochrons may be regarded as level contours of new phase ϕ' plotted above the (ϕ, M) plane. They therefore describe a screw surface winding up around the critical stimulus (ϕ^*, M^*) as observed in the several biological oscillators tediously overworked in this volume.

The isochrons may also be regarded as the successive positions of a wave front, the loci of $\phi' = 0$ on the cylinder. The wave front circulates around the base of the cylinder, on ring $M = 0$, at the same time as it rotates, pivoting about (ϕ^*, M^*).

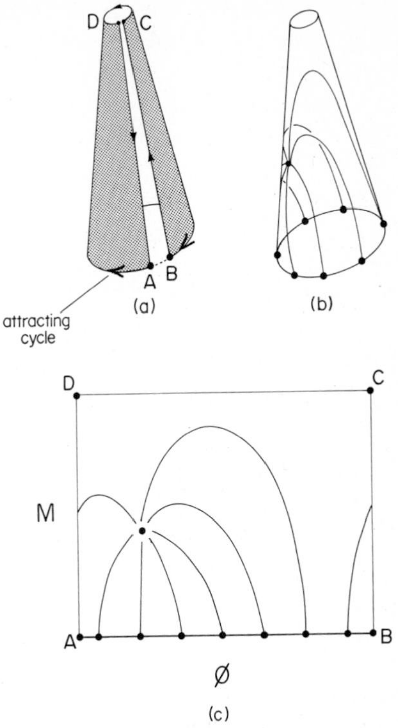

Figure 11. (a) The conical image of the cylinder after a graded stimulus (from Figure 10). Locus ab-dc is the cylinder element abdc in Figure 9. (b) Part (a) is redrawn to show intersection loci of the cone with the isochrons and their phaseless manifold. (c) Parts (a) and (b) unrolled and straightened into the rectangular format of $\phi \times M$ (cf. Box B and Figure 13 of Chapter 2, Figure 13 of Chapter 4, Figure 13 of Chapter 5, Figure 6 of Chapter 6, Figure 18 of Chapter 7, Figure 2 of this chapter and Box B of Chapter 12).

This (ϕ^*, M^*) remains stationary once all oscillators have independently recovered nearly to the common attracting cycle, because then phase advances by the same increment everywhere in each increment of time.

We've been through this argument before, perhaps too often. In each previous context the object of interest was the single oscillator. Here it is a population of oscillators, specifically a two-dimensional continuum. The argument is the same because the area elements of this continuum are here assumed not to interact.

This idealization has sometimes been adopted as a useful approximation to the mechanism of periodic patterning in two-dimensional films of malonic acid reagent (Chapter 13). Thoenes (1973), Smoes and Dreitlein (1973), and Smoes (1976) have even gone so far as to suggest that all the wave phenomena, spirals included, can be so interpreted. According to this interpretation every volume element is at some phase on the attracting cycle. The two-dimensional wave pattern is then a contour map showing where phase = 0 at the moment. On account of the phase singularity implicit in every rotating wave, this description would be tenable only if it were acceptable to suppose (as Smoes's model does) that arbitrarily small adjacent volume elements can remain at finitely different phases i.e., that the state variables of the oscillator do not diffuse. But they do, and in fact diffusion of $HBrO_2$ and Br^- is the driving principle behind one kind of wave propagation in this medium (Field and Noyes, 1972; Murray, 1976a; Tyson, 1976). These diffusion-coupled waves are typically a few millimeters apart, or less, and travel relatively slowly, in the order of several millimeters per minute. These "trigger waves", first described by Zaikin and Zhabotinsky (1970), are probably what Busse (1969) and Herschkowitz-Kaufman (1970) mistook for stationary "dissipative structures", and what Beck and Varadi (1971), Varadi and Beck (1975), and Beck, Varadi, and Hauck (1976) mistook (I suspect) for pseudowaves or kinematic waves. In contrast, the waves observed by Zhabotinsky (1968), by Thoenes (1973), and by Kopell and Howard (1973a) were spaced apart by centimeters or more, and moved at speeds in the order of meters per minute. I think these probably *were* the pseudowaves or kinematic waves expected theoretically by Beck and Varadi (1971, 1972), Winfree (1972c), Kopell and Howard (1973a), and Thoenes (1973) i.e., phase gradients in a spatially distributed oscillating medium. So far they have been examined in detail only in one-dimensional context. See Box C of Chapter 13.

The Ascomycete Frontier as a Two-Dimensional Population of Independent Attractor Cycle Oscillators

We turn back to the fungus *Nectria* for an example of attractor cycle oscillators arranged in a two-dimensional spatial continuum. In Chapter 4, Section C the colony's frontier was idealized as a ring of ring devices. You may have wondered at the time what is to be done about the interior of the colony which, after all, constitutes both its overwhelming bulk and the whole of the area in which patterning is clearly visible. We mapped the frontier continuously onto the phase circle, confident that this was only an act of selective attention, that the frontier map was only a restricted part of the whole disk's continuous map onto the phase circle.

1. A Point of Ambiguous Phase.   But a paradox lurked in the circumstance that *Nectria* makes not only ring-shaped formations but, alternatively, spirals. The frontier map for spiral morphogenesis necessarily has nonzero winding number. It is impossible to continuously map a disk onto the circle in such a way that its boundary has nonzero winding number.

Must the map then be literally discontinuous? Very steep concentration gradients pose no challenge to the imagination but an actual discontinuity would be hard to accept in a fine-meshed mycelium of richly interconnected hyphae. Moreover, there is typically no hint of phase discontinuity at any time during the mycelium's growth, as recorded in the visible zonations left by the moving frontier. Is this discontinuity always kept away from the frontier? Even in the tiny young mycelium? How could a tiny web of cytoplasm, less than a millimeter across, harbor the fierce discontinuity of phase implicit in spiral zonations?

Rather than answer such riddles, we could abandon the single-variable simple-clock model for a more complicated, but more realistic vision of oscillator dynamics: an attracting cycle in a state space of two or more variables, as in any of the last decade's reviews of biochemical oscillations (Higgins, 1967; Hess and Boiteux, 1971; Nicolis and Portnow, 1973; Goldbeter and Caplan, 1976). In such reviews, there is not one ring device to be found. In a previous section of this chapter our map of the mycelium's frontier onto the phase ring was accordingly revised, becoming a map into a space of at least two dimensions in which phase is a *function* of state as in Chapter 6. We look to this state space of at least two dimensions for the state of each tiny patch of tissue, small enough to be considered homogeneous, but large enough to have a steady average composition.

2. Mapping $\mathbb{R}^2 \to \mathbb{R}^3$. Now it is time to quit restricting our attention to the one-dimensional frontier. Figure 12 depicts a snapshot of the whole two-dimensional mycelial disk mapped into a three-variable state space. I use three dimensions instead of two because three is the most we can easily visualize, and restriction to two implicitly assumes some rather special features that the fungus has not yet exhibited. I call it a "snapshot" because Figure 12 only depicts each cell at its instantaneous combination of chemical concentrations, and those concentrations are all thought to be changing rhythmically in time.

The heavy ring represents the conjectured attracting cycle, and the finer curves show how concentrations tend to that cycle "in the limit" of long enough time.

The chief assumption implicit in this picture is that concentrations vary smoothly in space, i.e., nearby hyphae map as neighbors in state space. The frontier maps wholly onto the attracting cycle. If its winding number is 0, then the whole disk can map wholly onto the cycle. But Figure 12 shows a case with winding number $W \neq 0$. Where is the continuity problem? Instead of a discontinuity of state and the implicit infinite concentration gradient somewhere inside the mycelium as required by the simple-clock conjecture, we have here a perfectly smooth variation of concentrations across the mycelium.

If we insist on assigning a phase to each point in the mycelium, there still has to be a phase singularity. But that no longer forces us to accept a singularity in the biochemical state of the mycelium unless we insist on adherence to the attracting cycle everywhere. The mapping to state space can be, and presumably is, perfectly smooth. It is the *further* map from state space to the ring of latent phase values that harbors the singularity, as explained in Chapter 6. And that poses no paradox: There is no reason to expect an arbitrarily chosen function of state ("latent phase") to be free of pathological behavior. The fine trajectories of Figures 5, 6, 7, and 12 show how most sets of initial conditions lead onto the attracting cycle. The cycle's

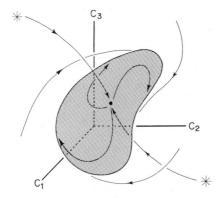

Figure 12. An image of a disk of oscillating medium in state space, supposing its border has winding number $W = 1$ around an attracting cycle.

attractor basin is necessarily shaped like a doughnut surrounding the cycle. Even if the doughnut is puffed up to enclose as much of composition space as possible, a part of its boundary (what bounded the inner hole) must thread the cycle. In the less graphic terms of topology: Any cap bounded only by a cycle must contain at least one point (an initial state) from which trajectories never return to the cycle. Guckenheimer (1975) further showed that trajectories from nearby initial states reach the attracting cycle at every phase. This point is the phase singularity first intimated by the simple-clock model of Chapter 4. This point is a part of the "phaseless manifold", the locus of at least $D - 2$ dimensions from which you cannot get to the attracting cycle by spontaneous processes without help from outside. In Figure 12, $D = 3$, so the phaseless manifold is one-dimensional. Specifically, it consists of the two trajectories labeled with an asterisk.

3. The Germinating Spore.   A mycelium with nonzero winding number constitutes such a cap, so somewhere inside it is a patch of tissue with ambiguous or indeterminate phase. Following this line of argument back in time to the very young and very small mycelium, scarcely more than a germinating spore, we infer that the single spore from which emerged a spirally-patterned mycelium must have been in or very close to the phaseless manifold.

In principle a spore could contain any concentrations of the relevant substances and still not oscillate until a parameter change during germination disinhibits their chemical interactions. In such a case, the initial concentrations would determine the latent phase of the whole mycelium, which would accordingly go onto the attracting cycle homogeneously. A homogeneously oscillating mycelium makes concentric rings. This is what most rhythmic ascomycetes do.

Therefore any fungus that spontaneously makes spirals must have had its initial conditions near the phaseless manifold. That means little in terms of concrete biochemistry until the oscillating mechanism is chemically isolated, but it does predict a somatic polymorphism, as noted in the previous section and as observed in *Nectria*.

So if spiral morphogenesis is observed, then genetically identical spores, cultivated under identical conditions, must be expected to develop into mycelia bearing rings, spirals of left and right handedness, double spirals, etc. This is just what happens in *Nectria*. The other two ascomycetes known to make spirals (*Penicillium diversum*, *Chaetomium robusta*) also make rings, but further details of their behavior have not been reported.

Note that the pattern originates in every case from the relatively homogeneous initial conditions required by our assumption of smoothness of concentrations in a tiny mycelium. Such homogeneity in a simple-clock model would imply homogeneity of phase and therefore concentric circular zonations. But the phaseless manifold of the attractor cycle model gives us the possibility of a *phase* singularity without the biologically impossible *state* singularity. The state is initially relatively homogeneous, but the divergence of trajectories from the phaseless manifold gives us a deviation-amplifying device that soon puts points of the frontier at widely different states on the cycle.

Box C: Testable Implications

(1)   We suppose that the angular distribution of phase around the frontier of a small mycelium is achieved by small random displacements from an initial state. If this initial tiny ring of states cannot adopt a variety of winding numbers, then it must not be close to the $D - 2$ dimensional locus (where the isochrons converge) about which the winding number is defined. Being tiny, it therefore does not enclose that locus, its winding number is $W = 0$, and its patterning therefore consists of concentric rings. In other words, this model requires that a fungus which routinely restricts its rhythmic patterning to one winding number must make closed rings, as is most common among rhythmic fungi.

(2)   A disturbance which alters the phase of a biochemical rhythm does so by temporarily changing the rates of reactions involved in the oscillation. Only in the exceptional case that all rates change in the same proportion is the geometry of flow in composition space unaltered. Otherwise not only rates but also directions of flow are altered. Thus trajectories previously constituting the boundary of the phaseless set during unperturbed oscillation are replaced during perturbation by trajectories which cross that boundary. A ring of oscillators encircling the phaseless set is then blown across it and ends up lying to one side, no longer encircling the phaseless set (Figure a). Its winding number has changed to $W = 0$, so morphogenesis thereafter changes to concentric closed rings.

(3)   The same principles allow transition from rings to spirals, but only if the stimulus is exquisitely timed. Suppose a frontier with winding number $W = 0$ is exposed to the perturbation diagrammed in Figure a. Suppose the stimulus begins when the ring of initial conditions is exactly "upwind" from the phaseless set. Let it continue just long enough to take part of the ring through the phaseless set, changing its winding number from zero. If the stimulus terminates at this time, spiral morphogenesis must ensue. If the stimulus continues longer then the winding number reverts to zero as in (2). However, if the stimulus is applied at the wrong time or in a mycelium initially too homogeneous in phase, then the winding number remains $W = 0$.

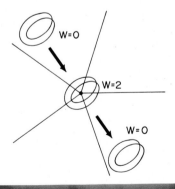

Figure A.   Similar to Figure 8, but the ring-shaped image of the frontier is moving along trajectories of the perturbed dynamic. As it moves, its winding number changes discretely. The five radial lines are supposed to be isochrons.

4. Consequences.   Three predictions follow from the foregoing (see Box C):

1.   If a fungus makes patterns of only one winding number, then that winding number must be zero and the pattern must consist of equispaced disjoint rings. For example, there should be no mutant that makes only clockwise two-armed spirals.

2.   A fungus with spiral polymorphism should go on to make only rings if shocked during germination. (A shock is anything that disturbs the oscillator's mechanism by the criterion that it inflects a phase shift on a rhythmic mycelium).

3.   A fungus which mainly exhibits rings should sometimes convert to spiral morphogenesis following a suitably timed shock.

3. Potential Difficulties.   There exist alternatives to the central conjecture elaborated above that each tiny patch of mycelium harbors a continuous attractor cycle "clock" which functions essentially independently in each patch and whose state varies continuously across the organism.

For example, more might be made of the inevitable interactions between nearby patches of tissue: The model elaborated above achieves its seductive simplicity by ignoring molecular diffusion and cytoplasmic mixing. Particularly in very young and small mycelia, spatial coupling of oscillators could dominate the launching of morphogenesis into concentric ring patterns. In the next section of this chapter, a beginning is made toward incorporating such interactions.

What about the many other internally rhythmic ascomycetes that do not make spiral zones, but only concentric rings? Why the anomalous ability of *Nectria*, *Penicillium*, and *Chaetomium* to make rings *and* spirals? Suggestions:

1.   Conceivably the others, if they harbor biochemical oscillators at all, have "simple clocks". The simple clock can only have winding number $W = 0$ around a very tiny mycelium, and this can only initiate morphogenesis in rings.

2.   The critical feature of the attractor cycle model for spiral polymorphism is that the steady-state changes from attractor (in spores and old, crowded, or damaged hyphae) to a repellor (in rapidly growing hyphae) while staying fixed at the same concentrations. For if it moved in concentration space, then all the cells previously attracted would soon find themselves to one side of the repellor and go into oscillation synchronously. Nearly synchronous oscillation means $W = 0$. It might seem a bit too ad hoc to propose a model in which the steady-state changes type, but does not move, while the metabolic rate increases. But this idiosyncrasy is strictly necessary to account for spiral polymorphism in terms of an oscillator model. It might thus account for the phylogenetic rarity of the phenomenon.

Neurospora crassa (band, csp-1), which makes only rings spontaneously, exhibits type 0 resetting, so (1) is excluded for *Neurospora*. Dharmananda and Feldman (1979, Figure 2) report that the banding rhythm usually starts up from germination within a predictable third of the cycle. Thus the initial steady-state lies off to one side from the convergence of isochrons in the mature dynamics. So (2) seems likely for *Neurospora*.

D: Attractor Cycle Oscillators Interacting Locally in Two-Dimensional Space

To ignore interaction among neighboring area elements of a reacting continuum is to make a gross approximation. Sometimes that is exactly what we want to do in order to abstract an essential principle out of an incidentally more complicated context. The appropriateness of such an approximation sometimes depends on the scale of time and space we are concerned with, in relation to the time and space constants of reaction and of local interaction. But not here. In dealing with spiral patterns we confront a topological difficulty that cannot be resolved without invoking some kind of dynamical interaction between adjacent patches of tissue. Why is that? According to the hypothesis of independent kinetics, every patch of tissue promptly moves into the attracting cycle. The only exception is that single cubic micrometer that initially happens to be poised exactly at equilibrium. If all its neighbors soon begin to oscillate on the cycle, then this neighborhood experiences oscillating concentration gradients of startling intensity. Unless the tissue be divided up into largish chunks by nearly impermeable membranes, there will be a rhythmic exchange of materials between differently phased neighbors, and corresponding modulation of local kinetics.

Tying the Mycelium Together

Accordingly, we now add neighbor interactions. Specifically, let each area element not only go about its own business according to its local kinetics but also let it tend to adopt a state midway between its immediate neighbors, as in the last part of Chapter 4. In terms of equations this is represented by adding a Laplacian operator $\nabla^2 c$ to the rate equation of each state variable c. Physically it is equivalent to introducing molecular diffusion between adjacent area elements. Geometrically it corresponds to conferring on the image of the medium in state space a coherence or integrity which for many purposes resembles an elasticity (Winfree, 1974b,f; 1978a and Box D). Viewed in any of these ways, the essential result is that the area elements can no longer depart freely from their neighbors in composition space. Nearby points initially have, and continue to have, nearby compositions. Composition varies smoothly because extreme concentration gradients do not long survive unless aided and abetted by an extreme irregularity in the kinetic rate equations.

With this understanding, Figure 12 contains less to disturb the skeptical reader. Without local exchange of reactants, the center of the mycelium's image in state space would be stretched infinitely thin and taut as almost all volume elements independently approach the attracting cycle. With diffusion restraining the divergence of neighboring states, the map stays smooth as sketched. In the absence of neighbor interactions, area elements initially near the phase singularity would independently diverge, tearing a hole in the image of the continuum or stretching it infinitely thin. Only that infinitesimal area element that was initially exactly on the equilibrium state would remain inside the cycle. But the effect of molecular

Box D :   The Medium's Image in Concentration Space Behaves Like an Elastic Membrane

As noted on page 127, the parabolic partial differential equation of reaction and diffusion has the same form as the equation describing the movements of a simple kind of elastic membrane immersed in a viscous flow. To make this metaphor explicit imagine the membrane as a rectangular gridwork of springs connected at four-way junctions (Figure a). I wish each spring to develop mechanical tension in direct proportion to the vector separating its endpoints. These forces add at each junction of four springs:

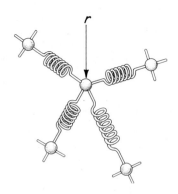

$$F = k \sum (r_{NBR} - r)$$

Now suppose this bedspring is immersed in a viscous medium moving with velocity $v(r)$ at each point r. So let each junction r be subject to an additional viscous drag force due to its motion relative to the local flow:

$$F(r) = k \sum (r_{NBR} - r) + \mu(v(r) - \dot{r})$$

If both bedspring and surrounding fluid be essentially massless, then there are no significant inertial forces F to balance the spring forces and viscous drag. Then we can write

$$\dot{r} = v(r) + \frac{k}{\mu} \sum (r_{NBR} - r).$$

Now call $k/\mu = D$ and interpret r not as $r = (x, y)$ in physical space but as $c = (c_1, c_2)$ in concentration space. Interpret $v(r)$ as the reaction flow $R(c)$. And recall that $\sum (c_{NBR} - c)$ is the discrete approximation to the Laplacian:

$$\dot{c} = R(c) + D\nabla^2 c.$$

To put it another way: the fundamental equations of chemical reaction, considered in terms of molecular collisions, are hyperbolic and isomorphic to those of a web of springs and balls. The more familiar parabolic reaction/diffusion equation is used only as an excellent approximation that overlooks transients of duration in the order of 10^{-10} second (Othmer, 1976b). This approximation amounts to neglecting the mass of each ball.

> I find it easiest to "intuit" the solution of such equations by thinking of them as motions of a thin elastic membrane caught up in flowing molasses. The flow has just the phase portrait of the local kinetics.
>
> Note that this metaphor is accurate only if the diffusion matrix is a *scalar*. Thus it is close-to-accurate only in reactions involving molecules of comparable mobility and in biological systems involving random bulk transport of all reactants, as in cytoplasmic translocation in fungi.

exchange is to preserve continuity. The image stretches toward the attracting cycle and thins out in the middle, but not forever. Beyond a certain point the concentration gradients near the phase singularity are steep enough (area elements have moved far enough apart in state space) that molecular exchange prevents further divergence. The oscillation remains thus attenuated near the phase singularity, as it would not were every area element independent. The image of the medium continues to rotate and every area element oscillates synchronously. But full amplitude is achieved at only some distance from the phase singularity. Area elements near the phase singularity are in close contact by molecular diffusion with area elements at all phases of the cycle so that they scarcely vary in composition.

Other Possible Outcomes and Interpretations

As described in the foregoing paragraph, this is a purely heuristic and conjectural story. Its support comes mainly from numerical computations of specific examples (Winfree, unpublished, 1973 and 1974; Yamada and Kuramoto, 1976; Erneux and Herschkowitz-Kaufman, 1977; Cohen, Neu, and Rosales, 1978).

There are three main alternatives to the qualitative pictures presented above. The first is that the whole situation could be subject to instabilities of a nonintuitive sort in which the medium's image, initially girdling the cycle, pulls across the convergence of isochrons, defying the local kinetic trajectories. After this, all rings have winding number $W = 0$, the whole medium oscillates synchronously, and its entire image contracts to a point circulating on the attracting cycle. This happens in my unpublished numerical solutions if the physical medium is too small or (the same thing) if the reaction is too sluggish relative to the homogenizing influence of molecular exchange.

In the case of one-dimensional media, the mathematics of Kopell and Howard (1973b), Othmer (1977), Ashkenazi and Othmer (1978), and Conway et al. (1978) can be invoked to show that in the presence of molecular diffusion (or random cytoplasmic streaming), nonzero winding number of phase is unstable on too small a physical ring, though it is perfectly stable on large rings [and this is borne out quantitatively in numerical simulations (Winfree, unpublished)]. The critical size depends on the ratio of the diffusion coefficient to a quantity related to the vigor of radial divergence from the phaseless manifold. Given $W \neq 0$ in such an unstable situation, the rate at which it evolves towards a switch to winding number 0 depends on these two factors and on the frontier's circumference, which is meanwhile increasing as the mycelium grows, and might increase enough to restore stability before winding number changes. But if it doesn't then only concentric rings will be observed.

At present, we lack figures to test this conjecture for *Nectria*. But estimating the effective diffusion coefficient at 10^{-5} cm^2 per second and assuming kinetics such that the attracting cycle would be approached with a time constant in the order of an hour, the minimum stable size for spiral morphogenesis should be in the neighborhood of 1 cm of circumference. In *Nectria*, this represents an age of about 50 hours. So it might be that in *Nectria*, unlike Ascomycetes that embark only upon ring morphogenesis, intramycelial coupling is restricted during the first 50 hours; or it might be that the instability doesn't develop to completion within that time; or it might be that the clock process is not disinhibited until at least 50 hours after germination. In fact, this region never does develop clear banding. I personally doubt that coupling by diffusion plays any important role in *Nectria*, other than ensuring reasonable smoothness of the bands. But the foregoing possibilities remain unexcluded alternatives to interpretations based on strictly parochial oscillation.

The second conspicuous alternative to the simple ideas about *Nectria* underlying Figure 12 is that while $W \neq 0$ *is* preserved, the medium's image does continue to rotate, but not rigidly: Area elements near the phase singularity dart this way and that irregularly, always tethered inside the attracting cycle by the requirements of continuity, but not stably so. This makes a rotating wave whose apparent "center" is forever meandering irregularly. Oscillating versions of the malonic acid reagent exhibit such meandering centers (Winfree, 1973c; Rössler, 1978, Rössler and Kahlert, 1979).

Kuramoto (1978) points to still more ghastly irregularities of timing that can arise in a field of oscillators coupled by diffusion. He calls these "phase turbulence" and "amplitude turbulence". Like the "diffusive instabilities" pursued by the Brussels School (Turing, 1952; Nicolis and Prigogine, 1977), these pathologies can only occur where diffusion coefficients differ sufficiently. Mutants of *Nectria* which typically exhibit irregular periodicity might conceivably represent this parameter range.

Third, it could well be that only the frontier oscillates. Hyphae left behind the frontier in crowded conditions on staled nutrient medium might not be oscillating at all. If the pertinent metabolic parameters differ so drastically from place to place, then it is wholly inappropriate to map all those places onto a common dynamic flow as in the foregoing diagrams. However, at least in the case of *Neurospora*, Dharmananda and Feldman (1979) recently succeeded in showing experimentally that the oscillation does persist, only slightly altered in period, at points sampled well behind the frontier.

My fourth conspicuous alternative occurs first to most people, i.e., that the local kinetics (neglecting transport) does not oscillate at all, but that the growth rate at the frontier "stutters" due to an instability of the balance between inward diffusion of nutrients and production of "staling" byproducts that locally inhibit

Box E: Chains and Sheets of Coupled Oscillators (Bibliography)

The behavior of oscillating continua has been of vital concern for over a century in connection with mechanical vibration, acoustics, and optics. From such studies of crystalline solids we inherit the geometric wonders of Brillouin zones, Fermi surfaces, and such useful technology as laser optics and semiconductor electronics.

On a more nearly macroscopic scale than that of atomic lattices, spontaneously rhythmic living tissue and oscillating chemical reactions have recently acquired new interest. What will come of it cannot be guessed as yet, but there seems good reason to anticipate insights of comparable scope and applicability in the area of biochemical dynamics. For the reader's convenience I have gathered here a representative collection of recent research papers under three headings.

The Malonic Acid Oscillator
Zaikin and Zhabotinsky (1970)
Field and Noyes (1972, 1974b)
Busse and Hess (1973)
DeSimone et al. (1973)
Tatterson and Hudson (1973)
Kopell and Howard (1973a, 1974)
Murray (1976, 1977)
Stanshine (1976)
Hastings (1976)
Troy (1977a,b)
Tyson (1977a)
Reusser and Field (1979)

Biological Situations
(a)   Electrical rhythms of intestine:
Nelsen and Becker (1968)
Diamant et al. (1970)
Sarna et al. (1971, 1972a,b)
Specht and Bortoff (1972)
Brown et al. (1975)
Linkens et al. (1976)
Patton and Linkens (1978)
(b)   Ecological cycles geographically coupled by diffusion:
Murray (1975, 1976b)
Jorne (1977)
Wilhelm and van der Werff (1977)
(c)   Extremely precise chemical oscillation in *Physarum*, coupled by flow of endoplasm in a loose two-dimensional network:
Durham and Ridgway (1976)
Yoshimoto and Kamiya (1978)
(d)   Monolayer of slime mold amoebae, considered as limit-cycle oscillators coupled by molecular diffusion:
Novak and Seelig (1976)

General Diffusion-Coupled Oscillations
Gmitro and Scriven (1966)
Scott (1970a,b)
Goldbeter (1973)
Kopell and Howard (1973b, 1974, 1977)
Howard and Kopell (1974, 1977)
Nazarea (1974)
Ortoleva and Ross (1973, 1974)
Aris (1975)
Erneux and Herschkowitz-Kaufman (1975, 1977)
Kuramoto (1975)
Kuramoto and Yamada (1975, 1976)

Pavlidis (1975)
Guckenheimer (1976)
Greenberg (1976)
Hastings (1976)
Kuramoto and Tsuzuki (1976)
Ortoleva (1976)
Ross (1976)
Yamada and Kuramoto (1976)
Zwanzig (1976)
Cohen, Hoppensteadt, and Miura (1977)
Nicolis and Prigogine (1977)
Cohen, Neu, and Rosales (1978)
Greenberg (1978)
Greenberg and Hastings (1978)
Ashkenazi and Othmer (1978)
Kuramoto (1978)
Neu (1979a,b,c)
Grasman and Jansen (1979)

For updates subscribe to Joseph Ford's "Non-linear Science Abstracts" c/o School of
Physics, Ga. Inst. of Tech., Atlanta, Ga. 30332.

growth. With the proviso that any such model must involve at least two mor-
phogens in order to accommodate patterns other than concentric rings (Winfree,
1970a, 1973d), this seems to me as plausible as the "local clocks" model. The
question simply awaits someone to explicitly formulate a plausible model, verify
that it exhibits the observed delicate polymorphism in two dimensions, and point
to a distinguishing observation.

And here we arrive at the badly frayed end of this story. There are good problems
here for those with a flair for mathematics. (See Box E for collected references
to the research literature.) As the takehome lesson from this chapter, I emphasize a
conjecture: that the usual mode of organization in a field of independent oscillators
is a wave rotating around a central pivot (a point in two dimensions or a curve in
three dimensions); and that this remains so if neighboring volume elements are
coupled by balanced diffusion. The following chapter asserts that it even remains
so if we now break open the attracting cycle so that the medium is merely excitable.
There we entertain kinetic schemes similar to the hourglass kinetics considered in
Chapter 3 in which oscillation is not spontaneous. Such media of course do not
support pseudowaves, but given neighbor interaction they do support real propagating waves, even pivoting rotating waves similar to those we encountered
above. Some promising starts have been made in the last few years toward identifying the properties of these waves both by numerical experiments in digital
computers and by observing biological media of practical interest such as brain and
heart muscle.

9. Excitable Kinetics and Excitable Media

The beauty of life is, therefore, geometrical beauty of a type that Plato would have much appreciated.

J. D. Bernal, *The Origin of Life*

Chapter 3 on ring devices is followed by Chapter 4 on populations and communities of such single-variable units: both simple clocks and the nonoscillating hourglasses. Chapter 6 on oscillators with more than one variable of state is followed by populations and communities in Chapter 8. What about nonoscillatory kinetics with more than one variable? That case is taken up briefly here, together with consequences of interaction in spatially distributed communities. The upshot is a new kind of oscillator and a new kind of phase singularity, both of which are apparently exhibited in diverse chemical and physiological systems. Even though no isolated piece of it may oscillate, an excitable medium can organize itself spatially in a way that stabilizes oscillation at a characteristic period. Architecturally, this configuration more resembles a clock than anything encountered in previous chapters: It consists of crossed concentration gradients, any one of which might be taken as the clock's "hand", a pointer that physically rotates about a fixed pivot once in each cycle of oscillation. At the pivot, nothing changes; the pivot is a phase singularity and all the rest is built around it.

A: Excitability

Clocks and Hourglasses Again

In Chapter 3 we encountered the simplest form of smooth oscillator, the simple clock. We saw that a tiny, local change in its rate law converts it from a spontaneously recycling clock into an hourglass. An hourglass is a system which rests poised at a delicate equilibrium, waiting to be helped across a shallow barrier to complete one cycle on its own. Both simple clocks and hourglasses are ring

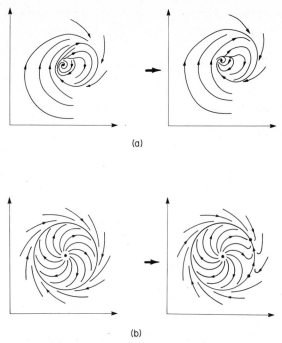

(a)

(b)

Figure 1.   Two ways for an attracting cycle to change locally, resulting in excitable kinetics without spontaneous oscillation. Case (a) is always an excitable oscillator; case (b) is not initially excitable. These are sketches, not computations.

devices. Any ring device is *excitable* if a tiny but finite disturbance triggers it into a big excursion followed by spontaneous approach toward the original state. If the recovering excitable system sticks at an attracting equilibrium, awaiting a stimulus to trigger the next excursion from that original state, then we have an hourglass. If it never quite stops but just creeps on to go spontaneously into another excursion, then we call it a simple clock.

I called both by the name "ring device" to emphasize their distinguishing feature, that the state space is a one-dimensional loop. Real kinetic systems, however, can seldom be realistically approximated in terms of a single degree of freedom. So in Chapter 6 we generalize the ring device to a dynamic system with an arbitrary number of degrees of freedom D. In such a higher dimensional dynamic system, all trajectories can eventually funnel onto an attracting closed trajectory like the unique cycle of a simple clock. The same sort of generalization is appropriate in the case of an hourglass. In this larger world of D-dimensional composition space, attractor cycle kinetics and hourglass kinetics can be closely related through a small local change in the directions of trajectories. It is common for model biochemical schemes or more abstract dynamic models such as the Hodgkin-Huxley equations to exhibit either behavior, depending only on fine adjustments of some parameter (Figure 1). Box A gives a bibliography of examples in biological and biochemical contexts.

Box A: Oscillation in Excitable Kinetics

Oscillatory kinetics is rendered additionally excitable by any adjustment that makes the flow very much slower in a limited arc of the attracting cycle. This happens, e.g., if the equilibrium point is moved very close to any point on the cycle.

Non-oscillating excitable kinetics is easily converted to spontaneous oscillation. Verbally, it requires only a reduction (not necessarily to zero) of the threshold beyond which a small displacement from equilibrium will grow enormously before the system again approaches equilibrium. In the world of biophysics, this transition to or from spontaneity has been prominent for the longest time in the study of electrically excitable membranes. It is a conspicuous feature of most neurodynamic models (See Katchalsky and Spangler, 1968; and Jack et al., 1975 for reviews and particular references).

More recently studies of biochemical kinetics have produced spontaneous oscillators, including the excitable kind that readily convert to mere excitability without oscillation. For theoretical examples see Rössler (1972b,c, 1974a,b), Hahn et al. (1974), Karfunkel and Seelig (1975), Othmer (1975), and Sanglier and Nicolis (1976). The cAMP kinetics of social amoebae exhibits this transition in real life (Cohen, 1977, 1978; Goldbeter and Segel, 1977; Goldbeter et al., 1978); and so might the phosphofructokinase oscillator of sugar metabolism (Goldbeter and Erneux, 1978).

A conceptual model convenient to the themes of this volume is implicit in the radially symmetric oscillator used in Chapter 6, Section A. There an attracting cycle was approached by two variables according to kinetics described in polar coordinates as $\dot{\phi} = 1$, $\dot{R} = KR(1 - R)$. With large K, this is practically a simple clock as noted in Box A of Chapter 6. Adding a uniform rate of increase of one concentration (e.g., of $x = R \cos 2\pi\phi$, $dx/dt = C \sim 2\pi$) as in Figure 1(b), the attracting cycle is scarcely deformed, but ϕ changes to $1 + f(C/K) \sin \phi$ as in the perturbed ring devices of Chapter 3. Above a critical C/K, the cycle is interrupted by an attractor and repellor just as in the single-variable models.

Another class of conveniently simple two-variable models are piecewise linear with the two pieces of state space separated by a threshold level of one variable. This kinetic scheme has been used for a long time to caricature nerve membrane (Offner et al., 1940; Cohen, 1971; Rinzel and Keller, 1973), clocks (Andronov et al., 1966), and excitable reactions (Winfree, 1974b,f; Zaikin, 1975). The computation shown in Section C used such a kinetic equation. With slight adjustment of the threshold level, it oscillates spontaneously. By playing about with this model and with the malonic acid reagent I believe I have shown that the wavelike phenomena described up to now in this reagent came from a property quite independent of its oscillation, namely its excitability alone.

Excitable Media

Consider now an extended line, surface, or volume, each point of which harbors the excitable dynamic system. Neighboring volume elements are coupled so that each can excite the next. Such a continuum is called an excitable medium. Just as in the case of the excitable continuum of ring devices discussed in Chapter 4, these dynamically more elaborate media turn out to support waves. Analytic solutions are more challenging than when only a single rate equation is involved, but some examples have been solved explicitly. The most thoroughly explored equations of course exploit some particular simplification. The commonest are

1.   Elimination of all but one degree of freedom (examples in Chapter 4).

2.   Incorporation of all the nonlinearities into a single threshold (piecewise linear models: Offner et al. (1940), Minorsky (1962, Chapter 31), Andronov et al., (1966, Chapter 8), McKean (1970), Rinzel and Keller (1973), Winfree (1974b,f) Zaikin (1975), Zwanzig (1976). The models of Tyson (1977a, Figure 3) and of Zaikin and Kawczynski (1977, Figure 2) and of Goldbeter et al. (1978, Figure 1) are geometrically essentially the same as the piecewise linear models.

3.   Segregation of the dynamics into one or two slowly changing variables and one other that moves much more quickly (Minorsky (1962, Chapter 26) and Andronov et al. (1966, Chapter 10)). Cells are commonly coupled only through this one variable [Hodgkin and Huxley (1952), Zhabotinsky and Zaikin (1973), Karfunkel and Seelig (1975), Ortoleva and Ross (1975), Yakhno (1975), Hastings (1976), Troy (1977), Zaikin and Kawczynski (1977), Collins and Ross (1978), and Fife (1979)]. This class includes catastrophe theory models of excitable media and their waves (Zeeman, 1972; Winfree, 1973c; Tyson, 1976; Feinn and Ortoleva, 1977; Schmitz et al., 1977; and Rössler, 1978).

A variety of wavelike phenomena have been derived or simulated in these approximations. Some of these waves propagate through the medium by means of an effect of one area element on the next, like the trigger waves first encountered in hourglass media in Chapter 4, Section D. A trigger wave can travel stably as a single isolated event, tending to a standard waveform and velocity. In a trigger wave, each volume element's local dynamical excursion is triggered by encroachment of the wave through neighboring volume elements. The triggering event propagates like a grassfire, behind which the grass grows back after a while, restoring readiness to conduct another trigger wave should one come along.

Rotating Waves

In Chapter 4, Section D, we saw that, given a suitable spatial distribution of activity, nearest-neighbor interactions can gloss over the slight difference of geometry (Figure 1a) that determines whether or not an excitable kinetics will cycle spontaneously. The result is that a ring of potentially quiescent hourglass material is enabled to sustain rhythmical activity. Looked at another way, this is not remarkable. The period is just the circulation time of excitation propagated as a wave on a ring of excitable medium.

Something more interesting arises when we carry over the same principle to two-dimensional media. This was impossible in Chapter 6 because a rotating wave has a nonzero winding number about an interior disk, and that implies a phase singularity inside the disk. Phase singularities are verboten in continua composed of coupled ring devices because they require infinite gradients of chemical composition whereby all the different phases confront one another at the singularity.

As we saw in Section D of Chapter 8, no such dilemma arises in continua with two or more dynamic variables. A two-dimensional rotating wave arises in every pinwheel experiment, and its geometry can remain much the same when the adjacent oscillators are coupled by diffusion. It is not hard to imagine the same thing happening in media whose isolated volume elements do *not* spontaneously cycle, if excitability is the dominant principle. Imagine how Figure 12 of Chapter 8 might change while we change the local dynamic as suggested in Figure 1 of this Chapter. The alteration affects only a tiny domain of composition space, occupied at any moment by a small patch of the physical medium (the wave front, roughly speaking). Because the medium is cohesive, adjacent patches help along the patch that is temporarily tempted to backslide into equilibrium. It is pulled away from equilibrium and stimulated into another cycle of excitation. Thus the whole medium perseveres in rhythmic activity almost as though the local kinetics were spontaneously oscillatory.

At the center of this rotation is a new kind of singularity. It is closely related to the phase singularities previously described in two-dimensional continua of attractor cycle oscillators such as the rhythmic fungi, circadian clocks in plants and flies, and oscillating glycolysis. This new kind of singularity differs from those encountered before in that it fundamentally involves coupling between volume elements by molecular diffusion. It arises in excitable media such as the malonic acid reagent (Chapter 13), heart muscle (Chapter 14), and slime mold (Chapter 15). I call it a "rotor".

B: Rotors

A rotor is the self-maintaining source of a rotating wave. It arises where the phase singularity arose in the final section of the previous chapter. Initial conditions for a rotor can be set up by executing the pinwheel experiment on an excitable medium whether or not it happens (as in that section) to oscillate spontaneously. This has not yet been attempted in the direct and obvious way, using a graded stimulus transverse to a phase gradient or a traveling wave. Let me show how it *has* been done and then show a few more pictures describing rotors in composition space. Then we will turn to a brief romance about rotors in three-dimensional excitable media.

The Basic Anatomy of a Rotor

What are the essential qualitative features of the situation in Figure 12 of Chapter 8? I see two:

a.   A piece of the physical medium maps smoothly across the interior of the usual cycle realized elsewhere, and

b.   (Which is the same thing) within that piece the level contours of one state variable are not parallel to those of another state variable. At the moment captured

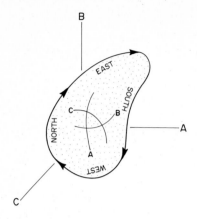

Figure 2. As in Figure 12 of Chapter 8 but geographic labels are affixed to the image, rotating with it. Three level contours are also indicated: one for A, one for B, and one for C. These are fixed in concentration space and therefore rotate in relation to the geographic labels in physical space.

in Figure 2, A increases from north to south within the rotating center of the disk's image. B increases from west to east and C increases from southeast to northwest. All three gradients rotate together.

I now argue backward to this situation from:

1.   The observed existence of a spiral wave (see below)

2.   An experimental method of creating a rotor by abutting a wave onto another piece of medium not containing a wave (Winfree, 1974d, in malonic acid reagent), and

3.   An experimental method of creating a rotor by shearing a preexisting wave (Winfree, 1972c, also in malonic acid reagent)

1: *To show that the existence of a spiral wave implies features* a *and* b.   Suppose a wave rotates about a hole. On each ring concentric to the hole, nearly the same sequence of states is realized in clockwise order around the ring. In other words, each ring maps with winding number 1 onto a common ring of states in composition space. Now plug the hole with a disk of the same medium and assume that the wave continues to rotate around the former border of the hole. Suppose this central disk, this plug, adopts a distribution of compositions compatible with the wave. How does the central disk map into composition space in such a

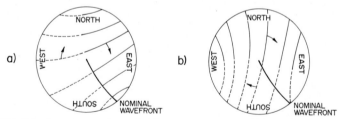

Figure 3. The disk whose image appears in Figure 2. In (a), contour lines of A rotate clockwise. At this moment maximum A occurs in the southeast and minimum A occurs in the north. In (b), contour lines of B rotate synchronously with those of A. At this moment maximum B occurs in the east and minimum B occurs in the west. The indicated wave front is the locus of A maxima on concentric rings. Alternatively, a wavefront could be a certain level of A or B, etc.

Figure 4. If to each *A* level there corresponds a unique *B* level, then the image in concentration space of any ring, and in fact of the whole two-dimensional medium, would be one-dimensional.

way that its boundary continuously joins the border of the hole? It has to map across the interior of the cycle in composition space as in Figure 2. This is feature a.

Now let's do it in terms of level contours of concentrations *A* and *B* in the physical medium (forgetting *C* for simplicity). Traveling clockwise around the border or around any inner ring concentric to it in Figure 3a, *A* must fall and rise back to its maximum. Call the maximum of *A* along any ring the wave front. Contour lines of *A* are indicated. There are two branches of each level contour: one branch (dashed) where concentration is falling clockwise and the other (solid) where it is rising back again. *A* and *B* do not vary together along any ring, otherwise their cycle in composition space would be flat as in Figure 4 and there would be no asymmetry to distinguish clockwise from counterclockwise circulation of the wave. *B* rises and falls around each ring, but out of phase with *A*. Thus the central disk is seen to contain a concentration gradient of *A* pointed north-south, and a concentration gradient of *B* pointed east-west. It contains crossed gradients of these two essential state variables. This is feature b.

In contrast, a disk of medium entirely on the attracting cycle (therefore necessarily with winding number $W = 0$) has all its concentration gradients parallel. This can be seen as follows. Each point necessarily has some phase on the cycle, so that contour lines of uniform phase can be smoothly drawn without singularities on a snapshot of the disk. All concentration gradients are perpendicular to those contours. Thus they are locally parallel.

2: *To show that a method of reliably creating a rotor is compatible with features a and* b. The method I have in mind starts with a wave in a two-dimensional piece of medium and a second piece of identical medium containing no wave. The two are shoved together as in Figure 5. Concentrations initially vary discontinuously across the dotted seam. How do these discontinuities resolve themselves into a continuous gradient? We depict the level contours of *A* in the upper left half of Figure 6 and of *B* in the lower left half. The right half shows how the concentrations would be expected to spread into the unexcited medium in the first instant of contact, prior to any significant degree of reaction. Figure 7 superimposes these two right halves. Gradients *A* and *B* are seen to criss-cross through the cycle average values of each state variable in the stippled region. This is feature b.

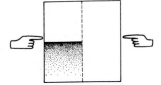

Figure 5. Two rectangles of medium are pushed into contact. The one on the left bears a wave moving from bottom to top.

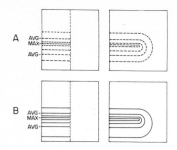

Figure 6.   (left) *A* and *B* levels at the moment of contact in Figure 5. (right) Shortly thereafter, when initial discontinuities of concentration have been resolved: equal levels (marked "AVG") behind and in front of maximum connect together.

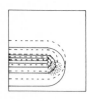

Figure 7.   The right side of Figure 6, combined to emphasize criss-crossing of the average levels of *A* and *B* in the stippled region.

Now let's look at the same manipulation in terms of the images of the two blocks of medium in composition space. The left half block maps degenerately along a one-dimensional trajectory from near-equilibrium through excitation and back to near-equilibrium (Figure 8). This is the locus of states realized during the traveling wave cycle. The right half block maps even more degenerately to a single point, the equilibrium. When the two are joined along the seam, each pair of abutted points becomes the center of a little composition gradient stretching in a straight line in composition space from the composition of the left block at that point along the seam to the (uniform) composition of the right block. Thus area elements from the inner edges of both blocks are pulled into the interior of the cycle. This is property a. This procedure is implemented experimentally in Winfree (1974d) and is implemented computationally in Winfree (1978a). It seems to bear an intriguing analogy to the regeneration of limbs on animals manipulated surgically in analogous ways. (See Chapter 16.)

3: *To show that the "shearing" method of creating a rotor in an excitable chemical medium is also compatible with features a and b.* Start a solitary wave in a thin

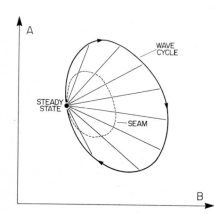

Figure 8.   *A* is plotted against *B* along the wave moving up the left side of the squares in Figures 6 and 7. The entire right half of each square is initially at steady-state (called "equilibrium" in the text). At the moment of contact each point along the seam abuts (radial lines) a point at steady-state. The seam then comes to adopt a midway composition (dashed).

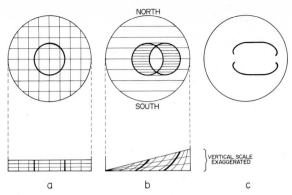

Figure 9.   (a) A cylindrical wave propagates away from the center of a shallow dish of liquid excitable medium; (b) the dish is tilted, shearing the cylinder; (c) only the north and south sides of the cylinder survive, while the east and west sides are thoroughly mixed with liquid at steady-state. The result is four free end points of wave front!

layer of the malonic acid reagent. The wave propagates outward as an expanding ring. Gently tilt the dish containing this liquid reagent, shearing the upper layers of fluid relative to those closer to the bottom of the dish as in Figure 9. The east-west shearing has little effect at the north and south edges of the wave because there the fluid flows parallel to the wave front. But at east and west the tilting stretches each cylinder of uniform composition between closely overlying and underlying layers with composition much closer to equilibrium. Figure 10 shows the northeast quadrant (shaded) of the circular wave before and after the operation of Figure 9. Along the three radial loci the chemical state initially varies from equilibrium through a rise of A and then of B (the wave) and back to equilibrium (before, right). But after the spatially graded mixing is carried out, the changes along path 3 are a lot more attenuated than along path 1. The image of the shaded region, formerly one-dimensional on account of the polar symmetry of the original wave, has been made two-dimensional by the stimulus gradient. Because the stimulus consists of a mixing of states originally present on the wave cycle (before, right), the new image (after, right) fills the inside of that cycle.

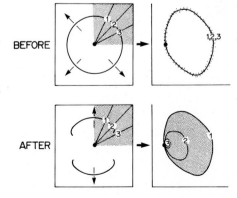

Figure 10.   The circular wave of Figure 9(a) and (c) is mapped into the concentration space of Figure 8.

The image thus shows property a. The preceding paragraphs have repeatedly shown that property b is equivalent. The result in this experiment is a rotor in each of the four quadrants of Figure 9(c), with appropriate mirror symmetries.

Bibliography of Examples

Experimentally, rotors in excitable media have been on record at least since 1963. Suzuki et al. (1963) had established such waves in a two-dimensional grid of iron wires in nitric acid. This is an electrochemical system which, in its simpler one-dimensional arrangement, has been studied as an analog to electrical excitation in nerves. Unfortunately, their Japanese publication and sequelae were never translated to English until 1976 (Suzuki) and came to my attention only in 1978. With the exception of Yoshizawa (below), all subsequent rediscoveries in diverse experimental systems were apparently also made in ignorance of the Nagumo result: Rotors in excitable media were found by Gerisch (1965) in the social amoeba (see Chapter 15), by Yoshizawa et al. (1971) in an electronic network, by Zhabotinsky (1970) and Zhabotinsky and Zaikin (1973) in the malonic acid reagent (see Chapter 13), by Allessie et al. (1973) in rabbit heart muscle, by Petsche et al. (1974) in rabbit cortex, by Shibata and Bures (1972, 1974) in rat cortex, and by Martins-Ferreira et al. (1974) the retina of the eye (see Chapter 14).

Using various mathematical approximations to an excitable but not oscillatory continuum, stable rotors were produced numerically, analytically, and/or by graphical arguments by Selfridge (1948), Buerle (1956), Farley and Clark (1961), Balakhovskii (1965), Krinskii (1968), Gulko and Petrov (1972), Winfree (1974b and f, 1978a), and Karfunkel (1975); and by Reshodko and Bures (1975), Greenberg and Hastings (1978), and Greenberg et al. (1978) who work with a reticulated medium like the one introduced by Moe et al. (1964) for computer simulations of fibrillating heart muscle. Pursuing a more mathematical analysis, Greenberg and Hastings (1978) ingeniously exploited a discretized analog of the winding number that plays such a conspicuous role in association with phase singularities of continuous media.

To give only one example of a rotor, I choose my own computations (op. cit.) using piecewise linear kinetics. This dynamic invokes only two variables often supposed to represent measures of excitation and refractoriness in nerve mem-

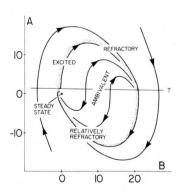

Figure 11.   A specific implementation of excitability using two variables named A and B as in Figure 8. The fine trajectories sample the vector field of Equation (1), which is linear above and below threshold T, The stippled ring with electrophysiological jargon on it is the cycle roughly followed in waves of excitation.

brane. In that application, adjacent cells are coupled by diffusion only through the first variable (voltage). I took the same kinetics to represent the nerve-like behavior of the malonic acid reaction (Chapter 13), and so coupled adjacent volume elements by diffusion equally through both variables (departures of the concentrations of small molecules from an equilibrium level).

Diagrammatically, the local kinetics (without diffusion) is as portrayed in Figure 11. In terms of numbers,

$$\frac{dA}{dt} = -A - B \quad (+20 \text{ if and only if } A > 1)$$

$$\frac{dB}{dt} = \frac{A}{2}.$$

(1)

In one simulation, a 1-mm square was divided into 61×61 cells, each harboring this kinetics. Each interior cell was connected to its four neighbors by Fick's law of diffusion, according to which A and B flow between adjacent cells at a rate proportional to the instantaneous concentration differences, $A - A'$ and $B - B'$, respectively. (Boundary cells have three or two neighbors only.) Figure 12 shows the result of a prolonged computation in which a rotor formed and stabilized. Each area element is here mapped into (A, B) composition space. All continually revolve along paths parallel to the heavy rings. A central area element (darkened) holds fast, its local kinetics exactly balanced by diffusion from its neighbors.

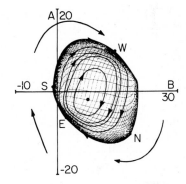

Figure 12. A square of excitable medium bearing a rotor is mapped onto the dynamical portrait of Figure 11. This is a moment in the periodic steady-state of a computation using Equation (1) together with no-flux boundary conditions and crossed-gradient initial conditions. Area elements in this 61×61 grid deviate from the trajectories of isolated homogeneous reactors: They hang together, circulating along paths such as the five concentric rings shown.

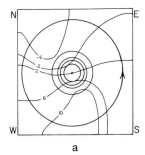

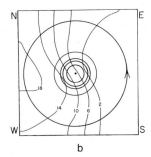

a   b

Figure 13. Figure 12 is replotted in dual representation as level contours of A (left) and of B (right) on the 61×61 cell grid of excitable medium. The five circles are the ring paths of Figure 12.

Box B: Spiral Waves in the Sky

In the physical sciences we are accustomed to waves of diverse sorts, emitted as concentric closed shells from a radiating source. But where is one to find a *spiral* wave? It turns out that they are not quite so uncommon as one might have supposed. In the simplest case, a rotating source of particles wraps itself in a spiral locus; this is apparently the typical configuration of the solar wind, e.g., winding through several turns as far as the orbits of the middle planets (Gosling and Hundhausen, 1977). Though the physical principles involved are a bit more subtle, much the same geometry arises in the magnetic field surrounding a rotating dipole. This is thought to be a prominent feature of pulsars (Ostriker, 1971). On a still grander scale, the process of blue star formation is thought to be "contagious", resulting in solitary waves of creation within the gas that constitutes most of any galaxy's mass (Mueller and Arnett, 1976). Behind the wave of creation lies a wave of supernovas and behind that lies a refractory zone that persists until gas again wanders into the relative vacuum. The simplest models of discrete-state excitable media (Wiener and Rosenblueth, 1946) predict involute spirals rotating about any long-lived holes in the susceptible continuum. Later refinements (e.g., Krinskii, 1968) allow spirals without holes. But even without these effects, galaxies develop spiral structure through the continual shearing of waves by the faster rotation of the galactic core (Mueller and Arnett, 1976; Gerola and Seiden, 1978).

The less illuminating but more familiar dual representation appears in Figure 13. Here we see the square covered by level contours of concentrations A and B. This pattern rotates counterclockwise about the central pivot cell at constant angular velocity, as shown in Figure 14. The corners, given compass labels N, S, E, W, are indicated by heavy dots in Figure 12.

Spiral Waves in Two Dimensions

In the preceding biophysical contexts among others (see Boxes B and C) a rotating source emits a wave into the surrounding medium, which conducts the disturbance at essentially uniform velocity. The resulting wave resembles a spiral with fixed radial spacing between its successive turns. Along any radial

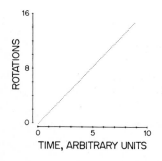

Figure 14.   The angular orientation of the $B = 7$ contour line of Figure 13, plotted at 1-second intervals spanning fifteen 12-second rotations.

Box C: Pastureland as an Excitable Medium

A "fairy ring" is a visibly distinct locus in a pasture or on a big lawn of grass that rather resembles a circle or fragments of a circle. It turns out to indicate where a certain basidiomycete fungus has recently been active. Parker-Rhodes (1955) wrote:

> A large grass field carrying a system of fairy rings can be visualized as like a pond in a light shower of rain. Each ring grows outwards at a constant (or irregularly fluctuating) rate, and new rings are added to the system with approximately constant frequency. When rings of the same species meet, then, unlike ring waves on water, their intersected portions are obliterated; but as between different species either or both may survive the intersection. The important thing is that no ring ever remains still, but only stops growing when it dies. However, all these processes take place with extreme slowness; in the rain-drop analogy the rings grow at a few feet per second, and new ones appear at the order of ten per square foot per second, whereas fairy rings grow at a few feet per year and appear less often than ten per hundred acres per year.

Ritzema Bos (1901) used the analogy of a grassfire in describing how fairy rings constantly advance, sweeping around obstacles, mutually annihilating one another in collisions and leaving a swath refractory to infection by the same species for some distance behind the advancing wave.

Parker-Rhodes's description might be taken almost word for word to describe the bull's-eyes of concentric circular waves in the malonic acid reagent (Chapter 13), except for one detail. The source, being the germination of the single spore, is not periodic. But he goes on to describe the effect of a breach in the wave front, such as might be induced by temporarily sterilizing a few yards of soil:

> Let us suppose that by some means . . . a length of the perimeter of a growing ring is abolished. The subsequent growth of such a ring should ideally proceed with two incurving horns which, when they meet, will reconstitute the closed perimeter and at the same time abstrict a T-shaped portion from which a new ring could grow inside the parent. Such a secondary ring will belong to the next impingent group after that of its primary; the points of origin of secondaries will lie at a more or less definite distance within the effective perimeter of the primary.

This is indeed a very close description of what happens in the malonic acid reagent when a solitary ring is broken by temporary blockade of a small portion of its perimeter.

Continued propagation of an interrupted ring cannot heal the two open end points. They should become centers around which spirals form. Fusion of adjacent spiral arcs makes rings.

Each end point becomes a rotor and waves from the two adjacent mirror-symmetric rotors periodically collide as diagrammed. Unfortunately, I have not located any scientific account of such fairy spirals, but I believe I have *seen* them in the lawn surrounding Salisbury Cathedral in England.

path out of the source one observes a periodic wave train. Along any closed ring around the source, cutting each radius once, the phase of this rhythm undergoes a net change of one cycle: The winding number of phase around the ring or of the ring's image around the circle of phase is $W = 1$. (See Figure 26 of Chapter 2).

To plot the time evolution of such a wave we might stack up a series of snapshots taken at unit intervals of time. The whole pattern rotates about its pivot as time advances upward, sweeping out a screw-shaped surface (Figure 15). This surface could also be constructed by plotting vertically the recurrence times of excitation of each point in the plane or by plotting the phase of the rhythm at each point in the plane as in the pinwheel experiment in Chapter 2. Any horizontal cross-section through this surface is a snapshot of the spiral wave rotating in the plane.

So far as algebraic description is concerned, Archimedes' spiral is the simplest. This is the shape of a neatly coiled rope and of the groove in a phonograph record. In normalized polar coordinates ($0 < \theta < 1$, unit spacing between turns) it is described by

$$\theta = r.$$

This is the locus of any fixed phase ϕ_0 taken to mark the wave front. To be more accurate, this is the locus of a stream of particles emitted radially from a uniformly rotating source like a garden sprinkler. Regarded as a wave, this locus has the peculiarity that its radial velocity rather than its velocity perpendicular

TIME

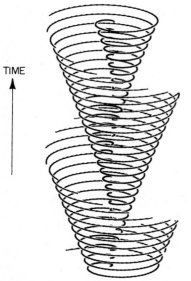

Figure 15.   Plotting wave arrival time above the plane of a wave emitted from a rotor is equivalent to stacking up successive snapshots of the wave. The surface so constructed resembles a screw. Its singularity is above the pivot.

to the local wave front is everywhere the same. This difference is conspicuous near the center where the perpendicular to a wave front is far from radial.

Another description of a spiral wave emitted by a rotating source is obtained by strict adherence to the assumption that the wave front locally propagates at unit velocity perpendicular to itself. This idea is embodied in Huygen's principle for conduction of light or sound or any other disturbance in a uniform medium with uniform propagation velocity. Taking the most symmetric case for illustration, one might simultaneously adopt this velocity rule and a second requirement, i.e., that the spiral wave's advancement in time be indistinguishable from a rotation. Taken together, these two rules imply

$$\left(\frac{d\theta}{dr}\right)^2 = 1 - \frac{1}{(2\pi r)^2}.$$

(See Winfree, 1972c, for a derivation in context of chemical waves.) This is to be contrasted with the previous case, which might be rewritten:

$$\left(\frac{d\theta}{dr}\right)^2 = 1.$$

The difference is conspicuous only at small r. Note the curious dilemma: Unlike Archimedes' spiral, this curve cannot continue toward the center beyond $2\pi r = 1$ because then $d\theta/dr$ turns from real to imaginary. This conundrum draws attention to the ultimate incompatibility of our two rules: Close enough to the pivot of a rotating wave, movement must be slower than any chosen speed unless the rotation period decreases toward zero.

This is the "involute" wave first described by Wiener (1946) as an approximation to waves of excitation on the surface of the living heart (Figure 16). The involute idea was later elaborated by Stibitz and Rytand (1968) in context of experiments on animal heart and by Durston (1973) in context of aggregation waves in slime mold. These models implicitly assume that the underlying kinetics has only a single variable of state as in a ring device, and they assume an unflinchingly steadfast velocity. On both counts such models conjure unreasonable paradoxes at the pivot. Other reasons that an involute wave cannot be exactly right are

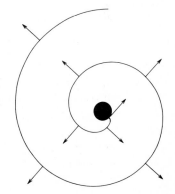

Figure 16. An involute of the circle can be drawn with a pencil tethered to spool of thread, and unwinding it. If every element of arc along this locus moves perpendicular to itself at the same speed (arrows) then the locus rotates rigidly.

Figure 18.   A locus of fixed composition, determined by at least two independent quantities, can cut a ring just once and then vanish (as $A = 1$ and $B = 1$ diverge).

So when my student André Laszlo brought to my attention the recent publications by Zaikin and Zhabotinsky (1970), we thought we understood why only ring-shaped waves were reported there. To check this argument, it seemed imperative at least to prepare the Zaikin and Zhabotinsky reagent, and to try various tricks which might induce spiral waves if the argument were somehow mistaken.

It came as a bewildering surprise, on October 10, 1970, to behold several perfectly stable spiral waves sedately rotating in a dish of this chemical reagent. This saved the oscillator interpretation of ascomycete morphogenesis. But what was wrong with the topological argument? The error lay in confusing *composition* with *concentration* in the second paragraph.

That identification is valid if and only if chemical composition, as it affects fungus growth or chemical wave conduction, is determined essentially by some single concentration. Figure 18 shows that if composition varies in two or more ways, a locus of fixed composition (for example $A = 1$, $B = 1$) can cut a ring around the pivot at an odd number of places while the concentration contour lines $A = 1$ and $B = 1$ each cut the ring an even number of times, as they must. Thus, spiral morphogenesis in fungus growth or in the malonic acid reagent is a mystery only in terms of single-concentration interpretations. The alert reader may retort, "Okay, it can cut *a* ring at an odd number of places. But to salvage the topological argument above requires that it cut *any* ring around the pivot an odd number of times. It can't do *that* unless the contour lines join at a sharp angle at the pivot dot, which is physically unreasonable. Moreover, only one composition locus could do so, but we're implicitly asked to believe that most, if not all, compositions achieved along this spiral follow spiral contours."

That's right. The topology of a spiral cannot be maintained all the way into the pivot. In fact, the "pivot" is an idealization, like the position of an electron, which serves a purpose only when not examined too closely. The practical essence of the spiral is that there are an odd number of intersections along any ring big enough to securely enclose the pivot, but the pivot cannot be localized within a minimum area. To think realistically and precisely about the structure of this critical disk requires that we turn inside out our habitual way of thinking about spatial distribution of reactions, thinking of the physical medium mapped into a concentration space rather than concentrations mapped onto the physical medium.

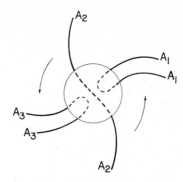

Figure 19. A wave is bounded both fore and aft by the same A concentration level, although the composition differs fore and aft because B's level contours are not parallel to A's (cf. computation, Figure 13).

In terms of composition involving two or more concentrations, there is no equivalent to the notion that what goes up must come down. *It need not*, so that an odd number of intersections is feasible. An odd number of crossings is also feasible in models based not on concentration (topologically, a point on the real number line $\mathbb{R}^1$), but on a phase of a simple-clock oscillation (topologically, a point on the circle $\mathbb{S}^1$). However, in this case the unavoidable implication is a discontinuity (e.g., a phase singularity) inside any such ring. This is physically unrealistic in any continuous, simply-connected medium spatially coupled by molecular diffusion. So simple-clock models are excluded in all these phenomena, just as are single-concentration models.

Another way to resolve the dilemma posed by a dish full of spiral waves is to recognize that although a concentration level contour may be a wavefront, the converse is not true: A wavefront is not a concentration level contour. It is *part* of a concentration level contour, the rest of which goes around behind the wave front to cut the boundary again, thus ensuring an even number of intersections, and removing the paradox (Figure 19; cf. Figure 13).

The notion that the contours of chemical composition must be rings or pieces of rings seems to have deep intuitive roots, perhaps because we take our cues from the simplest conjecture that composition equals the concentration of a single important substance. This notion impeded until 1969 the recognition that "rings" in fungi are often spirals (and there is a huge literature of rings, going back to pre-World War I German botanists). It impeded the recognition of spiral *Liesegang* rings for 30 years; in fact the cover photo of a book on these "rings" of chemical precipitation is in actuality (Figure 20) a photo of six parallel spirals! While studying electrical waves on the beating heart, Wiener and Rosenbluth (1946) conceived an otherwise fruitful single-variable model of excitation which led him to the mistaken conclusion that spiral waves could not exist. Selfridge (1948) refined the argument to show that they could exist but are unstable. Balakovskii (1965) and Krinskii (1966) and Kuramoto and Yamada (1976) argued further for stability, ignoring the implicit discontinuity at the spiral's source. All these problems evaporate when we quit restricting our thoughts to single-variable dynamics.

A similar assortment of riddles present themselves in context of morphogenesis in animals. Locke (1960), Lawrence (1970, 1971), and Lawrence et al. (1972)

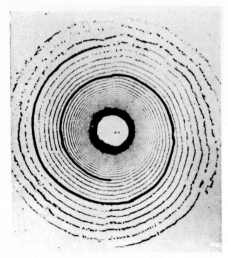

Figure 20.   Six spirals (one highlighted for this figure) from Rothmund (1907, Figure 1), as pointed out by the photographer Liesegang (1939).

perceived the wrinkle patterns on insect cuticle as contour maps of the concentration of a single morphogenetic substance. If a morphogenetic gradient were ever found to run in a closed ring, a fundamental challenge would seem to be posed for the notions that the gradient is a substance gradient and that local polarity is its direction of dilution.

This is exactly what has occurred in the case of limb regeneration in insects. The clockface model of French et al. (1976) and of Glass (1977) meets the challenge of an ostensibly continuous circular gradient by entertaining the notion of a morphogenetic phase variable. A phase gradient can close in a ring. To my mind, a tidier and more conservative approach would be to extend the substance gradient paradigm to encompass *two* substances whose concentrations jointly determine the local cell type. This would eliminate the phase singularity. Chapter 16 elaborates this approach in context of the experimental literature, in essentially the style of this chapter and the previous, but without invoking any time periodicity. In more mathematical terms, essentially the same response has independently appeared in a recent paper by Cummings and Prothero (1978).

C: Three-Dimensional Rotors

Isochrons and Phaseless Sets Again

An excitable continuum inhabited and organized by a rotor is driven periodically at the rotor's period of rotation. Every volume element can thus be assigned a phase relative to some reference clock oscillating at the same period. What do the loci of uniform phase look like? The wave front is one such locus. Its earlier positions mark other such loci. We might call these loci *isochrons* (loci of same time) even though in present context the isolated volume element is not an oscillator. The isochrons are thus defined only in real physical space, not in composition space. Goodwin and Cohen (1969, appendix) adopt this descriptive

Figure 21. A rotating wave in the malonic acid excitable medium (Chapter 13) is almost an involute spiral. Parallel involutes drawn 400 μm apart by computer approximately mark past positions of the same wave. They are thus isochrons. They cannot be reliably followed inward past the evolute circle but appear to be converging to the pivot.

convenience in using the eikonal equation[1] to assign a phase to each cell in an organism whose development is supposed to be governed by periodic waves.

In the case of a rotating wave in two dimensions it is clear that the isochrons necessarily converge somewhere near its center. If the wave strictly rotates about a pivot point, like a picture engraved on a rotating turntable, then the isochrons converge toward the pivot (Figure 21). Inside a central disk of circumference ≈ 1 wavelength, recognizable phases are not realized, owing to the mixing of states through molecular diffusion. But for purposes of gross description, I want to ignore the finite dimensions of this disk, and now pretend that isochrons come all the way in to the pivot point. Phase can be defined operationally on any circle concentric to the pivot by simply watching the periodic fluctuations of any concentration at any point. Whenever such a concentration reaches its maximum, call that phase zero and measure off a full cycle of phase behind it around the circle through the pivot. Concentration fluctuations on the tiniest circle will not amount to much, so the chemical state realized at each phase on a tiny circle is not identical to the state at that phase on bigger circles where the full-amplitude cycle of excitation is realized. Nonetheless, these isochrons permit us to deal with timing relations throughout the whole continuum. (We can do this only if concentrations vary in a strictly periodic way. They do in computer solutions of some equations, but not in others, nor in malonic acid reagent. In the latter cases the periodicity approximation gets worse closer to the pivot. I wish here to sweep such matters under the rug.)

The isochrons so defined converge to a pivot which is therefore a phase singularity. This simply means that there is no discernible timing relation here relative to the reference clock for the very good reason that concentrations do not change in time at the center of a rotating pattern.

What becomes of such a point in the larger context of a three-dimensional continuum? Like the intersection of isochrons in state space, the zero-dimensional

---

[1]   The eikonal equation formalizes the notion that every piece of wave front advances parallel to itself at constant speed. See Keller (1958).

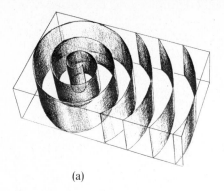

(a)

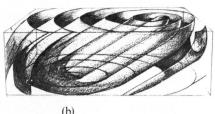

(b)

Figure 22.   Scroll-shaped waves have a spiral cross-section perpendicular to the pivotal axis from which they emerge. In (a) the axis is a short upright line segment. In (b) it slants from floor to ceiling. In (c) it curves around in a U. From Winfree (1973b) Copyright 1973 by the American Association for the Advancement of Science.

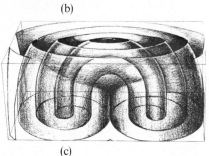

(c)

convergence point of isochron lines in two-dimensional physical space becomes a one-dimensional convergence line of isochron surfaces in three-dimensional physical space. The simplest three-dimensional extension of the spiral wave is a scroll (Figure 22). Its rotation axis is the three-dimensional equivalent of the pivot point in two dimensions. It is a one-dimensional phase singularity. This rotation axis threads its way through the three-dimensional continuum as a filament of ambiguity, so far as phase relations are concerned. Clark and Steck (1979) argue that this actually happens in the migrating slime mold slug.

Scroll Rings

How does it end? It doesn't. Apart from the possibility that the medium itself might end (at a glass wall in the case of malonic acid reagent), its usual behavior is to make a ring [Figure 23(b)]. This may seem strange. Isn't it rather unlikely that a one-dimensional locus meandering through three-space should exactly close on itself? Certainly so if that were the real structure of this situation. Were we

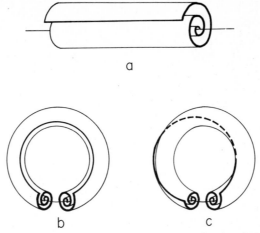

a

b   c

Figure 23. A scroll ring may be envisioned by joining the ends of an imaginary cylinder containing a scroll wave (a). This can be done as in (b) or as in (c), imparting a twist of 360° (or multiples thereof) to the cylinder before joining its ends.

dealing with a continuum of ring devices, that indeed would be the case: The isochron surfaces would be arbitrarily positioned, and so would be their convergence, so that the locus of convergence would not typically ever intersect itself or close into a ring. But then neither would there be a convergence locus: A singularity is a serious matter, strictly an impossibility, for ring devices because it implies finite differences of state between volume elements arbitrarily close together.

So let's go one step further toward realism. A biophysically realizable excitable medium typically consists of interacting chemical reactions. Its state is not determined unless at least two concentrations have been specified. The organization of reaction in space is a mapping of the physical continuum $\mathbb{R}^3$ or I^3 into a composition space such as $\mathbb{R}^2$. A whole one-dimensional set of the three-dimensional continuum thus maps onto each composition. In particular the pivotal composition is realized along a one-dimensional set: a thread in the physical continuum.

Looked at another way, the pivotal composition is a particular combination of at least two concentrations. Call them A^* and B^*. Let's think about that. The locus along which $A = A^*$ is some two-dimensional surface in the three-dimensional continuum. The locus along which $B = B^*$ is some other two-dimensional surface. If a composition (A^*, B^*) is realized, it is an intersection of those two surfaces. So it is a one-dimensional locus. Though such a locus might end at the boundaries of the medium (e.g., a glass wall), its typical configuration is a ring as in Figure 24.

These arguments are hardly scholarly. They obviously become sticky when one thinks of the composition space as involving three or more reactants. And the geometry of mapping I^2 or $I^3 \to \mathbb{R}^3$ with a singular locus closed in a ring is quite difficult to visualize. It is not easy to bring our three-dimensional geometric intuition to bear on problems involving five or more dimensions.

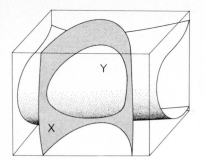

Figure 24. If two non-self-intersecting orientable smooth surfaces intersect in three-dimensional space, they typically intersect along a ring. The ring may be knotted.

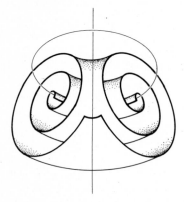

Figure 25. A scroll ring, sliced open along diameters (similar to Figure 22(c) but viewed from below).

Nonetheless let's forge ahead as best we can. If the scroll's axis typically closes in a ring, then scroll rings should be easy to demonstrate in three-dimensional excitable media such as the malonic acid reagent. This has been done. Scrolls of 1-millimeter wavelength have been found in malonic acid reagent, with the axis closed in rings of any length exceeding about 1 millimeter. This was demonstrated in serial sections of waves chemically fixed in a three-dimensional block of reagent (Winfree, 1974d). The photographically enlarged sections, reconstructed on stacks of plexiglas sheets, resemble Figure 25.

Fancy Rings

With that much established as fact, we are encouraged to ask the next question: What other modes of ring closure might occur? For example, could the ring be knotted? Or could the spiral cross-section be gently twisted through $360 \times N$ degrees along the length of the closed ring? [Figure 23(c)]. Or combinations of both? These are all feasible geometries so far as the surfaces of uniform concentration are concerned. At least the once-twisted, unknotted ring can be realized in terms of concentration gradients from source loci without cutting or otherwise doing topological violence to the continuum (Winfree, unpubl. sculpture). Are such structures stable then? I think probably yes, at least as stable as the experimentally demonstrated spirals, scrolls, and scroll rings. Any instability would have to be based on long-range interactions, since the scroll is locally stable.

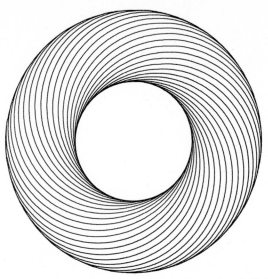

Figure 26.   Circular loci of fixed phase on the surface of a doughnut. Only half of each circle is visible. One of these circles is the edge of the wave in Figure 23(c). The others are its earlier and later positions. Any disk plugging the hole touches these isophase circles in order along its boundary. Phase thus increases clockwise through a full cycle around this boundary.

Long-range here means several millimeters. In the case of chemical scroll rings in the malonic acid reagent, the typical time scale for long-range communication through diffusion is in the order of L^2/D. For a scroll ring of diameter 2 mm, with diffusion rates typical of small reactants in water, this is about 30 minutes. This is much greater than the 1-minute rotation period of a scroll emitting the typical 1-mm waves.

So far as I am aware no one has looked into this matter computationally or experimentally. Some peculiar topological consequences invite testing by such experiments. For example, in the case of the once-twisted scroll, phase runs through a full cycle along any ring around the axis perpendicular to the plane of the ring ambiguity (Figure 26). Waves emitted from this ring therefore *converge* in a spiral *toward* that perpendicular axis (Figure 27). Along any two-dimensional surface bounded by a ring interior to the phaseless ring (the scroll axis), phase has winding number $W = 1$. How is phase organized along such a surface? The answer is a map of the two-dimensional disk to the phase circle. But we know that no continuous map exists unless $W = 0$ along the border. Therefore, at least one phaseless point (or locus) lies within every such surface. Since the three-dimensional medium is composed of a continuum of such surfaces, this argument indicates a new phaseless locus of at least one dimension threading the previously established phaseless ring. Does this new locus typically close in a ring, linking the first ring? Does it emit a spiral wave? Do all twisted scroll rings necessarily come in linked pairs?

Such questions abound in the study (not yet undertaken, so far as I know) of three-dimensional continua of oscillators. We confront such media in the malonic

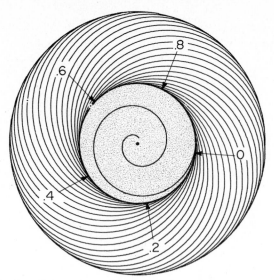

Figure 27.   The twisted scroll ring of Figure 26 is repeated to show the progression of phase through one cycle around an equatorial ring. Every disk bounded by that ring must therefore contain a phase discontinuity or singularity, presumably located along the hole axis. The spiral wavefront shown in that disk is converging toward the center.

acid reagent (Chapter 13), in cardiac muscle and in brain tissue (Chapter 14), in aggregations of *Dictyostelium* cells conducting pulses of cAMP (Chapter 15), in suspensions of yeast cells rhythmically processing sugar to make ATP (Chapter 12), in suspensions of mitochondria rhythmically making ATP by oxidative phosphorylation (Chance and Yoshioka, 1966), and in three-dimensional tissues composed of cells rhythmically dividing (Chapter 22) or harboring circadian clocks (Chapter 19). Any investigation of the geometric and topological constraints characterizing waves in such media would probably encounter something new almost immediately.

However, there are few true firsts in science. Questions about rotating patterns of activity, the threads about which such patterns rotate, their closure into rings, linkage among the rings, and interaction among such rings (attraction, repulsion, etc.) were addressed a century ago by Herman Helmholtz (1867) and Sir William Thomson (1867) (later Lord Kelvin).

Vortex Atoms

Helmholtz, Thomson, and contemporaries were concerned with the aether, then considered as an incompressible three-dimensional fluid, a continuum whose motions are organized by vortex lines not unlike the axis of our scroll waves. The idea was entertained that atoms are indestructible vortex rings. According to this vision, molecules are assemblages of vortex rings, linked, knotted, knitted, and otherwise bound together. Thomson even went so far as to calculate their hydrodynamic interactions in quantitative terms and to show that the principles so discovered lent themselves to quantitative interpretation of binding energies in reactions fashionable to the physical chemistry of his day.

Thomson's vortex atoms died a quiet death, forgotten rather than discredited by experiment (apart from the experiments that undermined belief in palpable aether). In excitable media we may have a new context in which something like the vortex atom theory can live again, strangely transfigured.

10. The Varieties of Phaseless Experience: In which the Geometrical Orderliness of Rhythmic Organization Breaks Down in Diverse Ways

A cautious man should above all be on his guard against resemblances; they are a very slippery sort of thing.

Plato, "The Sophist", translated on
p. 180 of F. M. Cornford,
Plato's Theory of Knowledge, 1935

In Chapter 1 we dwelt on the notion of smooth maps from one space to another. In Chapter 2 it emerged that certain kinds of mapping involving circles cannot be contrived smoothly. As an application we saw that certain kinds of experimentally observed continuity and smoothness involving measures that are periodic in space or time inescapably imply an unobserved (but observable) discontinuity. A phase singularity is one way to resolve this crisis of continuity implicit in the observation of nonzero winding number.

The six chapters 3, 4, 5, 6, 8, and 9 elaborated a variety of dynamical schemes commonly employed in biological time keeping. Those included were chosen for two essential features: periodicity and smoothness of mechanism. In each case it emerged that the topological crisis encountered in Chapter 2 has a parallel in the more concrete terms of mechanism and structure.

This feature acquires a disproportionate interest on account of the thoroughness with which our sciences are permeated with the idea of smooth maps. An assumption underlying almost all science is that similar causes have similar effects. It is this rule of thumb that we implicitly invoke in supposing that our deliberately simple-minded "explanations" of idealized aspects of things are useful approximations to the real causes of real-world messy phenomena. Without that faith, there would be no laboratories in which especially simple versions of phenomena are contrived for study and there would be no explanations of phenomena conceptually isolated from all the goings-on that surrounded them.

If science is mostly approximation and approximation means a smooth map from cause to effect then exceptional interest inevitably attaches to those isolated states near which the cause-to-effect mapping loses its smoothness, even its continuity. There is little we can do to satisfy curiosity about such things when the discontinuity is as abrupt as an earthquake, a mutation, or a miracle. But if the discontinuity is surrounded on all sides by smoothness (e.g., the trajectories and phase patterns of Chapter 7) and is even implicit in that smoothness, then we can creep up on the exceptional event gradually and know why it has to be there. This is the beauty of Thom's catastrophe theory, of the physical theory of state transitions, of the mathematics of singular perturbation and, I would add, of the singularities implicit in the geometry of biological time.

We have looked at the smooth geometric patterns of timing (or more abstract phase) that surround singularities, by which their existence is inferred. This we did both abstractly in Chapters 1 and 2, and in terms of component processes in Chapters 3, 4, 5, 6, 8 and 9. We now fix our attention steadfastly on the organizing center of those patterns: the singularity itself.

In doing so Sections A and B below will again follow the outline of mathematically oriented Chapter 2: Section A consists of three intuitively familiar cases, followed by six cases involving biological clocks not necessarily involving physical space, and six more critically involving spatial periodism; Section B concerns three cases in which there may exist a physiologically important phase singularity, but if so it cannot be inferred just from the geometry of smooth timing relations elsewhere. I hope you will find it useful in connection with many of these brief statements to read the corresponding chapter of experimental background in the Bestiary.

A: The Physical Nature of Diverse States of Ambiguous Phase

Example 1. Color Vision.   A short list of the most tantalizing mysteries prominent in daily experience could hardly exclude human color vision. Even the simplest aspect, the classification of homogeneous colors according to hue along a color wheel, has no parallel in psychophysics unless it be that we classify musical tones along a cycle by perceiving a repeat at octave intervals. This analogy, together with the color sensations induced by suitably timed rhythms of black and white, has prompted suggestions that the neural coding of color is somehow basically rhythmic, involving periodic patterns of firing by visual neurons (see Box B of Chapter 1). Unlike disparate musical tones, two different colors combine to make a new color. Implicit in this fact is a singularity of hue: a color of no hue or of ambiguous hue. Obviously this is merely gray, obtained by mixing complementary colors.

In present context the most important fact about gray is that it is a singularity only from the limited viewpoint of hue measurements. When one takes account of *saturation* as well as hue, gray no longer seems discretely different from nearby colors. Suppose there were mutants who perceived only hue and intensity, but not saturation. To such a person the singularity of hue would present a striking phenomenon. But when colors are distinguished according to saturation no singularity is perceived. It is evaded by the saturation going to zero at gray, making all hues indistinguishable: Grays slightly tinged by any hue are all perceived as similar, not as disparate points on a color cycle.

Example 2. Navigation in a Rowboat.   Here again the singularity is an artifact of an artificially restricted viewpoint. If we attend only to direction of movement, neglecting speed, then the zero of velocity is a singularity. A boat rowed directly upstream so as to remain at rest with respect to the shore has all directions or no directions, as you like. An arbitrarily small perturbation directs the rowboat in any direction. It is much the same in correcting the compass with the two little magnets attached to screws mounted in the case (Chapter 1, Section C, Example C): At a certain setting, the earth's magnetic field is exactly neutralized and any further adjustment, no matter how small, can give the needle a preferred orientation in any direction. Only the delicacy of its mounting hides from us the fact that the magnitude of the orienting field is passing through zero at the singularity of direction. In the same way the directional singularity implicit in wind patterns revolving about barometric highs and lows resolves itself simply in a calm at the center. Even in a tornado, the center is becalmed so far as compass direction of flow is concerned. The conservation of mass is satisfied by evoking an overlooked dimension; the wind roars straight upward.

So it is with singularities in general. As noted in Chapter 2, Section C, the ostensible singularity is always evaded by bringing into play a degree of freedom that remained inconspicuous everywhere else. This comes as a surprise only if you really didn't think there *was* another degree of freedom. The way in which that evasion comes about can be most revealing, a viewpoint which fanciers of black holes will warmly endorse, and which I also adopt in connection with biological clocks.

To return to the rowboat and its compass, nothing of much interest transpires at these singularities except that some vector passes through a reference zero in a perfectly smooth way. These singularities amount to little more than artifacts of the choice of coordinate systems and consequently have no further interest.

Example 3. What Time Is It?   On the scale of global time keeping, just as in the eye of a hurricane or a tornado, a singularity is inevitable: It is a pivot, a point of zero amplitude rotation, an axis about which the whole earth turns. Viewed from this axis the sun neither rises nor sets each day but simply boxes the compass, invoking a new degree of freedom, an "amplitude" of the day-night cycle which is zero if the sun keeps constant altitude.

Example 4. Timing the Tides.   An equally smooth attenuation of amplitude presumably explains the amphidromic points of the global tide charts (though to be strictly correct, I don't think anybody has ever identified such a point and monitored the sea level there). The tidal range falls to zero at each such point, with nearby points of the ocean surface slightly rising and falling in sequence around the horizon. It is worth noting that this vanishing of amplitude pertains only to one harmonic of the tide. It is not a vanishing of fluctuation because the other harmonics generally have their amphidromic points elsewhere. In this respect the singularities of the tidal rhythm resemble those of attractor cycle oscillators. Putting such an oscillator into its phaseless set typically ensures only that the subsequent trajectory, wherever else it might go, does not lead back to the standard cycle. This phaselessness amounts to absolute timelessness only in the special case $D = 2$, which in the case of the tides corresponds to recognizing only a single harmonic.

Example 5. The Fruitfly's Body Clock. Does any of the foregoing provide a useful clue to the meaning of phaselessness in circadian rhythms? And what has the fruitfly to tell us of such matters? First of all it does not seem likely that phaselessness in the fruitfly represents phase scattering in a population of simple clocks, if only because it can be achieved in mere seconds even under dim light. It seems unreasonable to suppose that ring devices could zip through half of a 24-hour cycle in so short a time. However a population of clocks each having two or more variables of state, one of which is photolabile, might easily achieve the large phase shifts required in an arbitrarily short time.

The Clock's Singularity

But setting aside the possibility that the singularity is an artifact of physiological incoherence, the first question to ask is whether the singularity might simply be the steady-state of a biochemical oscillation. Before we tackle that question, let's make sure not to beg the question by perpetrating a pun. As noted on page 73, the word *singularity* means one thing to mathematicians and physicists, and another to engineers, and both, as well as other things, to biologists. The phase singularities inferred on topological grounds throughout this book are singularities in the sense of mathematicians and physicists. They are discontinuities at which physically important variables change infinitely abruptly. This is the sense of Penrose and Hawking in their study of black holes as space-time singularities (see Misner et al., 1973), of Guckenheimer (1975) in his study of the phase singularities of attracting cycles, and of René Thom (1975) in his study of catastrophes as singularities of maps. The "physically important variable" in present context is the phase of the subjective daily cycle.

In contrast, the singularity commonly referred to in engineering literature is the critical point, equilibrium, or steady-state of a set of kinetic equations. This is the sense of Higgins's (1967) analysis of biochemical oscillations, of Pavlidis's (1967) analysis of circadian rhythms, and of Kauffman and Wille's (1975, p. 50) study of the dynamics of the cell cycle.

Unfortunately, for my present attempt to distinguish usages of the word, I used the term *singularity* in both senses simultaneously in the first three papers describing *Drosophila*'s phase singularity. These two senses do coincide in the limited context of two-variable models. I argued explicitly from experimental data that the phase singularity *is* a reaction steady-state in a state space of dimension $D = 2$ (Winfree, 1968). That argument could have been correct and these two senses of "singularity" could be reasonably taken to coincide, as Pittendrigh and Daan (1976c) choose to emphasize:

> [Winfree] has accomplished the delicate feat of "stopping" the *D. pseudoobscura* pacemaker by driving it onto the "singularity" of its phase plane where it is, apparently, stably motionless . . . . [Winfree] has done the same to oscillations in the redox state of NAD which he generates by disturbing the steady-state of yeast glycolysis . . . . In all oscillating systems capable of self-sustained limit-cycle movements, no matter what their physical nature, there is, as analytic necessity, a point equilibrium or singular state on the phase plane . . . . The only empirical question is whether or not that singularity is stable—and how stable.

Unfortunately, there is no such analytic necessity, nor do we know that we are working with a limit cycle process. The statement quoted is true only on the assumptions that a steady-state exists, that it is accessible, and that the only acceptable models are ordinary differential equations involving exactly two variables. Some models proposed in the circadian rhythm literature are of this sort but many others are not; e.g., simple clock models (Chapter 3), independent-oscillator population models (Chapters 4 and 6), time delay models (Johnsson and Karlsson, 1971, 1972, Engelmann et al., 1973, 1974, 1978), singularity-free discontinuous oscillators (e.g., Pavlidis, 1967a), differential equations involving three or more variables (Chapter 6 and Cummings, 1975), and coupled oscillator models (Pavlidis, 1976). It *could* be that $D = 2$ and if it is then the data plots of Chapter 7 acquire much more vital impact. But I no longer find the evidence persuasive on this point. (See the end of Chapter 7.)

The Clock's Steady-State

Following the tradition of enquiry into chemical oscillators (Chapter 12), oscillations in sugar metabolism (Chapter 13), neural rhythms (Chapter 14), and rhythms of cell division (Chapter 22), almost every conceivable conjecture has been offered about the steady-state of circadian clock.

1. It doesn't exist. The circadian cycle could be an oscillation with no steady-state and only slightly adjustable amplitude, as in Danziger and Elmergreen's (1957) models of the female cycle applied to the circadian cycle by Strahm (1964) and Pavlidis (1967a). Or it could be a simple clock (Swade, 1969), even a physical ring of DNA traversed periodically by a polymerase (Ehret and Trucco, 1967), or a physical rotor like the turbine wheel of bacterial flagellae. In such a model perturbations only vary the rate of change of phase. Nonphase states make no sense.

2. It does exist but is so violent a repellor that for all practical purposes it is inaccessible, and amplitude regulates completely back to the standard cycle from almost any subnormal amplitude within a day (Pavlidis, 1967b, 1968, 1978a quoted on p. 422).

3. It exists and repels (i.e., amplitude regulates away from 0), but not so quickly. The near closure of trajectories during the two days monitored after

perturbation to lower amplitude (Chapter 7) could be interpreted as lethargic regeneration of amplitude (Pavlidis, 1976).

4. The clock has more than the two variables assumed under 2 and 3. For example, Cummings (1975) invents a three-variable biochemical oscillator and Pavlidis (1976) supposes an oscillator composed of two or more oscillators coupled together, each contributing two variables. Thus, if a phase singularity is found, it is probably not because clock kinetics was forced to a steady-state, but rather because it was tricked into some other dynamic mode (possibly leading to the steady-state) conspicuously different from the usual 24-hour oscillation.

5. The steady-state exists and *attracts* from nearby states (Kalmus and Wigglesworth, 1960; Pavlidis and Kauzmann, 1969). Starting from below a critical amplitude, such an oscillation is unstable and degenerates to stationarity. Such a situation could even result from coupling many oscillators whose steady-states individually repel. Below a critical amplitude of collective rhythmicity, mutual synchronization is unattainable so that the population remains incoherent and any slight initial collective rhythmicity degenerates (Winfree computer experiments privately circulated, 1967).

6. The question cannot be answered from the observed behavior of whole animals because multicellular clocks consist of many cellular clocks functioning independently for all short-range purposes. In such a case the vigor, amplitude, or intensity of circadian oscillation in the whole organism, being a sum over many separate oscillators, would mainly measure the coherence of their phases. Any stimulus that further scatters phases within the population would thereby attenuate the collective rhythmicity, and if interaction were weak, then coherence would not soon recover. Without an independent assay of phase scatter among the oscillating units, little would be revealed about each unit's amplitude.

In using a succession of models, theorists have thus covered every base.

My favorite conjecture is 6, that the circadian clock is actually a composite of two or more independently competent oscillators and that the whole-organism rhythm reflects some kind of average over this population. In such composite clock models, the facts are explained by incoherence in the population. I suggested a model involving only these essentials, not even invoking coupling among the component clocks. This simplest, though perhaps least general, interpretation accounts for all the *Drosophila* data quantitatively (unpublished computations). The quantitative fit can probably be retained even with weak synchronizing interactions between the individual clocks (Pavlidis, 1976) or even with strong synchronizing interactions (Aldridge and Pye, 1979b). Any number of conjectures might suffice. The essential qualitative point is that the phase singularity is interpreted in terms of incoherence in a population rather than as a steady-state of a reaction. Given any such composite clock interpretation, the phase singularity could be a steady-state only if all the clocks were perturbed to their individual steady-states simultaneously. The fact that a phase singularity is obtained by adjusting only two stimulus parameters makes this an implausible interpretation.

In short, whatever evidence points to a composite structure of circadian clocks in multicellular organisms also points away from interpretations of the phase singularity as a reaction steady-state. A few experiments are suggested at the end of Chapter 7 which might help to resolve this ambiguity.

The Singularity in Development

A related question dealt with in more detail on pages 417–420 et seq. concerns the arrhythmicity of eclosion in naive pupae who have never experienced a change of lighting or temperature by which to start or set their clocks. Are they in the same state as are "rhythmically experienced" pupae returned to arrhythmicity by a singular stimulus? They *are* by the criterion that both populations become indistinguishably rhythmic after exposure to light or a temperature shock, though evidence from more discriminating comparisons would be reassuring. Until such further comparisons reveal unforeseen distinctions, it seems a plausible guess that the singular state in *Drosophila*'s clock is its primordial state, as assembled in the maturing egg prior to activation by an external stimulus.

Example 6. A Mutant Fruitfly's Body Clock.   *Drosophila melanogaster*'s clock has been probed nowhere near so intensively as *D. pseudoobscura*'s. But neither have any conspicuous differences been detected to distinguish the 19-hour mutant *melanogaster* from the wild type *pseudoobscura*, apart from the 20% difference in their periods.

Example 7. A Fly that is 1000 Times Less Sensitive.   *Sarcophaga* has not been examined in respect to its singularity, except to demonstrate its existence by the screw-shaped pattern of rephasing around it and the relative arrhythmicity of eclosion near it. Saunders (1978) believes that some forms of arrhythmicity in *Sarcophaga* represent incoherence among circadian rhythms within the individual fly. But the question whether this is true of arrhythmicity elicited by a singular pulse has not yet been posed experimentally.

Example 8. Resetting a Flower's Clock.   As noted in Chapter 21, Engelmann et al. (1978) believe that the arrhythmicity induced by a critical stimulus in the individual *Kalanchoe* flower represents a steady-state of the circadian oscillator throughout the flower, possibly because the clock's biochemical steady-state is locally an attractor. If this inference withstands the further tests in process in Engelmann's laboratory, it may provide a unique opportunity to implement the following experiment proposed by Winfree (1970b) and by Tyson et al. (1976).

On the assumption that diverse physical and chemical influences alter clock dynamics in diverse ways, then, applied to a clock at its steady-state, they should induce rhythmicity at diverse phases. If time be allowed for relaxation to the steady-state after the critical stimulus then the dependence on stimulus magnitude of the phase of renewed rhythmicity after a second stimulus should characterize the site of interference in clock dynamics. Thus, suitably selective interfering agents can be classified according to their sites of action (supposing they don't all act at multiple sites). The sites of action can thus be enumerated. The chemical nature of each site may be suggested by chemical properties common to those agents which apparently act at that site. There might also be a minimum dose required to elicit any rhythmicity, revealing that the arrhythmic state is a local attractor and thus confirming that it is not a mere artifact of incoherence. Engelmann has had experiments of this sort in progress since 1977 (personal communication). There seem to be technical difficulties in reliably inducing and maintaining arrhythmicity in the controls.

In the case of *Avena*'s 30-minute transpiration rhythm, the experiments of Johnsson (1976, Figures 4 and 8) and of Johnsson et al. (1979) gave particularly clear-cut results: The singular stimulus kicked the seedling into a locally attracting steady-state of water transport.

Example 9. A Phase Singularity for Beer Drinkers. The chemically well-studied oscillation of anaerobic glycolysis in yeast cells provides a nice opportunity to simultaneously test the above experiment in a better understood system. A start has been made through the independent experiments of Winfree (1972d), of Greller (1977), and of Aldridge and Pye (1979a,b). In these three investigations it was found that oxygen, acetaldehyde, and calcium elicit rhythmicity from previously arrhythmic cell suspensions respectively at the phase of NADH maximum, one-quarter cycle later, and one-half cycle later.

The nature of arrhythmicity in this system has not been clearly established. However, Greller (1977) and Aldridge and Pye (1979a,b) argue cogently both

1. that the individual oscillator has an attracting steady-state in addition to an attracting cycle, and
2. that the cell population is made inhomogenous by near-critical stimuli which divide it into a synchronously oscillating subpopulation and another subpopulation trapped within the steady-state's attractor basin.

These inferences may eventually be confirmed by direct microfluorimetry of individual cells.

There are now five ways to drive glycolysis into its phaseless manifold:

1. by mixing differently phased cell populations,
2. by applying a suitably timed critical dose of oxygen,
3. or of acetaldehyde,
4. or of calcium,
5. or of EGTA (which chelates calcium).

If glycolysis has a unique attracting steady-state, then all five techniques presumably direct cells eventually to that state, even if the five techniques enter the phaseless manifold at different points. But no one has yet excluded the alternative possibility that *different* arrhythmic states are reached by these five techniques. The question could be substantially resolved by reinitiating rhythmicity with agent *i* after annihilation with agent *j*. If the new phase depends only on *i*

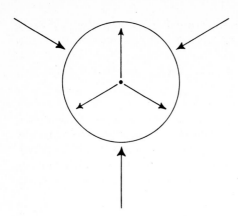

Figure 1. A bounded divergent flow in two-space necessarily includes a cycle (of "most-proximal" cell types encircling the "most distal" type).

of a process involving at least two mutually interacting factors). I am accordingly skeptical of theories which take the prominently featured ring as a cue for seeking causal answers in terms of oscillators, circulating waves, or singular points. Needless to say, I would rejoice to find this skepticism unfounded.

I think there *is* one way in which my pessimistic outlook could soon prove wrong. The fact seems increasingly well established that cells do not acquire their type specificities in a passive way by smoothly bridging the differences between their neighbors, but are rather invested with some dynamic. Goodwin (1976) was the first, I think, to suggest attributing to new cells a tendency to diverge from more distal types toward more proximal types in a way that might be summarized geometrically by a vector flow diverging from a point in state space. This divergence must be bounded and its boundary is topologically a ring (Figure 1). This ring presumably consists of the most proximal cell types, and the center of divergence is the most distal cell type. Thus it is at least conceivable that there exists a meaningful natural origin for a polar coordinate system which can be identified with the singularities implicit in the clockface scheme. To establish this as a fact would require monitoring the changing fate of proliferating cells.

Example 12. The Cuticle of Arthropods, A case in which the singularity even more clearly turned out to harbor only a mathematical artifact concerns the regular plywood-like laminations of skeletal material in arthropods. The passage of a microtome knife through cuticle elicits a pattern of periodic banding that might at first suggest rhythmic secretion by the cells that constructed the cuticle. (It did to me.) The observed bands would then be seen as markers of biological time, like archaeological strata, indicating the horizontal distribution of phase of those vertically secreting clocks. The wonder of it was that in many cases, most conspicuously in the lens cuticle of the compound eye, a ring could be drawn on the cuticle round which the phase of this putative "secretory rhythm" had nonzero winding number. The implicit singularity was plainly visible in the microscope as the inner end point of a spiral of relatively more opaque cuticle. But here for sure my enthusiasm to find biological expressions of phase singularities

---

Box A: A Structural Phase Singularity in Muscle Fibers

In iron whiskers, tobacco mosaic virus, and bacterial flagellae, a strictly structural rhythm is so arranged in space (around a screw dislocation) that a point of ambiguous phase is implicit. In the following case it also turns out to be biochemically unique.

The individual fibrils of vertebrate skeletal muscle show conspicuous bands a few micrometers apart along their whole length of some millimeters or centimeters. This astonishingly precise periodism reflects the muscle's underlying molecular architecture. Myosin filaments alternate with overlapping actin filaments in repeated structural units called sarcomeres. The sarcomeres are joined end-to-end in thousands of repeats like the unit cells of a crystal. A muscle fiber is composed of many such fibrils lying side-by-side with their sarcomeres exactly in register. The fiber thus gives the appearance of a stack of flat disks, each one sarcomere long, monotonously repeated.

With this architecture in mind it comes as a surprise to note that many kinds of *in*vertebrate muscle fiber are *helically* striated, like a barber pole. In such fibers the molecular filaments of actin and myosin are still aligned parallel to the fiber's long axis and they still repeat with exquisite periodism in this direction. But adjacent filaments are slightly out of register in a systematic way. Proceeding once around the fiber following the edge of a helical band, one advances along the fiber by some integer number W of unit cells. Or girdling the fiber at a fixed level, one crosses W helical bands. In other words the phase of the banding rhythm has winding number $W \neq 0$ around the fiber.

So much for the visible surface of the fiber. What about its interior? In vertebrate striated muscle ($W = 0$) each band visible on the surface is only the border of a disk extending smoothly through the interior. Can a disk have a helical boundary? It cannot. A helical boundary cannot be continued into the interior of a cylinder with creating a singular axis along which phase must be ambiguous. Does anything of physiological interest transpire there? In this case, it *does*: helicoidally striated muscle fibers have a central cytoplasmic axis utterly devoid of actin and myosin filaments! (Lanzavecchia, 1977 Figures 16 and 21.)

---

ran aground on the facts. It turned out, thanks to a remarkable geometric insight by Yves Bouligand (1972), that these "rhythms" derive from the bent-liquid-crystal ultrastructure of cuticle. As elaborated in Chapter 17, Gordon and I (1978) were able to argue that the bands of darkness are created by the influence of the microtome knife itself on the periodically oriented chitin fibrils of which cuticle is composed. The phase of this rhythm of banding is determined geometrically. It becomes ambiguous where the local screw axis of the liquid crystal is perpendicular to the plane of the cut. Mathematically it is indeed a phase singularity, but all it represents physically is a critical orientation relative to the knife. No rhythm of secretion, no pattern of phases is revealed. (See Box A.)

Example 13. The Cellular Slime Molds. What then of the phaseless center from which wave trains of cAMP release emerge in spiral pattern on a monolayer of *Dictyostelium* cells? Is that point different in any significant way from all other points, apart from being the locus of a mathematical phase singularity? It certainly becomes so soon enough. All cells within reach of those ripples of cAMP move toward the source and eventually accumulate into a multicellular organism at that site (see Chapter 15). But was there initially any special kind of cell at the phaseless origin of the rotating wave? I know of no reason to think so. In continuous media dynamically similar to the cellular medium of a *Dictyostelium* monolayer, spiral waves are created by temporary interruptions of continuity. A wave passing such a gap acquires two free end points, each of which becomes the source of a spiral wave if continuity is reestablished before the end points find one another again. In media composed of wandering cells such temporary interruptions presumably come and go all the time as local cell density fluctuates. In continuous media the source points so formed exhibit a unique interplay between reaction and diffusion by which a state is created and maintained which never occurs anywhere else. But in media composed of discrete cells communicating by pulses, the arguments of continuum mechanics and of topology have no clear application. I see no reason to believe that the phase singularity in *Dictyostelium* initially has any locally detectable unique properties. (But once again, being the center toward which cells collect, it soon *acquires* unique properties.)

Example 14. Pathological Rotating Waves in Heart, Brain, and Eye. On a grosser scale, waves geometrically similar to the slime mold's temporally organize the repetitive firing of excitable membranes in vertebrates. These patterns were initially discovered in the early 1970s, in every case by severe artificial stimulation. This only shows that neuromuscular tissue is capable of persistent rotating self-excitation. The normal role of such rotors, if any, remains to be discovered. In all cases but one the singularity is not available for examination, being lost in a central patch of tissue damaged by the stimulus. (See Chapter 14 for details.) In that one case (Allessie et al., 1973) the phaselessness of the singularity is nothing so simple as attenuation of amplitude to zero. The pivotal patch of membrane is not held stably at a voltage (and at ion permeabilities) intermediate between those realized in the wave circulating around it. Instead the rotor's central patch is irregularly invaded by sizable "wavelets", so its phaselessness consists of irregularity rather than of timelessness. Nonetheless it may appear as timelessness in a grosser macroscopic description of the phenomenon (see Box B).

It seems natural to interpret atrial flutter or ventricular flutter in terms of one or more rotors. What about fibrillation, the higher frequency, less regular mode of arrhythmia? One line of thought suggests that fibrillating tissue is possessed by many tiny rotors, continually being created and extinguished within a two-dimensional medium of spatially inhomogeneous refractoriness (Krinskii, 1966, 1968; Krinskii and Kholopov, 1967a,b; Krinskii et al., 1967). But a slightly earlier and apparently independent literature places less stress on the "tiny rotors" idea and more on the "many tiny" idea. These are the papers of Moe and collaborators (Moe and Abildskov, 1959; Moe, 1962; Moe et al., 1964) which I suspect were imaginative and deep beyond their time, yet remain today undeservedly neglected. Moe's

> variant of the circus movement theory proposes that fibrillation is maintained
> by the irregular wandering of numerous wavelets generated by the fractionation

Box B: The Human Heartbeat's Phase Singularity (?)

The figure shows two kymograph tracings recorded in 1913 by a young physiologist who, a few months after submitting them for publication, accidentally killed himself in electrical experiments on his own heartbeat. These tracings were made by a rabbit's heart whose regular oscillation was occasionally perturbed by an electrical stimulus of suitable intensity (Mines, 1914). As you can see following each dot on the tracings, most such stimuli induced at most a transient upset without resetting the sino-atrial pacemaker. But stimuli delivered at a critical phase instead stopped the heartbeat for a long time. Presumably this singular result requires not only the exact timing determined by Mines but also suitable stimulus intensity, in that a much weaker stimulus would have little effect and a much stronger one would resynchronize the heart muscle and SA node altogether. But when administered at this phase, the intensity employed induced fibrillation: microscopically a chaotic jumble of circulating waves, but macroscopically a phase singularity in the behavior of an elaborate partial differential equation oscillator.

So far as I know this is the first recorded encounter with a physiological phase singularity. The whole subject might have received much earlier development had not its discoverer's enthusiasm led so promptly to stopping his own heart.

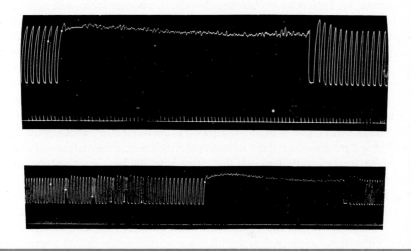

of a wave front passing through tissue in a state of inhomogeneity with respect to excitability and conduction velocity. The arrhythmia is assumed to sustain itself when the number of wavelets is so great that coalescence is improbable. The number of wavelets which coexist in the tissue should be directly related to some function of the mass of the tissue, and inversely related to the duration of the refractory period and to the conduction velocity.

This sounds superficially like Krinskii et al. seven years later, but Moe's computer simulations show very little that suggests locally regular rotation. Rather the process more resembles the "chemical turbulence" or "diffusion-induced chaos" first noticed by Zhabotinsky (1968, p. 90) in the homogeneous malonic acid reagent, and more recently studied by Rössler (1978), Kuramoto (1978), and

others in homogeneous media. As Moe described his fibrillation, spatial inhomogeneity of parameters is the *sine qua non* of fibrillation. When refractoriness was
suddenly made spatially uniform in his model, the previously turbulent activity
simplified, sometimes to one or more rotors. My bet is that the role played by
inhomogeneity of one parameter in Moe's essentially single-variable model of his
local kinetics is served by the second or third degree of freedom required for
chaotic behavior in partial differential equations and in ordinary differential
equations, respectively. Thus I think the issue of inhomogeneity is a red herring
and that the theories of chaos and turbulence arising in the late 1970s may in the
end have more to do with fibrillation than will theories about rotors and phase
singularities.

Example 15. Self-Organizing Patterns of Chemical Reaction.   A strictly continuous analog of the excitable media discussed in the two preceding examples is
presented by the malonic acid reagent (Chapter 12). In this case one might expect
the volume element at the center of symmetry to be held at time-independent
average composition. This would seem a reasonable consequence of diffusion
from volume elements only a small distance away which are at all stages of the
periodic cycle of excitation. In fact though, the center is subject to irregular color
changes. Tracing the movements of the indistinct end point of the spiral wave of
oxidation, one obtains a jumble of loops each of length about one wavelength, all
consistently clockwise or anticlockwise, mostly contained within a central disk
of diameter about one-half wavelength. (The wavelength is the spacing between
turns of the spiral.) This area is certainly phaseless but by no means time-
independent in its activity. The cause of the irregularity has not yet been determined.
As noted in Chapter 9, some computer simulations of idealized excitable media
show similar effects, but in others the patterns pivot rigidly about a fixed center.
In each case the amplitude of the unit period fundamental falls to zero at a central
fixed point but what determines the amplitude pattern at other frequencies,
harmonic or otherwise?

Setting aside that question, it remains of interest to inquire about the size of
the central region within which the fundamental is substantially attenuated,
whether or not other frequencies substantially increase their amplitudes there. I
call this finite disk surrounding a singularity a "rotor". In the malonic acid reagent
it is about one-third the wavelength in diameter. This is the practical dimension
of the singularity. In three-dimensional experiments the rotor is a thread of about
that thickness wandering through the reagent like the core of a tornado. Rotors
are unstable within confines smaller than this dimension, or closer than this to an
impermeable barrier. It is my impression that rotors are attracted to walls and to
counter·rotating rotors which come this close. This also happens in computer
simulations of the reaction-diffusion equation used in Chapter 9. Whether adjacent
rotors of the same handedness also repel, and if so whether the effect varies according to their phase difference, remain to be looked into. Such effects could have
much to do with the allowed topological changes undergone by the rotor thread
in three dimensions, such as decay into little rings, knotting or unknotting, linkage
or unlinking of rings.

Another way to characterize the rotor leaves aside consideration of phase and amplitude and instead focuses attention on concentration gradients. In any plane wave, all level contours of chemical concentration are parallel to the wave front. Gradients are normal to the front, pointing along the direction of propagation. In its outer parts, the spiral wave looks like a plane wave and its concentration gradients are nearly parallel. To measure departures from this parallelism in computer simulations, I used the magnitude of the vector cross-product of the gradients. As one follows any gradient inward toward the source of rotation, this measure increases. It increases slowly at first and then quite abruptly increases 1000-fold within a short range of the singularity. The width of this bell-shaped peak, taken between its inflection points, is once again about one-third wavelength (Winfree, 1978a).

In geometric terms the region of crossed gradients is the part of the medium's image in state space that departs substantially from the cycle, comprising the cap or diaphragm bounded by the cycle. This is the part that produced time singularities in state space analyses of the pinwheel experiments and which produces distal organs (limbs) in state space analyses of regeneration experiments in Chapter 16. In terms of developmental biology, this chemically structured region would be called the "organizer", or a "regulatory field". It imposes a spatial and temporal organization on the surrounding medium, maintains its own dynamic structure against small perturbations, but vanishes altogether if too big a piece of it is removed. It behaves as do limb fields in surgical regeneration experiments in respect to the number and handedness of new rotors (limbs) created. In other words, it satisfies the formal rules of French et al. (1976) and of Glass (1977) for reasons which I believe derive from little more than the continuity of maps.

B: The Singularities of Unsmooth Cycles

A theory has only the alternatives of being right or wrong. A model has a third possibility: it might be right but irrelevant.

M. Eigen

The alert reader may have noticed something missing from discussions of the malonic acid oscillator: There has been no mention of resetting curves and no use of the pinwheel experiment argument that type 0 resetting implies rotating waves. That is because no such measurements have been reported. And I suspect that when they do appear, anything of topological interest about them will be well hidden in practical discontinuities.

The adjective "practical" is important here. Chemical kinetics follow continuous differential equations which, in the case of the malonic acid reaction, describe an attracting cycle. Chemical perturbations being modifications of the dynamic equations, continuously map points initially on the cycle to new latent phases, effectively mapping the cycle to itself with some integer winding number. All the reasoning used in previous chapters must apply . . . in principle. But maybe not in a practical way. This remains to be seen.

An analogous case is seen in the slime mold, whose cAMP oscillator also has a very abrupt jump in its biochemical cycle. Nonetheless it is in principle continuous and type 0 resetting has been measured by exposing these cells to cAMP pulses. It is not yet clear how the singularity implicit in this observation relates to the singularity geometrically observed at the pivot of the spiral wave in a monolayer of cells.

Spontaneous rhythmicity in nerve membrane presents another case of this sort. The Hodgkin-Huxley equations and their various adaptations to neural oscillators are continuous. With most practical settings of parameters, their solutions include jumps so abrupt that simpler discontinuous models are more often used than the full continuous equations. Nonetheless the full continuous equations, if not the simpler discontinuous models, have a phaseless manifold and in it a dynamic steady-state. Moreover, experiments on pacemaker neurons published as long ago as 1964 show what has more recently been interpreted as type 0 resetting (Winfree, 1977). It follows that a phaseless set should be accessible to a sufficiently attenuated stimulus of the same sort, suitably timed. Best (1979) demonstrated this computationally using the Hodgkin-Huxley equation for rhythmically firing squid axon, current-biased to make the steady-state a local attractor. In this way he expanded the phaseless manifold from its minimum dimension $(D - 2)$ to its maximum dimension (D), turning this least promising of neural oscillator models into a sure-fire demonstration. As in James Thurber's "The 13 Clocks",

> . . . he tampered with the clocks to see if they would go, out of a strange perversity, praying that they wouldn't. Tinkers and tinkerers and a few wizards who happened by tried to start the clocks with tools or magic words, or by shaking them and cursing, but nothing whirred or ticked. The clocks were dead . . .

Since then Guttman et al. (1979) and Jalife and Antzelevitch (1979) have confirmed the computational results experimentally. A suitably timed voltage pulse of just the right size kicks the membrane off its attracting cycle into its phaseless manifold wherein it promptly retires to the attracting steady-state. This effect was also observed by Teorell (1971) in a 2-variable model of a sensory pacemaker neuron.

I wonder if a similar phenomenon may await discovery in other biological clocks? Section B of Chapter 2 described three counterexamples, viz neural pacemakers, the cell division cycle, and the endocrine oscillator that times ovulation and menstruation. Because of their characteristic near-discontinuities, these clocks render only disrespectful and scarcely recognizable homage to the topological principles that supposedly govern biological clocks. They probably *have* phaseless manifolds and, within them, steady-states, but these are evidently such asymmetrically violent repellors that only discontinuities are evident in coarse measurements. This violent repellingness rather than nonexistence of the singularity probably underlies the unsatisfying outcome of Kauffman and Wille's pioneering investigation of the mitotic cycle in *Physarum* (Chapter 22).

The example of neural pacemakers raises the question whether the steady-states of the female cycle and the steady-state of the mitotic cycle might not also become local attractors in the right hormonal environments. In fact, we know they can, but with the techniques currently in use ("the pill", antimitotic cancer drugs) they become *global* attractors, leaving no attractor basin for the cycle. As everyone realizes, there are medical side effects connected with such severe alteration of the normal parameters. Is it possible that the steady-state can be given an adequate but much smaller attractor basin by using far less of the medications already in use? No one would ever have noticed a small attractor basin by accident without looking for it. It might be there. The way to look for it is by a pinwheel experiment, in the singularity trap format of Chapter 2, Example 6. While maintaining the treatment intended to alter the steady-state's stability, one varies the two parameters (T and M) of a fleeting perturbation. After each stimulus the resetting of the continuing rhythm is measured. Those measurements can be put together to construct the resetting maps, isochrons, and winding numbers in such a way as to guide the search for the singularity. When it is found it will prove either attracting or repelling . . . hopefully attracting as in Jalife's and Guttman's experiments and in Johnsson's (1976) and Johnsson et al.'s (1979) experiments with the oat sprout's transpiration oscillator.

In the cell cycle this would switch off proliferation, either permanently or until another sufficient perturbation reinitiates it. In the case of unicellular organisms that reproduce by cellular fisson, this would also be arrest of the life cycle. In larger animals, e.g., female humans, it would mean arrest of ovulation. One might switch it back on by administering another stimulus to kick the system back out of the steady-state's attractor basin or by removing the relatively mild chronic therapy that keeps the steady-state attracting.

Obviously this is an ambitious if tedious program. But it seems to me likely to work, and more monotonous searches have been successfully conducted in the past. As the Hobbit was once overheard to remark, "There's nothing like looking, if you want to find something".

This sort of phenomenon may be more common than we realize. Richter (1965) has collected myriad examples of biological rhythms, mostly of a pathological sort from his clinical experience, which seem to be turned on and off unexpectedly by fleeting perturbations.

Returning from clinical speculation to a somewhat broader context, it seems to me remarkable that in this whole catalog of examples there is not one clear case of a smoothly cycling biological clock with a violently repelling phaseless set. This is in striking contrast to the models most commonly elaborated by theorists. Wiedenmann's cockroach shows promise of providing the first example of a strongly repelling phaseless set. Attempts to locate its singularity produced unpredictable rephasing but never a sign of arrhythmicity in the individual animal (Wiedenmann, 1977).

C: Transition to Bestiary

This ends the predominantly theoretical section of the materials I have gathered between these covers. Probably you have already delved extensively into the more experimental material that follows in the distressingly one-dimensional order inevitable to books. Nonetheless, this spot marks a transition and seemed to me the best place to articulate a few caveats explicitly. As Hermann Hesse says in "The Steppenwolf Treatise",

> All attempts to make things comprehensible require the medium of theories, mythologies, and lies; and . . . [an author who respects his readers] . . . should not omit at the close of an exposition to dissipate those lies as far as may be in his power.

Some authors of a theoretical bent, in neglecting to do this, have brought down around their necks the wrath of the Eumenides, manifest most notably of late as Kolata (1977), Pearson and McLaren (1977), and Zahler and Sussmann (1977). I don't think that much in the way of disclaimers will be required for *this* book, as no one is likely to mistake my pictures and guesses for mathematical proofs of anything. There is much room for proof in many of the more theoretical chapters where remarks are made about the anticipated behavior of idealized model systems. Such remarks are mostly "proofs by example" buttressed by geometric diagrams and computational results. Perhaps some reader of mathematical inclination, recognizing a favorable case, will pick up one of these remarks and establish its exact domain of validity as an exercise. For myself, the pictures are of value mainly in a heuristic way in connection with laboratory work. I am not persuaded that much is to be gained for empirical science by distinguishing, for example, exactly which kinds of dynamical system give phase singularities inside a ring of nonzero winding number, and which others give discontinuities of other sorts. I adopt this rather cavalier attitude partly because topological notions are in principle incapable of rigorous application to empirical science: sufficient quantitative distortion would (and does) alter any topological structure beyond hope of practical recognition; it can make any topological structure look like any other, so far as experiments of finite accuracy can resolve. My aim is more nearly to discover patterns in the sense of the quote opening the Introduction to this book than to prove that they had to be there all along and that it wasn't really necessary to look. I believe that I have found some, and that pleases me profoundly. As to their meaning in terms of particular mechanisms, I follow George Polya's recommendation in "How to Solve It":

> Quite often it matters little what your guess is; but it always matters a lot how you test your guess.

The matter preceding this page largely concerns guesses. The matter following largely concerns testing.

11. The Firefly Machine

The Glowworms represent another shew, which settle on some Trees, like a fiery cloud, with this surprising circumstance, that a whole swarm of these Insects, having taken possession of one tree, spread themselves over its branches, sometimes hide their Light all at once, and a moment after make it appear again with the utmost regularity and exactness, as if they were in perpetual Systole and Diastole.

Kaempfer, 1727

A: Mechanics

A machine once existed (Figure 1) in which 71 flickering neon lamps were each coupled electrically to all the others. (There were 71 because out of 100 constructed, 29 drifted outside the intended range of autonomous period during initial "wearing in".) The purpose of building this machine was just to "look and see what would happen", on a hunch that groups of oscillators might synchronize together in fleeting alliances. One hope was that by plotting the output of this population of interacting oscillators in the same format as biologists use to plot activity rhythms of multicellular animals, enough resemblances might be noticed to suggest some interpretation of the tantalizingly complex biological records.

The neon lamp operates as follows (General Electric Glow Lamp Manual, 1963). Current from a fixed voltage power supply—V, in Figure 2—trickles into a capacitor C, steadily raising its voltage. The capacitor's voltage appears across two metal electrodes separated by neon gas N. The gas has some slight electrical conductivity, and passes a current from one electrode to the other. For reasons that have to do with the kinetics of ionized atoms in electric field, this current depends in an S-shaped way on the electric field maintained between the electrodes (Figure 3). At least, this is how the current-voltage relationship would look under steady-state conditions. However, conditions are not steady because the capacitor voltage rises as current trickles into it. When it reaches $E2$, the glow discharge ignites and the current suddenly rises to $I2$. This current short-circuits the capacitor so that E very quickly falls down to $E1$ and the glow discharge ceases. The current

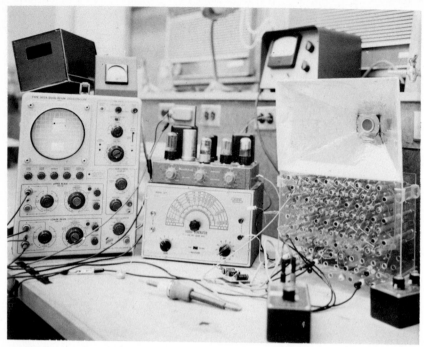

Figure 1.   On the right, mounted in a plexiglas frame, the array of 71 neon oscillators, individually shielded in brass tubes. The silvered funnel on top swings down to expose a photodetector to the 71 flickering lights. The oscilloscope is painting a pattern like Figures 4 through 7. Equipment borrowed from the lab of Victor Bruce, Princeton, 1965.

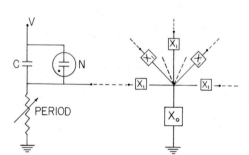

Figure 2.   A single neon oscillator (left) is one of many connected through impedances x_1 and grounded through common impedance x_0. The impedances used were resistive or capacitive.

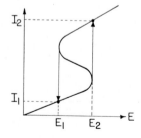

Figure 3.   The steady-state current-voltage curve of a neon glow tube. The negative-slope branch represents unstable states, so current jumps to a stable branch at E_1 and E_2.

drops abruptly to $I1$. From here E slowly increases again. The neon oscillator was originally chosen for its simplicity and reliability and for the analogy it presents to a rhythmically flashing firefly, a heart cell or other pacemaker neuron (see Chapter 14), the cAMP oscillator in a slime mold cell (Chapter 15), or any other biological oscillator based on a quantity like membrane voltage which accumulates to a threshold and then discharges abruptly.

The several-millisecond period of the cycle can be made quite stable by "aging" the lamps at high current for several days, cleaning the glass bulbs against conducting films of humidity, using an individually trimmed trickle resistor, and encapsulating the whole circuit in a grounded metal cylinder to protect against extraneous electric fields and especially against visible light. In the machine, the distribution of periods was tuned to a standard deviation about ten percent of the mean. The oscillators were all coupled equally to one another by connecting each to a common terminal through a small capacitor. This terminal was grounded through a larger variable capacitor so that the magnitude of mutual influence could be adjusted.

A signal depicting the aggregate activity of the 71 oscillators was derived electronically or through a photoelectric monitor watching a subset of the flickering lamps. The signal was used to modulate the brightness of an oscilloscope display, just as in a television receiver. The horizontal sweep was tuned to twice the mean period of the oscillator population, thus painting a horizontal strip of bright or dark segments corresponding to activity above or not above a chosen threshold throughout two consecutive activity cycles (essentially the same as one cycle repeated for easier visualization). The vertical scan rate was much slower so that these "daily records" were painted one above the next on the oscilloscope screen, stacked some hundred cycles deep in compliance with the custom of circadian physiologists. Thus, if all 71 oscillators flash in synchrony, the horizontal sweep can be tuned to (half) that frequency and the display presents a (repeated) solid vertical column of horizontal bright segments. If the population period lengthens, then the formerly vertical column drifts to the right as it goes down, as each day's activity is a little later than on the day before (taking the horizontal scan period as two conventional "days") (see Figure 4).

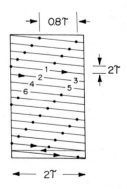

Figure 4. The time axis is broken into stacked segments (slanting) of length 2τ. An event repeats (dots) at intervals 0.8τ in this idealized example.

B: Results

What, in fact, happens? First of all, with no coupling all oscillators run at their own periods. Though initially synchronous, phases spread around the whole cycle within 20 cycles after starting. During these 20 cycles, the activity bar widens from a point and its mean level drops proportionately, eventually falling below the display threshold. Figure 5(b) shows a population in this condition, displayed as described above. The horizontal axis is two "days" or two cycles at the population's mean period. Vertically the recording covers several hundred cycles. A black gap appears in the record whenever an oscillator discharges. Vague trails of gaps slant across the display as little groups of oscillators gain phase (to the left)

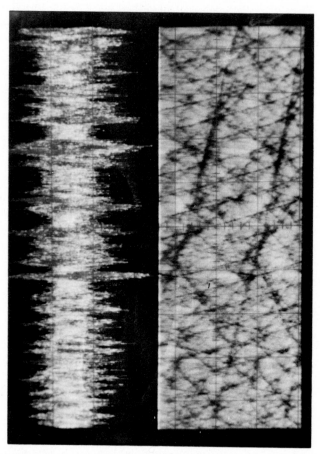

Figure 5.   (right) In the double-plotted format of Figure 4, spanning several hundred cycles from top to bottom, sudden changes of voltage appear as black dots arranged in criss-cross trails. Each trail represents a fleeting alliance of some few synchronous oscillators. (left) The corresponding voltage waveform is plotted on the same vertical time axis, increasing downward, with collective voltage at the common node plotted to the right.

or lose phase (to the right) relative to the scan period. Mutual coupling is so weak that only the most fleeting alliances occur. Figure 5(a) shows a simultaneous tracing of a voltage to which all 71 oscillators contribute equally: Its amplitude is quite low and fluctuates irregularly as oscillators drift in and out of synchrony.

With resistive coupling (C_1 and C_0 replaced by resistors), no matter how strong, the result is about the same: random phases. Thus, it appears that simply admitting mutual influence is not sufficient to guarantee mutual synchrony in a population of "clocks". This result goes somewhat against intuition, especially for oscillators which easily lock together in pairs, given almost any kind of coupling of intensity commensurate with the difference between their natural periods. The reason for this failure of synchronization in groups seems to be a phase compatibility problem, as touched upon in Chapter 4, Section B and in Chapter 14, Section B. Near the moment of the nearly synchronous discharge, the phase adjustment inflicted on each individual depends on that individual's exact phase in such a way that the spread of phases in the population is increased rather than decreased.

However, with capacitive coupling mutual synchrony *is* achieved abruptly at a threshold intensity of mutual influence. In this situation the intrinsically slower oscillators are imposed upon to speed up a bit, and the faster ones slow down, so all stay in step at one common frequency. But the exact mechanism of mutual synchronization was never determined quantitatively in this machine. Figure 6 is analogous to Figure 5 but has its vertical time axis much expanded to reveal individual cycles. Figure 6(a) shows the sawtooth voltage waveform to which all 71 oscillators contribute equally. As groups of sawtooth oscillators drift out of synchrony the collective waveform develops multiple zigzags in each cycle and decreases its peak-to-peak amplitude. Each zigzag appears as a whiter patch (because it is traced twice, not just once) of width proportional to the number of oscillators that discharged at that time. The phase relations among these groups are displayed more conveniently in Figure 6(b) [(with the same vertical scale as Figure 6(a), i.e., much expanded relative to Figure 5; the horizontal scale also happens to be compressed relative to Figure 5(b)]. The scan is solidly interrupted (black gap) when many oscillators discharge simultaneously. The discharge of smaller subpopulations appears as a proportionally smaller interruption. When two major groups are about one-half cycle apart the waveform in Figure 6(a) exhibits about half its maximum amplitude.

With still stronger coupling (smaller C_0) the synchronous population abruptly splits in two, with a slight change of period (Figure 7). This spontaneous schism doubles the frequency of the aggregate rhythm. Formally, this is a familiar bifurcation in the behavior of coupled ordinary differential equations (as one might view the coupled oscillators). I found its mechanism in detail in the case of smoothly varying simple-clock oscillators (Winfree, 1967a). Pavlidis (1971, 1973) derived similar behavior in oscillators involving two variables. Durston (1973) reports a similar abrupt frequency doubling in the similarly pulse-like rhythmic coherence of social amoebae (Chapter 15), though he suggests quite a different mechanism for it. I expect to see more studies of mutual synchronization in populations of equally coupled cells of this creature in the near future, thanks to the suspension

basis. By the time this much was accomplished (1965) it had become clear that computer simulation offered a more controllable model and a more flexible format for observation and output. Moreover, in this same year, Farley (1965) published a computer simulation of spatially coupled, idealized neurons which exhibited synchronization, waves, and even pinwheel-like rotating waves. In the same year, Gerisch (1965) published an account of nerve-like excitability and rotating waves in populations of social amoebae. Also in the same year, Balakhovskii (1965) published a geometric model of excitable media showing wave propagation and rotors. Then came the discovery that yeast cell suspensions are populations of mutually synchronizing oscillators (Chapter 12), and Zaikin and Zhabotinsky's (1970) discovery of a repeatedly excitable chemical medium (see Chapter 13). Then even computer simulation seemed inconvenient as a way to investigate mutual synchronization and the spatial activity modes characteristic of such media. So the neon device was never wired for neighborhood interaction and such behavior as had been recorded under conditions of promiscuous interaction was never rationalized in quantitative terms.

12.   Energy Metabolism in Cells

The want of which incomparable Artifice (microscopes) made the Ancients' . . . erre in their . . . observations of the smallest sort of Creatures which have been perfunctorily described as the disregarded pieces and huslement of the Creation . . . . In these pretty Engines are lodged all the perfections of the largest animals . . . and that which augments the miracle, all these in so narrow a room neither interfere nor impede one another in their operations. Ruder heads stand amazed at prodigious and Colossean pieces of Nature, but in these narrow Engines there is more curious Mathematicks.

Henry Power, physician, *Experimental Philosophy*, 1663

A :   Oscillators

Cells have three alternative means of procuring energy for digestion and biochemical synthesis, for maintaining concentration gradients, for muscular contractions and cell division, and for maintaining body heat:

1.   Photosynthesis: The chloroplasts of green plants capture photons to convert ADP to ATP. Water is split to reduce NADP to NADPH, releasing oxygen.

2.   Respiration: The mitochondria use that oxygen and convert ADP to ATP. In the process, NADH is oxidized to NAD and water.

3.   Glycolysis: Lacking illuminated chloroplasts or lacking oxygen, cells metabolize sugars by fermentation to make a little ATP from ADP. Historically, this was probably the first way to make the high-energy pyrophosphate bond of ATP. All cells maintain this pathway. Most cells fall back on glycolysis only when they have no better alternative, but it is common to have no better alternative. The microorganisms of yogurt, sauerkraut, gangrene, and food poisoning, for example, subsist wholly on glycolysis, as do faculative anaerobes such as intertidal bivalves (e.g., oysters) and parasitic helminths (e.g., schistosomes) and diving vertebrates (e.g., green sea turtles) during their prolonged periods of contented abstinence from respiration. Red blood cells have no other energy supply. Poorly vascularized tissue such as the cornea of the eye, compact tumors, and embryos rely heavily on glycolysis for their energy needs. For shorter intervals, so do the aching muscles

of any animal in violent exercise, especially those of diving mammals such as the porpoise and whale. The rabbit heart muscle used by Allessie (Chapter 14, p. 332) to support a rotating wave presumably drew its energy predominantly from glycolysis.

The first three biochemical oscillators discovered (all since 1955) were in these three sources of energy: photosynthesis in chloroplasts, respiration in mitochondria, and glycolysis in the cytoplasm of yeast cells (and later in more dilute cell-free extracts of yeast, heart muscle, and skeletal muscle). See reviews by Hess and Boiteux (1971), Nicolis and Portnow (1973), by Goldbeter and Caplan (1976), and by Goldbeter and Nicolis (1976). According to a proposal by Wagner et al. (1975, 1976), circadian rhythmicity also arises from the control dynamics of energy metabolism, regulated adaptively in relation to diurnal changes in light intensity.

B: The Dynamics of Anaerobic Sugar Metabolism

At present, the most vigorously and successfully studied biochemical oscillator is oscillating glycolysis in yeast. In fact, biochemistry originated in studies of the mechanism of glycolysis in brewer's yeast, which makes ethanol. The chemistry is consequently well understood. Yet it wasn't until 1957 that Duysens and Amesz obtained the first hint that glycolysis does not always produce energy in a steady DC mode: it sometimes goes AC, the cells oscillating like little dynamos with a period on the order of a minute (Figure 1). With this realization came a new emphasis on the second and less well-developed aspect of biochemistry: Not only must we understand the sequence of molecular transformations among reaction pathways, but we must also understand the regulation of *rates* of transformation. Molecular transformations are mediated by enzymes. The enzymes' catalytic rates are governed by the concentrations of pertinent substrates and cofactors, most of which are themselves produced in other enzyme-catalyzed reactions of the same pathway. The discovery that such a system can spontaneously vary its rate by

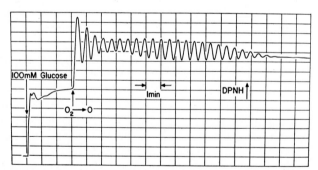

Figure 1.   NADH fluorescence in a suspension of metabolically oscillating yeast cells (from Pye in Biochronometry (1971) page 627 with the permission of the National Academy of Sciences, Washington, D.C.). Note that *other* variables change in a much more abrupt fashion. People committed to describing this oscillation as "pendulum-like" or "relaxation-oscillator-like" would be divided into mutually antagonistic schools according to their choice of a state variable to watch.

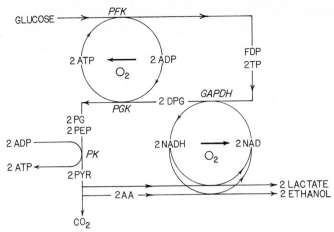

Figure 2.   A schematic of the sequence of transformations undergone by a glucose molecule becoming ethanol or lactate. These changes are wrought by some dozen enzymes, of which only four are indicated here in italics. The bold arrows indicate the immediate effect of oxygen: increasing ATP and NAD.

factors of 10 to 100 in a regularly repetitive way seemed at first to offer the experimental biochemist a particularly favorable opportunity to study regulation. And it seemed to offer a particular challenge to those who would formulate the mathematics of regulatory kinetics. But it turns out that as far as investigations of mechanism are concerned, the fact of oscillation in itself has provided no particular experimental opportunity. It also has provided few constraints by which to eliminate what proved to be a plethora of possible mechanisms. Far from requiring very special arrangements, oscillation turns out to be a common feature of even very simple regulatory kinetics, a feature that rapidly becomes all but unavoidable as reaction complexity increases (Goodwin, 1963; Higgins, 1967; May, 1973).

In the particular case of glycolysis, then, the problem is not to think of a possible mechanism, but to determine by ingenious experiment which of the many potentially oscillatory aspects of that pathway in fact assume dominant importance under physiological conditions. This job was carried out primarily by Higgins in Philadelphia, working with biochemists brought together by B. Chance, and by Selkov in Moscow, working with biochemists brought together by Hess in Dortmund and by Betz in Stockheim.

Glycolysis consists of a chain of a dozen enzyme-catalyzed transformations of a six-carbon sugar (Figure 2). The sugar is twice phosphorylated, then split into two three-carbon fragments, each of which gets phosphorylated again, gives up both phosphates to ADP, and ends up as pyruvate. On the way to pyruvate, each gets oxidized, simultaneously reducing NAD[1] to NADH. In oxygen-starved muscle a final reaction converts three-carbon pyruvate to three-carbon lactate, a cause of general anxiety, of charley-horses, and of labor pains. In yeast, a preliminary step cleaves off carbon dioxide, making the broth fizzy and leaving two-carbon

---

[1]   Before 1970, NAD was DPN.

acetaldehyde (AA) to be reduced. The result is then grain alcohol, a cause of highway disasters for drinkers and their victims. Either way, the final reaction oxidizes the NADH accumulated earlier back to NAD. This NADH, as it happens, fluoresces blue-green in ultraviolet light. Following the innovation of Duysens and Amesz (1957) one can monitor the flux of material through the pathway by watching this glow. This works because the reactions increasing and decreasing NADH happen to fluctuate out of phase with each other. Like the NAD-NADH couple, an ATP-ADP couple also links remote reactions in this pathway (Figure 2). But here the books are not kept so neatly balanced: Each triose that runs this gantlet uses and restores one ATP but additionally converts one ADP to ATP in shedding its final phosphate group to become pyruvate. And here we come to the crux of the matter: Glycolysis is *for* making ATP, and is regulated accordingly.

C: The Pasteur Effect

The rate of glycolysis is regulated by the cells' balance of ATP and ADP. When there is more ADP (less ATP), glycolysis runs faster, converting ADP to ATP at the expense of glycogen or starch reserves. When there is less ADP (more ATP), e.g., when there is enough oxygen for operation of the mitochondria, the cell's main ATP factory, then glycolysis shuts down. This regulation is called the Pasteur effect.

How is this trick managed? It turns out that it is managed mainly by one enzyme, the one that puts the second phosphate on the sugar. This is phosphofructokinase (PFK). When there is enough ADP around, the concentration of AMP is also high and some of it sticks onto a special receptor site on the PFK molecule. This changes its shape so that it works more quickly, admitting a swift flow of sugar through glycolysis to convert the excess ADP back to ATP. The activity of PFK in fact changes almost 100-fold during each cycle of oscillation (Hess et al., 1969).

But here lurks a paradox. PFK is a phosphorylating enzyme. It uses ATP, degrading it to ADP. Thus it is activated by its immediate product. A potential for instability lurks here, not far beneath the surface. As sugar passes PFK, ADP accumulates and enhances the rate of sugar passage and of ADP accumulation, unless that ADP is consumed at a correspondingly increased rate further down the pathway. If there is any delay (and there is) in the arrival of phosphorylated sugar further down the disassembly line, then the pulse will grow until its front does reach the later reactions. Then so much ADP is converted to ATP that PFK shuts off the input altogether.

Verbally, this is how the oscillation arises. But as Hermann Hesse observes in *The Steppenwolf Treatise*, "Things are not so simple in life as in our thoughts, nor so rough and ready as in our poor idiotic language." All such verbal arguments about cycles are specious, a realization aptly articulated by Wojtowicz (1972, p. 65) and by Aldridge (1976, p. 2). One knows where to make such an argument only by the hindsight acquired through biochemical experiments and through solving kinetic equations. The problem with verbal rationales lies in their inability to decide whether

1.   The verbally articulated mechanism can or cannot settle down to a nice balanced rate at which ATP is produced exactly as fast as it is consumed.

2.   The regularly periodic oscillation is approached spontaneously from the steady-state.

However these questions have been resolved in a variety of other ways. Most recent, and probably the most successful quantitatively, is the work of Goldbeter and collaborators, whose kinetic analyses we will also meet in Chapter 15 in connection with *Dictyostelium*'s cAMP oscillation. Goldbeter's analysis (Goldbeter and Lefever, 1972; Boiteux, Goldbeter, and Hess, 1975; Goldbeter and Nicolis, 1976) derives much of the observed behavior from an attracting cycle of allosteric activation of PFK by ADP. This is not to deny that other reactions are involved, but this reaction asserts itself especially conspicuously under common operating conditions.

D: Goldbeter's PFK Kinetics

Though it is not glycolysis, the single enzyme oscillator built around PFK is sufficient to account for many of the facts of oscillating glycolysis. Not only sufficient, it may be a necessary part of the glycolytic oscillator. This is revealed by the following fact. It is possible to bypass the first steps of the pathway by starving the cell-free extract for sugar, then dripping into the reacting extract any substrate that normally appears after the PFK step. With PFK thus bypassed, no oscillations have been observed (Hess and Boiteux, 1968).

Higgins and Chance's early recognition of the pivotal roles of PFK and of the ATP-ADP couple also rested heavily on the fact that the many substrates of glycolysis oscillate in essentially four groups. All those prior to PFK oscillate synchronously. All those after PFK and before GAPDH also oscillate as a group, but 180° phase-shifted from those preceding PFK, as though PFK were the fulcrum of a teeter-totter. Those from GAPDH to PK are synchronous, though somewhat delayed relative to their precursors. And all substrates after PK are 180° opposite to that phase. The two enzymes PFK and PK mainly seem to be gating the flow. Whenever mass flow through one of these bottlenecks increases, downstream concentrations rise and upstream concentrations fall.

An essential third clue implicating the ATP-ADP balance in controlling the oscillation came from phase-resetting experiments. An injected pulse of ADP was found to profoundly affect the oscillation in cell-free extract (Chance, 1965; Pye, 1969, with an even smaller dose). Not every substrate does this. In particular, Pye (1969) found that FDP has no such effect. FDP is the other product of PFK, and like ADP it also activates PFK. The demonstration that FDP has no effect and that ADP has a profound effect led to Goldbeter's updating of the earlier models of Higgins (1964) and Selkov (1968a,b).

Caveats

Every theory of the course of events in Nature is necessarily based on some
process of simplification and is to some extent, therefore, a fairy tale.

Sir Napier Shaw

Without going into much more detail here, it seems appropriate to note that
glycolysis is really much more complicated than its simplest adequate model. In
particular:

A.   Phase shifting in extracts requires about 30 times more ADP (about
12 μmole/g of cells) than would be expected for the 0.4 μmol/g swing of its concentration in the oscillation. The strong phase-shifting by oxygen requires the
equivalent of about 1 μmol/g of ADP. Thus one might guess that the effect of O_2
is not mainly by depletion of ADP and that ADP is not a primary state variable
in the oscillation mechanism. In fact the other adenosine phosphates play weightier
roles in controlling PFK activity. All three adenosine phosphates are in equilibrium
through myokinase, so from a functional point of view, *any* one may be considered
the control. Selkov and Betz (1968) found that, under conditions allowing oscillation, PFK is controlled almost exclusively by AMP. Chance et al. (1965a) found
that cyclic AMP phase shifts the oscillation about 100 times more efficiently than
does ADP.

B.   At least two other reactions play important roles in the oscillation:

1.   Pyruvate kinase (PK), another phosphorylating enzyme, is sensitive to
ADP levels and rhythmically gates the flow of material through glycolysis almost
as much as PFK does.

2.   Glyceraldehyde phosphate dehydrogenase (GAPDH), regulated by fluctuating NADH levels, likewise plays a modulating role. In fact, 3-PGK and
GAPDH alone, without PFK, seem to constitute a separate autonomous oscillator
under some conditions (Chance et al., 1967; Dynnik and Selkov, 1973). The
dynamics of two interacting oscillators is much richer than that of one alone. This
lower part of glycolysis is not so thoroughly studied as is the upper tract, and may
yet contain many secrets. The effect of cyclic AMP (see A above) seems to be on
an enzyme after 3-PGK.

C.   All the critical control points of glycolysis change according to the cell
type and its physiological state (determined by dietary conditions and hormone
levels). Substrate concentrations and cofactor concentrations determine the
operating points of the enzymes and so determine what quantities their rates react
to most sensitively. Selkov (1968a,b), Garfinkel et al. (1968), and Higgins et al.
(1968) have given myriads of different approximations to the complete kinetics of
glycolysis, each appropriate under different operating conditions.

D.   Glycolysis can be reconstituted in glassware from crystallized enzymes,
substrates, cofactors, and an ATPase to simulate "the rest of the cell". It works
fine and it oscillates (Hess and Boiteux, 1968). But this trick has not yet been
achieved using any less than all the enzymes from hexokinase through alcohol
dehydrogenase.

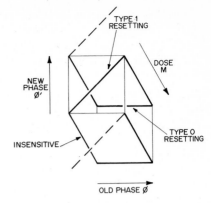

Figure 3.   Piecing together observations of phase resetting at high dose, low dose, and insensitive phase, we obtain a helical boundary on the surface that describes new phase as a function of old phase and dose.

E: Phase Control of the PFK-ADP Oscillator

The resetting data of Chance (1965) and of Pye (1969) are fragmentary but they suggest type 1 resetting in response to a 0.7 mM ADP increase and type 0 resetting in response to a 2.5 mM ADP increase. They also indicate that an injection of ADP during about half of the cycle has little lasting effect on the oscillation's phase. Taking these observations at face value, they describe a helical boundary on the three-dimensional graph of the glycolytic oscillator's new phase as a function of the size of an ADP dose and the old phase at which it was administered (Figure 3). Interpolation into the middle range of doses and phases in Figure 3 leads us to anticipate a helicoidal resetting surface inside the helical boundary (see Chapter 1, Section C and Chapter 2, Box B). We are driven to this surface by assuming a smooth dependence of phase on the stimulus parameters. Yet such a surface necessarily contains a discontinuity. The most compact discontinuity we can get away with is a phase singularity.

If the unavoidable discontinuity manifests itself in this form, then a phase singularity should result from perturbing glycolysis by a suitable change in the ATP-ADP balance, neither too drastic nor too subtle, at just the right phase. What would happen to cells so treated? Their new phase should be unpredictable and irreproducible. We might also observe some effect on the amplitude of oscillation. The magnitude and nature of such effects could tell us whether the glycolytic oscillation has a smooth attracting cycle or more nearly resembles an hourglass-like relaxation oscillation, whether it has one or more steady-states, and whether they are repellors or attractors.

F: More Phase-Resetting Experiments

I went to E. K. Pye's lab in Philadelphia to try this experiment in 1970. Without his making every service available to me, it never would have worked in the single week I had scheduled (in which the daylight hours became unexpectedly busy because Nixon chose that week to invade Cambodia).

As a novice, leery of cyanide-inhibited respiration and of homogenized cells, I preferred to use intact living cells. So I chose to perturb the ATP-ADP balance indirectly, using an oxygen pulse. This was necessary because phosphorylated intermediates do not penetrate the cells' plasma membrane,[2] but oxygen does. By injecting a tiny amount of buffered water containing dissolved oxygen, the cells' mitochondria are stimulated to quickly deliver a measured burst of ATP, whereupon the corresponding amount of ADP vanishes within the cells. During the few seconds required to consume it, the oxygen concentration was well above the 0.4 μmolar concentration required to saturate mitochondrial activity (Chance, 1965). In other words, I used the Pasteur effect as a perturbation. Glycolysis was allowed to run in its usual anaerobic oscillation and was then switched to aerobic metabolism. Aerobic metabolism operates at a high steady ATP level with very slight flux through the pathway. But the switch to aerobic dynamics was scarcely begun when the small pulse of oxygen ran out.

G: Results: The Time Crystal

In 88 separate experiments, I gave various doses of oxygen at various times in the cycle, all the while monitoring the NADH fluorescence rhythm. The results appear in Figure 21 of Chapter 2, and, in stereo without artwork, as Figure 4 of this chapter. Each NADH maximum is plotted as a dot in three dimensions along three axes:

1.   Vertically downward, the time of the maximum, measured from the time of oxygen injection

2.   Horizontally to the right, the time of the injection, measured from the time of the prior NADH maximum

3.   Horizontally in depth, the amount of oxygen given

Sure enough, the data points fill out a screw-like surface in each 30 seconds × 30 seconds unit cell of the time crystal. Its singularity lies a few seconds before the NADH fluorescence maximum (which is also the ADP maximum) at about 20 μl of oxygen-saturated buffer. These 20 μl contain 28 nmole of oxygen which, if completely used by the cells' mitochondria, would deplete 170 nmole of ADP in the 160 mg of suspended cells. So we're talking about an ADP depletion of about 1 μmol/g of cells. As noted above, this is at least double the whole range of fluctuation of ADP in these cells. Thus, some doubt is raised about the simple interpretation I use throughout this chapter in terms of the ADP-PFK mechanism considered in isolation. The *phase* of the singularity seems right, though. A coordinated increase of ATP and depletion of ADP from the ADP maximum on the anaerobic cycle would move the oscillator toward its stationary state.

---

[2]   They do in *Aplysia* neurons, whose electrical rhythms are regulated by PFK activity (Chaplain, 1976, and Chapter 14).

Box A: How to Look at Figure 4

The stereo picture in this chapter, the three in Chapter 20, and the one in Chapter 21 are intended for viewing with the help of a mirror. (This is my alternative to the eye-crossing method, which requires practice, and to the red and green spectacles method, which involves the publisher in color printing.) A first-surface mirror works best, but ordinary silvered glass will serve if you can ignore the duplicate reflection off the glass front surface. Mount the figure directly in front of you. Hold the mirror in the symmetry plane of your head so that it touches your nose and the paper and reflects the right image into your right eye. Look at the left image with your left eye. Adjust the mirror to align the two images. The graph should now snap into three-dimensional perspective.

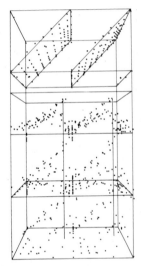 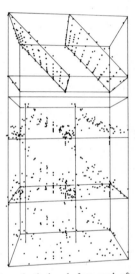

Figure 4.   Stereographic views of 342 measurements of NADH peak timing before and after an oxygen pulse. (For viewing, see Box A.) Coordinates and perspective as in Figure 21 of Chapter 2. The two interior uprights depict the presumed singularity, a pole around which data points climb in cork-screw fashion. The 10 dots stacked up in middle background are unperturbed controls. All data are double-plotted to the right.

Amplitude effects were a little harder to evaluate because, being a tyro in this business, I prepared rather poor cells. Their oscillations ran down exponentially, each maximum being only about half the size of its predecessor after the cells start at high amplitude upon running out of oxygen and going anaerobic. [Later experiments by Aldridge and Pye (1976) show that I had too great a density of cells in suspension.] In experiments right at the singularity of the phase surface, the

Box B: The Fluorescent Pinwheel Experiment

As in the case of other phase-resetting experiments explored in this volume, the Pasteur-effect rephasing by oxygen pulses applied to oscillatory energy production in yeast cells can be described in terms of one single experiment conducted in two-dimensional space. Imagine a thin layer of yeast cells spread out on a table top, conducting aerobic metabolism. Now we begin to cover them with a glass sheet or some other surface capable of excluding oxygen. Slide the shield from east to west at a constant velocity sufficient to reach the cells furthest to the west within a minute or two. As the cells are covered and consume a residue of dissolved oxygen, they enter upon anaerobic metabolism and begin oscillating. By grading from east to west the times at which this happens we establish an east-west phase gradient covering two to four cycles. Now to administer a transversely graded stimulus, we slide the oxygen barrier northward then return it immediately to its original position. The southernmost cells are thus exposed (at various phases along the east-west gradient) to oxygen for a relatively long time, while those at the northern extremity are exposed for only a moment or not at all, at the extremity of the oxygen shield's movement. We have thus assembled in an orderly way in two dimensions all the individual resetting experiments that can more conveniently be carried out one by one in the fluorometer. This experiment presents in a continuum all combinations of (phase, oxygen dose) within the chosen limits. Our 88 fluorometer experiments constitute 88 samples within one cycle of phase in this idealized two dimensional experiment. The results were plotted in Figure 4 and in Figure 21 of Chapter 2 in the format of a time crystal. Each unit cell of that plot contains a screw surface whose altitude at each point is the time of the NADH fluorescence maximum at that particular combination of initial phase and dosage of oxygen administered at that phase. A contour map of this surface consists of curves, each of which indicates a locus of constant altitude on this surface (Figure a). Constant altitude means constant time of the fluorescence maximum. Each such contour could be obtained by taking a cross-section at fixed altitude through the time crystal. The time crystal is a stack of serial sections of this sort or, equivalently, a set of contour lines as sketched in the adjacent figure. In these contours we see the result of the phase-resetting experiment reassembled in two-dimensional format as a pinwheel experiment. At any given moment, one of those contour lines marks the locus of brightest blue-green fluorescence. A few seconds later the next contour line is brightest and a few seconds after that the next, and so on. Thus

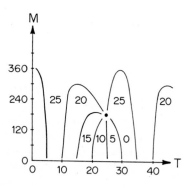

Box B Mean loci along which NADH fluorescence peaks up at the indicated number of seconds after an oxygen dose of M nanomoles per wet gram of cells applied T seconds after an NADH maximum. (Sketched from serial sections of the data cloud in the time crystal of Figure 4, and Figure 21 of Chapter 2.

we see that the glow is a wave which sweeps 'round and 'round a central pivot at the singularity of the screw surface.

This rotating wave is only a "pseudowave", in the sense elaborated in Chapter 4. However, if the two-dimensional experiment were conducted in a small enough physical space, let's say 1 centimeter by 1 centimeter, then the molecular interactions which normally synchronize nearby yeast cells would be reaching across cells of significantly different phase in the cycle in times comparable to one cycle duration or less. Without solving the equations of kinetics and diffusion, one doesn't know what would become of this situation, but similar initial conditions in the malonic acid reaction (Chapter 13) and in *Dictyostelium* (Chapter 15) result in stable rotation of a spiral wave around that fixed pivot. To actually see this in yeast would require the services of a scanning fluorometer not unlike the scanning electron microscope or a television camera.

According to Goldbeter and Erneux (1978), the glycolytic oscillator can be "excitable" in a physiologically reasonable range of parameters, in the same sense as the malonic acid reagent and *Dictyostelium* are excitable. On this basis, too, one might reasonably anticipate metabolic wave propagation, rotors, and scroll rings in one-, two-, and three-dimensional media composed of yeast cells.

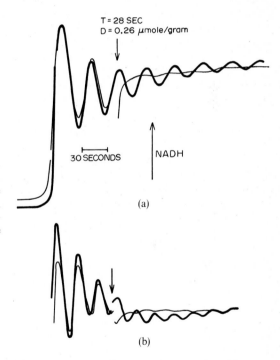

Figure 5. (a) Two NADH fluorescence traces superimposed. The heavy trace is an unperturbed control. In the thin trace, a critical dose of oxygen was administered at the critical phase (arrow), quenching rhythmicity (from Winfree 1972 Figure 11). (b) As above but from Greller, 1977, Figure 29, using acetaldehyde instead of oxygen. (The time scale is somewhat different.)

cell suspension nearly quit oscillating [Figure 5(a)], but I saw no conspicuous effects on amplitude otherwise. A more careful reexamination of the data years later [prompted by Aldridge's (1976) resetting experiment using Ca^{2+} pulses and by Greller's (1977) resetting experiments using acetaldehyde pulses] shows that there are effects on amplitude following nonsingular perturbations. Specifically, the damping is precociously advanced the closer the stimulus was to the critical phase and critical dose. This is what would be expected of the ADP-PFK mechanism. Goldbeter's equations show trajectories winding many times around the steady-state before the oscillation recovers most of the way back to the attracting cycle, after an increment of adenosine phosphate moves its state into the inside of the cycle.

Aldridge and Pye (1979a,b) envision a slightly more exotic mechanism of amplitude adjustment. They propose that the steady-state and the limit cycle are *both* attractors. A near-singular perturbation then deposits *some* cells inside the steady-state's attractor basin and they stay there, contributing little or nothing to the aggregate amplitude. It is argued from numerical studies that this situation is stable even in the face of the strong mutually synchronizing interaction known to be at work in yeast cell suspensions of the density used.

H: A Repeat Using Divalent Cations

Essentially the same phase and amplitude behaviors as noted above were elicited by Aldridge and Pye (1979a,b) using pulses of Ca^{2+} or Mg^{2+} concentration. The effect of *removing* ions from the extracellular medium was qualitatively the same as the effect of *adding* a comparable amount a half-cycle earlier or later. The mechanism by which these sudden changes of ion concentration affect the kinetics of glycolysis has not been resolved. Activation of membrane-bound enzymes that process ATP, ADP, and/or AMP seems a likely possibility.

I: A Repeat Using Acetaldehyde

There are basically two good ways to deliver a swift kick to glycolysis in intact cells. One is to transiently kick glycolysis indirectly, by activating mitochondria with oxygen or by afflicting the membrane with ions as above. The other is to transiently kick glycolysis itself directly with one of the nonphosphorylated substrates that occur late in sugar degradation. Pyruvate (PYR) and acetaldehyde (AA) are the obvious candidates (Figure 2). And it turns out that both cause large phase shifts at submillimolar concentrations. As with ADP, AA has been used to measure type 1 (weak) resetting (Betz and Becker, 1975b; Greller, 1977) and, at the same dose, type 0 (strong) resetting (Greller, 1977). An AA concentration of 50–60 μM seems to be the dividing line between weak and strong resetting. Greller's data, like Aldridge and Pye's, once again describe a helical boundary for the surface within each unit cell of the time crystal. Greller found the expected singularity at M* about 50 μM AA, at a phase ϕ* about one-third circle before the NADH peak [Figure 5(b)]. AA presumably (Figure 2) strikes first at lower glycolysis, in the less studied GAPDH/3-PGK oscillator. Its immediate impact is on NADH, oxidizing it and thus further diminishing NADH concentration. The measured ϕ* is hard to interpret on this basis because the ϕ* is already near the minimum of NADH fluorescence. From there, a further decrease of NADH leads outside the cycle and so *cannot* lead back toward the stationary state if only two variables are importantly involved. Unless there is somehow a delay of one-half cycle, this observation would seem to preclude any such simple tidy interpretations as attempted above for the oxygen singularity in terms of ADP and ATP. Greller points out that a model involving at least three variables seems to be required. This might be a reflection of the fact that this experiment, by striking low in glycolysis, does indeed implicate additional variables in the disruption of the oscillator.

Mutual Synchronization

Another point of interest in AA as a perturbing agent is that AA has been considered a candidate for the coupling agent diffusing between cells to keep them synchronous. Without such an agent, a cuvette full of yeast would not long exhibit coherent oscillation. But with a steady trickle of sugar input, von Klitzing and Betz (1970) kept it running for days with roughly normal periodism. As in the yeast mitotic cycle (Halvorson et al., 1971) and in *Dictyostelium*'s cAMP pulsations in cell suspensions (Gerisch and Hess, 1974; Malchow et al., 1978), it is evident that mutual synchronization is occurring. It is probably mediated by a small ion. AA (Pye, 1969) and PYR (Betz and Becker, 1975b) have been nominated as candidates.

On the other hand, Ghosh et al. (1971) show that no loss of synchrony follows addition to the extracellular medium of enzymes that tried to remove PYR and AA. This experiment might be refined, possibly to show that neither substrate alone is the intercellular messenger. The candidacy of AA is also weakened by Betz's observation that its concentration in and between cells (summed) doesn't fluctuate during the cycle. These hints are corroborated by an implication of the following experiment (according to my interpretation only).

The strength of the mutually synchronizing interactions was measured in a dilution experiment by Aldridge and Pye (1976). If cells are far enough apart, their mutual coupling should be negligible. At some point in the dilution of a cell suspension, they will fail to synchronize their metabolic rhythms, and the collective amplitude of the population will fall drastically. Aldridge and Pye (1976) found this threshold[3] at a dilution of about one cell volume per 10,000 surrounding water volumes. This indicates a startling specificity of the coupling agent. Assuming a flux of one-half μmole of fructose per gram of yeast per second (Betz, 1966), at most 1 μmole of AA can be secreted per second into the 10,000 grams of water. Thus its concentration could increase by only 3 μM per 30-second cycle. To maintain sychrony, cells must adjust phase substantially during each cycle. A dose of even 10 μM produces scarcely noticeable phase shifts, even if applied at the

---

[3] Othmer and Aldridge (1978), and Aldridge and Pye (1979b) favor an alternative interpretation, i.e., that *each* cell quits oscillating when too isolated from its neighbors by dilution. This possibility can be checked by spectrophotometry of single-cell NADH rhythms (Chance et al., 1968) but definitive results have not been reported.

optimal phase (Greller, 1977). Moreover Betz and Chance (1965b) find that AA concentration

1.   Does not fluctuate during the cycle, but only climbs steadily and

2.   Is three orders of magnitude lower than the above maximum estimate, presumbly because ADH promptly changes it all to ethanol

Thus AA seems after all a poor candidate.

PYR fluctuates by 0.2 mM/g during the cycle. A substantial portion of this amount appears extracellularly (Betz and Chance, 1965b). In 1:10,000 suspension, this amounts to 0.02 mM changes. Since PYR causes large phase shifts only at 70 mM concentration, this 0.02 mM variation does not seem adequate to maintain synchrony.

Aldridge and Pye also found an upper limit to cell density, at volume ratios in the order of 1:5, above which the rhythm dies out. Presumably the oscillation is suppressed in every cell by excessive exposure to its own and its neighbor's diffusible products. The density of 1:12 used in my experiments and in many others' is near the upper end of this permissive range. The commonly observed damping of oscillations at such high densities in unreplenished medium might represent accumulation of those products. The whole question about the role of cell density subverts Chance et al.'s (1968) estimation of the range of single-cell natural frequencies in a cell suspension: Their calculation was based on comparison of observed damping rates of single cell rhythms and population rhythms, but the cell densities in those two cases were not the same and were not reported.

J :   Phase Compromise Experiments

The vigor of cell-cell interaction is shown in another series of experiments, all done at the same cell dilution (1:12) as in my phase-setting measurements using O_2 and in Greller's using AA. In these experiments two equal suspensions of cells at different phases in the cycle are poured together. They quickly establish a phase compromise, as in *Physarum* (Chapter 22) and resume normal but now synchronous oscillation (Pye, 1969). The design of this kind of experiment has a striking symmetry, and that symmetry has a curious mathematical consequence (Chapter 1, Section C). The consequence is that the experimental result (a compromise phase) cannot depend continuously on the phases of the two parent populations at the moment they are poured together. This statement has nothing to do with any interpretations of the mechanism of rhythmicity. However, the nature of the discontinuity might reveal something about the mechanism. For example, a smooth attractor cycle mechanism would localize its discontinuity as two isolated phase singularities, supposing perfectly instantaneous mixing of all reactants and neglecting the exceptional case of perfect symmetry (Chapter 6, Box F). On the same basis, a discontinuous relaxation oscillator would have a phase discontinuity varying in magnitude along a pair of perpendicular lines at the jump phase. In fact, glycolysis in yeast shows neither pattern clearly. At first sight, it shows almost no pattern at all, according to the experimental results of Ghosh et al. (1971). With a little imagination (Winfree, 1974c), the data can be seen as a noisy approximation to a two-screw pattern surrounding two isolated phase singularities (Chapter 2, Example 9). This is what would be expected from such attractor cycle models as those of Higgins (1964, 1967; et al., 1968), Selkov (1972), and Goldbeter and Lefever (1972).

13. The Malonic Acid Reagent ("Sodium Geometrate")

A new chemical reaction with either excitable or periodic dynamics appears every month in the theoretical journals. But only one has been widely studied experimentally in ways that reveal wave-like organization in space. There are already two entire books about it: Zhabotinsky (1974, in Russian) and Tyson (1976b).

This is a catalytic oxidation of an organic fuel to carbon dioxide in water solution at room temperature. As far as that much goes, it sounds almost biochemical, especially as three of its possible fuels (citric acid, oxaloacetic acid, and hydroxylsuccinic acid) are prominent in the Krebs cycle, one of them (citric acid) inhibits PFK (see Chapter 12), and another (malonic acid) is a specific inhibitor of the Krebs cycle. The catalyst is typically an iron ion held in an organic ring complex called "ferroin", not unlike the heme group of oxidative enzymes. But the resemblance to cell metabolism ends about there. The organic fuel must be easily brominated and can be any one of many water soluble organics. The ferroin can be replaced by ruthenium bipyridyl, cerium, manganese, or cobalt and presumably by other transition metal ion complexes. In fact, oscillation with *no* catalyst has been reported (Kuhnert and Linde, 1977; Körös and Orban, 1978; Orban and Körös, 1978).

The chief indispensible feature is an acid solution rich in bromate ions. Much of the chemistry seems to rest on peculiarities of bromate, bromide, and more exotic bromine-based ions. Its essence is oxidation of an organic substrate by bromate in water at low pH.

Exotic though some of the elementary reactions may be, the reagent is easy to prepare and it works reliably. The recipe I prefer is one adapted from Zaikin and Zhabotinsky's version of 1970, which was itself an improvement on Zhabotinsky's 1964 recipe, in which the oscillations were enhanced by substituting malonic acid for Belousov's original citric acid. The 1970 recipe substituted the colorful ferroin for cerium, to serve double duty as catalyst and redox indicator following Busse's (1969) introduction of a trace of ferroin to improve color contrast in the cerium recipe (see Boxes A and B).

Box A: Recipes

To prepare this reagent, the following aqueous solutions are to be made from anhydrous reagents:

To 67 ml of water, add 2 ml of concentrated sulfuric acid and 5 g of sodium bromate (total 70 ml). To 6 ml of this in a glass vessel, add 1 ml of malonic acid solution (1 g per 10 ml). Add 0.5 ml of sodium bromide solution (1 g in 10 ml) and wait for the bromine color to vanish. Add 1 ml of 25 mM phenanthroline ferrous sulfate and a drop of Triton X-100 surfactant solution (1 g in 1000 ml) to facilitate spreading. Mix well, pour into a covered 90 ml Petri dish illuminated from below (Figure 5, top). Bubbles of carbon dioxide can be removed every 15 minutes by stirring the liquid: It turns blue then reverts to red and the color changes begin anew. The reagent will turn permanently blue after about 45 minutes at 25°C. Amounts are not critical except that no more than 4 mM chloride can be tolerated; in other words, use distilled water and keep your salty fingers off any surfaces that will contact the liquid. Avoid using the European Merck ferroin, which contains 75 mM chloride. I have usually found it necessary to recrystallize the sodium bromate, even from Analytical Reagent grade stocks. Leaving out half of the recommended sulfuric acid will suppress spontaneous oscillation, at least in thin layers exposed to the air (Figure 5, bottom). In this condition, the blue, oxidizing phase propagates like a grassfire when started but never occurs spontaneously.

The easiest way to watch the chemical waves is to pour a 1-mm depth of reagent into a very clean dish resting on top of a lightbox. Disposable plastic tissue culture dishes are good for this purpose. They are optically perfect and free of dirt and scratches, which otherwise serve both as nuclei for carbon dioxide bubbles and as pacemaker nuclei that engender a discouraging profusion of waves. Waves may start spontaneously. If not, a touch of a hot needle will help. Setting the dish over a second dish filled with blue copper sulfate solution (to which a few drops of sulfuric acid are added for clarity) both prevents the rapid destruction of waves by convection currents and greatly enhances visual contrast between the vivid blue wave front and its orange-red background. Be sure to cover the reagent dish: The slightest air current disrupts the perfection of the wave fronts. Escaping byproducts will give you a headache as a remainder. You must be very careful not to bump, vibrate, or tilt the dish, or liquid flow will destroy the waves.

Two ml of collodial silicon dioxide can be used to gel 1 ml of reagent. Waves propagate in this peanut-butter-like gel exactly as in the unbound fluid. This makes it easy to insert barriers to block a wave, or to shear waves in a controlled way, or to cut out the center of a spiral or ring source, etc.

DeSimone et al. (1973) showed another way to eliminate possible complications from hydrodynamic movements, by binding the ferroin catalyst-indicator in a thin film of "collodion" cellulose esters. This film is then wetted by floating on a solution of all the other ingredients of the malonic acid reagent. I find it more convenient to prepare the complete reagent and then drop a Millipore filter disk onto the liquid's surface. Millipore is a paper-like material made of cellulose esters or of vinyl chloride pervaded by submicroscopic tunnels less than a micrometer in diameter. The reagent quickly soaks into this ultra-sponge and is immobilized by its own viscosity. Thus the Millipore can be handled freely without disrupting wave patterns. However, once again, beware of salty fingertips and iron forceps: Use nylon or stainless. I have not found any other kind of micropore filter that works well. Not all versions of the reagent propagate in

Millipores, but the recipe given above does when suitably protected from air to retard evaporation and exclude oxygen. I sandwich the filter between plastic sheets or stick it to the inside of the lid of a closed plastic dish, or drop it in oil. Carbon dioxide passes off without forming bubbles and, in such a thin layer, reaction heat passes off efficiently enough that all the phenomena observed can be reasonably thought of as isothermal.

For demonstration purposes, it is sometimes convenient to have a supply of dry Millipores which need only be wetted to start the reaction. These make excellent waves that continue to move about for half an hour at room temperature. Prepare 1 Molar ammonium malonate[1] in water. Mix this with an equal volume of standard 25 mM ferrous phenanthroline. Combine this mixture with an equal volume of 4 M ammonium bisulfate. Float a Millipore on this solution, remove surface liquid by dabbing with filter paper, and allow it to dry on a plastic surface. Don't let it dry *completely* before removing it from the plastic surface, lest it crack when removed. Stored dry in a plastic baggie in the refrigerator, these have shown no sign of deterioration in a year. To produce waves, wet the filter by floating it on $\frac{1}{3}$ M sodium bromate in water. (0.015 ml/cm^2 is the right amount per unit area of dry Millipore.) Protect the wetted filter from oxygen and drying between two sheets of plastic. Blue dots soon appear and grow into waves propagating at a few millimeters per minute in ring and spiral configurations.

---

[1] In a previous publication (Winfree, 1978a) I gave this recipe with ammonium *bromo*malonate. It turns out that this is not necessary, which is just as well since bromomalonate is hard to come by.

A: Mechanism of the Reaction

The mechanism of this chemical oscillation has been dissected in fine detail by many investigators, but most notably by Field, Körös, and Noyes (reviews in Eyring, 1978). In gross outline (see Tyson, 1976b, and Figure 1), bromate is continually converted to molecular bromine, which continually attacks malonic acid and brominates it while inhibiting a second reaction. The bromate is decomposed by either of two routes, depending on the bromide concentration. At high bromide, a series of oxygen atom transfers is involved, which incidentally consume bromide. When bromide falls below a critical concentration, a free radical mechanism (the second reaction) comes to dominate. As bromide falls below this switching threshold, $HBrO_2$ appears explosively (increasing 10^5-fold within much less than a second) by a reaction which abruptly consumes the last vestiges of bromide. In this transition, the red ferrous ferroin in oxidized to the blue ferric form which, in turn, oxidizes the accumulated bromomalonic acid. As it does so, it turns back to the red form and bromide and carbon dioxide are released. When enough bromide accumulates, control is handed back to the first reaction.

This reaction is spatially coupled by molecular diffusion. According to Field et al.'s analysis (1972, 1974b, 1979), things would probably go quite differently if $HBrO_2$ were somehow prevented from diffusing. Physical movement of other reactants is probably less important, though for conceptual convenience equations are commonly written as though all diffusion coefficients are exactly equal.

Box B: Chemical Oscillations, Practical Jokes, and Chaos

The first version of this reaction was encountered by Belousov in 1950 while trying to assay citrate quantitatively in an acid mixture containing bromate, bromide, and a standard reagent for oxidation-reduction titration, ceric sulfate. Needless to say, the fluctuating composition of the mixture occasioned initial consternation. I found this useful as a practical joke during recalibration of a repaired Cary spectrophotometer. A cuvette of pale yellow malonate reagent (using the original cerium instead of the visually more colorful ferroin) is placed in the sample beam, with an equivalent cuvette lacking malonate in the reference beam. Ordinary reactions show a monotone approach to a stable optical density at each wavelength. But with the malonic acid reagent, large irregular excursions of optical density in the ultraviolet continue for half an hour while the visible yellow scarcely changes (Figure a). The perplexity of an unsuspecting expert in electronics can last longer. (The irregularity is due to carbon dioxide bubbles and spatial differentiation of phase in the unstirred oscillation.)

E. K. Pye, while a research student of biochemistry at the University of Manchester in 1963, encountered similar vexing "instabilities" while spectrophotometrically following sugar metabolism in yeast by NADH fluorescence. Measurements were simply irreproducible. Every other possible source of error was proven fault-free before testing the outrageous conjecture that NADH concentration was fluctuating with a regular 30-second rhythm (See Chapter 12).

Many phenomena, including vital clinical measurements, in which irreproducible readings have been tolerated as "biological variability", have been found to be regularly periodic. For an especially nice example, see Young (1978). This fact provides the foundations for the whole field of circadian physiology (see Chapters 19, 20, and 21).

Of more recent vintage, a deterministic interpretation of irregularly variable measurements has been found in the concept of "strange attractors". In such cases, the ostensibly chaotic fluctuations of chemical concentration are traced, not to probabilistic causes, but to lawful variation through known reactions. The malonic acid reagent exhibits such behavior in an appropriate range of oxidant concentrations. (See Schmitz et.al., 1977; Rössler and Wegmann, 1978; Wegmann and Rössler, 1978; Tyson, 1978; Yamakazi et al., 1978, 1979; Hellwell and Epstein, 1979; for other examples of chemical chaos see Olsen and Degn, 1977; and Schmitz et al., 1979.)

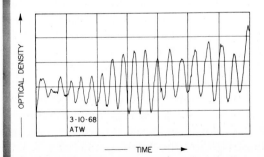

3-10-68
ATW

OPTICAL DENSITY →

TIME →

Optical density in the ultraviolet, as traced by a spectrophotometer look-looking through 1 cm of cerium-catalyzed malonic acid reagent. This was in the days before ferroin. The irregularities are presumably due to spatial gradients of phase and slow fluid convection.

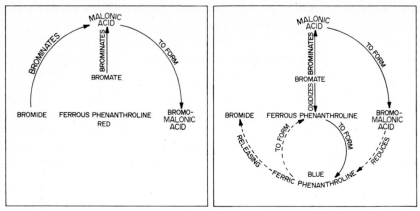

Figure 1.   Two sets of reactions can account for the oscillation of the malonic acid reagent from red to blue and back to red. The mechanism can be followed in skeleton form by describing reactions among bromide, bromate, malonic acid, and iron phenanthroline, which serves double duty as a catalyst and as an indicator dye. The concentration of bromide determines which of the two sets of reactions will dominate in a certain region. In the first set (left) the bromide and bromate both brominate (add bromine to) the malonic acid to form bromomalonic acid. During this process the phenanthroline dye is red, with its iron atom in the ferrous form. If the concentration of the bromide drops below a threshold level, then the second set of reactions (right) starts to dominate. The last vestige of bromide is consumed and the bromate takes over the bromination of the malonic acid. Simultaneously it oxidizes the iron atom in the indicator dye, changing it from red ferrous phenanthroline to blue ferric phenanthroline. Accumulated bromomalonic acid later (dashed arrows) reduces ferric phenanthroline back to red ferrous phenanthroline, thereby releasing bromide and carbon dioxide. High concentration of bromide shuts off this reaction sequence and restarts the red stage. Other substances such as oxygen in the air are also involved but they are not shown.

B: Wave Phenomena

This chemical reagent provides very nearly what I aspired to create by wiring together a lot of neon glow tubes in Chapter 11, what Franck and Mennier (1953) created by wiring together electrochemical oscillators, and what Diamant et al. (1970), Sarna et al. (1971), Brown et al. (1975) and Patton and Linkens (1978) approximated by wiring together electronic oscillators (Chapter 14). It is a continuum of mutually coupled relaxation oscillators. However, they are not all indiscriminately coupled: Nearest neighbors are coupled most strongly. Thus the sought-after mutual synchronization phenomena acquire *spatial* organization. It was in thin layers of such a ferroin-malonic acid reagent that Busse (1969) and Zaikin and Zhabotinsky (1970) discovered sharp bands of vivid blue cruising through the motionless red liquid. The behavior of these waves has stimulated much curiosity and investigation. Two extremes of wave behavior warrant separate attention, though it should be remembered that they are only extremes of a continuum which grade into each other through intermediates (see Box C).

At one extreme, the malonic acid reagent exhibits "pseudowaves". These are periodic patterns of color which appear to move through the liquid, not by virtue of any physical conduction or propagation but as a natural consequence of a

Box C: Pseudowaves

As the papers of M. L. Smoes testify (1979 and submitted), I have introduced some confusion into the research literature of chemical waves along with my notion of pseudowaves (Winfree, 1972c and this book pages 124–127, and 215–220.) I believe the problems stem from a mistake I presented in December 1973 at the Faraday Society meetings of the Royal Society (Winfree, 1974d).

This twofold error consists firstly of calling something a pseudowave that was not definitively shown to be such (and, as I now perceive, mostly likely was not one); and secondly in falsely equating pseudowaves with wave trains of time period equal to the spontaneous oscillation's period in an isolated volume element.

Strictly speaking, pseudowaves exist only in imaginary continua of completely independent identical oscillators. Their spatial integrity gradually decays, being provided only by the initial conditions. Any real medium is a poor approximation to that ideal, especially in that adjacent volume elements are inevitably tightly coupled by diffusion. However, phenomena do occur in real oscillating reactions which strikingly resemble pseudowaves. These occur where the local oscillation is not quite synchronous across a wide space. Such waves were first reported by Zhabotinsky (1968, p. 90). They are distinguishable by their wide spacing (wavelength at least several centimeters) and high velocity (at least a millimeter per second). Such waves occur only as periodic wave trains. They pass each point at the same period, indistinguishable from the period of oscillation in an isolated volume element. Their apparent velocity thus necessarily varies locally. They are not noticeably disturbed by impermeable barriers. The only documentation of this effect is in Kopell and Howard (1973a), whose wave is only one-dimensional and runs along a *gradient* of oscillation period. In this context Kopell and Howard call it a kinematic wave.

The notion of a pseudowave thus seems at present primarily a theoretical construction, introduced primarily as a contrast to "trigger" waves. A trigger wave moves with spatially uniform velocity. It can occur as a single unrepeated wave front.

Now, what I showed at the Faraday Society meeting was an oscillating reagent in which passage of a trigger wave train left the reagent behind it without any source of excitation save its own spontaneous oscillation. That oscillation was phased in each volume element by the prior passage of the last trigger wave. Every volume element turned blue at the fixed intervals of local oscillation after passage of that trigger wave. Thus even if every volume element were completely isolated from its neighbors, they would still turn blue in succession as observed, and a train of pseudowaves would follow the receding trigger wave forever at exactly its velocity. That is what I saw and I called it a pseudowave. But it couldn't have been.

It could not have been a pseudowave because the volume elements were not mutually independent. Neither could the concentration gradients have been shallow enough that diffusion had negligible impact on the course of local reactions. On the contrary, all the diffusion gradients were as steep as in the initial trigger wave because whenever a volume turns blue it initiates a trigger wave like the first one. In a slightly shallower phase gradient, volume elements would turn blue in succession *faster* than the trigger wave propagates and this spontaneous bluing would outrace true propagation. The continually triggered wave would be continually anticipated by the scheduled local oscillation in the volume elements ahead. But here the steepest sustainable phase gradient was established by a moving trigger wave, so that the pseudowave had no greater speed than a trigger wave.

I think what I observed was in fact a series of trigger waves, each nearly a replica of the first one used to establish the phase gradient. Each was initiated at the extreme point of the phase gradient, when that volume element turned blue in its spontaneous oscillation. But away from that point diffusion could not be ignored. The fact that the waves passed each volume element at the period of the homogeneous oscillation does not mean that the wave *is* only a phase gradient along a continuum of independent local oscillators (a pseudowave) but only that it was triggered by the local oscillation at the origin with that period. Had all volume elements been magically uncoupled, a true pseudowave of much the same appearance and timing would have continued. But the real wave was not a pseudowave because volume elements *were* distinctly coupled, and diffusion was *not* negligible, and the phase gradient was *not* infinitely shallow.

I suspect that Beck and Varadi's (1971, in Hungarian) introduction of the same notion fell prey to a similar error. They rightly noted that Busse's (1969) "dissipative structures" probably were traveling waves. But they were probably not pseudowaves and probably *did* critically involve diffusion, being so close together. I think they were trigger waves.

Recent calculations by Reusser and Field (1979) show a gradual transition from trigger-wave-like velocity to pseudowave-like velocity over the critical range of initial phase gradient steepness. They clearly show that the waves I contrived at the steep end of that range were like pseudowaves only in their periodicity and that "true" pseudo-waves are revealed only in the limit of infinitely shallow gradients.

parochial oscillation whose phase varies in space. In such a case, the successive color changes of adjacent volume elements give an appearance of motion.

At the other extreme let the spatial phase gradient be very steep. The waves will then creep so slowly and be so close together that molecular diffusion between adjacent volume elements has a substantial impact on the local reaction within each volume element. This will occur when the emigration of molecules from a volume element proceeds faster than local synthesis can replace them and when their immigration is correspondingly faster than local degradation can consume them. When the waves move this slowly and come this close together, the blue oxi-dizing excitation actually propagates like a grassfire. Volume element after volume element is triggered to a pulse of oxidative activity by the activity of its neighbor, and in turn conducts the impulse to its neighbor on the other side. In this extreme, we have what Zeeman (1972) called a "trigger wave". Such waves are characterized by a reasonably uniform velocity, though it depends somewhat on wave spacing. Pseudowaves, in contrast, are characterized by a uniform periodicity in time (Figure 2). The pertinent physical quantity which determines which limit the system falls into is a ratio between a measure of diffusion flux and a measure of the reaction's sensitivity to concentrations. To be specific, the measure of diffusion is the second derivative of concentration in space. This determines the rate of accumulation or loss of substances from a given volume element. The measure of the reaction's sensitivity is the differential rate of change of reaction rate as local concentration is changed. These conditions are specified quantitatively in mathe-matical papers by Kopell and Howard (1973b), Othmer (1977), and Conway et al. (1978).

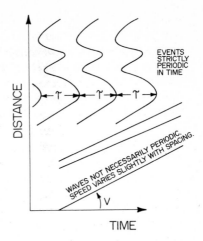

Figure 2. (above) Pure pseudowaves. Only in a linear gradient of phase do such waves all have the same apparent speed. (Period = horizontal spacing and speed = slope on this diagram.) (below) Real propagating waves. Only when far apart do such waves all have the same period.

Trigger waves persist when the chemical composition of the reagent is so altered that its oscillation period becomes very long or ceases altogether, that is, when the clock-like behavior is turned into hourglass-like behavior (Chapter 3). This can be done either gradually, following the wave behavior and bulk oscillation period as the reagent ages by consuming its substrate (Winfree, 1974d; see Box C), or discretely, by preparing a reagent that doesn't oscillate even in the beginning (Winfree, 1972c).

C: Excitation in Non-oscillating Medium

I contrived a non-oscillating version of the malonic acid reagent in 1970 in order to demonstrate abolition of waves when the isolated volume element's limit cycle was abolished. That seemed a reasonable expectation at the time, since oscillation and periodic wave propagation were the main features distinguishing this reaction. At the time, I was not thinking "excitable media" such as nerve membrane or layers of social amoebae, which are capable of recurrent wave propagation whether or not they support a local oscillation. But by early 1971 my reagent bottles, facetiously labeled "1 normal solution of the wave equation", had to be relabeled "aqueous solution of the Hodgkin-Huxley equation" because the *non*-oscillating reagent proved to conduct waves almost indistinguishable from those seen earlier in oscillating reagents. Rössler and Hoffmann (1972) were, I think, first to note that the qualitative parallels to electrophysiological waves are quite striking. This analogy has been pursued heuristically in Winfree (1974e) and in mathematical detail by Troy (1978). Troy and Field (1977) have assembled an analysis of excitability without spontaneous oscillation, based on the original oscillator mechanism. [Since then Goldbeter et al. (1978) and Goldbeter and Erneux (1978) have achieved the same for the oscillatory biochemistry of *Dictyostelium* and of yeast glycolysis, respectively.] Tyson (1977a) has reduced the chemical equations to the mathematically typical phase portrait of an excitable medium. In this approximation, the malonic acid reaction's kinetics resemble the still simpler caricature I used (Winfree, 1974b,f) to isolate the essential geometric principles of stably rotating patterns of reaction and diffusion in this reagent.

It is my impression that the main phenomena of wave-conduction and pattern-formation in this chemical reagent derive not from its oscillation, but from its completely independent property of excitability.

The non-oscillating reagent presents us with a second surprise. Its quiescence turns out to be perfectly stable in the *un*stirred liquid, as well. No diffusive instabilities occur. As in other excitable media, the spatially uniform steady-state is stable. To initiate oxidative activity (red-to-blue color change), a perturbation is required, and it must exceed a threshold magnitude. This fact about the malonic acid reagent is persistently overlooked by theorists of "dissipative structures" who see waves in this reagent as examples of structures arising from diffusive instabilities [Ross, 1976; Defay et al., 1977, p. 507, who also, following Beck and Varadi (1972), mistook the waves for a· *surface* phenomenon; Cottrell, 1978].

D: Wave Pattern in Two-
and Three-Dimensional Context

A flask of liquid malonic acid reagent, at first glance, looks like effervescent deep purple Kool-Aid. The bubbles are carbon dioxide. Closer inspection reveals that the color is mottled red and blue in bands a millimeter or less apart which constantly move. Turmoil is guaranteed by the occasional ascent of a bubble from the vessel wall to the upper surface and by the convective turnover of the reacting liquid as it warms itself by several degrees Celsius. This liquid proves to be as full of surprises as any of the organisms I expected it to resemble. Latent in this reagent, waiting for the right stimulus to conjure it into vivid red and blue animation, is a three-dimensional self-regenerating pattern of activity quite unlike anything seen before in a chemical solution.

In sufficiently thin layers of reagent, frame by frame analysis of movies reveals this pattern in two-dimensional cross-section. It is a rotating spiral wave.

(a)   Each piece of the spiral wave is propagating at fixed velocity perpendicular to the local wave front, as though following Huygen's principle (1690).

(b)   The whole wave is also pivoting rigidly as though engraved on a rotating disk.

As noted on pages 244–247, one waveform is compatible with both descriptions. This is the involute of a circle, i.e., the curve traced by the end of a taut string winding around a circle of circumference equal to the standard wave spacing. One property of such a curve is that perpendiculars to it are all tangent to that central circle. Figure 3 shows this construction on a typical snapshot of a spiral wave.

As Eugene Wigner once noted, "Every empirical law has the disconcerting quality that we do not know its limitations." Some of the limitations of approxi-

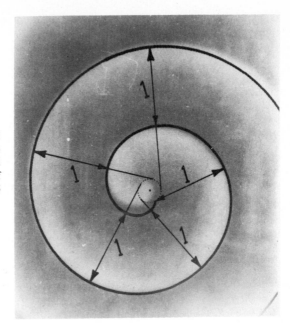

Figure 3. An involute spiral is plotted atop a photograph of the inner two turns of a spiral wave in malonic acid reagent. Its evolute is a central disk of unit circumference. (The unit is 1.85 millimeters.)

mation (a) plus (b) are exposed in the following four points:

1. Experimentally, wave velocity is not quite uniform: The malonic acid reagent conducts an excitation somewhat more slowly if it has done so in the recent past. In consequence, the wave velocity increases somewhat with wave spacing (and a variety of situations do admit local variations in wave spacing, e.g., the sudden extinction of a nearby source).

2. The experimentally observed wave does not truly pivot. Its inside end more nearly "meanders" through a central disk of circumference about equal to the standard opening.

3. Mathematically, the involute solution terminates on the boundary of the central disk. Inside that disk, the solution becomes imaginary, signalling that our two descriptive approximations a and b are incompatible too close to the pivot.

4. According to the involute approximation (a and b together), all the contour lines of uniform reactant concentration on a snapshot would be parallel involutes. But this is impossible, since on a trip around any ring-shaped path the concentration of each substance must go up and back down, so we run across each concentration an even number of times. The contour lines thus come in pairs that must join together on the inside. They cannot all just dive into the pivot because an infinitely steep concentration gradient would then appear at the pivot, but this is impossible in the face of molecular diffusion. Instead, each concentration contour line connects onto its mate as in Figures 13 and 19 of Chapter 9, violating the exact involute description. Guckenheimer (1976), Hastings (pers. comm.), and

Greenberg (pers. comm.) all independently proved mathematically that concentration contours cannot all be parallel spirals without violating the basic equation of reaction and diffusion. But far from the core, they can approximate an involute spiral quite well (see Tyson, 1976b, pp. 98–103) and "far" can be quite near if the kinetics are nearly discontinuous.

Details aside, we do see a spiral wave in thin layers. Note that the thinness required to justify a two-dimensional description is about the diameter of the troublesome central disk of the involute approximation. This disk is the source of the outer wave and is the only region in which something more interesting than straightforward wave conduction is taking place. Experimentally, the rotor's core cannot be confined in a box smaller than about the diameter of the model's central disk. It is unstable in smaller confines: The reacting liquid lapses into uniformly red stationarity. This critical dimension gives us a practical definition of the topological idealizations "zero-dimensional", "one-dimensional", "two-dimensional", and "three-dimensional".

How does the two-dimensional rotor fit into the more richly structured three-dimensional picture? The first clue is that the period of the three-dimensional structures is nearly the same as the period of the two-dimensional rotors. (There also exist structures of quite distinctive appearance with other periods, called "pacemakers" or "leading centers". See below.)

The second clue is that the two-dimensional layer is not really two-dimensional. It has a finite thickness and within that thickness, the spiral wave resembles a very narrow scroll, like a watch spring. And that scroll has an axis, a short vertical line seen in vertical projection as the pivot point. What if it were longer? What if it were tilted over, not perpendicular to both interfaces? Figure 22 of Chapter 9 sketches such situations. It turns out that the structures seen in somewhat thicker layers can be interpreted in some detail in terms of such curvy tilted scrolls. There is a depth threshold at which simple spiral patterns give way to the more complex patterns which I interpret as three-dimensional structures seen in projection. This depth turns out to be just what is required to admit a horizontal scroll axis. Beyond this depth, the scroll axis can meander to arbitrary distances between the liquid's upper and lower interfaces.

The scroll axis has some intriguing properties. Particularly notable among them is its indecisiveness. Every other volume element in the reacting fluid cycles regularly through red and blue stages with a universal period. Only at the scroll axis is such regularity impossible. It is impossible precisely because of the regularity elsewhere. On any cross-section through a scroll axis, the contour lines of fixed phase in the red-blue cycle are all spirals converging to the scroll axis. Near the scroll axis, all phase values appear in order around any tiny encircling ring. Around this ring, phase has winding number $W = 1$. At the scroll axis, all phases converge, making the scroll axis a point of ambiguous phase. In three dimensions, it is a one-dimensional locus, a *filament* of ambiguous phase. Here the chemistry necessarily becomes time-independent or varies in a way that has no component at the otherwise universal period (see Chapter 2, Example 15 and Chapter 9, Section B).

Another peculiarity is that the scroll axis is physically so long and narrow, unless confined in a tiny reaction vessel. Laterally, its dimensions are in the order

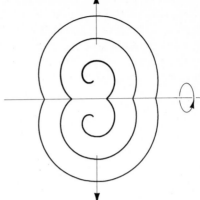

Figure 4.   A scroll ring is cut open along a plane through its axis of rotational symmetry. The inner wave is topologically a disk bounded by the scroll ring. As this rotates about the scroll ring, it buds off tomato-shaped waves that become more nearly spherical as they expand.

of tenths of a millimeter while in length it may commonly extend many centimeters before running into a wall or closing in a ring.

Furthermore exact closure in rings in typical. This feature initially seemed quite surprising in view of the narrow cross-section of this filament of ambiguity. In retrospect, and in view of the interpretation of the rotor's core given in Chapter 9, it seems less surprising. According to that view, a rotor appears when concentrations of ceric ions and bromide ions cross through a critical range as transverse concentration gradients in space. The scroll axis is seen as the intersection of two surfaces of critical concentration in three-dimensional space: one of ceric ion concentration and one of bromide ion concentration. The intersection of two curving surfaces in three dimensions is typically a ring (Figure 24 in Chapter 9). Thus the typical configuration in which an excitable reaction organizes itself in time and space is a scroll *ring* (which appears within the confines of a small container as a scroll axis bounded by walls). At any point in the volume dominated by any such wave (except along a ring-shaped phase singularity from which the wave emerges), excitation recurs at the same fixed interval of time. The volume thus dominated is bounded by surfaces along which outgoing waves collide with waves emitted by another source. Such surfaces have fixed shape and position if the other source is also a scroll ring and therefore emits waves at the same period. Because scroll rings emit waves all along their length, there may be complicated collision surfaces even within the domain of a single ring. Outside the smallest sphere completely enclosing the simplest kind of scroll ring, the waves are topologically equivalent to disjoint concentric spheres; inside that sphere the scroll ring wave is one single continuous surface (Figure 4).

Finally, the scroll axis is stable, once created. In this respect, it resembles the vortex line of classical hydrodynamics, but the scroll axis differs from a vortex line in that there is only one allowable vorticity: All scroll axes rotate with the same period. A scroll ring also differs from the quantized vortex ring of liquid helium in that the ring can have any length, above a minimum required for stability.

There remains at present a conspicuous shortage of physical insight into the mechanism of these rotating diffusive structures that generate spiral waves in two dimensions and generate scroll rings in three dimensions. Any chemically realistic picture of this tiny chemical vortex must be built out of the continuity properties of reaction kinetics in time and the continuity properties of concentration patterns in space. These matters are briefly dealt with in terms of models in Chapter 9.

Even more mysterious is the appearance of spontaneous pacemakers. These may represent local violations of continuity, e.g., dust motes.

E: Pacemakers

As noted in Figure 4, a rotationally symmetric scroll ring is a periodic source of concentric spherical waves of very short period.

There is another mechanism by which longer-period concentric spherical waves are instigated. It gets short shrift in this volume because it does not involve a phase singularity, but since it occurs in the malonic acid reagent, in the cellular slime mold, and in heart muscle, it deserves at least passing mention. This is the pacemaker point.

A pacemaker can be made in oscillating reagent by simply phase-advancing a spot. Busse and Hess (1973) did this by briefly irradiating a spot in a dish of malonic acid reagent with ultraviolet light. That spot turns blue ahead of surrounding volume elements in the next cycle and thereafter, so that a ring of blue is seen expanding out of that spot.

By another trick involving a scroll ring (Box C), a radial phase gradient can be established. Its central extremity becomes the origin of a "bull's-eye" of circular waves radiating away at intervals equal to the period of oscillation in spatially uniform reagent.

Mathematical analyses by Ortoleva and Ross (1973) and Greenberg (1978) still leave in doubt the stability of such phase gradients.

If a spontaneously oscillating medium has any slight parametric inhomogeneity (e.g., nonuniform temperature), then there is a point of locally minimum period. The phase of the oscillation advances faster here than elsewhere, so a spatial phase gradient develops. When the phase gradient becomes steep enough, waves from the center reach outer points before the liquid out there spontaneously completes its own cycle. (If the local gradient of phase is not initially that steep, it will eventually become so, since the short-period center is constantly advancing in phase relative to its surroundings.) The waves emanating from such a source are of course concentric spheres. All points within that source's domain of influence cycle at that same period *without* exceptions for any locus of ambiguous phase. As with scroll rings, the domain of influence is bounded by impermeable barriers or by surfaces along which waves collide with those from other sources. In the latter case, the collision surfaces are constantly changing shape unless both sources should happen to have identical periods, a situation guaranteed if and only if both sources are pieces of scroll rings. If the periods differ, then the collision surface constantly moves toward the longer period source. This is because each collision wipes out a pair of waves. Of the two waves next in line, the one from the short-period side will reach that prior collision point after a short period and the one from the long-

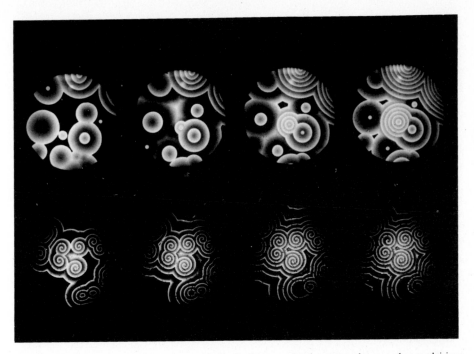

Figure 5. (top) Circular waves emerge periodically from randomly scattered pacemaker nuclei in a 1.5 millimeter thickness of reagent. (A few pacemakers lie outside the field of view.) These blue waves of malonic acid oxidation cancel each other where they collide. Each successive collision occurs closer to the lower frequency pacemaker until it is entrained. These four successive snapshots by transmitted light were taken 60 s apart. The field of view is 66 millimeters in diameter. Wave velocity is about 8 millimeter/minute. Pacemaker frequencies range from about one to three rings per minute. The tiny circles are growing CO_2 bubbles. (bottom) Everything is as above except that the first two intervals are 90 s and the reagent is less acidic (to minimize pacemaker activity and to slow wave propagation to 4 millimeter/minute). This reagent, when conducting ring-shaped waves, was deformed by gentle tilting. (See Chapter 9, Figure 9.) Segments of blue wave revert to red and the surviving pieces begin to pivot around their free ends, winding into paired involute spirals. Each spiral rotates once each minute. Notice the widening space between the outermost waves: Waves move a little faster into virgin territory than into reagent still recovering from the passage of a piror wave.

period side would reach it after a long period. Thus, the next collision occurs after an intermediate period closer to the long-period source. Each pacemaker has its temporary domain of control, but shorter period domains expand at the expense of longer period domains (see Figure 5, top).

In non-oscillating media, heterogeneous nuclei seem to induce local oscillation with much the same consequences. In the malonic acid reagent, these nuclei seem to be dust particles and/or surface defects in the container (Rastogi et al., 1977). In heart muscle they are called *ectopic foci*, which include patches of tissue irritated by abrasions, or by chemicals, or by local defects of blood circulation. In cellular slime molds, they seem to be single cells that have shorter period than most others.

Numerous models have also been proposed whereby non-oscillating excitable media may generate concentric ring waves of periods greater than the rotation period of scroll rings *without* the help of heterogeneous nuclei (see Shcherbunov et al., 1973; Zhabotinsky and Zaikin, 1973; Yakhno, 1975; and Zaikin and Kawczynski, 1977). The mechanisms invoked there suggest another application of the "wave broom" principle (Chapter 4, Box G), i.e., substances can be transported toward a source of waves. The transported substance would only have to encourage spontaneous cycling, or shorten its period, in order to make the origin of a first wave into a permanent pacemaker.

In no case has a pacemaker been found with a period less than the rotor's. The reason for this peculiarity seems clear only in terms of relatively naive models that involve the notion of an absolute refractory period. In such a case, the minimum supportable period is the refractory period, and this is realized by a rotor, so no pacemaker can do better. Unfortunately, such models imply a discontinuity of concentration within the rotor's core. This may be acceptable in cellular media such as heart muscle and social amoebae, but it is definitely not acceptable in the malonic acid reagent. When we turn to strictly continuous models of excitability, however, there seems a possibility for the rotor's period to exceed the minimum stable period of plane waves. Whether it can also exceed the minimum stable period of waves emitted from point source remains to be determined.

14. Electrical Rhythmicity and Excitability in Cell Membranes

Every cell has a plasma membrane. The plasma membrane is a thin film, less than a hundred angstroms thick, which maintains a difference between inside and outside by gatekeeping the passage of molecules and ions. Every cellular membrane is freely permeable to some substances (e.g., water) and essentially impermeable to others (e.g., proteins and certain ions). Nerve cells and some secretory cells are distinguished from most other kinds of cell chiefly in that the selective permeability of their plasma membranes depends sharply on an electric field. All cells experience an electric potential difference between inside and outside, ultimately because amino acids bear an ionic charge and, once polymerized inside the cell, they can't get out.[1] This potential difference is typically about one-tenth of a volt, so the thin plasma membrane is stressed by an electric field in the order of 10 million volts/m. In nerve cells, molecular anatomy within the plasma membrane is believed to readjust when this field is reduced to less than a certain threshold. With its selective permeability altered, the membrane passes certain ions that it had formerly restrained, resulting in a further decrease in the field maintained, and a self-catalyzing breakdown of membrane potential quickly ensues. But things are so arranged that a recovery promptly follows in which electrical imbalance is restored.

The kinetics of this "action potential", as it is called, has been a favorite object of study among biophysicists. The first thorough quantitative description of the major ionic events was given by Hodgkin and Huxley (1952) on the basis of their electrical experiments using a squid's giant axon, the high-speed conductor for the squid's escape reflex.

Hodgkin and Huxley clarified the mechanisms of nerve excitability by establishing five facts in the squid axon:

1. Excitability is a membrane phenomenon. If the cytoplasm is involved, it is involved only as a dilute solution of the several essential ions (and, in the long term, as the biochemical support of the membrane).

---

[1]   This potential difference would persist without expenditure of energy were the membrane perfectly impermeable to select ions. But it leaks a little, so the observed voltage must also be attributed to ion "pumps" in the membrane.

2. Fast changes of membrane potential V derive from changes in selective passive permeability to ions, notably sodium, potassium, and chloride (and calcium, in many of the pacemaker membranes of concern in this chapter; but that is a more recent appreciation).

3. The rates of these permeability changes are governed instantaneously by V.

4. A standard linear electrical equation describes the rate of change of V in terms of the membrane's capacitance, its permeability to ions, and ion currents driven by the potential difference across the membrane.

5. Active pumps in the membrane compensate for imperfections of its selective permeability.

For at least a decade this much had been explicitly formulated, though not yet universally accepted (e.g., see Offner et al., 1940). The achievement of 1952 lay in contriving ingenious experimental methods to actually measure the expected functional dependencies and in showing numerically that they suffice to rationalize the familiar neuroelectric phenomena. Hodgkin and Huxley summarized the kinetics of voltage and permeabilities in a four-dimensional local differential equation augmented by diffusion terms representing the spread of current in space. The best description of the equation's behavior, in both topological and biophysical terms, is due to FitzHugh (1960, 1961). A variety of simplifications in current use today retain the essentials of interest in present context. So do other modifications which introduce altered conductivity functions or a fifth variable quantity expanding application of Hodgkin and Huxleys' paradigm to encompass myelinated nerve, Purkinje fibers of the heart, pacemaker sensory neurons, etc. The books of Cole (1968); of Jack et al. (1975, especially Chapter 8); and of Kuffler and Nicholls (1976) give excellent and readable accounts of experimental and theoretical aspects of neuroelectric dynamics. Some fascinating historical background is summarized by Harmon and Lewis (1966) and in the book of MacGregor and Lewis (1977).

My purpose in this chapter is not to dwell on membrane biophysics or differential equations but to describe four classes of neuroelectric behavior that relate to the themes of this book through their phase singularities and their rhythmic interactions:

1. Rephasing schedules of pacemaker neurons
2. Mutual synchronization in the zero-dimensional case
3. Waves in one dimension
4. Rotating waves in two dimensions

This material is intended to provide the physiological context that readers untrained in biology may require in order to evaluate the significance or triviality of phase singularities described in the main text.

A: Rephasing Schedules of Pacemaker Neurons

Experimental Results

Some kinds of nerve cells fire with impressive regularity at a rate determined by environmental conditions. The most sensational examples I know of are the stretch receptor cell in crayfish muscle (Firth, 1966), the pacemaker in the brain of the fly *Calliphora* (Barneveld, 1971; Leutscher-Hazelhoff and Kuiper, 1966), the similar neuron in the brain of the housefly *Musca* (Hengstenberg, 1971), and the pacemaker in crayfish optic nerve (Nudelman and Glantz, 1977). Under constant conditions, firing intervals vary by only a couple percent of the mean, in these cases. Many other kinds of pacemakers also fire with sufficient regularity to suggest description in terms of phase in a cycle. Perkel et al. (1964) and Schulman (1969) measured the rephasing schedules (alias resetting maps) of a stretch receptor cell in the crayfish, which serves as a strain gauge on the muscles: Its firing frequency is modulated by mechanical tension. At a fixed tension its electrical cycle is regular but it is transiently upset by arrival of impulses from a synapsing inhibitory neuron. Upon recovery of regular firing (essentially within one cycle time), the rhythm continues as though it has been displaced along the time axis. (For details about systematizing such observations, see below. If you read the first half of this section easily, then skip the second half.) A replotting of their data reveals type 0 resetting

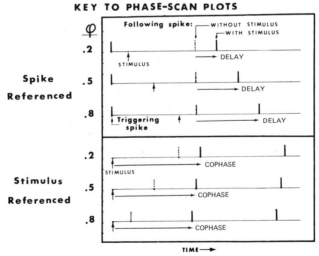

Figure 1. The firing pattern of spikes (solid vertical bars) following some triggering event (at left margin of plot) is represented as a function of phase within the normal interspike interval. The phase ϕ of the perturbation (arrow marked "stimulus") is translated into vertical level in the plot ($\phi = 0$ at top). Three representative phases are shown in the figure for purposes of illustration. The plot is triggered by the spike preceding the stimulus (spike-referenced plot) or by the stimulus itself (stimulus-referenced plot), and the following spike pattern in time is plotted as tics (vertical bars) to the right of the time origin. For comparison, the plot patterns for a null stimulus (one producing no pattern perturbation) are indicated with broken bars (adapted from Hartline, 1976).

Resistance Capacitance time constant (1–10 ms) and roughly matches the period of the highest frequency pacemaker activity. Possibly, random fluctuations of some sort occasionally direct the membrane to recycle along the subthreshold route. Wilson and Wachtel (1974), Gola (1976), Connor (1978), and Gulrajani and Roberge (1978), working with pacemaker neurons in molluscs and crustacea, trace their rhythmicity to a negative-resistance segment of the membrane's current-voltage characteristic. Connor and Stevens (1971a,b,c), Plant and Kim (1975, 1976), and Plant (1976, 1977), and Carpenter (1979), among many others, also trace rhythmic modulation of firing rate to the Hodgkin-Huxley equation with suitable modifications. This approach quite satisfactorily accounts for electrical rephasing of pacemaker rhythms and control of their frequency (or even suppression, leaving only the rhythmic "slow wave") by a steady biasing current.

Nonetheless it is not yet clear that the whole cause of rhythmically varying conductivities (etc.) is to be sought within the membrane itself in all cases. Both et al. (1976) and Chaplain (1976) postulate a separate metabolic oscillator whose activity affects membrane permeabilities and potential. They emphasize the role of glycolysis and in particular of phosphofrutokinase (PFK), suggesting some relationship to the glycolytic oscillation in yeast, as explored in Chapter 12. Rapp and Berridge (1977) also suppose a separate metabolic oscillator but put more weight on observations implicating calcium ions and cAMP. They suggest a relationship to the cAMP oscillations in the cellular slime mold, explored in Chapter 15. (Note that Goldbeter's kinetic model, Chapter 12, allegedly serves for both PFK oscillations and cAMP oscillations.) Rapp and Berridge (1977) and Pollack (1977) even go so far as to argue that the ultimate source of rhythmicity in the heartbeat may be a Ca^{2+}-cAMP oscillator, rather than an instability of the membrane potential.

Formats for Rephasing Measurements

Perhaps the single most conspicuous fact about most biological rhythms is our present ignorance of their chemical mechanisms (for exceptions, see Chapter 12). In many cases, we do not even know for sure whether or not our chosen observables themselves participate in the oscillatory mechanism. In some cases, it seems likely that they are causally "downstream" from the undiscovered source of periodicity, being only indirectly affected by it, and not affecting it in return. Despite extraordinary advances in biophysical measurement since 1950, the mechanisms even of electrophysiological rhythms still challenge the most resourceful investigators.

In this situation, workers in several areas of experimental physiology (cell cycles, circadian rhythms, biochemical oscillations, and neural and cardiac rhythms) have frankly accepted the challenge of indirect inference. One can ask a question of the mechanism by probing it with discrete stimuli. One sure indicator of impact (whether direct or indirect) on the mechanism generating a rhythm is alteration of the rhythm's phase, period, or shape *as measured under standard conditions*. Thus the imposed disturbance must be temporary, e.g., a current *pulse*, not a step up or down. If the measurement were conducted under conditions other than those in which its prior phase was measured, then it would be hard to know what the new phase measurement means in terms of the old phases. The significance of the experiment would be obscured to the extent that we are ignorant of the changes imposed by the changed conditions on the structure of the rhythm-generating mechanism.

In the cases of concern here, ephemeral disturbances usually have little lasting effect on a rhythm's period. But the detailed time course of the rhythm may be altered, most conspicuously by a permanent phase shift. The phase of an oscillator is the main feature we can measure reliably by watching an arbitrarily chosen rhythm without necessarily knowing anything about how the hidden oscillator imposes its rhythmicity on this feature that we find it convenient to monitor. We measure the phase some time after a fleeting disturbance, after transient effects on the rhythm's shape or period have died away. One can discriminate between alternative conjectures about an oscillator's dynamics on the basis of the distinctive regularities in its patterns of phase adjustment in response to the timing and magnitude of such a stimulus.

Thus biochemists have probed reaction sequences by looking for an advance or delay in rhythmic NADH fluorescence following injection of a consumable substrate (Chapter 12); behavioral physiologists have revealed the causes of seasonal flowering and hibernation by examining the phase control of circadian rhythms by suitably timed exposure to visible light (Chapters 19–21); cell physiologists have examined the "excess delay" of cell division following some physiological shock to a regularly dividing cell suspension (Chapter 22); and electrophysiologists have studied motor coordination by systematically measuring how far a post-synaptic potential resets the phase of a pacemaker neuron.

Such data are now widely available. They exhibit certain regularities which may be diagnostic of various classes of mechanism, permitting us in some respects a "sneak preview" of what will not be properly understood in terms of molecular mechanism for a very long time yet. But it is troublesome that resetting data are available in such an imaginative array of formats. (See Box C of Chapter 4.)

My purpose here is to redescribe the formats in use among neurophysiologists in relation to the format I've chosen as a common standard for the diverse rhythmic systems and conceptual models gathered in this book.

A physiological rhythm consists of either a discrete conspicuous event repeated rhythmically (e.g., a spike discharge) or smoother rhythmic variations in some quantity (e.g., membrane potential). In the latter case, a "phase reference event" is chosen on the smooth rhythm to serve as a discrete event (e.g., a local maximum or minimum or an upward or downward zero crossing). In the former case, the system has already chosen for us: The discrete event was presumably triggered by some underlying smoothly varying quantity reaching a threshold.

It is usual to plot the rhythm as a trail of dots, each marking the occurrence of the chosen event, along a time axis extending horizontally to the right. It is usual to stack up such lines vertically in order of the time when a stimulus of fixed quality and duration was given. Here two choices present themselves. In depicting them, I use the notation of Hartline (1976), who electronically automated an on-line display of a pacemaker neuron's resetting, simultaneously in both formats.

Format 1: Spike referenced or control referenced (Figure 1). Experiments are plotted on horizontal time axes stacked vertically with the prestimulus rhythms aligned; hence the name "control referenced". In Hartline's cases, the data points outlining the rhythm consist of neuroelectric spikes, hence the name "spike referenced". On each successive line, vertically, the stimulus comes later, so that the vertical downward axis is the old-phase axis.

In this format, data points in type 0 resetting fall along curves wriggling about a parallel to the stimulus diagonal. Type 1 resetting data wriggle about vertical lines parallel to the controls. This format has the advantage that deviations of the poststimulus rhythm from periodism in both horizontal and vertical dimensions are plainly exposed, and that if one wishes to simplify, omitting those features to plot only a phase *shift*, then this is easily done by clipping out a square unit cell. Rotated 90° and flipped over, this is a phase-resetting curve (PRC). It plots vertically the *change* of phase expected if an impulse arrives at the phase plotted horizontally. I dislike this operation because it breaks up the continuity of the data: Type 0 resetting acquires a spurious full-cycle "breakpoint" which corresponds to no comparable discontinuity in the data or in the physiology of the oscillator (e.g. see Figure 1 of Chapter 19).

Presenting results in a format reduced to a single unit cell not only violates the continuity of experimental data and shows only one square out of a larger pattern, but it also commonly entails presenting the difference between a perturbed rhythm and an unperturbed control rather than presenting the unprocessed direct observations. This procedure might eliminate correlated sources of error, but in my experience those are slight compared to the independent sources of deviation from perfect rhythmicity. So, in taking the difference, data variance is increased. Moreover, the difference must be taken between a chosen event in the perturbed rhythm and a chosen event in the control rhythm. Which ones are chosen? There are several different equally valid conventions in use, and if the result is to be confined to the interval $\pm\frac{1}{2}$ cycle, then it usually requires a change of convention at that point.

However, this format is most convenient in arguments about entrainment, which typically deal with small phase shifts repeatedly inflicted during a standard cycle by stimuli of one standard duration. It is also convenient in arguments where the *process* of phase change is of less interest than the end effect: a small advance or delay in each cycle of the stimulator. Perkel et al. (1964), Walker (1969), Stein (1974), Taddei-Ferretti and Cordella (1976), Pinsker (1977), Scott (1979), and Jalifé and Moe (1979) have presented their experimental results in this format.

Format 2: Stimulus referenced (Figure 1). This is the same as spike referenced except that the successive horizontal lines are sheared over 45° to vertically align the previously diagonal trail of stimuli. "Cophase" or "stimulus latency" is measured to the right from stimulus time, so that the poststimulus horizontal axis becomes the cophase axis in all experiments. Cophase being complementary to phase, the horizontal time axis is also the new phase axis, running from right to left. In type 0 resetting, data points lie along a curve wriggling rhythmically about a vertical, parallel to the stimulus column. Type 1 resetting data wriggle about diagonals parallel to the control diagonal.

If the stimulus referenced plot is rotated 90° and flipped over, then we have the old phase axis horizontal to the right (previously vertically down), the new phase axis vertically up (previously horizontal to the left), and the cophase axis vertically down. This is the format of Schulman's (1969) and Harcombe and Wyman's (1977) presentation of pacemaker neuron resetting, and the format of all my publications, including this book. Schulman's silent period D is exactly my cophase θ, supposing exactly normal rhythmicity follows the first poststimulus spike. This format has the advantage of presenting all the experimental data plainly while portraying the phase transition curve, alias resetting map, in each square unit cell. This curve is convenient for analysis of entrainment in terms of phase just before and just after the successive entraining pulses (Perkel et al., 1964). It also lends itself to analysis of the *process* of phase resetting (during the stimulus, as its duration increases) in terms of the phase reached, or the internal state of an oscillator at each moment. Such analyses are more cumbersome when conducted in terms of the phase *shift*, the changing difference between the phase reached and whatever the initial phase may have been.

B: Mutual Synchronization

Mutual synchronization is an interesting problem, and one that all societies have to solve. It can be solved dictatorially, every individual being entrained to an external rhythm (a common *zeitgeber* in the jargon of circadian physiology) or to a unique pacemaker individual (a "master clock"). Or it can be solved democratically, each individual contributing equally to an aggregate signal, a virtual pacemaker that serves the same purposes as a zeitgeber or master clock, holding all entrained in synchrony.

In the case of nerve-like individuals, the spike discharge is presumably the most influential part of the signal. If so, then mutual synchrony requires that whether the spike be nominally excitatory or inhibitory, each cell should advance (or delay) slightly if it encounters the aggregate spike slightly before (or after) its own was scheduled to occur. Ideally the advance or delay would be exactly equal to the phase discrepancy. Taking the output spike to mark phase $\phi = 0$, the ideal response would consist simply of resetting the phase to 0, i.e., $\phi'(\phi) \equiv 0$ independent of ϕ. This is the extreme form of type 0 resetting. However all that is really *required* is that $-1 < d\phi'/d\phi < 1$ near $\phi = 0$. This is observed for example, in the cases of rhythmically flashing fireflies (Figure 18 of Chapter 4; Hanson et al., 1971; Buck and Buck, 1976; Hanson, 1978) and of chorusing crickets and cicadas, phase-shifted by hearing their own collective chirp (Walker, 1969; Figure 16 of Chapter 4).

If $\phi' \neq \phi$ at $\phi = 0$, then each time a population fires, it advances or delays itself. If $|d\phi'/d\phi|$ exceeds 1 near $\phi = 0$ (as in Winfree, 1977, Figure 1A, for example), then entrainment is unstable (Perkel et al., 1964). Thus such patterns of response would bring about interesting consequences, such as a general speeding up or slowing down, or fragmentation of the population into two or more subgroups (see Buck and Buck, 1968, on alternation synchrony in chorusing grasshoppers), or even

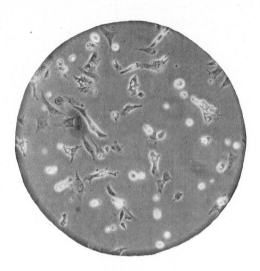

Figure 3. Single cells of dissociated chick heart creeping across the floor of a tissue culture dish in summer 1966. The rounder brighter cells are about to divide. The field of view is in the order of 0.1 millimeter in diameter.

active resistance to mutual synchronization such as we saw in context of simple clocks in Chapter 4, Section B.

Mutual synchronization is especially important among the cells of the sino-atrial pacemaker node of any mammal's heart (Jongsma et al., 1975; Clay and DeHaan 1979), and among all the cells of certain less differentiated hearts, e.g., in the hagfish (Jensen, 1966). The process of mutual synchronization is directly observable in a culture dish of individual pacemaker cells taken from the enzymatically dissociated heart of a four-day-old chick embryo (Figure 3). The cells visibly beat as they creep about underwater. They encounter one another, form electrical junctions, and begin to beat synchronously (DeHaan and Sachs, 1972). Eventually large clusters of electrically joined cells are seen to twitch rhythmically as waves of contraction sweep repeatedly from a leading center. It was my good fortune to work under Robert DeHaan's direction in summer 1966 while trying to measure the phase resetting of a pacemaker cell in response to injected current pulses. I wasn't successful, but Stephen Scott was (1979 thesis). His rephasing curves (ϕ' vs. ϕ) do show the features required above. See also Figures 16–18 in Chapter 4, Jalifé and Moe, 1976, Figure 9, and Sano et al., 1978, Figure 5 for cells of the Purkinje fibers and sinoatrial node.

There exists a tantalizing prospect that in some cases mutual synchronization among nerve-like cells might not be based so much on exchange of electric spikes as on electrotonic influences or chemical interactions among the smooth metabolic rhythms that time the spikes or slow waves. Watanabe et al. (1967) found that mutual synchronization among the heart cells of a large shrimp is mediated both by spikes and by "slow waves" (oscillatory pacemaker potentials), both of which independently affect the phasing of the slow waves in the cells they reach. The measurements of Jalifé and Moe (1976), of Sano et al. (1978), and of Scott (1979) also suggest that mutual synchronization may be mediated by electrotonic interactions. It is not entirely clear as to whether the slow wave, like the action potential, originates in the ionic mechanisms of the membrane or has some other metabolic basis.

Whatever its mechanism, local synchrony does not necessarily guarantee long-range synchrony. One possible outcome of extension in space is propagation of a wave, as in a row of upright dominos. Buck and Buck (1968) mention waves of synchrony in swarming fireflies. We have seen this in layers of pulsing *Dictyostelium* cells (Chapter 15) and in oscillating chemical reactions (Chapter 13). It is equally familiar in neuromuscular contexts.

C: Waves in One Dimension

Hodgkin and Huxleys' analysis of a quarter-century ago admirably fulfilled their aim to calculate the squid's traveling action potential and to rationalize its constant velocity in terms of local permeabilities. A similar wave sweeps across the bulging surface of a human's full atrium to squeeze the contained blood into the ventricle. A wave propagating along the Purkinje bundle, arborizing down into the much thicker muscle of the ventricle, initiates its almost synchronous contraction. But the wave-like aspect of cardiac contraction plays little functional role in either the atrium or ventricle alone. Both in the heart and in the squid's motor neuron, things would go even better with instantaneous transmission of each impulse within each compartment.

Wave behavior plays a more essential functional role in vertebrate intestine where it ushers food along the route from stomach to anus. The intestine is a long tube enveloped by a lining of smooth muscle. There is a layer of longitudinal smooth muscle and a layer of smooth muscle of quite different electrophysiological and chemical properties engirdling the intestine in circular bands. Each of these rings contracts homogeneously around its circumference like a sphincter. In an isolated segment of intestine, these contractions are regularly rhythmic. No one knows how the electrical signals for contraction originate, but they are apparently *myogenic* in that they originate in the longitudinal muscle itself, not in the associated nerve cells. Spontaneous rhythmicity is apparently the rule for smooth muscle in bladder, ureter, vas deferens, oviduct, uterus, scrotum, penis, and all the veins and arteries, as well as in the gastro-intestinal tract (Golenhofen, 1970).

In the latter case, the native frequency of this rhythm is highest nearest the stomach (~ 20/minute) and falls off in a monotone way toward the ileum, where the native frequency is lowest (~ 10/minute). All this is most commonly observed electrically, by way of chronically implanted electrodes (in dogs and cats). The visible muscle contraction follows in consequence of the electrical rhythm crossing a threshold of depolarization which elicits action potentials which in turn evoke the mechanical response (Holaday et al., 1958).

The word *wave* enters this story because intestine does not normally come in disjoint segments, but in one long continuum. Local electrical activity stimulates a similar response in adjacent muscle membrane, which responds vigorously enough to conduct a propagating wave. The one-dimensional gradient of native frequency is apparently responsible for the unidirectional bias of movement in the small intestine from stomach toward anus, a feature of considerable adaptive importance. The waves of electrical depolarization are typically several minutes and several centimeters apart. By stimulating circumferential contraction, they divide the small intestine into moving beads, each encouraging the conveyance of a bolus of digested food, as in a peristaltic pump. (However, as in almost everything biological, it isn't really that simple. Superimposed "segmentation" contractions squish food in both directions. See Davenport, 1977, Chapter 3.)

The smooth unidirectional gradient of native period encourages unidirectional movement of peristaltic contractions. But it does not guarantee *smooth* movement because the frequency gradient covers a *range* of native frequencies spanning a full octave. To a first approximation, in any intact segment of intestine, the high-frequency end serves the functional role of a pacemaker. At any point in that segment there are as many waves per unit time as there are pacemaker cycles. Downstream, these periodic disturbances stimulate electrically more lethargic membranes. At *some* distance, the local membrane has become too sluggish to respond as often as waves arrive from the high-frequency pacemaker. So an occasional wave encounters refractory membrane, is ignored, and gets deleted. This might be looked upon in two ways:

1. A local oscillator is driven by a high-frequency stimulus. If the frequency discrepancy is too great, then entrainment is imperfect and an occasional pulse is lost, or the slow oscillator simply runs at its own native period, somewhat perturbed by impinging electrical disturbances. But strict entrainment is maintained within the segment up to that critical frequency. This segment might be looked upon as a population achieving mutual entrainment through electrical interactions. Only a certain range of frequencies can thus cohere, so the intestine breaks up into a series of domains, each operating at its own frequency. Figure 4 shows some of these frequency plateaus and the temporally disorganized transition zones between them. Models of this situation have been contrived and studied in relation to real intestine. These models are basically anterior-to-posterior chains of van der Pol oscillators, each coupled electrically to its posterior (Diamant et al., 1970) and anterior (Sarna et al., 1971) neighbors. The most advanced modeling of this sort that I know of is in Brown et al. (1975) and in Patton and Linkens (1978). See the bibliography in Chapter 8, Box E.

2. An excitable membrane is refractory for some time following each response to an adequate stimulus. During this time, a bigger stimulus is required to elicit another response, and the response may even then be only half-hearted. So if a

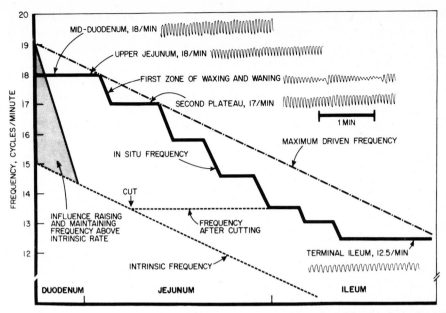

Figure 4.   Frequency of the basic electrical rhythm of the small intestine of the anesthetized cat. The stepwise heavy line shows the observed frequency, 18 per min in the duodenum and upper jejunum and descending to 12.5 per min in the terminal ileum. Segments of the electric record from which the frequency was measured are shown. The upper slanting dashed line shows 1/refractory period. The lower slanting dotted line shows the intrinsic frequency displayed by short, isolated segments of the intestine. If the intestine is transected at the point labeled CUT, the frequency distal to the cut falls to that of the intrinsic frequency at the point of transection. [From Davenport, H. W.: PHYSIOLOGY OF THE DIGESTIVE TRACT, 4th edition, 1977, Year Book Medical Publishers, Inc., Chicago. Used by permission. (Adapted from Diamant and Bortoff, 1969. Electric records supplied by A. Bortoff.)]

second wave arrives during the refractory interval, it fails to stimulate and is blocked at that point as surely as though a gate had been lowered. Arshavskii et al. (1964) contrived an analogous situation in a long nerve from an earthworm by laying it on a temperature gradient. Waves initiated electrically at the hot end get only so far before occasional deletions are inevitable. The surviving waves readjust their spacing and proceed as far as the next critical temperature, at which once again some are deleted (and so on).

I suspect that the differences between these two approaches are in part semantic differences and that a quantitative, rather than a verbal, biophysical model may behave much the same under either description. However, the differences of concept can be exposed by changing the model's parameters. I think so because that's how it is in the malonic acid reagent (Chapter 13), in which the analogous situation can be set up (see Box A) and in which models similar to those of electrophysiology have proven useful (Rössler Hoffmann, 1972; Winfree, 1973c, 1974b,e

Box A: A Biochemical "Analog Computer"
for Intestinal Peristalsis

It might be amusing to use our model excitable medium ("aqueous solution of Hodgkin-Huxley equations", Chapter 13) to mock up a length of intestine, or Arshavskii's nerve (1964). I think this is what Diamant, Sarna, Daniel, Brown, and the others might have done instead of running van der Pol equations in the computer for a couple dozen discrete segments of intestine, had they access to such an analogous continuum of excitable and rhythmical material. I tried it crudely. By mounting a copper bar between hot and cold water baths a temperature gradient was established along its length. Then I laid along the bar a strip of chromatography paper soaked in malonic acid reagent, so that the warmer end alternates between red and blue more frequently than the colder end. Each time the warmer end turns blue, that blue wave is added to the train of waves propagating toward the colder end. Now imagine any intermediate point and ask, "What is its oscillation frequency?" Is it the local oscillator frequency corresponding to the temperature at that position? No, it turns blue more often than that, to be exact, as often as a blue wave arrives from the hot end. So far as electrode recording goes, you'd say it adopts the frequency of the hot end, as does every upstream point that those waves passed.

But what becomes of these waves as they propagate into naturally slower regions? Eventually they must get to a place where the reagent cannot respond as often as waves arrive: Its refractory period is too long. So a wave gets deleted. From there on, "frequency" is lower, and remains the same at every station along the chromatography strip up to a point where a next wave is deleted. Now I can't say that this process unfolded itself before my wondering eyes with perfect clarity in that one first crude attempt, but something like this was happening, rather like the plateaus of synchrony in gut (see Figure 4). There is a region of so many waves per minute, a disorganized transitional region, a region of fewer waves per minute, another disorganized transitional region, and so on. And all these plateaus are rather labile, their boundaries shifting back and

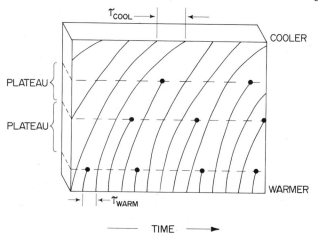

A strip of oscillating medium, seen at time zero on left end of box, conducts horizontal waves vertically from warmer end (below) toward cooler end (above). Waves are shown at front edge of strip as time increases to right. Shorter period waves from below must be deleted as they ascend into longer period regions. Deletion levels (dashed) mark boundaries of frequency plateaus in Figure 4.

forth as wave spacing varies, resulting in critical spacing and deletion of waves at different points along the full length. (These phenomena pose intriguing mathematical problems.)

What the analogy to excitable media suggests, then, is simply that a slightly different emphasis is possible in which the critical parameters would be not local oscillator frequency but rather local refractory period. It seems to me likely that the local capacity to oscillate is not very important in intestinal smooth muscle, except at the most anterior end. (And perhaps elsewhere to provide a local "backup" pacemaker in case waves from the anterior end should be somehow blocked, e.g., by ulceration or by surgery.) Local rhythmicity might be important if local excitation were not so grossly anticipated by an arriving wave; but if it is, then it hardly matters whether the local membrane would have recycled spontaneously somewhat later or would have lingered at resting potential until excited. In either case what matters is not native frequency, but refractory period. Specht and Bortoff (1972) and Sarna and Daniel (1974) did emphasize refractory period. This change of emphasis has proven critical for a number of phenomena (for example the rotor) in excitable films of malonic acid reagent and of slime mold cells. One way to find out whether it is important in smooth muscle might be to inhibit spontaneity of firing without greatly altering the refractory period, perhaps pharmacologically [as was done chemically in the malonic acid reagent (Winfree, 1972c)]. A direct measurement of refractory period along the intestine should allow explicit prediction of the positions of plateau boundaries and of the step-down in frequency from one plateau to the next by deletion of waves rather than by mutual synchronization of local oscillators. For a theoretical treatment of wave propagation along a gradient of refractoriness, see Krinskii and Kholopov (1967).

Troy, 1977b, 1978). But a systematic study of such dynamical systems has not yet appeared in the literature. It might be a useful exercise.

D: Rotating Waves in Two Dimensions

The wall of the intestine is treated as a one-dimensional continuum because our interest in it in this chapter is limited to a single layer of muscle (eliminating the depth dimension) which acts much more nearly synchronously around each circumference than along its length. But if we follow the intestine anteriorly, we discover that it abruptly flows out into a much more symmetric bag called the stomach. The smooth muscle lining of the stomach behaves much as does the intestinal smooth muscle. The lower half of it is spontaneously rhythmic in a way that suggests to Sarna, Daniel, and Kingma (1971, 1972a,b) and Sarna and Daniel (1973, 1974) and Kingma and Min (1975) a two-dimensional continuum or network of electrophysiological oscillators beating several times per minute.

The local coherence and larger scale wave-like organization of the stomach activity keeps food properly churned up and eventually forces it toward the upper bowel via the pylorus. As in the intestine, there are marked gradients of native frequency, the most anterior muscle having the highest frequency. Surgical excision of ulcerated or cancerous parts of this gradient can result in peculiar alterations of wave behavior, by juxtaposing regions of dissimilar native frequency. Diamant et al. (1973) remark that the postoperative side effects of operations for peptic ulcer often include diarrhea and anomalies of stomach movement which are usually attributed to damaging the vagus nerve, but might really reflect altered coupling in a sheet of distributed pacemakers. Specht and Bortoff (1972) and Sarna and Daniel (1973) explored this problem somewhat, both with dogs and with networks of van der Pol oscillators. They believe electronic gastrointestinal pacemakers may have medical uses comparable to the now-conventional cardiac pacemaker (see Sarna and Bowes, 1976).

Apart from derangements of the normal gradient, are there new geometric possibilities for wave conduction in the more expansive context of stomach wall? Reshodko (1974) obtained rotating spiral waves in models of smooth-muscle sheets implemented in a digital computer. The data of Sarna et al. indicate that real stomach is characterized by steep gradients of excitability and of native frequency in the spontaneously rhythmic region. I find it hard to imagine stable rotation of a wave in such a situation, but the question remains to be studied empirically.

The situation is quite different in the excitable tissue of the heart and of the brain. In both cases, the theoretically predicted possibility of perniciously rotating excitation has been demonstrated experimentally during the 1970s.

Rotors in Heart

We first examine the heart muscle, in which the demonstration is clearest. For almost a hundred years it has been unoriginal to speculate that some forms of high-frequency irregularity of heart beat (e.g., atrial flutter) might represent the effect of one or more circulating waves. Details of this hypothesis have varied and the most ingenious evidences have been brought to bear, but the following papers comprise the main contributions I have found to this story.

The basic idea is that an excitable medium that can fire rhythmically forever should also be able to conduct a circulating wave, if the medium is made into a ring and asymmetrically stimulated. This was shown in jellyfish (Mayer, 1908, 1914; Kinosita, 1937), in strips of heart muscle (Mines, 1913, 1914; Garrey, 1914), and in earthworm nerve (Arshavskii et al., 1965). In fact, Pastelin et al. (1978) showed that one mode of flutter is essentially one-dimensional in this sense: It originates in pulses circulating along three loops of specialized conduction bundles.

The equivalent experiment in a two-dimensional sheet of excitable medium would be to punch out a hole or to inactivate a disk whose periphery may then serve as the required ring. According to the simplest models, this would only work if the circumference of the ring or central obstacle were adequate to accommodate the shortest periodic wave that the material will support; or otherwise put, if it is long enough so that the circulation time of a pulse will exceed the material's absolute refractory period. The hole might even be the aorta or superior vena cava in a large vertebrate's heart, such as man's. Stibitz and Rytand (1968) attempted to demonstrate such an excitation whirling around inlet and outlet ducts of dog heart and of human heart, using the published data of several research groups.

Box B: The Iron Wire Model

William Ostwald (1900) was first to notice that iron wire in nitric acid exhibits an electrochemical surface phenomenon quite similar to the action potential in nerve. It was vigorously investigated with this in mind for decades (Suzuki, 1976; MacGregor and Lewis, 1977 for reviews). A *two* dimensional surface (a ten-inch iron sphere) behaves in many ways like a human heart, even "fibrillating" when made too excitable or stimulated too frequently (Smith and Guyton, 1961). A gridwork of iron wires also supports waves. The first publication of this experimental system (Suzuki et al., 1963) exhibited both circular waves radiating from a point source of excitation *and* spiral waves rotating loosely about one endpoint of the wavefront (Figure a). Unfortunately for many in the West, this remarkable and thorough study was published only in Japanese in a series of papers that passed almost unnoticed in English-speaking countries.

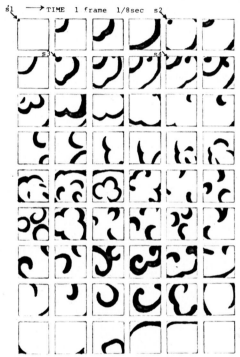

Pencil tracings at 1/8 s intervals (left to right, then down a row) from photographs of a 30 cm × 30 cm grid of 26 × 26 iron wires in nitric acid. Stimuli introduced at S_1, S_2, S_3, and S_4. Spontaneous activity persists for a while in the form of waves irregularly pivoting about moving end points. (From Nagumo et al., 1963, Figure 15, with permission.)

Balakovskii (1965) and Krinskii (1966) argued that a hole might not even be necessary. Farley and Clark (1961), Farley (1964), and Moe, Rheinboldt, and Abildskov (1964) had observed such rotors in computer simulations of nerve-like excitable media without holes. See Box B. In 1965, Gerisch published a photograph of a rotating wave of excitation in an ostensibly uniform field of *Dictyostelium* cells (see Chapter 15). Rozenshtraukh et al. (1970) attempted to demonstrate such a wave using the atrium (alias auricle) of a frog's heart stretched out on a frame in Ringer's solution. He may well have had such a wave, but the demonstration was unconvincing without an adequate number of simultaneous recordings from multiple electrodes scattered over the muscle surface. Gulko and Petrov (1972) and Shcherbunov et al. (1973) computed rotors in perfectly homogeneous tissue using heart-like versions of the Hodgkin-Huxley equation. In 1970 in Chicago and Puschino, Zhabotinski and I found stably rotating waves in a perfectly uniform and continuous excitable medium that resembles heart muscle in several qualitative essentials (see Chapter 13).

In sufficiently inhomogeneous media a more aggravated condition arises in which no macroscopic periodism has been detected. This "fibrillation", as it is called, has been explored by Gordon Moe and collaborators in an exciting series of experiments and computations emphasizing the role of microheterogeneity of refractory periods. See Chapter 10, Example 14.

The most conspicuous features of flutter (and of fibrillation) that suggest a sourceless rotating wave are the facts that

1.   High-frequency "tachycardia" has a period limited only by the refractory period and is thus faster than any other pacemaker. This is potentially troublesome medically because a focus throwing out waves more often than other sources is able to expand its domain of control with each mutually annihilating collision of two waves.

2.   A minimum area is needed, as in the malonic acid reagent. In Allessie's experiments (see below) this is about 40 mm². Flutter is not stable in smaller hearts and is not observed at all in much smaller hearts. Fibrillation is never seen in frog heart or house cat atrium. Cat ventricle (which is bigger than the atrium) can fibrillate transiently, but spontaneously recovers. So does dog atrium. Dog ventricle, however, fibrillates persistently, as does human atrium.

3.   Convulsive synchronization restores control to the sinoatrial node of the heart. I've watched this happen during open heart surgery at Johns Hopkins Hospital. The heart starts flopping more and more irregularly, and less and less coherently until it is all ajumble. Synchrony is restored with a pair of smooth, flat electrodes, making the whole heart simultaneously refractory: The next pacemaker discharge restores coherent contraction. This erasure of the pathological wave would not work if its source were a little nucleus of irritable membrane, because the nucleus would still be there after synchronization destroyed the preexisting wave pattern.

4.   Gulko and Petrov obtained a rotating wave in a computer by interrupting an established wave front with a temporarily inhibited zone, such as might be produced by an untimely local stimulus (Figure 5). [The procedure is similar for the malonic acid reagent's rotor and for the rotating wave of spreading depression (see below). All three were independently discovered in formally similar excitable media within a span of one or two years. No one has yet reported attempts to deliberately induce a rotor in *Dictyostelium*.]

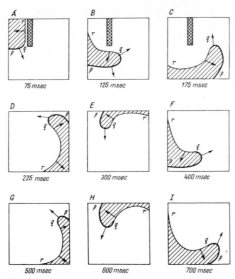

Figure 5.   Computer simulation of an action potential wave in nerve membrane. Its progress to the southeast is obstructed by cross-hatched barrier (artificially held at resting potential), which is later removed. The wavefront pq continues to revolve about the middle of the square. (From Gulko and Petrov, 1972, Figure 5.)

All these observations set the stage and enhanced anticipation for the experiment of Allessie et al. (1973). They used a sheet of rabbit atrium, once again without an anatomical obstacle, stretched on a frame perfused with an oxygen rich solution at 37°C. To initiate tachycardia, they stimulated electrically at one point at regular intervals of time. Every so often, they also stimulated just behind that wave and then waited to see if spontaneous high-frequency excitation persisted in the absence of further stimulation. Sometimes it did (Figure 6). The effect is not consistent nor is the mechanism of action very clear. The procedure originated from two observations:

1.   Wiener and Rosenblueth's (1946) original theoretical study of wave conduction in heart muscle argued (incorrectly: see Durston, 1973) that it should be possible to initiate a rotor in this way. The corrected version of that argument requires an anatomical obstacle or inhomogeneity to wipe out one of a pair of symmetric waves thus initiated. Allessie and friends believe that they are taking advantage of such inhomogeneities in the refractory zone just behind the wave initiated by the first of their two pulses.

2.   According to Katz and Pick (1962), periods of tachycardia or fibrillation in man almost without exception are preceded by a premature beat.

In any case, once tachycardia started, Allessie et al. used an array of 10 electrodes, moving the array to as many as 30 different positions during a period of persistently sustained activity in the rabbit's atrium. From those hundreds of recordings, they were able to build up a composite two-dimensional picture of the rotor. For example, Allessie et al. (Figure 6) show how the time of pulse arrival

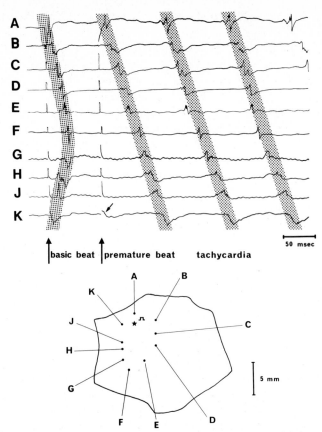

Figure 6.   Initiation of tachycardia by a premature beat. Surface electrograms recorded at 10 sites on the left atrium (A-K). The atrium was driven regularly with a basic interval of 500 ms. The first arrow indicates the last basic stimulus, applied near A and spreading toward F. The second arrow indicates the premature stimulus (interval 81 ms). The sketch at the bottom shows the sites at which the electrograms were recorded; the site of stimulation is represented by the asterisk. Note that conduction of the premature impulse failed in direction KJHG. This area was activited from site F in a retrograde fashion. Sites A, B, C, etc., were then reexcited from site K. This pattern persisted during the ensuing tachycardia. (Adapted from Allessie et al., 1973, Figure 4, with permission from the American Heart Association, Inc.; see also the idealization in Figure 24 of Chapter 2.)

lags progressively along the perimeter of a circle of 10 electrodes around the putative pivot. The winding number of phase of this periodic excitation is thus $W \pm 1$ around the ring, and so the ring must contain a phase singularity. In their Figure 5 the wave front is shown in 17 successive positions as it pivots around this singularity. Allessie et al. are careful to point out that this demonstration does not prove that such rotors occur in living animals, nor that they are the only mechanism of flutter even if they do. But this is the first mechanism that has actually been verified physiologically.

One of its most intriguing features is of course the phase singularity. What can be going on there, electrophysiologically? This is not even clear in Gulko and Petrov's computations, which serve up every variable to the experimenter for precise observation. Their wave does not exactly pivot around a fixed point but seems to wobble, to meander, as can be seen in Figure 5 by superimposing tracings of the successive panels. So the central patch of membrane, while not participating in the regular rhythm expressed everywhere else, is also not electrically stationary. A similar perplexing wobble occurs in the malonic acid reagent's rotor. Rössler (1978) and Rössler and Kahlert (1979) believe that it may be an instance of "chemical turbulence"; in the heart we would have to call it "electrophysiological turbulence". There is also turbulent activity, not quiescence, at the center of the rotor in rabbit atrium. The wavefront seems to run through the center irregularly. Given many such rotors packed close together we might see only a writhing mass of short-lived wave-icles not unlike the turbulence described by Moe et al. (1964) when their electrophysiological equivalent of a critical Reynolds number was exceeded in computer simulations.

Rotors in Brains

We are accustomed to think of nerve membrane as an electrically excitable medium along which a sharp impulse can propagate at a steady high velocity. There are also lower velocity waves of collective firing in brain tissue (Verzeano, 1963; Petsche and Sterc, 1968, Petsche et al. 1970). Petsche et al. (1974, Figure 6) show that these waves can stably rotate at 10 cycles per second around little patches of rabbit cortex a few millimeters square.

Not everyone knows that there is still another mode of electrical wave activity in masses of nerve cells, a mode that is insensitive to tetrodotoxin, i.e., a mode which does not at all require the sodium-based electrical action potential. It is induced by applying any of a variety of mild stimuli to any of a variety of cortical structures of any of a variety of different animals. In this mode, cells depolarize and leak potassium ions from their potassium-rich insides, thus further depolarizing themselves and their neighbors, who therefore also leak potassium and so propagate the condition. A droplet of potassium chloride solution is thus a reliable stimulus for starting it. After a few minutes, metabolic potassium pumps restore normal ionic conditions and the cells return to their usual function. There remains only a ring-shaped frontier of newly depressed cells which will in turn soon recover. In this way the frontier moves outward [Tuckwell and Miura (1978)]. This phenomenon, called "spreading depression", was discovered by Leao during his Ph.D. thesis research during World War II (Leao, 1944). Leao was studying epileptic seizure by deliberately initiating the bulk excitation that causes it in the cortex of a rabbit. It turned out that weak mechanical or electrical stimuli applied to the exposed cortex can also trigger a depression of activity that lasts locally for some minutes and spreads as an expanding ring. In fact, this phenomenon commonly follows epileptic seizures anyway. During this depression, all neurons are turned off within a 2-mm zone as the wave goes by. Accordingly, the EEG is locally attenuated. The ring's rate of expansion is determined by the diffusion rate of potassium in the extracellular medium, the rate of potassium leakage from depolarized neurons, and the rate at which cells bail themselves out after ceasing to leak potassium. In the presence of obstacles (e.g., the edge of the tissue, or a knife cut, or a bit of thermally coagulated tissue), the wave propagates around the obstacle according to Huygens' principle exactly as in *Dictyostelium* or the malonic acid reagent (page 245). It can be induced to *rotate* around a big enough obstacle by blocking one of the two waves into which the obstacle splits an oncoming wave front. The surviving half-wave then circulates for as many as 50 rotations of approximately constant period (Shibata and Bures, 1972, 1974). These waves travel about 3 mm per minute, comparable to the cortical disturbance associated with Jacksonian epilepsy. [In this connection it is intriguing to note that although spreading depression is supposedly a local condition, propagated by diffusion of potassium ions and maybe neurohormones, Leao found that when only one hemisphere is stimulated, spreading depression arises also at the mirror image spot on the other hemisphere (presumably by way of the corpus callosum). Jacksonian epilepsy also involves alternation of excitation between left and right sides.] Could the coma following concussion be a spreading depression rotor?

The conduction speed of spreading depression is also quite similar to that of the scintillating blindspot associated with migraine headache (Lashley: see Bures, 1959, p. 245; Richards, 1971; Martins-Ferreira et al., 1974). Gouras (1958) discovered that spreading depression also occurs in the retina of the eye. In this case, one does not need electronic machinery to observe it because there is a visible change of light scattering in the retina itself as the wave goes by. As in the cortex, spreading depression can be induced to circulate around an obstacle for six to seven hours, completing as many as 50 rotations of strikingly uniform period (Martins-Ferreira et al., 1974).

After Shibata went back to Japan, Reshodko came from Kiev to Bures's laboratory in Prague to undertake computer simulation of the rotating wave and of the mechanism of its induction. Basically, his model is similar to Wiener and Rosenblueths' (1946) and the other two-dimensional excitable media discussed in this book. As in those cases, Reshodko and Bures (1975) found that the central obstacle is not necessary. A stable spiral wave was found to pivot around a point in a sheet of several thousand cells (in the computer). This has not yet been observed in real nerve tissue; however, I see no good reason to discount the possibility that it will be. Allessie et al. (1973, 1976, 1977) found that it is tricky to start a rotor in heart muscle unless there is a substantial obstacle. It can be done, but the mechanism relies on unpredictable inhomogeneities in the excitable medium, such as might not have been achieved in trials on cortex or retina.

15.   The Aggregation of Slime Mold Amoebae

Two kinds of slime mold play central roles in this book. Later on we will meet the "true" slime mold (*Myxomycetes*), an acellular jelly remarkable for the regularity and synchrony of mitosis in its many nuclei. Topologically, the true slime mold is one single monstrous cell. But in the present chapter, our concern is with the cellular slime molds (*Acrasiales*), the best studied of which is *Dictyostelium discoideum* (Bonner, 1967; Gerisch, 1968). This creature is more conventional in its cellular structure but is equally astonishing topologically in that its cells wander independently, like the individual workers of an ant colony. Like the ant hive, *Dictyostelium* is a "superorganism", a genetically homogeneous being composed of autonomous individuals, nevertheless organized altruistically for the collective good. The life cycle runs as follows.

A: The Life Cycle of a Social Amoeba

Spores borne on wind and rain arrive at a substrate that encourages their germination. Following much visible turmoil within the ovoid spore capsule, a crack appears and the vibrantly active little blob of jelly escapes, leaving behind the transparent shell. This 10-micrometer-diameter single cell, a "social amoeba" as it is called, wanders about the substrate voraciously consuming bacteria. As they feed, the amoebae divide by fission. Eventually, the population outruns its food supply. During the next few hours, an internal transformation takes place by which these formerly independent cells become more alert and responsive to their neighbors. Depending on conditions not yet clarified in detail, a cell may become spontaneously active whereupon it emits a pulse of cAMP every five minutes, or it may only emit a pulse of cAMP when triggered by its neighbors. This principle of pulse relaying was discovered by Shaffer in 1957. The fact that the substance involved in the pulse is cAMP was discovered by Konijn et al. (1967) at Princeton. As in other systems of similar biochemical dynamics, a hairline difference in membrane properties is believed to distinguish these two modes of behavior: spontaneous activity (as in pacemaker neurons, spontaneous ovulators, or oscillating malonate reagent) or mere excitability (as in quiescent neurons, reflex ovulators, or the non-oscillating version of malonate reagent). This developmental change is believed to be spontaneous, in large measure. But it is also accelerated by exposure to cAMP, especially if the cAMP comes in pulses (Gerisch and Hess, 1974). Darmon et al. (1975) and Gerisch et al. (1975) also showed contact site formation to be accelerated by cAMP pulsing.

Relaying behavior results in wave conduction if the cells are sufficiently well coupled, i.e., if a cell receiving the chemical impulse passes it on, amplified up to a saturation level. If cells lie close enough together, then the amplified impulse reaches enough unexcited cells before it is attenuated too much to trigger them. So they relay it and a wave of cAMP release propagates across the field of cells at 2–3 mm per hour. The passage of a cAMP wave is made visible by the slightly later passage through the field of cells of a geometrically identical wave of altered refractivity (Alcantara and Monk, 1974). These waves were discovered and filmed by A. Arndt in 1937. As in heart muscle (Chapter 14) and in malonic acid reagent (Chapter 13), two kinds of wave occur: concentric rings emitted from a homogeneous pacemaker, and spiral waves emitted from a quiescent center. Figure 1 shows a photograph of a *spiral* wave, first published by Gerisch in 1965. As in heart muscle and in the malonic acid reagent, these rotating waves resemble involutes of a circle whose circumference equals the wavelength. See Figure 2; cf. Figures 2-26, 9-15, and 13-3. The wavelength as well as the rotation period are both shorter than in concentric ring waves. Apart from their sources and their distinctive shapes, the two kinds of wave (rings and rotors) share the same mechanism and properties.

Nonetheless, *Dictyostelium* is not heart muscle and *Dictyostelium* is not the malonic acid reagent. There are some differences. For example, *Dictyostelium*'s cells move around. Shortly after it emits its reflex pulse of cAMP, each cell elongates and moves about 1 cell diameter in the direction from which it was triggered. In consequence of the repeated movement toward the source of the waves, cells

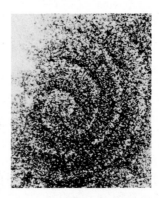

Figure 1. The first photograph of a rotating-spiral wave in a layer of aggregating *D. discoideum* amoebae. The wave spacing is about $\frac{1}{5}$ mm. at velocity 43 microns/min. Spacing and velocity as much as 10-fold greater are commonly observed in less dense monolayers. From Abb. 7 of Gerisch (1965) with permission.

accumulate at the center. When enough have gathered, a "tip" differentiates and secretes cAMP continuously, thus ending the spiral wave. The ring of nearby excitable cells is triggered to release cAMP as often as it recovers from the prior release, and the subsequently emitted waves form a nest of concentric rings (if they hadn't already adopted that configuration around a pacemaker cell).

To finish the life cycle, cells continue to accumulate until the tip is raised aloft on a substantial column of cells which eventually topples over. This little blob of transparent jelly, called a "grex" or "slug", is motile, perhaps due to the caterpillar-like waves of contraction rippling down its length from the tip at about five-minute intervals. It crawls off toward whatever light and warmth may be nearby, stops, and rights itself again. Then the tip is shoved aloft by a fountain-like streaming of cells, in which about one-third of their number sacrifice themselves to become a hard dead stalk. The remainder turn into spores in a tiny ball atop this erect column. Eventually, the spore capsule ruptures and the life cycle is complete.

B: Questions of Continuity

Because the slime mold consists of discrete cells, it is not obvious *a priori* that continuum models are appropriate vehicles for understanding its wave behavior. In fact, Cohen and Robertson's (1971a,b) quantitative interpretation of aggregation waves takes a radically different approach. It is based on a discretized approximation in which each cell is a point in a lattice (betraying Cohen's alter ego as a solid state physicist), capable of responding to stimulation after a fixed delay by emitting a discrete pulse of cAMP. Following this it is completely refractory to stimulation for a fixed interval and then once again returns to excitability. A medium of such cells might conduct waves not smoothly but by discrete leaps at fixed intervals of time. Alcantara and Monk (1974) observe just such leaps of 60 μm every 12 seconds. Also according to this view, the source of the rotating spiral wave is not a phase singularity but a discontinuity. The geometric model of Balakhovskii (1965) makes this discontinuity a line of length equal to half the spiral's wavelength, along which cells at all stages of excitation and refractoriness abut. In this respect *Dictyostelium*, as a collection of discrete cells like a nerve network (Chapter 14), may differ qualitatively from continuous excitable media such as the malonic acid reagent (Chapter 13) and associated partial differential equation models.

The chief alternative to this supposition that adjacent cells can differ discretely in their biochemical states is to suppose that rotors appear only where a cAMP pulse can circulate around a physical *hole* in the "continuum" of cells. I doubt that a sound case can be made for this alternative. In many cases, a distinct doughnut-shaped stream of connected cells clearly supports a circulating impulse, and its center is empty [e.g., Gerisch, 1961 and computer-simulations by Mackay (1978) at low cell density]. But in other cases, a thick droplet of cells, many layers deep, plainly supports a rotating wave, pivoting within the pile of cells (Gerisch, 1971, Figure 9). Neither the original photos of Gerisch (e.g., Figure 1) nor a more recent one published editorially in *Nature* (12 June 1975, p. 522) shows a hole in the field of cells near the spiral's core. Moreover the spiral waves seen in early aggregation fields all have nearly the same period and wavelength. That would seem to impose upon the hole model the awkward *ad hoc* assumption that every hole must have the same circumference viz the observed common wavelength. Although waves *can* also be made to circulate around the edge of a hole, and the source of the spiral wave *can* sometimes be a pulse circulating around the hole's perimeter, my inference is that this is not a necessary or even typical mechanism.

Whether or not biochemical continuity is commonly maintained in two-dimensions in the film of cells, their discreteness and motility no doubt do make *Dictyostelium*'s connectivity at least precarious in some situations. As the pictures of Gerisch (1961) and the simulations of MacKay (1978) both show, cells *are* further apart in some regions than in others. A sufficiently low-density region may serve *functionally* as a hole. These inhomogeneities presumably have much to do with the early appearance of rotors and the spiral waves they emit. Rotors never arise spontaneously in the malonic acid reagent because the chemical solution remains almost perfectly homogeneous until a suitable local irregularity is deliberately introduced. But in a certain *Dictyostelium* mutant (aggr 52) which makes no pacemaker cells, rotors arise in abundance (Gerisch, 1971). The cause may be that local density fluctuations ensure that the functional equivalent of holes continually appear and vanish again. The Russian literature of excitable media is replete with mathematical analyses of such inhomogeneous media (e.g., Krinskii, 1973, 1978). These papers are directed toward fibrillation in cardiac muscle, but the formal results might as well describe *Dictyostelium*. The prediction for both media is that waves of minimum supportable period should arise and multiply. In the normal case, this would go on only until each differentiates into a pulsing "tip". However, Gerisch's mutants do not form tips so they support rotating waves for a very long time.

Gerisch (1971, Figure 9) has also seen tiny blobs of cells possessed by rotors with two or more arms. Such multiarm rotors seem to be unstable in other excitable media, for reasons as yet unknown but possibly related to the difference between continuous and discrete media.

C: Chemistry in the Single Cell

In regard to the single cell's biochemical dynamics, it seems worthwhile to mention two special features of *Dictyostelium*'s spatial coupling mechanism.

1.   Although released cAMP is the only known coupling agent, it is not transported as such inward across the receptor cell's plasma membrane. Rather, cAMP affects membrane surface receptors which in turn affect intracellular reactions.

2.   The influence of cAMP seems more nearly proportional to the rate of change of extracellular concentration than to the absolute amount released.

What goes on inside the cell? For purposes of qualitative analogy we might think of the cell as a triggerable switch that can be at rest, excited, or refractory, as in the discrete-state model of Cohen and Robertson (1971a,b). But suppose we want a continuous biochemical characterization of the process of excitation and of its spatial distribution? Goldbeter (1975) put forward the first detailed quantitative model of this kind. The model consists of three rate equations for enzyme-mediated reactions involving cAMP synthesis and degradation.

1.   Energy is provided by a constant intracellular production of ATP, which is consumed by the following reactions.

2.   At a rate enhanced by cAMP, ATP pyrophosphohydrolase degrades ATP to 5'-AMP, which is constantly removed by a nucleotidase, and

3.   At a rate enhanced by 5'-AMP, adenyl cyclase turns ATP to cAMP, which is constantly degraded to 5'-AMP by a phosphodiesterase.

In another version of this model (Goldbeter and Segel, 1977; Goldbeter et al., 1978) reaction 2 is ignored, and is replaced by an isomorphic expression for extra-cellular cAMP transport. The same equations are saved under a renaming of variables to replace internal 5'-AMP by external cAMP [there being currently some difficulty in confirming the Rossomando and Sussman (1973) measurement on which the original interpretation was based (Klein, 1976; Roos and Gerisch, 1976).] Either way, behavior strikingly like that of *Dictyostelium* cells is obtained when the pertinent biochemical parameters are assigned plausible values. Depending on the exact values, the equations either oscillate, thereby producing a cAMP spike every several minutes, or respond with a cAMP spike only upon stimulation by extracellular cAMP.

Essentially the same kinetic equations were previously deployed, under another exchange of names, to mimic the mechanism of glycolytic oscillations in yeast cells, based on PFK inhibition (Goldbeter et al., 1978; Goldbeter and Erneux, 1978). As yeast has been shown to support a rotor in the pinwheel experiment without diffusion (Chapter 12), much the same might be expected of *Dictyostelium*, even with diffusion added.

Marcus Cohen (former student of the Morrel Cohen referred to above) undertook a graphical and mathematical analysis of kinetic schemes of this sort, analogous in many ways to formal models of excitability in nerve membrane and in the malonic acid reagent (Chapter 13). Cohen's (1978) model seems to account for the development of excitability and other aspects of social dynamics during *Dictyostelium*'s life cycle.

Using digital and analog computers, kinetic schemes of a qualitatively similar nature have been diffusion-coupled in two dimensions through one reactant, as in *Dictyostelium*:

1.   Gulko and Petrov (1972) obtained a rotating action potential using three equations of membrane conductivity coupled only through membrane voltage (see Chapter 14, Figure 5).

2.   Karfunkel (1975) obtained a rotating biochemical wave using enzyme kinetics coupled only through one substrate, and

3.   Winfree (1974b,f) obtained rotating waves in a hypothetical excitable medium borrowed from nerve modeling, coupling volume elements through diffusion of all reactants (Chapter 9). It works nearly indistinguishably when

coupling is only through the variable that represents cAMP or membrane voltage (unpublished).

Presumably the same will be shown in this case, coupling cells by way of cAMP.

The various models are susceptible to biochemical testing, thanks to the development by Gerisch and Hess (1974) of techniques for observing synchronous biochemical activity in aqueous suspensions of many *Dictyostelium* cells. At this writing, it has been found that adenyl cyclase activity fluctuates as in the postulated kinetics, but the predicted intracellular ATP fluctuations are not yet measureable. [Roos et al., 1977; Geller and Brenner (1978)].

D: Phase Resetting by a cAMP Pulse

It is possible to rephase the spontaneous oscillation by injecting cAMP into a cell suspension. The new phase varies according to the old phase at which the

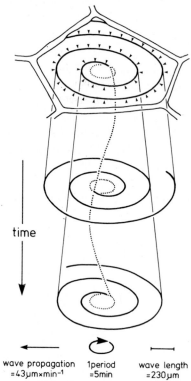

time

wave propagation   1period   wave length
=43µm×min⁻¹   =5min   =230µm

Figure 2.   Diagram of the spiral pattern of chemotactic signals in an early stage of cell aggregation. *Top*: Cells are attracted towards their inner neighbours (arrows) along a curved line which forms a spiral. The zones where such a spiral wave meets the neighbouring ones become sharp boundaries between adjacent aggregation territories. From these boundary zones the cells are withdrawn by chemotactic attraction towards either side.

From top to bottom the movement of the origin of the spiral around a central area (dotted) and the propagation of the spiral wave over the aggregation territory is shown. The data on bottom apply to dense cell layers. In more dilute cell populations the speed of wave propagation is higher (Alcantara and Monk, 1974). (Data from Gerisch, 1965). From Fig. 1 of Gerisch 1978 with permissions.

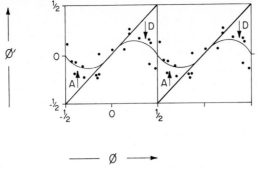

Figure 3.   The experimentally-determined new phase ϕ' reached by rhythmically pulsing *D. dis-codieum* cells when stimulated by a cAMP pulse at phase ϕ. The data are replotted from Figure 5 of Malchow et al. (1978), which plots phase *shifts* (advances A, delays D). The spontaneous cAMP pulse occurs at $\phi = 0$. The second cycle of ϕ here is merely a duplicate of measurements reported in the first cycle.

cells are exposed to cAMP. Using a cAMP pulse much smaller[1] than what the suspended cells provide themselves every eight minutes, Malchow et al. (1978) have found type 0 resetting (Figure 3). Their resetting curve lingers along the diagonal (new phase-old phase) near the usual phase of cAMP release. Thus a slow cell gets advanced and a fast cell gets delayed by each collective pulse of cAMP, so that synchrony is maintained. However, there are also effects on the *amplitude* of the cAMP rhythm following these pulses, so the effects of repeated pulsing cannot be evaluated simply by repeated application of this curve. It should be possible to compare these results with similar experiments on the Goldbeter model and others. However, it is first necessary to verify that *all* cells are oscillating in the cell suspension and not just a small minority with the more numerous remainder responding after a delay.

E: Historical Note

Before leaving the subject of periodic wave-like activity in slime mold mor-phogenesis, I interject a point of historic comparison with other experimental systems in this Bestiary.

1.   The first quantitative analytical attack on these intriguing problems (Keller and Segel, 1970) was based on a linearized continuum approximation. This invoked diffusion-induced instabilities of a spatially uniform steady-state. This approach proved to be in error:

A.   Experimentally, the uniform steady-state is stable so the phenomenon is intrinsically nonlinear.

---

[1]   By "smaller" I mean the total amount released. The cells might be more concerned about the peak *rate* of release.

B. The linear analysis produces only periodic solutions, but *Dictyostelium* is capable of propagating a solitary pulse (Robertson et al., 1972).

C. The linear analysis allows arbitrary superposition of whatever eigenpatterns have positive growth rates. But in fact only pure rings and pure standard-pitch spirals are observed. Also of interest in this connection is Nanjundiah's quantitative argument (1973, last page) that continuous secretion as postulated by Keller and Segel would be much less efficient as a detectable chemotactic signal than would be a periodic pulse of the same substance. Pulsatile morphogenesis may have been adopted by cellular slime molds for reasons of economy.[1]

2. The same thing happened in the first attempts to rationalize ring and spiral patterns in the malonic acid reagent (e.g., Desimone et al., 1973). The appearance of periodic patterns reminded theorists of the instabilities of linear kinetics coupled spatially by sufficiently asymmetric linear diffusion. Because the mathematics was well developed, it seized the imagination, squeezing other observations out of consciousness. For examples: the same patterns arise in reagent with a *stable* uniform steady-state, and spirals do not arise spontaneously, and colliding waves completely vanish. To my mind such phenomena make "diffusive instabilities" a far-fetched metaphor. Yet so tenacious is the appeal of a beautiful idea that theorists continue to interpret the real phenomena in these terms (Ross, 1976; Defay et al., 1977; Thompson and Hunt, 1977).

3. The similar concentric ring and involute spiral patterns in *Nectria* fungi (Chapter 18) predictably served as a conspicuous target for application of the same models (various people's preprints, withdrawn before publication), despite the equally conspicuous fact that no hint is ever observed of the pattern superposition expected of linear models (unless by exquisite tuning of parameters all wavelengths but one are given negative growth rates).

I think it is important for research students to be aware of such trends in the elaboration of theory, and doubly important for those whose mathematical aptitudes make their minds susceptible to infection by whatever beautiful and sophisticated models chance to be currently in fashion. As Sherlock Holmes once remarked, "It is a capital mistake to theorize before you have all the evidence. It biases the judgement" (*A Study in Scarlet.*) My impression is that after the facts are collected with an eye receptive to surprises new biological phenomena almost always suggest quite different models than whatever first came to mind.

---

[1] Although Keller and Segel's original interpretation of slime mold aggregation seems incorrect for *Dictyostelium*, practically the same equations seem to account nicely for the moving bands of chemotactic bacteria (Keller and Segel, 1971a,b). This is not at all an unusual pattern of discovery. For example, Newton's equilibrium theory of the tides is utterly mistaken for seawater, but it served beautifully for tides in the atmosphere when they were discovered.

16.   Growth and Regeneration

> One of the principal objects of theoretical research in any department of knowledge is to find the point of view from which the subject appears in its greatest simplicity.
>
> J. Willard Gibbs, 1881

Many kinds of living organisms regrow appendages that are crushed or torn off in the mishaps of an active life. People have scarcely any abilities of this sort, a fact which contributes to their jealous curiosity about the mechanisms of regeneration in more resilient organisms. This curiosity runs deeper than mere jealousy would motivate because regeneration in many ways resembles the initial normal development of an animal's structures. Normal development plus regeneration, collectively called morphogenesis, presumably operates by some general rules that we might at least elucidate empirically as a prelude to ferreting out deeper mechanisms. Yet for all the imaginative and meticulous efforts of at least four generations of developmental biologists, few general rules have stood the test of time. If principles of widespread applicability exist, they remain tantalizing obscure.

A: The Clockface Model

New hope in this area comes from a recent synthesis by French et al. (1976). They draw on experiments with the fruitfly larva's "imaginal disks" (the precursors of the adult fly's various appendages) (P. Bryant), with the legs of cockroaches (V. French), and with amphibian limbs (S. Bryant). Their paper is an extremely impressive tour de force of focused experimentation and organization of data. They come up with two principles which together account for an extraordinary diversity of peculiar experimental results. Let me first remark that what one takes to be "the principles" is presently subject to so much subjective interpretation that there are as many versions as there are people retelling this story. I will retell it in four principles, in such a way as to draw out the aspects of greatest interest in context of mappings and phase singularities. The reader will want to examine French et al. (1976), Goodwin (1976), Bryant, et al. (1977), Glass (1977), French

(1978), and Cummings and Prothero (1979) for other slants and for more detailed references to the original experimental papers.

The main point introduced by French et al. (1976) is that each tiny patch of tissue is somehow labeled permanently with an unchanging quantity that behaves like a phase or an angle in that it denotes a point on some abstract ring of states.

French et al. (1976) place the digits 1–12 as indicators of local tissue specificity around the circumference of a limb or other developmental field. Many of their diagrams thus resemble the face of a clock. Although this notation gives the model its name, it unfortunately also requires a numerical discontinuity (12/1) where none is intended biologically. As we saw in Chapter 1, compass directions or hues of color provide more apt quantitization of a ring than do numbers borrowed from the real line. Whatever the notation, we cannot yet imagine what this circular state space might correspond to biochemically, but the following simple rules make use of it and suffice to systematize a lot of otherwise very perplexing experimental results. At an appropriate stage in regeneration, tissue will develop structures (sensory hairs, muscle attachments, color patches) corresponding to their phase labels. This is an application of Wolpert's (1969) principle of separation of tissue specificity, encoded in some kind of chemical gradient, from the cell's *interpretation* of that "positional" specificity.

Rule 1: Each little patch of cells is labeled with a phase which is part of a smooth phase gradient across the tissue. Thus there exists a smooth map from the tissue to an abstract ring of biochemical specificities. Once established, this map does not change. (Note 1: Some independent second label is implicitly assumed, whereby to distinguish cell types in two dimensions.) Rule 2 introduces an exceptional point near which this map is not smooth.

Rule 2: In the normal limb, the phase values are supposed to run one full cycle around the limb axis (the azimuthal direction) as in Figure 1. If a limb were a hollow cyclinder, this would pose no problem. But a limb has a foot or a hand at the end. (Note 2: Being a three-dimensional volume, it also has insides, a fact that we find expeditious to sweep under the rug.) Again we invoke the much-used theorem that there is no smooth map from a disk to a ring unless the rim of the disk maps with winding number zero onto the ring. Here the winding number is ± 1 so we see that there must be a discontinuity. French (1978) showed that there is

Figure 1. The circular dimension around an appendage is believed to be encoded in the cells by a time-independent phase-like quantity which increases smoothly around the appendage. For the sake of graphic clarity I sketch human-like appendages, even though mammals do not regenerate whole limbs. Most of the experiments used the less familiar-looking appendages of insects.

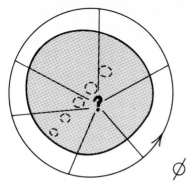

Figure 2. The epidermis of the appendage in Figure 1 is stretched out flat. Six idealized phase contours are indicated. The dashed rings indicate fingernails.

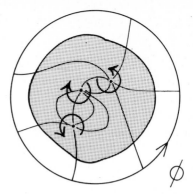

Figure 3. Any odd number of counterrotating singularities would satisfy the boundary conditions as well as the one singularity of Figure 2.

no discontinuity along some "seam" running the length of the leg (the proximodistal direction) like an International Date Line. The discontinuity must therefore be more compactly localized than that. It might be an isolated phase singularity, as depicted in French et al. (1976, their Figure 1 = my Figure 8) and in Figure 2. Alternatively, there could be several isolated phase singularities, e.g., three, of which two have opposite handedness and cancel out (Figure 3).

Rule 3: (French et al.'s (1976) first rule, the rule of intercalation.) So long as the phase gradient is shallow enough (if adjacent cells are sufficiently similar in phase), cells divide only to replace those that happen to perish. But if cells with normally nonadjacent phases are juxtaposed (e.g., by cutting off a leg, rotating it, and sticking it back on), then those cells begin to divide in earnest. (Note 3: An exception apparently occurs at the singularity, where cells do *not* always divide although the phase gradient is infinitely steep there.) As proliferation continues, the new cells take on phase values intermediate "between" their immediate neighbors. Thus the phase discontinuity is soon bridged through newly regenerated tissue. (Note 4: There are two spans of phases, two arcs, "between" any pair of points on a ring. The new cells may "decide" to populate either arc. Which one? We evade this question for now as it will turn out to be superfluous.) Proliferation continues until it has restored the initial shallowness of the phase gradient in space. (Note 5: Some kinds of discontinuity *cannot* be smoothed over by intercalation of intermediates, e.g., a phase singularity such as French et al. (1976) postulate at each limb's distal tip.)

Rule 4: (The second rule of French et al., (1976) the "complete circle" rule of distal transformation.) A new limb will grow out wherever there is a phase singularity. It is left- or right-handed according to whether the winding number of phase around the singularity is $+1$ or -1. Multiple limbs of appropriate handedness are possible, and as we shall see, occur in fact. Within a patch of tissue whose border has winding number W around the phase ring, there will emerge R right limbs and L left limbs and $R - L = W$.

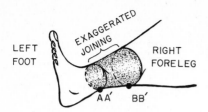

Figure 4.   In the stippled cylinder of regenerating and healing tissue the transition is made from the clockwise polarity of the left-handed graft (foot) to the counterclockwise polarity of the right-handed host (foreleg). A human limb is sketched only because its left and right aspects are familar to most readers; humans do well to regenerate even as much as a finger.

Applying the polar coordinate rules by way of illustration, consider the regeneration of a severed foot. A layer of epidermis, a *blastema*, grows from the cylindrical edge to cover the stump. By rules 1 and 3 and Figure 2, it must contain a phase singularity of the same handedness as the original foot. So by rule 4 a new foot of the same sort replaces the old. In terms of winding numbers, $W = 1$ (or -1), so in the simplest case $R = 1$ and $L = 0$ (or $L = 1$ and $R = 0$).

Consider a second example in which a right foot is cut off and replaced by a left foot severed from the other leg. The cylinder of new tissue proliferating in the junction according to rule 2 is bounded by two complete circles of phase, as indicated in Figure 4. This two-part border has winding number $W = 2$ around the cylinder of skin it encloses. This is shown in Figure 5 by slitting the stippled skin along the line AB and laying it flat: Along path ABB'A'A the phase increases by $0 + 1 + 0 + 1$ cycle. Thus we expect two additional new right feet to emerge, and possibly any number of left-right pairs. In animals capable of regeneration, the actual result is indeed two additional right feet. [A Chinese woman suffered this very operation 1973, following piecemeal destruction of both limbs in a railroad accident (Mobile Medical Team of Wuhan Fourth Hospital, 1976). The inability

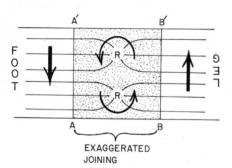

EXAGGERATED
JOINING

Figure 5.   The stippled cylinder of Figure 4 is slit open along the line from AA' to BB'. Lines A'B' and AB are the same. Contour lines and arrows show the smoothest way to join the oppositely oriented phase circles AA' and BB': It requires two singularities of right-handed orientation.

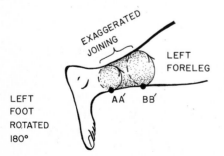

Figure 6.   As in Figure 4 but the left foot is only rotated 180° rather than suffering removal to a right stump.

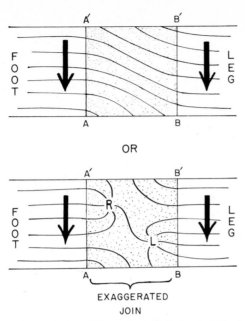

Figure 7.   As in Figure 5 but now the transition can be made by just twisting the contour lines through 180° without invoking a singularity (a). In (b) an alternative construction invokes a pair of complementary singularities. Any number of such pairs (here none above, one pair below) would satisfy the experimentally imposed boundary conditions but some choices are smoother than others, depending on the topology and metric of the underlying state space.

of human beings to regenerate limbs presumably spared all concerned an embarrassment.]

As a last example consider a left foot cut off and replaced (Figure 6) 180° rotated. By the same argument, we expect no supernumerary limbs, or else any number of left-right pairs (Figure 7). A common result is a right foot and a left foot. The other common result is no new limbs.

B: An Alternative Description

Superficially, the data compiled by French et al. (1976) would seem to argue forcibly that phase singularities do play a central role in the morphogenesis of higher animals, not only in the social amoebae (Chapter 15) and the ascomycete fungi (Chapter 18).

This is a distinct possibility that I personally find very exciting, but it does have difficulties. These arise when we ask, as in the other organisms exhibiting phase singularities, what happens *at* the singularity? There is no abrupt discontinuity in cell type, no unique structure, and in apparent violation of rule 2, cells do *not* keep proliferating indefinitely. All this suggests that the proximodistal aspect of tissue specificity interacts in an essential way with the azimuthal tissue specificity represented on the phase ring. Though French et al. (1976) do not dwell on this

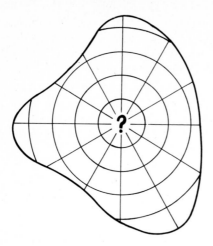

Figure 8. On a piece of tissue, level contours of proximodistalness are drawn as rings concentric to the most distal point. Level contours of the complementary phase-like tissue specificity are drawn perpendicular to these rings. The ? denotes ambiguity of the phase-like quantity.

interaction, it is implicit in their polar coordinate diagram (Figure 8) in which the proximodistal level is represented radially with the most distal parts (toes, fingers) at the center.

If cells have two interdependent aspects of tissue specificity then one must ask what motivates the choice of one coordinate system over another on this two-dimensional state space. Of course, one would prefer a coordinate system natural to the topology of the state space. In the published experiments cited in French et al. (1976) I find no reason to postulate any exotic topology (a cylinder, a toroid, a Moebius band, a Klein bottle, or a sphere) for the two-dimensional state space. A plain piece of paper seems to have the right topology. One might also prefer coordinates natural to a dynamic implicit in observations that states tend to evolve in a certain direction. The evidence for a dynamic would be that homogeneous patches of tissue change their specificity autonomously and predictably. I know no certain evidence of such changes after the earliest stages of growth. (However, see implication 1 below.) Also one might prefer a coordinate system that gives a preferred role to any unique tissue types (e.g., an indispensible organizing center such as the grey crescent of amphibian eggs or the pulsating tip of the slime mold slug). But I am not aware of any special places or tissue types revealed by regeneration experiments. There is no natural origin for whatever coordinate system we will use to describe the type-determining state of a cell. In fact, since we know nothing of the biochemical basis anyway, we might abandon coordinate systems altogether. By dealing in a coordinate-free representation, we isolate the essential facts necessary for understanding the empirical results.

C: Redescription in Terms of a Map Without Singularities

A coordinate-free representation may be undertaken as follows. Let us suppose a tissue specificity space (TSS). We endow it with enough dimensions (two) to distinguish cell types on a two-dimensional surface such as the surface of a leg. We depict as a position in TSS the latent tissue specificity of a cell or a patch of cells.

To each region in TSS there corresponds a type of structure that cells in that region will make when they mature. As we have done so often before, let us map the creature, organ, or tissue into TSS. There is nothing new in this procedure. This is only drawing a fate map, inside out as it were: Instead of drawing the organism in the real world and writing tissue names on an overlay, we write the tissue names at fixed places in tissue specificity space and then draw the organism as an overlay, distorted as required to put places on their corresponding names. For example, a bilaterally symmetric organ has a folded map so that two patches of cells symmetrically disposed to the left and right of a mirror plane map to the same place in TSS. The mirror axis maps to a fold line (Figure 9). Now consider two pieces of an organism, normally not adjacent in the intact, mature stable organism. These appear in TSS as two islands of tissue. If they are now physically juxtaposed, this does not initially affect their tissue specificities, but dotted lines should be used

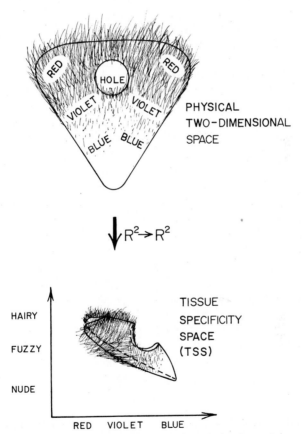

Figure 9. An imaginary organism (top, in physical two-dimensional space) is mapped to a two-dimensional space of tissue specificities (bottom, TSS). Each tissue is characterized by color and hairiness. The organism's bilateral symmetry corresponds to folding of its projection into a two-dimensional TSS. The discrete cell types of a given genotype would occupy tiny polygonal regions of this plane. If there be discontinuities of "positional information" at the boundaries of compartments or fields, then replace "organism" by "organ" throughout.

to connect the cells that are physically adjacent along the surface of contact (Figure 10). Unless TSS has some topology more exotic than $\mathbb{R}^D$, it should be possible to make these dotted lines straight by suitable local stretching and twisting of whatever map we should first plot. The new map is as good as the old in the absence of any definite operational significance attached to the implicit coordinate axes, and it is better in that "in-betweenness" now acquires a definite geometric interpretation.[1] We rewrite the rules now in four parts as above, using this geometric language:

Rule 1: Each little patch of cells is labeled with a state which is part of a smooth gradient of states across the tissue. This state is a point in a two-dimensional space of biochemical specificities, homeomorphic to the plane $\mathbb{R}^2$. Once established, this smooth map does not change.

Rule 2: No exceptional point or state exists.

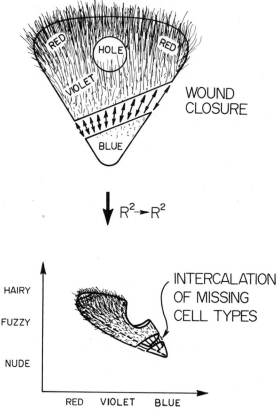

Figure 10.   A strip of tissue is cut out of the organism and its image, without affecting adjacent tissues. Physical apposition of the two surviving pieces is symbolized by fine lines in TSS connecting the cells that come in contact. Daughter cells arising as the wound heals adopt tissue specificities intermediate between their progenitors' (along the fine lines).

---

[1]   The alert reader may, with good cause, feel dubious about the foregoing. Once again I find myself "The Man Without a Metric" (p. 163). Some work is needed here.

Rule 3: Cells proliferate at a rate determined by the local gradient of tissue specificity, i.e., determined by the distance between adjacent cells in TSS. If they are initially quite far apart (e.g., cells on the interface, connected by dashed lines in Figure 10), they begin mitosis. The tissue expands in physical space while its image in TSS remains unmoved but becomes denser with cells. Proliferation stops when the density of cells throughout the image has risen to a local norm, i.e., when the local gradient of tissue specificity has diminished to normal for each kind of tissue. There is no problem about endless proliferation at an unremovable phase singularity; No singularities occur. Just as in the original rule 3, new cells take up tissue specificities intermediate between their physical neighbors. This is the intercalation rule exactly as above, but now in a two-dimensional TSS homeomorphic to a plane. No ambiguity arises about intercalating along the "shorter" or "longer" arc between two cell states. This is geometrically quite different from defining in-betweenness independently both on a phase ring and on an implicit proximodistal axis. The latter approach is equivalent to defining in-betweenness on the cylindrical product space.

Rule 4: No separate rule is needed for induction of a limb. The image in TSS of any ring of tissue is necessarily a ring, and by rule 3 the tissue inside the physical ring must acquire the tissue specificities inside its image. At an appropriate stage in development cells will develop the structures by which their positions in TSS were named. If those should happen to include the distal parts of a limb (fingers, for example), then such structures arise when tissue specificities are interpreted biochemically, and we have a "limb".

D: Applying the TSS Image Rules

If the most distal structures are normally internal to more proximal structures in TSS, then amputation in the real world corresponds to ablation of a disk in TSS. Blastema formation followed by wound healing corresponds to stitching across the empty disk dotted lines along which new cells take up the missing tissue specificities, restoring the more distal organs (Figure 11). Note that this picture in TSS is the same for right or left limbs.

Replacing a left hand (L) by a right hand (R) on a left stump corresponds to cutting out the disk in TSS and replacing it with an identical disk. But remember that because of the inevitable L-R mismatch in physical space, tissue connections from the (R) central disk reach across to the opposite side of the (L) hole (Figure 12). We thus have a three-layered map across the distal core of TSS. The three layers are the existing hand and the cylinder of wound along the wrist that joins hand to forearm. The two new layers (the cylinder of wound proliferation) have the same orientation, so that we get two replicate left hands.

Or imagine that the arm is amputated at *both* ends. Both the proximal and distal blastemas must then span the middle ("hand") region of TSS. So mirror image hands must emerge (and do) at each end. This is the simple geometric essence of the rule of distalization. According to this rule more distal structures regenerate from any stump even at the proximal end where, if regeneration were naively

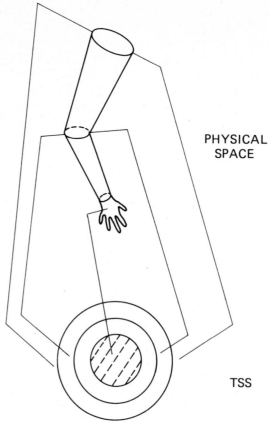

PHYSICAL
SPACE

TSS

Figure 11.   Following French et al. (1976), the parts of a limb are supposed to be so disposed in TSS that more distal structures are concentrically interior to more proximal structures. Amputation at any level thus deletes a central disk. Contact of "wrist" tissues across the stump is indicated by dashed lines spanning the ablated tissue specificities.

expected to restore missing parts in an anatomically faithful way, one might have expected a shoulder to grow.

And so on. All the diagrams of French et al. (1976) can be executed in this format.

The phase circles of Glass (1977) can be constructed by drawing circles on the physical tissue and then following the circle images in TSS. The number of singularities corresponds to the winding number of the image about an arbitrarily chosen origin for the angular coordinate system. The usual choice of origin is the distal-most tissue, but I see no biological reason to prefer such a choice to any other. From my point of view, the arguments of French et al. (1976) and of Glass (1977) about circles, phase maps, and winding numbers amount to using a circle embossed on the organism as a means of bookkeeping the folds and rotations of the regenerating tissue's image in TSS. Glass's calculational procedure is very useful here because it takes practice to visualize handedness in TSS maps. But so far as I can tell, the predictions all come out the same with any origin for the polar coordinate system or (as done here) with no origin at all.

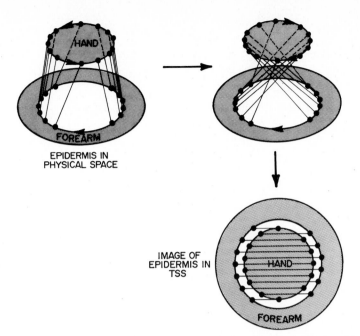

Figure 12.   The experiment of Figures 4 and 5 is represented, not as there by joining up phase contours as smoothly as possible, but by examining the locus of wound tissue in TSS hung between the surgically established boundary conditions. In physical space the left hand's epidermis is seen as a disk, like the disk of latex rubber from which a glove is molded. Its polarity runs counter to that of the forearm to which it adheres through a cylinder of "new wrist". To plug them into Figure 11, hand and forearm must be oriented to a common polarity, which means flipping the hand's image. In TSS the hand's physical attachment to the forearm therefore twice crosses through the central (most distal) range of specificities.

Two new hands of polarity opposite to the graft are thus induced in the new wrist.

E: Experiments Needed

> False facts are highly injurious to the progress of science, for they often endure long; but false views, if supported by some evidence, do little harm, for everyone takes a salutary pleasure in proving their falseness.
>
> Charles Darwin

Certain implications of the geometric viewpoint adopted above may allow critical testing:

1.   This style of mapping physical media or organisms into a state space or a TSS differs from our previous examples [ascomycetes (Chapter 8) and rotating diffusive structures (Chapter 9)] in that no dynamic is postulated in state space. Applied to nondividing or relatively slowly dividing tissue, this is the assumption of *epimorphosis*. But dynamics may play an essential role in more rapidly dividing young or regenerating tissue. As Goodwin (1976, p. 175) was first to emphasize, this is a deficiency that needs attention. For example, consider the following:

a.   Although French et al. (1976) address themselves exclusively to epimorphosis (old cells retaining their specificity), some amount of morphallaxis (re-specification) may commonly occur in regenerating limbs. According to Maden's interpretation (1977), this fact is assimilated by simply allowing our Rule 3 to govern wounded tissues even prior to cell division. This scarcely alters our diagrams. However experiments to elucidate the following two points are potentially more subversive.

b.   Every tissue starts as a tiny patch of indistinguishable cells which, as they grow, acquire divergent specificities. Thus at least at some stage of development there is a dynamic in TSS. Duranceau (1977) shows this empirically as a changing fate map of the immature imaginal disk. Kauffman et al. (1978) provide an alluring linear theory of this dynamic. Bunow et al. (1980) critically refines their contribution by pointing to the structural instability of linear models. The essence of reliable pattern formation must then come of non-linear amendments that remain to be specified.

c.   Peter Bryant's observations on transected imaginal disks of the fruitfly argue for a strong tendency of tissue specificity to evolve more proximally. The qualitative observation is that both edges of the cuts produce the same new structures: one piece exactly duplicates and the complementary piece regenerates

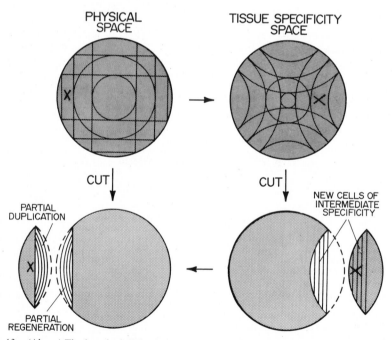

Figure 13.   (Above) The imaginal disk of a fly's appendage is painted with reference circles and rectangles (left) then mapped into TSS (right). The central (more distal) tissue is imagined to occupy disproportionately little space in TSS as though chemical concentrations change more slowly near the middle of the disk than near its edges.
(Below) Cutting off a chord in physical space (left) severs a corresponding lens-shaped piece in TSS (right). Both pieces fold to close the cut edge and recover tissue specificities intermediate between apposed boundary cells (fine vertical lines). Mapping the regenerated tissue (stripes back into physical space (right to left) we find some *but not all* of the missing tissue types restored.

completely. The knife line in Figure 13, though straight in the real world, is generally curved on the image of a disk in TSS, TSS being defined in such a way that in-betweenness follows straight lines. No matter *how* the cut edge heals onto itself, the proliferating cells will adopt tissue specificities intermediate (parallel lines) between points on the knife cut's image in TSS. These intermediates cannot include the whole domain occupied by the excised piece if the disk's image is convex in this TSS—and it must be, else the intercalation rule could produce tissue types not normally present. So transection of the disk leaves each part completely without access to some region of TSS occupied by the other part (Figure 13). Without a dynamic, one piece must fail to *fully* regenerate while the other fails to *fully* reduplicate. Transection experiments may thus provide the means to verify the existence of a dynamic and to quantitate the rates at which tissue specificity can change during normal growth and during regeneration.

2.   According to this coordinate-free representation, ablation of a feature on the symmetry plane of a bilaterally symmetrical organism (a tongue, a nose, a penis, a tail) corresponds to cutting a hole out of the fold of the two-layered image in TSS (Figure 9). Such a hole can heal in two quite distinct ways (Figure 14):

a.   Closing horizontally (top-to-bottom contacts), the hole in TSS would be filled again. Thus we would anticipate complete regeneration.

b.   In contrast, a vertical closure (side-to-side contacts) would join mirror image tissues. Thus, no proliferation would ensue. Even if it did, the missing tissue specificities would not be recovered.

c.   The results of diagonal closure must vary between the extremal results a and b according to the angle of the diagonal. This series of experiments may help to distinguish between geometric interpretations of tissue specificity.

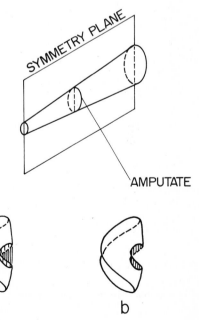

Figure 14.   As in Figures 9 and 10, a bilaterally symmetric appendage maps as a folded disk into TSS, unlike the unfolded disks of asymmetric appendages in Figures 1–3 and 11–13. Amputation ablates an internal disk straddling the fold line in TSS. Wound closure might achieve intercalation as in (a), regenerating, or as in (b), without issue.

$(x + iy)$ to a two-dimensional tissue specificity space $(u + iv)$. Here $u = f(x, y)$ and $v = g(x, y)$ as in Figure 5 of Chapter 1 but with the additional constraint that the gradient of v is assumed perpendicular to and proportional to the gradient of u. This makes u and v vary in space like steady-state solutions of the heat equation or the equations of electrostatics. This both guarantees smooth interpolation of tissue specificities and, as a special case of the diagrams more vaguely sketched in this chapter, gives us access to an elegant body of classical mathematics as a quantitative model.

ACKNOWLEDGMENT. Conversation with Jay Mittenthal was particularly critical in thinking out this chapter.

17. Arthropod Cuticle

The shop seemed to be full of all manner of curious things—but the oddest part of it was that, whenever she looked hard at any shelf, to make out exactly what it had on it, that particular shelf was always quite empty, though the others round it were crowded as full as they could hold.

Through the Looking Glass

A: Rules for Development

Experiments with the limbs of amphibians, roach legs, and fly wing disks suggested to French et al. (1976) some simple rules governing growth and pattern formation in the insects and in higher animals. To apply these rules (Chapter 16), we must first find a point on a ring associated with each point on the animal's two-dimensional surface. French et al. argue that morphogenesis is conducted primarily within two-dimensional sheets of cells and that within these sheets cells know their identity in part as a point on a ring, which we might think of as an angle or a phase. The pattern of phase (together with a second, independent quantity) across the two-dimensional sheet determines the qualitative pattern of growth and differentiation. In particular, phase singularities play a crucial role, for example determining the number and handedness of limbs (Glass, 1977).

Convenient as the notion of developmental phase may be for conceptual purposes, we still lack any direct way to measure the phase of a cell. Two potential clues are provided by cuticle, the plastic exoskeleton that characterizes all the arthropods.

Cuticle is made of two substances, a hard chitin and a more rubber-like resilin. These substances are secreted from the underlying cells, not synchronously, but alternately. Thus, as the insect grows, the cuticle becomes laminated in alternating sheets of chitin and resilin, some micrometers thick. This alternation is governed by a circadian clock. In the normal light-dark cycle, the distance between successive sheets of chitin in this laminated cuticle represents 24 hours growth. In constant conditions in the laboratory the alternation continues but now with only *circadian* period. It is not yet clear whether this period reflects the oscillation of some central, perhaps neurosecretory, pacemaker or whether it is a local property deriving from the separate circadian clocks in each cell that secretes cuticle. Wiedenmann (1977) finds that the cuticle clock is at least independent of the brain clock that controls activity rhythms. Zelasny and Neville (1972), Brady (1974, Section 3.2.4), and Lukat (1978) suggest that insect epidermis is a two-dimensional sheet of oscillators, each capable of autonomous cycling, but each coupled to its immediate neighbors. Thus, local synchrony might be maintained even when cells are not all entrained by the external light-dark cycle. If this were so, then the insect epidermis would have many organizational properties in common with the models of local oscillation in two-dimensional fungus mycelia considered in Chapter 18. In particular, there might be phase gradients, even circular phase gradients en-

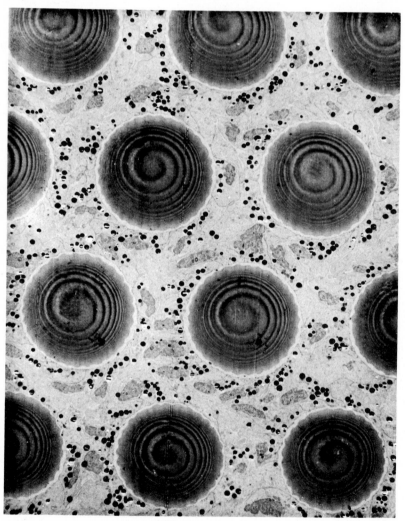

Figure 1.   A thin section through the hexagonal array of exocone lenses in a firefly's compound eye. Each lens is about 10 micrometers in diameter. From Wolken (1971, Figure 3.12) with permission.

closing a phase singularity. Whether such spectacles occur normally and whether such phase patterns as might exist bear any resemblance to those inferred by French et al. (1976), no one has yet reported. But the record of local phase is presumably there, preserved in the insect's hardened cuticle, awaiting inquiry.

B: Insect Eyes

A very suggestive lead appeared on the cover of *Science* for 7 Sept. 1968 (Figure 1). Similar photographs appeared in a few other publications, also without comment. These are all electron microscope pictures of sections through arthropod eyes in which alternating lamina of dark and light were quite conspicuous. The part in question, the corneal lens, is described in textbooks as a secretion from underlying cells, like other parts of the cuticle. The alternating lamina of dark and light cuticles are portrayed as tiny cups, all skewered concentrically on the optic axis of the lens (Figure 2). The cup-like shape of the layers of cuticle would be determined by the curvature of underlying cell surfaces and the relative phases of their secretory rhythms. Herman Gordon and I took a keen interest in this structure because sections (e.g., Figure 1) perpendicular to the plane of Figure 2

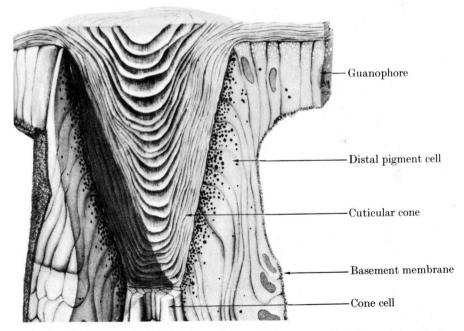

Guanophore

Distal pigment cell

Cuticular cone

Basement membrane

Cone cell

Figure 2.   The exocone lens (alias cuticular lens) from Fahrenbach (1969) with permission. The lens is depicted as a nest of concentric layers secreted by adjacent cells.

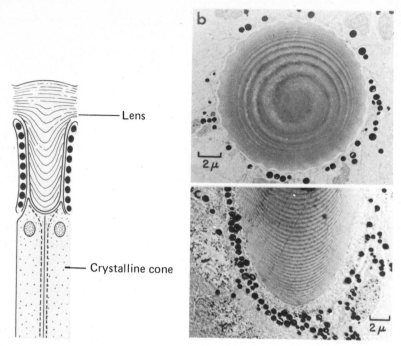

Figure 3.   (a) Schematic similar to Figure 2. (b) Cross-section perpendicular to plane (a) and to (a)'s long axis, as in Figure 1. (c) Cross-section in plane (a). From Wolken (1971, Figure 3.13) with permission.

showed not concentric *ring*-shaped zones of black and white as required by the textbook model, but rather a single-armed *spiral* winding into the center. Figure 3 shows both views. If the alternating black and white zones reflect alternating phases of a secretory rhythm, then the spiral pattern betrays a phase gradient in space which, in one full cycle of black to white to black, winds once around a symmetry axis. That means that the center of the spiral is a point of ambiguous phase, a phase singularity of the sort postulated by French et al. as an organizing center for discrete organs such as arms, legs, wings, antennae, and, in this case, perhaps the single corneal lens.

It turns out that this growth rhythm has much shorter than circadian period. The growth rate of lens cuticle is such that the period works out to minutes to an hour. We are not looking at Zelasny and Nevilles' sheet of long-period clocks. In fact, an entirely different interpretation comes to light with the realization that although in oblique sections the dark zones appear much more widely spaced, they are only a few tenths of a micrometer apart in three dimensions. It turns out that cuticle is not just an amorphous, isotropic plastic. It is richly structured, similar in architecture to cholesteric liquid crystals. We owe this remarkable discovery to A. A. Michelson in a remarkable paper written late in his dazzling career. There we read of the beetle *Plustiotis resplendens*, whose

> whole covering appears as if coated with an electrolytic deposit of metal, with a lustre resembling brass.

> Michelson, 1911

Finding that light reflected from this surface is circularly polarized, he conjectured,

> The effect must therefore be due to a screw structure of ultramicroscopic, probably of molecular, dimensions.

Michelson, 1911

This startling insight collected dust in libraries for the customary half-century before the riddle was solved in detail by the morphologist and geometer Yves Bouligand.[1] Bouligand (1965) brilliantly inferred that arthropod cuticle is composed of layers of parallel chitin fibrils whose orientation rotates as one penetrates deeper into the cuticle normal to the planes of parallelism. The fibrils' orientation turns through 360° in about a micrometer of this laminar material, thus completing a full cycle of semi-crystalline ultrastructure in about the distance occupied by a wavelength of visible light. This kind of architecture is common to aggregates of a wide variety of elongated macromolecules (Bouligand, 1978) including the collagen constituting the corneas of certain vertebrate eyes (Trelstad and Coulombre, 1971; Gordon, 1976); the DNA packed into cell nuclei (Bouligand, 1972a; Livolant et al., 1978) or into sperm heads (Sipski and Wagner, 1977); and the cellulose deposited in the walls of growing plant cells (H. Gordon, ms. circulated privately in 1976). As Michelson foresaw even before World War I, much of the iridescent decoration of insect cuticle derives from this helicoidal periodism in the molecular architecture of chitin (Neville, 1975).

C: Micromechanical Models

Bouligand (1972a) was the first to realize that the zones of light and darkness seen in microtomed sections of cuticle are not different materials at all, but artifacts of mechanical interaction between the moving microtome knife and the cuticle's periodic fibrillar ultrastructure.

Bouligand argued that the dark regions, as seen by the electron microscope, arose where the chitin fibrils had a certain three-dimensional orientation relative to the knife's path. But Bouligand's liquid crystal model of chitin implies a certain kind of mapping from any slice of chitin ($\mathbb{I}^2$) to the sphere of fibril directions ($\mathbb{S}^2$). According to the idea that certain fibril orientations result in local darkening, that region of orientations can be painted dark on $\mathbb{S}^2$, indicating that darkness is acquired by whatever parts of the slice ($\mathbb{I}^2$) map onto that region. It turns out to be implicit in the symmetries of liquid crystal architecture that the pertinent maps must be two layered, in such a way that if some locus in $\mathbb{I}^2$ acquires darkness, then

---

[1] If you love geometry, look at Bouligand's papers on dislocations in liquid crystals (*Journal de Physique*, 1972–1974). They are full of astounding singularities, including cusps, screws, and even Möbius bands. At least some of these have been identified in biological materials and may play essential roles in the mechanics of plant cell elongation and of chromosome condensation. The literature of liquid crystal spherulites (of polypeptides, DNA, etc.) likewise is a study in singularities, since there is no way to smoothly map a sphere's surface to the ring of possible compass orientations for molecules on that surface. See also Slonczewski and Malozemoff (1978) on orientational singularities in three dimensional magnetic bubbles.

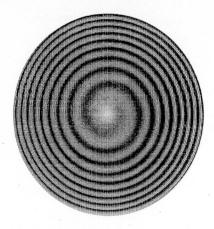

Figure 4. Computer simulation supposed to represent a cross-section similar to those in Figures 1
and 3(b). Each tiny dot is the endpoint of a chitin
fiber. The dots were initially uniformly dense, then
each was slightly displaced according to local fiber
orientation relative to microtome knife motion.
Orientations were described mathematically according to the observed curvatures of cuticle in the
exocone lens and Bouligand's liquid-crystal model
of chitin. The displacements create a single-spiral
locus of thickening and thinning.

so must another, symmetrically disposed, locus. So if there is only *one* spiral of
darkness then either Bouligand's liquid crystal architecture is wrong (a repugnant
notion) or (more likely) the spiral is not after all a microtoming artifact of the sort
postulated. Then might there still be something to discover in this phenomenon,
something about rhythmic secretion, phase maps of nonzero winding number,
and a morphogenetic phase singularity?

This was my hope. Unfortunately, however, Gordon and I managed to contrive
an alternative mechanical model by which the single spiral of darkness could
arise as an artifact of microtoming. According to this simple interpretation (as yet
untested by experiment), the moving knife compacts material in certain places by
inducing relative movement between chitin fibrils and their surrounding matrix
(Gordon and Winfree, 1978). Mathematically describing the helicoidal structure
of cholesteric liquid crystals in a computer and arithmetically displacing fibrils
according to this rule of movement, we obtained patterns of compression such as
in Figure 4. These look disappointingly similar to the patterns observed, for
example, in Figures 1 and 3.

Under this interpretation, the phase singularity at the center of the dark spiral
is nothing more than the point of tangency between a plane of parallel fibrils and
the cutting plane of the knife. Thus a phase singularity in the formal description
of an object does not necessarily indicate an event of singular interest.

In the next chapter we review some facts about the development of fungi which,
on a much larger scale, again display this same geometric pattern: a spiral derived
from spatial periodism in which the phase of a rhythm has a nonzero winding
number around a ring of tissue. Is anything of physiological interest associated
with the encircled phase singularity in this case?

18. Pattern Formation in the Fungi

Good problems and mushrooms of certain kinds have something in common: they grow in clusters. Having found one, you should look around; there is a good chance that there are some more quite near.

H. Polya, p. 65 of
How To Solve It

(See also Chapter 9, Box C.)

In Chapter 12 we dwelt on a biochemical clock in an ascomycete, the yeast cell. The familiar bread molds and their relatives are also ascomycetes. They are called *colonial* ascomycetes because of their habits of growth. Like animals, these fungi derive their energy by oxidizing organic fuels. Like plants, they grow where the seed falls and feed through roots. An ascomycete colony starts when a spore falls on a food surface. It germinates and extends a fine web of hair-like filaments, called hyphae, across the food as an expanding disk. This two-dimensional disk of hyphae is called a mycelium. It is not really a cellular organism since the septa dividing hyphae into tiny compartments typically have holes in them, so that cytoplasm flows freely between the compartments.

It is not unusual to find mycelia banded with concentric rings of heavy growth and/or sporulation. In many cases, these rings turn out to be 24 hours apart on the expanding disk. Each ring marks those regions that were in a certain range of ages when it was daylight, or hot, or wet (or whatever) in the diel cycle, much as rings of heavier growth mark the annual seasons of a tree.

Setting aside such cases of obvious environmental influence, there remain a vast array of papers describing spatially periodic growth in ascomycete mycelia. Independence of environmental rhythms can be verified in these cases by adjusting either the growth temperature or light intensity or nutrient conditions to change the period, or by using a transient stimulus of the same sort to inflict a permanent phase shift in the growth rhythm.

What might cause such banding? We may be short of explanations, but there is no lack of plausible conjectures and little reason to expect any one of them to cover all cases (see Winfree, 1973d).

By a daring leap of faith, we might for example conjecture that what is not driven by an external rhythm of environmental conditions is driven by an internal metabolic rhythm. Only recently has a single experiment been published showing repetitious activity of any sort in patch of hyphae [Dharmananda and Feldman (1979) in the case of *Neurospora crassa*].

An alternative seems equally plausible, i.e., that conditions on the frontier of growth locally determine the immediate activity mode of newly created cells, which in any case soon lapse into a standard metabolic condition as the growth frontier moves on. For example, frontier cells might be transiently disinhibited to "high" activity if and only if the nearest high activity event was a sufficient number of hours in the past and/or sufficient number of millimeters behind the growth frontier. In the latter case, some mechanism of transport, such as cytoplasmic streaming or diffusion must be invoked to separate events in space. Models of this sort have been exploited to good advantage in connection with periodic banding of bird feathers (Nickerson, 1944), Liesegang ring formation in inorganic gels (Stern, 1967), etc.

It seems appropriate to look at these phenomena with respect for the diversity of nature's inventions. We first cursorily survey similar phenomena in four ascomycetes and then stick to one for a more detailed correlation of facts.

A: Breadmold with a Circadian Clock

Neurospora crassa, the geneticist's favorite mistress before *E. coli*, exhibits several mutant varieties of circadian rhythm as well as banding rhythms of other periods. These currently offer some of the best clues to a circadian clock's bio-chemical mechanism. The typical circadian rhythm properties of temperature-compensated period, suppression by dim light, phase resetting by brief exposures to blue light, and clarity of period are best exhibited by the "band" mutant, usually grown on very long "race tubes." In this technique, a floor of nutrient agar is poured in a U-shaped glass tube with a wide flat bottom. The inoculum of spores or hyphae placed at one end soon elaborates a mycelium which propagates at a uniform speed of about 1 mm per hour across the sterile nutrient. After a weeks' growth in constant darkness at 25°C, six to eight bands are found to be placed at circadian intervals along the tube. Within each band, spores differentiate in the mycelium's luxurious aerial arborization (Figure 1). A brief exposure to light increases or decreases the interval to the next band, depending on the phase of the rhythm at exposure time and the amount of light given. After that phase shift, the rhythm continues at circadian intervals, but phase-shifted relative to the series of bands preceding the stimulus. Thus resetting curves can be measured as in any other circadian rhythm. Type 1 (weak) resetting is observed following brief enough exposures. Type 0 (strong) resetting is found after greater exposures (Feldman, Dharmananda, pers. comm.).

Dharmananda and Feldman additionally tested bits of mycelium *behind* the frontier for their phase. This is done by cutting them out and transplanting them to a new race tube to observe band phasing in the renewed growth, as Chevaugeon and van Huong (1969) had done previously with the rhythmic ascomycete

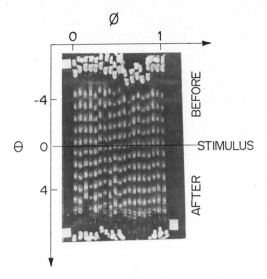

Figure 1.   Rhythmic banding in race tubes of *Neurospora* "timex" (= an invertase-deficient strain of "band") from unpublished experiments of W. Engelmann. This stack of tubes is presented in the format of Figures 31 through 33 of Chapter 1, calibrated in cycles of 21 hours. Growth is vertically downward. The stimulus was a cold shock which scarcely affected the rhythm.

Ascobolus immersus. However, unlike *Ascobolus*, it appears that each patch of *Neurospora* mycelium, even many days old, continues to keep time with the same circadian rhythm. The whole mycelium apparently remains nearly[1] synchronous in phase beneath the waves of growth and sporulation that record each patch's phase at that moment long ago when it passed a decisive age. Demonstration of a persistent *clock* in a fungus is not surprising in itself. The clocks of *Daldinia* (Ingold and Cox, 1955) and of *Pilobolus* (Bruce et al., 1960) were studied 20 years earlier, but this *is* the first demonstration that a clock underlies the *spatial* periodicity of a fungus.

B: Breadmolds in Two-Dimensional Growth

Although much less studied, *Penicillium diversum* also exhibits a rough 24-hour, temperature-compensated rhythm of growth and sporulation. In this case, however, no sensitivity to visible light has been shown. Because its frontier moves much more slowly, it is convenient to culture *Penicillium diversum* as a disk on two-dimensional agar plates, not only along essentially one-dimensional slants. In this configuration, Bourret et al. (1969) made a most remarkable observation: Though many mycelia are banded in concentric circles 24 hours apart, many also make a single continuous band winding round and round the inoculum as an Archimedian spiral (Figure 2). Along any radius (equivalent to a radial race tube), zonation has a 24-hour rhythm. But as one changes the azimuth of the chosen radius, the phase of that rhythm drifts, drifting through one full cycle as the imaginary race tube is rotated through 360°.

---

[1]   More exactly, Dharmananda and Feldman (1979) find a phase *gradient* of about $1\frac{1}{2}$ hours per day's growth, suggesting that young hyphae right on the contemporary frontier cycle that much faster than older hyphae left behind on staled medium.

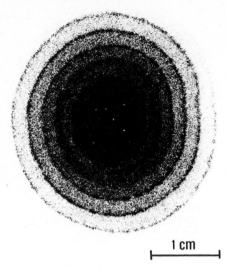

Figure 2.   A spiral locus of heavier sporulation on a disk-shaped mycelium of *Penicillium diversum*, from Bourret et al. (1969) copyright 1969 by the American Association for the Advancement of Science.

|⊢   1 cm   ⊣|

Archimedean spirals of zonation, as an occasional alternative to the more usual concentric rings, were also reported in *Chaetomium robustum*, another filamentous ascomycete with a 24-hour rhythm on nutrient agar (Kraepelin and Franck, 1973).

C: Pattern Polymorphism in Bourret's *Nectria*

Another ascomycete, identified as *Nectria cinnabarina*[2] exhibits a noncircadian 16-hour rhythm of growth and sporulation in slant tubes. It has the same habit in two dimensions (Figure 3). Its bands are approximately 1 millimeter apart at 20°C. By shaving all the surface growth off an agar plate like that of Figure 3, one obtains an agar layer containing bands of denser and lighter growth. Running this through a densitometer, and plotting the positions of successive density maxima along any radius, one obtains a distinctly periodic record. Figure 4 shows the distribution of distances (0.9 ± 0.2 mm) between 264 adjacent bands. This interval and its variability showed no conspicuous differences with age or distance from the inoculum in the 10 mycelia examined.

In two-dimensional growth, *Nectria*'s bands appear as distinct rings nominally 1 mm apart in about two-thirds of the colonies, and as clockwise and counterclockwise spirals in about two-thirds of the remainder. *Nectria* has also been caught making distinct two-armed and three-armed spirals. In such cases, each arm winds around at a pitch of 2-mm or 3-mm (respectively) per 360°. Thus bands along any radius occur at 1-mm intervals as usual. The difference is that the phase of the radial rhythm progresses through two or three cycles rather than one (as in one-armed spirals) or zero (as in disjoint rings) as the chosen radius is swung

---

[2]   Probably misnamed. It is not the same as the British *N. cinnabarina* (the "coral spot fungus") nor the same as the *N. cinnabarina* that decimated N.Y. State beech trees in 1975. According to the rather confused taxonomy of order *Hypocreales*, to which family Nectriaceae belongs, *N.c.* is the perfect stage of a fungus whose imperfect stage (fusarium) is called *Tubercularia vulgaris* (Bessey, 1950, p. 287). *Hypomyces* may be another name for *Nectria*.

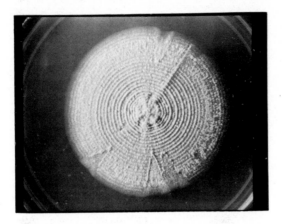

Figure 3. This disk-shaped mycelium of *Nectria*, printed at actual size bears two spiral ridges of spores. Note the glass coverslip inserted deep into the agar radially at 2 o'clock before the mycelium had grown big enough to touch the glass. It has no discernible effect on the sporulation pattern. The wedge-shaped patches are colonies of faster growing mutants. See also the cover photograph of *Science* for November 7, 1969.

Figure 4. This histogram classifies into 100-μm intervals 264 measurements of the distance between adjacent bands. The measurements were read by computer from densitometer tracings of hyphal density in shaved agar along radial tracks from the inocula of ten 8-cm mycelia.

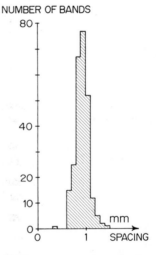

through 360° of azimuth. Figures 5 and 6 show digitizer tracings of $R(\phi)$, the radial distance of a spore band from the inoculum as a function of compass angle: Figure 5 is the two-armed spiral of Figure 3 and Figure 6 is a one-armed spiral. The spacing between successive bands is nominally 1 mm in both cases.

This situation lends itself to several different styles of description. For example, one might emphasize the global integration of pattern: Spores differentiate along an integer number of distinct rings or spiral arms, each of which either encircles the inoculum concentrically (in the case of ring morphogenesis) or (in the case of spiral morphogenesis) winds at mean pitch proportional to the number of arms. This point of view raises challenging questions about the large-scale integration of pattern. How are the distinct loci of sporulation distinguished? At various times

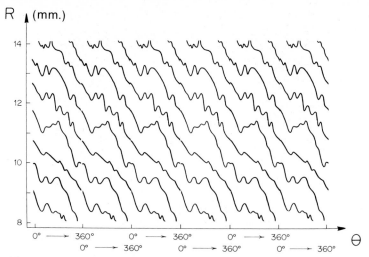

Figure 5.   Figure 3 was placed on a digitizer platen while the electronic cross-hair was moved by hand and eye along the locus of densest sporulation. Its position was continually plotted in millimeters from the mycelium's center (vertically) and in degrees of azimuth from north (horizontally). The first 360° panel was replicated 6 times so that both spirals can be followed through several complete turns without jumping back from 360° to 0°.

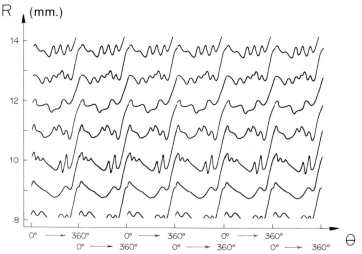

Figure 6.   As in Figure 5 but the sporulation locus is a single spiral of opposite handedness in a different petri dish. The radial spacing is about the same as in Figure 5.

during growth of the disk-like mycelium, each one is interrupted by the frontier at several places, yet as the mycelial disk expands, they grow together exactly. How do the segments of each successive ring "know" that they belong with the other segments of that ring and avoid joining up with segments of other rings currently under construction elsewhere? One is led by these questions to consider possible channels of long distance communication within the mycelium, such as the waves postulated by Goodwin and Cohen (1969) [and in fact *seen* in *Dictyostelium* (Chapter 15)]. But by implanting glass barriers, cutting out pieces, etc., it can be quickly shown that pattern formation is independent of spatial continuity and continuity (Winfree, 1973d). Integration of pattern is somehow managed on a local basis.

D: Integration of Pattern

Another clue is afforded by looking for the most predictably reproducible aspect of the observed patterns in hopes that it may suggest the appropriate language of description, which in turn may suggest a mechanistic basis for the terms used. For example, is the slope $dR/d\phi$ of the spiral reliably predictable? No. Neither the distance to the frontier from a spore ridge, nor its distance from the inoculum varies regularly with angle (Figures 5 and 6). The only sure thing is that the *average* slope on a full cycle of 360° azimuth will be an integer.

To cut a long story short, what is most reliably conserved is the nominal 1-mm periodicity along each radius (Figure 4). Even when the pitch of spirals or rings changes by local quantitative variation or by a global qualitative change of connectivity, still a 1-mm spacing is conspicuous in densitometer tracings made along every radius. The standard deviation of this spacing is fully 22% of the mean and shows no serial correlations (unpublished computations). This makes it about half again as unreliable as the notoriously variable menstrual period (Chapter 23). But the essential point in present context is that the spacing is not correlated with pattern topology: It does not double or halve in two-armed spiral patterns.

A second reliable feature is the propagation of pattern defects along the direction of growth. It commonly happens that rings do not quite close: There is a mismatch as in a lock washer. Maintained during cycle after cycle of growth, this sudden change of phase with azimuth becomes the site of a radial mismatch of many bands, resembling a zipper or an edge dislocation in a crystal. Only after many cycles do the adjacent hyphae synchronize well enough for bands to cross this radius smoothly. This feature points once againt to substantial local autonomy, certainly so far as interactions transverse to the mean direction of growth are concerned.

A third reliable feature is that zonation patterns are always concentric to the inoculum. The centers of rings and the pivots of spirals are always where the colony started.

If we choose to emphasize these clues over others, then the natural terms of description will feature the radial rhythm and the variation of its phase with azimuth. The clues that suggest local autonomy lead us directly to a simple interpretation in terms of such a local oscillation as has been widely assumed for many years and recently shown in *Neurospora*. The clues that point radially to the inoculum suggest that we seek the origin of the topologically distinct morphs in the initial condition of the germinating mycelium.

So my next experiment asked, "Does the shape of the inoculum determine the choice between the several patterns of differentiation?" I found no way to classify the microscopically observed shapes into groups that correspond to the eventual patterns that each develops. So my answer is "no".

Then "might the initial conditions be invisibly latent in the hereditary material?" The required experiment is to grow a mycelium from a single spore, collect its identical asexual spores, and plate these separately on agar, several to a dish. Each dish turns out to contain mycelia of the several various morphs even though each is started from genetically identical units and is grown under identical culture conditions. This is so even if the spores are floated on a Millipore raft on a sea of liquid nutrient to make culture conditions just as homogeneous as possible.

So it would seem that the difference between initial conditions is very subtle and delicate, though these differences do lead to qualitatively distinct patterns of differentiation into sporulating and nonsporulating hyphae. In other words, we are looking for some reasonable interpretation of a singularity. It turns out that the kind indicated is just the kind found to underlie the helicoidal resetting behavior of circadian clocks. (See Chapter 4, Section C and Chapter 8, Section C.)

But what about other internally rhythmic ascomycetes that do *not* make several alternative rhythmic patterns, but stick to one reliably? That one is always the concentric ring pattern, with rings parallel to the frontier and concentric to the inoculûm. When it occurs, spiral polymorphism reveals something intriguing about the mechanisms of differentiation and pattern formation, but why is it so rare? At present this question can only be answered by conjecture in terms of a proposed interpretation of *Nectria*'s biochemistry. Such conjectures are found at the end of Chapter 8, Section C.

19. Circadian Rhythms in General

Time sleeps quite naked, Anaitis, and, though it is a delicate matter to talk about, I notice he has met with a deplorable accident.

So that time begets nothing anymore, Jurgen, the while he brings about old happenings over and over . . . .

Jurgen, J. B. Cabell

One doesn't have to look at many living organisms before noticing that a lot of behavioral physiology is temporally organized in periodic patterns. In fact, if I had to decide what impresses me as the single most conspicuous feature of natural ecosystems, I would say that it is the daily and seasonal periodism and the consequent temporal organization of niche structure, food webs, and behavior. Of course, it would seem natural to assume that any given daily rhythm or seasonal rhythm is a response to the environmental cycle of days and nights, summer and winter. This is often the case. It is, for example, in the case of "deep scattering layers". These are layers of diverse fauna in all the world's oceans that show up very clearly on sonar. They go deeper in the daytime and rise back toward the surface by night. My job on an Indian Ocean Expedition cruise from Woods Hole in summer of 1964 was to study these deep scattering layers using sonar and tow nets during their diurnal up-and-down migrations. Investigators in Cousteau's diving saucer found that the deep scattering layer consists largely of small fish. They seem to simply follow light intensity, e.g., they come up during eclipses. Enright and Hamner (1967) later found a substantial role of endogenous rhythmicity in the vertical diurnal migrations of invertebrates in the scattering layers.

That opportunity to participate in field work in 1964 started me wondering about the dynamics of physiological rhythms in general. Probably the most pervasive physiological rhythms, phylogenetically speaking, are the ones with 24-hour period, a subject on which Colin Pittendrigh had delivered one of his typically enthralling lectures at Cornell shortly before my summer at sea. Thus I learned that 24-hour rhythms are commonly found to continue at about the same period even when apparently isolated from all diurnal influences. The period is often quite exact, but seldom exactly 24 hours. In other words, organisms have *endogenous* rhythms: circa*dian* rhythms as they are called from the Latin meaning

Box A: Possible Clinical Relevance of Circadian Rhythms in Humans

This Box provides an assortment of references in chronological order. The papers and books listed are only those that I have run across and found convincing or otherwise stimulating. The list is far from complete. Many of these papers report experiments not actually on man, but on laboratory mammals. I include them on presumption of similar effects in human physiology. Much of the work takes place in diel cycles rather than in the externally constant conditions required to observe endogenous oscillators. Thus, I am not distinguishing here between aspects of temporal organization imposed by an internal circadian clock (eg. the adreno-cortical system in man) and aspects imposed by external 24-hour clock (via light schedule, meal schedule and exercise schedule). At the time of writing, this literature is just beginning to grow. It will presumably be very much more extensive by the time you read this. The following bibliography might therefore be most useful as a source list for finding newer research papers and reviews in Science Citations Index. Also see Chapter 22 on circadian regulation of mitosis and Chapter 23 on circadian involvement in reproductive cycles.

Halberg et al.	1960	Haus et al.	1974
Hellbrügge	1960	Simmons et al.	1974
Pizarello et al.	1964	Stupfel	1975
Aschoff	1965a	Aschoff	1976
Scheving et al.	1968	Moore-Ede et al.	1976
Conroy and Mills	1970	Sollberger	1976
Kraft et al.	1970	Weitzman	1976
Colquhoun	1971	Halberg	1977
Luce	1971	Miles et al.	1977
Reinberg and Halberg	1971	Moore-Ede et al.	1977
Haus et al.	1972	Palmer	1977
Wongiwat et al.	1972	Sulzman et al.	1977a,b
Mills	1973	Fuller et al.	1978
Shackelford and Feigin	1973	Moore-Ede et al.	1978
Urquhart	1973	Young	1978

Most of the above concern normal rhythmicity of clinical norms and of the effects of drugs in man or similar laboratory animals on a 24-hour schedule. The following concern transient disruptions of the 24-hour schedule as in time-zone shifts by air travel and in unusual departures from 24-hour scheduling such as in space travel, submarine patrols, and polar expeditions.

Flink and Doe	1959	Wever	1973
Aschoff	1965a	Jouvet et al.	1974
Hauty and Adams	1965	Ehret et al.	1975
Aschoff et al.	1967	Klein and Wegmann	1975
LaFontaine et al.	1967	Wever	1975
Aschoff	1969	Aschoff and Wever	1976
Christie et al.	1970	Mills et al.	1977
Elliott et al.	1972	Moore-Ede et al.	1977

"roughly daily". The word has, unfortunately, become Americanized to cir-$c\bar{a}'$-dian, with the result that more than once I've been asked after a lecture on "cicadian rhythms" why I never got around to the subject of the 17-year cicada.

The observed circadian rhythms are said to reflect the functioning of a biological clock. It is called "biological" because it is apparently generated from within the organism, and it is called a "clock" because it measures time in a periodic way, sometimes with surprising precision, with period close to 24 hours (see Box A). Circadian rhythms have been found in every major group of animals and plants except the procaryotes, which comprise viruses, bacteria, and blue-green algae or cyanobacteria. There exist two reports of vague circadian rhythmicity in bacteria, but both await confirmation (Rogers and Greenbank, 1930, analyzed by Halberg and Connor, 1961; and Sturtevant, 1973).

In protozoa and algae we find 24-hour rhythms of capacity for photosynthesis, of mitosis, of motility, and of bioluminescence. In some fungi there is a daily hour of spore discharge. In plants there are diurnal rhythms of leaf movement and of flower opening and closing. Many kinds of insects periodically regulate the times of egg laying, egg hatching, and emergence from the pupal case (see next chapter). Vertebrates including man have conspicuous circadian rhythms of activity and sleep, of body temperature, and of hormonal activity. Menaker et al. (1978) provide an excellent recent review, focused on five specific experimental systems. Box A provides a bibliography of medically-pertinent literature.

Besides regulating the daily activities of organisms adapted to life on a rotating planet, circardian rhythms have become involved in *seasonal* adaptations to life in orbit around a star. Circadian clocks are involved in the measurement of daily photoperiod to govern flowering in agricultural plants; to regulate diapause in their pest insects; and to organize migration, navigation, and reproductive seasonality in the vertebrates.

The rest of this chapter is divided into three parts to review some general characteristics of circadian rhythms, to speculate "without fig leaves", as Bacon said, about their evolutionary origins, and to gather observations suggesting multioscillator organization of "the clock".

A : Some Characteristics of Circadian Rhythms

Circadian organization is one of the most pervasive facts of physiology and of ecology. The mechanism of endogenous temporal organization is unknown, but is surely mediated by different organ systems and biochemical pathways in men, fruitflies, paramecia, pigweeds, and bread molds (see Section B). Yet some aspects of function seem almost universal:

1.   A spontaneous periodicity of about 24 hours (Box B).

2.   Relative temperature independence of that period so long as the temperature is constant (see pages 379–381).

3.   Persistence of the rhythm even in the smallest unit of biological function, the single cell.

this point as though threading (nonexistent) data points on a very steep segment.[1] Discussions of this "phase jump" (example Wiedenmann, 1977) seldom make the critical distinction that it is really "phase *shift* jump" from a big delay to the complementary advance. This means that it is no phase jump at all unless one imagines that the clock reaches a given phase only by passing through all the intermediate phases, as in a ring device (Chapter 3).

Much of the literature gives the impression that "what goes up must come down" in the world at phase shifts: that if a system gives delays at one phase and advances at one later phase, then at some intermediate phase the curve must rise from delaying, through zero, to advancing. This is true of real numbers, i.e., points on the real line $\mathbb{R}^1$. But we are concerned with phases, i.e., points on the ring $\mathbb{S}^1$. On a ring there are *two* arcs "between" any pair of points, not only the one that includes zero. Thus one can go smoothly from the delayed semicircle of phases to the advanced semicircle of phases either through $0°$ or through $180°$.

In the case of a simple clock (the most popular, if implicit, mental model of a smooth physiological clock), the transition has to occur through zero. This is because the basis for advancing and delaying in a simple clock is a phase-dependent angular velocity which cannot go from minus to plus without passing zero unless the mechanism itself is discontinuous. Thus the "observed" data-free jump is sometimes interpreted (I believe incorrectly in most cases, though possibly correctly in some particular organisms) as a physiological jerk or discontinuity in the oscillator's mechanism or in its mechanism of response to the perturbation used. For example:

1.   It was apparently within this context that Frank and Zimmerman (1969) chose to determine action spectra for phase shifting at phases only two hours apart: an hour before and an hour after the phase jump. They sought different photoreceptors mediating advancing and delaying resets of the fruitfly's circadian rhythm. This is an important experiment and a very sensitive experimental design if one believes that the circadian clock is phase-shifted by advancing or delaying along its cycle, or that the phase jump represents a physiological discontinuity. However, no difference in photosensitivity was detected. This would be expected under *any* hypothesis about the photoreceptor if one interprets the adjacent big advance and big delay as two ways of plotting essentially the same physiological process.

2.   Zimmer (1962) obtained a smooth dependence of the phase of *Kalanchoe*'s flower-opening rhythm on the timing of a rephasing stimulus (see Figure 2 in Chapter 21). But the curve she sketches through the data (Figure 1 in Chapter 21) does not follow the smooth path: It leap across a cycle to join points in adjacent smooth trails of rhythmic data. The resulting discontinuity is preserved as a conspicuous feature of *Kalanchoe*'s circadian cycle when later writers plot the curve without data points in phase shift format.

---

[1]   There is one instance in which data points do appear on the very steep segment of a type 1 curve (Pittendrigh and Bruce, 1959, Figure 8). However, the raw data (Pittendrigh and Bruce, 1959, Figure 4) shows that these dots represent multiple recurrences of the monitored event during an interval of transient irregularities.

3. Karvé and Salanki (1964) follow the same procedures in repeating Zimmer's experiment in a different physiological context. Here a geotropic stimulus is used to upset a two-hour bending rhythm of a plant seedling. The resulting oscillations have their maxima at nearly the same interval after the stimulus, but the curve threaded through them is made to jump almost a full cycle between data points in order to preserve the appearance of type 1 connectivity. The phase jump thus imposed on the data is reckoned to be an important physiological event.

4. Hastings and Sweeney (1958) show smoothly varying data on the rephasing of *Gonyaulax* by a light pulse. But in Christianson and Sweeney (1973, Figure 3) the plotting is switched to phase shift format and a large phase jump is assumed in order to make the curve look like a magnified version of type 1 resetting. Sweeney (1974) suggests that the phase jump may reflect an abrupt change from active to passive transport in the plasma membrane of *Gonyaulax*. In my opinion the pervasive notion that there is a phase jump which needs to be accounted for derives less from any data than from interpreting them within the simple-clock paradigm. If one does so then it appears that the simple clock's cycle is not continuous but has a sharp break in it. This is accounted for by supposing that a chronogen accumulates during a prolonged tension phase up to a threshold concentration, at which an abrupt change in cell dynamics brings on the complimentary relaxation phase. This may be quick or protracted, but when completed it restarts the tension phase. This idea goes back at least to Rasmussen and Zeuthen's (1962) division protein model, to the mitogen accumulation model of Sachsenmaier et al. (1972) for cell cycle timing and to models of rhythmicity in pacemaker neurons in which a generator potential accumulates toward a threshold for membrane conductivity (Hill, 1933). In context of circadian rhythms Bunning (1960), Wagner and Cumming (1970), Wagner et al. (1975), and Wagner (1976) associate the chronogen with energy metabolism.

Though I look askance on such interpretations, it must be recognized that the sparseness of data points along the vast majority of circadian response curves precludes any very rigid inferences about the topological types of resetting thus observed. Type 1 has been widely assumed tacitly, but in only a few cases in the earlier literature [e.g., Figure 1, the *Drosophila* curve of Pittendrigh and Minis, (1964)] were the curves clearly enough resolved to make the reality of type 0 resetting arguable. (See Box C of Chapter 4.)

Advance vs. Delay Ambiguity

By classifying all phase responses as advances or delays we force ourselves to violate the topology of the phase ring as follows. On the phase ring a small advance and a small delay are nearly the same thing. Also any advance (or delay) and a slightly smaller advance (or delay) are nearly the same thing. A 90° advance is quite different from a 90° delay. But a 179° delay is not as-different-as can be from a 179° advance. They are in fact virtually the same thing, so far as the end result is concerned. The same is true not only of the end result but also of the *process* of

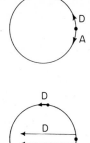

Figure 2. (a) Delay and advance are opposite directions along a cycle. (b) If perturbation is not restricted to a one-dimensional cycle then the largest delays and advances can be indistinguishable.

rephasing (except in the special case of simple clocks: Chapter 3). A small displacement of state by some amount in some direction (to the left in Figure 2) is interpreted either as a small advance or as a small delay, depending only on the prior phase of the oscillator. Large displacements of state that result in the biggest advances or delays not only change the state in the same direction by the same amount but also are very nearly identical in the sequence of states passed through. These two diagrams may look like mystic symbols here, but they acquire operational meaning in terms of the presumed multiplicity of equivalent oscillators in multicellular organisms in Chapter 4, in terms of hypothetical underlying biochemistry in Chapter 6, and in terms of experimental manipulations of the fruitfly's eclosion rhythm in Chapter 7.

The language of advances and delays suggests that advance and delay are opposites, but they are not, since a big enough delay is an advance. It suggests that advanc*ing* and delay*ing* are opposites, but they are not except in the special case of simple clocks. Moreover, an advance or delay can be obtained without advancing or delaying by simply freezing the oscillator for the duration of a prolonged stimulus. The very notion of advance or delay is ambiguous. The following three nonequivalent definitions are in current use:

1. If the perturbed rhythm leads the unperturbed control rhythm by less than one-half cycle, then one speaks of an advance, or if it lags by less than one-half cycle, one speaks of a delay.

2. A net delay can come about by accumulation of abnormally short intervals until the accumulated advance exceeds one-half cycle. Zimmerman (1969, pers. comm.) obtained a four-hour-delayed eclosion rhythm in the fruitfly by way of a progressive 20-hour advancing transient. In general, the resetting of a clock which can both advance and delay might register with the external world by way of a unidirectional process that can only be advanced or only delayed. See Christianson and Sweeney (1973, p. 98) for another example.

3. The net effect of a stimulus changes as its duration is prolonged. If $(\theta + M)$, the event time measured from the *start* of the stimulus, increases with M, the stimulus duration, then the stimulus is delaying the rhythm, and if $(\theta + M)$ decreases with M, then it is advancing the rhythm. Saunders (1976a) obtained both

advance and delay phase shifts by convention 1 but only delays according to this convention 3. And if one measures the new phase at the *end* of the stimuli, then all of Saunders's shifts are obtained by advancing.

For such reasons I have adhered to the conventions of full data presentation originally fostered by Hastings and Sweeney (1958, Figure 8), Pittendrigh and Bruce (1957, p. 95 and 1959, p. 486), and Zimmer (1962, Abb. 2). I avoid the more recent convention of data reduction in the notation of phase *shifts*. Neurophysiologists in the 1970s, encountering problems of phase control similar to those first tackled by circadian physiologists in the 1950s, follow similar practice (see Chapter 14, the second part of Section A).

Looking Behind the Adaptive Facade

If clues to dynamic organization are to be sought in behavioral experiments, it is essential to know something of the adaptive value of that behavior, and so about the evolutionary origin of the mechanism to be explored. Pittendrigh and Bruce (1959) succinctly stated the essential principle that features with adaptive value must be avoided:

> Again there are difficulties here which stem from the fact that diverse physical systems can be devised (or evolved) to behave in a formally similar way. In the comparison of living clock systems for instance, we can take neither temperature independence nor entrainability by light as evidence of a common mechanism; selection must have demanded both these features as functional prerequisites of clock systems and will, of course, have been indifferent to the particular physical mechanism by which the functional end is met. If we are to test the proposition of a common mechanism and use only formal properties, these must be of such a nature that selection can reasonably be dismissed as the agent responsible for their universal association with cellular clock systems; in short, the properties must lack adaptive value.

Or as Pavlidis (1971) put it:

> Therefore, their study can provide little insight into the structure of the system which regulates the daily rhythmicity of organisms. On the other hand, one may expect more from the study of features which do not seem to have any survival value and therefore may be indicative of the structural properties of the system. This point may be illustrated by the following example. Consider a passenger on a train who wants to find out whether the locomotive is steam, diesel or electric. (One may assume that the engine was attached to the train after he boarded it and therefore he did not see it). He may try to do so by observing the details in the changes in velocity while the train goes through curves or stops at stations. While it is theoretically possible to achieve identification in this way, it is quite difficult because all locomotives tend to show rather similar functional characteristics. On the other hand, the identification could be quite easy by observing non-functional features, like smoke, noise, etc.

Rephasing experiments probably serve this purpose admirably, so long as the stimuli used are quite unlike natural daylight photoperiods. For example we will see in the next chapter that the phase singularity of *Drosophila*'s clock can only be revealed by exposure to light whose energy is a million times attenuated relative to natural photoperiods, and only if it is given just once, not repeated daily as in nature.

But what are the evolutionary origins and purposes of circadian rhythmicity? The next section explores this question, leading to recommended experiments and observations.

B: Clock Evolution

I list here three alternative conjectures, all in need of development, about the evolutionary origin of circadian clock dynamics. These three make different assumptions about the selective pressures dominant at the time. The first starts with a timer selected to anticipate the dawn by degrading or accumulating a substance during the night. The second starts with the cells' daily cycle of activities, imposed by the daily cycle of light and dark, warmth and cold. The third starts with the cell metabolizing steadily in a constant environment, and asks how stable this situation might be.

First Dream: Dawn Warning

In the Archaezoic Seas.   Suppose it were necessary for cells in the primordial slime to coordinate their biochemical activities on a diurnal schedule. In a stimulating speculative essay, Pittendrigh (1967b) suggested advantages that would accrue to whatever mutants could best anticipate the rising of the sun:

> . . . it is a reasonable speculation that the daily alternation of light and darkness was the historical cause (selective agent) of circadian oscillations in the first place.
>
> This line of thought derives from recalling the prerequisites for organization in a chemical system. The principal of these is that the constituent reactions cannot proceed spontaneously at the prevailing levels of free energy. Thermochemically, this means of course that the reactions the cell employs involve energy barriers unsurmountable at prevailing temperatures; they proceed only on command which rests with enzymes and ultimately with the nuclear store of information. Little attention seems to have been given—in this general context—to the problem of visible radiation as an energy source that threatens organization.
>
> Of course, the fact is that the majority of the cells' constituents are colorless; and uncontrolled activation by the visible is thus excluded. It may well be that in the history of the cell there has been selection for colorless molecules, but if that is true (and it seems likely) it is a fact that for some functions, colorless molecular devices have not been found. The flavins and cytochromes are examples of ubiquitous fundamentally important molecules that are colored—and where color has no detected function.
>
> No attention seems to have been given to the consequences of illuminating these molecules whose color is without obvious function. At any rate it is surely reasonable to consider, at least, the likelihood that some subroutines in these cells' overall tasks are impaired by the activation of molecular piece-parts in the flood of visible radiation to which it is subjected each day. To that extent, the routine delegation of some chemical activity to the recurrent darkness of each night would be an obvious escape from the photochemical threat to organization.

If a timer is needed to anticipate the dawn, it seems reasonable to expect that the simplest timer would be invented first. This would presumably be a relaxation oscillator, the same as has been widely invoked to account for the timing of cell division (Chapter 22) and for the timing of rhythmic contraction in nerve and in muscle and in heart muscle (Chapter 14). Specifically, some byproduct would accumulate during the dark, up to a threshold. At threshold, a conversion is undertaken from metabolism appropriate to darkness to metabolism appropriate in the glare of sunlight. It is to be assumed that the sunlight and/or the cells' preparation for it turns off the accumulator and allows the product to dissipate until next darkness. (The reciprocal version is similar, in which a product decays in the night and is restored to a standard concentration by the end of day.) A sophisticated accumulator might even adjust the threshold or the rate of accumulation to improve timing tomorrow if the threshold was reached a little too early or too late today.

Conservative Evolution of a Simple Mechanism. If cells need to invent a timer, here is a simple beginning. Such a single-variable hourglass mechanism has appealed to many investigators of contemporary circadian rhythms. From the viewpoint of evolution, its most conspicuous features are its responsiveness to the environment and its one-shot relaxation character under hypothetical conditions of constant daylight or constant darkness.

Given enough mutants, it is easy to imagine the prosperity of the cell's descendents who have the most reliable hourglass and engineer the cleverest attachments to general metabolism. If all this confers a selective advantage, then woe be unto the apostate cell who attempts radical innovation or abandons altogether this valued heirloom. Inheritance may well be conservative and we could end up with a universal circadian clock or hourglass as faithfully replicated as the mechanisms of glycolysis, of oxidative phosphorylation, and of protein synthesis. In multicellular organisms, the indispensible timer might well be expressed in a specialized tissue that serves to coordinate the whole organism.

This case is the most favorable for scientists. An accumulator mechanism with a single most important variable, marked by sudden transitions once or twice a day, would be the easiest to pinpoint biophysically, especially if one is free to choose the most favorable organism in which to explore each aspect of the mechanism, and if it functions most conspicuously in a certain organ (let us say, the brain or the pineal gland).

On the Other Hand . . . . Unfortunately, I see rather little evidence to encourage this outlook. It is by now well established that many cell types in each organism exhibit circadian timing. Moreover, though there are exceptions (Lees, 1972), the timer's cycle most commonly does spontaneously repeat at 24-hour intervals despite the lack of obvious selective pressure for such a contrivance. A primordial hourglass easily could have mutated to turn itself over, but given the regularity of sunrise and sunset, no clear advantage would accrue to such cells. Saunders (1976b) notes of the fly's clock that "only when the night is extended to more than 24 hours is the oscillatory nature of the system revealed", and that seldom happens in nature. Remmert (1962) and Hoffmann (1971a, p. 199; 1976) are perplexed by the apparition of a spontaneously recycling internal timer. After all, organisms were selected only for ability to cycle *in a regularly cycling environment*. Enright (1970) in a particularly valuable essay addresses the question, "What selective advantages led to the evolution of internal timing mechanisms, and in particular to an internal timing mechanism which is rhythmic and self-sustaining rather than to a one-cycle timer which must be reinitiated by environmental stimuli each day?" At this time there seems to be no very persuasive answer. I suspect that this feature is not an adaptation at all, but an accident of neutral selective value. (For a contrary opinion, see Pittendrigh and Daan, 1976.) As such it may provide some clue that the origin and mechanism of the clock do not lie in this class of mechanisms. Finally, reinforcing this conclusion, I know of no indications of a physiological jump in the normal cycle, as assayed by resetting experiments.

Second Dream: Indirect Selection

Back to the Archaezoic Seas.   This one starts clock evolution not with the need for a single timer, but with a preexisting cycle of biochemical activity, driven by the daily alternation of light and dark, of warmth and cold. With many metabolic pools driven into reasonably predictable diurnal fluctuations, the cell could be regarded as executing a regular cycle of activities, e.g., photosynthesis by day, heat prostration by evening, perhaps mitosis by night, starvation by morning.

Given an alternating environment that intrudes forcibly into the cells' internal affairs and the consequently inevitable internal rhythms, a cell needs no additional clock. But it could improve itself by optimizing the transfer from one predominant metabolic activity to the next. To economize waste, to anticipate changes and bridge them smoothly, to resist diversions during vicissitudes such as an eclipse or the passage of a cold front, the successful cell would accumulate modifier genes to facilitate each handoff of control during this imposed cycle. Impelled by selection to gather more and more fail-safe backups to guarantee efficient state transition, the cell might surprise itself to discover, two billion years later when some scientist first puts it into constant conditions, that it shuffles its way spontaneously through almost the same cycle. Yet the ability to do so was never of any use and was never selected for.

A Critical Experiment.   If cells can actively internalize environmental rhythms in the way here vaguely suggested, then it should be possible to demonstrate the effect by bacterial selection experiments in a chemostat. By alternating the nutrient influx from glucose without oxygen, to oxygen without glucose, to alanine and oxygen, cells would be forced into a three-point metabolic cycle. With generation time adjusted to several cycle durations, mutants could be sought which prosper better in that milieu, and then tested for spontaneous cycling on nutrient agar with suitable color indicators. By reversing the order of the driving cycle, it should be possible also to select cells whose clocks run backward.

On the Other Hand .... This vision of the origin of clocks seems well fitted to models of clock dynamics in which there are no states accessible to the cell apart from one or a few discretely different sequences of transformations. Such one-dimensional models have enjoyed much success in connection with the cell cycle, which in fact might have arisen in circumstances similar to those portrayed above, the composition of cytoplasm being rhythmically changed by the recurrent replication and phased transcription of the nuclear genome (Goodwin, 1966; Ehret and Trucco, 1967). One-dimensional models have also heavily influenced the devising of experiments and interpretations in circadian physiology (see Chapter 3). However once again I doubt their suitability to account for circadian rephasing behavior: Type 0 resetting does not come naturally to such mechanisms. So let's at least think of one more class of plausible model for clock evolution.

Third Dream: Steady-State Start

The Statistics of Stability. This one starts not with an engineering goal, nor with a forced cycle, but with steady-state operation of cell chemistry, supposing perfect homeostasis in a constant environment. Our outlook here may be taken from Bunning (1973, p. 27), who cites Oatley and Goodwin (1971).

Statistical explorations of the kind of equations that describe reaction kinetics suggest a difficulty about steady-state operation (cited by Goodwin, 1963, May 1973). The more complex the reactions, and the more regulatory coupling to other reactions, the less the chance that any one steady state will be stable. Of course there may be more steady states in a complex kinetic scheme, but probably not enough to guarantee one being an attractor. In other words, mutants of complex regulated pathways almost all exhibit instabilities of the homeostatic steady state. Departure from a steady state need not lead to regular oscillations, but that *is* one possibility, and mutants of this sort must be expected in vastly greater abundance than mutants whose steady-state is well regulated.

Instability May be Desirable in Itself, Even in a Constant Environment. Are there intensive selective pressures *against* such mutations? I doubt it. There is presumably some advantage to homeostasis, but only to keep fluctuations within livable bounds. I see no virtue in perfect stationarity per se. In fact, several loose arguments suggest the contrary. Chemical engineers are finding more and more examples in which process control is improved by rhythmic departure from average conditions. This might be expected, in general: Deviation from monotony is one more degree of freedom in which to search for optimization. In reactions with sharply nonlinear rate laws, excursion to extreme conditions may "allow every dog his day": A cycle facilitates segregation *in time* of conditions conducive to incompatible reactions. Cells commonly accommodate such processes by compartmentation in space, but another possibility is to compartmentalize in time.

It seems worth noting in this regard that steady pressure on a nail accomplishes less than a rainfall of sharp hammer blows. A steady subthreshold level of sexual tension is less conducive to mating than the familiar build-up to a sudden release. Similarly nature has designed us to sleep and wake, but not to practice moderation in all things by groggily intermingling the two.

In short, there may be reasons why even if organisms didn't live in a fluctuating external environment, they might benefit by inventing one internally. The statistics of ordinary differential equations suggest that much more intense selection would be needed to avoid inventing one.

None of this specifically implies regular fluctuations, nor is regularity necessarily to be expected of kinetic schemes involving more than the two variables so familiar in theoretical journals (see Box B in Chapter 13). Nor is regularity widely observed in the physiological rates of organisms except in certain preferred selection windows along the frequency axis. But fluctuations that happen to occur regularly, and at frequencies that closely match dominant regularities of the natural environment, may have special uses. Having started clock evolution by simply not counterselecting the inevitable instabilities of homeostasis, we now turn to the rhythmic environment for refining selective pressures. For example, the effective value of any rhythmic environmental parameter can be adjusted to any value within the range of daily variation by locking the organism's experience of that parameter to a particular phase of the cycle. If an organism ventures from its underground burrow only by night, then daytime temperature is unimportant. This principle of phase locking has been put to spectacular use in the laboratory by scientists from Michael Faraday (item 33 of *Experimental Researches in Electricity*, to prove that voltaic and galvanic electricities are equivalent), to Henry Cavendish (the experiment to demonstrate gravitational attraction), to Richard Dicke (detection of the primordial fireball). It should be no surprise if organisms have exploited it almost since the fireball cooled down.

If so, then we may expect accumulation of modifier genes to augment the amplitude of such selected spontaneous fluctuations and to phase-lock them to the environment through a photoreceptor or other sensor.

From this perspective, one might anticipate a diversity of circadian rhythm mechanisms as great as the diversity of homeostatic feedback schemes ... even within the single cell. There may have been many independent evolutionary starts, and no *one* had to win out. Moreover with several independently competent systems oscillating within each cell, the clock is fail-safe, so each part is free to mutate without imperiling the function collectively subserved. Thus, no aspect of mechanism need be conserved during the radiation of species. On this basis, one would rather solve the mechanism of one cell completely, than to select clues to a universal mechanism from experiments on diverse cell types.

Conclusion

These three "dreams" are not mutually exclusive. They suggest three distinct principles that might have contributed to the origin of circadian clocks under different conditions. Which ones would work fastest under any given conditions? Did clocks of different sorts evolve under different conditions? We don't yet know.

A Critical Observation to Make. None of these three conjectures make any distinction between procaryotes ("lower" organisms, lacking subcellular organelles) and eucaryotes (made of one or more cells, with nuclei). Nor do they discriminate between cells with generation times exceeding 24 hours and those that divide more frequently. But a strong generalization among circadian physiologists has it that circadian rhythms occur only in eucaryotes with cell divisions more than 24 hours apart (Ehret and Wille, 1970). Counterexamples have been sought (Rogers and Greenbank, 1930; Halberg and Connor, 1961; Sturtevant, 1973) but not convincingly followed up. An implication of the conjectures risked above is that better counterexamples will be found. I would begin with a survey of the colonial eubacteria (Actinomycetes), e.g., streptomyces or the tuberculosis germ, mycobacterium. Other promising candidates among the procaryotes might be the nitrogen-fixing photosynthetic bacteria. If no circadian clocks are found, then the scenarios elaborated above have overlooked some major factor. For example, it might be that procaryotes typically enjoy much quicker biochemical responsiveness to environmental changes than do eucaryotes. If so then maybe no significant improvements of efficiency accrue to procaryotic mutants with long-period timers of any sort, so no single frequency band is favored among mutations to spontaneous oscillation.

The Importance of Choosing One Case to Solve. My own initial enthusiastic belief in a universal mechanism perished during the 1960s by counterexamples to almost every specific proposal. Some creature physically lacks a substance or structure postulated to be essential, or can be functionally deprived of it be a suitable poison, yet perseveres in its circadian time keeping. Like many others, I am a true believer turned apostate in this matter. But, disliking middle courses, I now believe at the opposite fanatical pole, i.e., that circadian mechanisms are as diverse as species, or even as diverse as their cell types, or even as diverse as the modes of operation in each cell in different hormonal and nutritive surroundings, or even more diverse, if there are several independently competent circadian oscillators within the cell's regulatory mechanisms under any given conditions.

It therefore seems to me of first importance to completely solve the mechanism in one cell type, forsaking for the present clues gathered eclectically from diverse phyla, and then to solve it completely in another cell type, or to look for involvement of the first found mechanism in another cell type. Thanks to Max Delbruck's advocacy of this strategem for virus genetics a quarter-century ago, surprisingly rapid and cross-fertilizing progress became the hallmark of that field. Why not in circadian physiology too?

Until a few years ago it was easy to indulge in speculation about the biophysical mechanism of any particular circadian clock because almost any particular physiological system X is involved: Either a disturbance of X disturbs the monitored rhythm or better alters its period or permanently resets its phase (perhaps indirectly), or disturbance of the clock alters the behavior of X. This game can be played both physiologically and genetically. What is needed and what has been lacking is a proof of both simultaneously: that X affects timing, and timing affects X, so that X is implicated as a part of the timer. But even given this, there remains one more necessary step: to show that X is an *indispensible* part of the timer.

These feats are not beyond the capacities of modern biochemical physiology, but they do require concerted effort focused on a single cell. The day is at hand when that effort will be made. From its origin in field naturalism and evolutionary biology, and from its confusion with astrology and biorhythm fads, the exploration of circadian rhythms has come a long way. Nowadays, even molecular biophysicists have heard of the "clock" and believe that it might have some definite universal properties. As indicated above, I think that is a mistake, if one means biochemical properties. But the point is that we are now in a position to find out.

It is my belief not only that one organism should be chosen, but additionally that it should be either a single cell or a cell population in which both autonomy of the individual and synchrony among individuals can be monitored (a big order). Obviously the dynamical behavior of a population of interacting oscillators can be quite different from the behavior of one alone. This point was belabored in Chapters 4 and 8. We also saw in those chapters that collective measures on a sufficiently incoherent population, even of noninteracting oscillators, can be very misleading about the basic type of dynamics underlying oscillation in each independent individual.

C: The Multioscillator View of Circadian Rhythms

This section is intended to gather indications that circadian rhythmicity may typically reflect the collective circadian activity of many independently competent sources, each oscillating at a circadian period with or without its neighbors. If this be true, then we have two new sources of dynamic complexity to explore above and beyond the behavior of the single oscillator:

1.   The interactions among such physiological oscillators and

2.   The manner in which their influences are pooled in a rhythm of activity, body temperature, or what-have-you.

Moore-Ede et al. (1976) provide a useful brief statement of the problem. In collecting the evidence, I omit two categories of information:

1.   Experiments showing that single-cell organisms exhibit circadian rhythms. Circadian rhythmicity does not require multicellular organization, but the fact argues neither one way nor the other in the question as to whether circadian rhythmicity is structurally and functionally composite in multicellular organisms. Experiments with unicellular organisms might ultimately contribute with a demonstration that behavior attributed to composite structure in multicellular organisms is found *only* in multicellular organisms (e.g., "transients"). But for the present, the most pertinent observations involve multicellular organisms and their isolated tissues.

2.   Experiments showing that multicellular organisms have many rhythmic activities which readjust to new phase at diverse rates. In this respect a multicellular

organism does comprise a population of transiently separable rhythmic functions but such indications do not in themselves argue for independently competent sources of rhythmicity.

I gather the evidence into two sets of experiments. The first set indicates that the clock is structurally composite in multicellular organisms. The second set of experiments indicates that it is functionally composite.

That "The" Circadian Clock is Structurally Composite

We turn now to experiments indicating that there are separate circadian oscillators physically distributed throughout the many organs and tissues of a multicellular organism.

1. In mammals, the clearest demonstration comes from Andrews' (1971) discovery that hamster adrenal glands, isolated in tissue culture, persist in distinct circadian rhythmicity of endocrine secretion for many days in ostensibly constant conditions. A similar effort by Tharpe and Folk (1965) to exhibit circadian rhythmicity in rat heart and in isolated rat heart cells in tissue culture produced only rather unconvincing evidence of rhythmicity. Hardeland (1972, 1973) reported clearcut rhythmicity of enzyme activities in suspensions of isolated rat liver cells, and Ashkenazi et al. (1975) have reported very conspicuous rhythmicities of enzyme activity in stored human red blood cells. No convincing results have yet come of attempts to repeat these experiments in other laboratories.

2. Aréchiga (1977) reported several experiments exhibiting circadian rhythmicity in isolated ganglia, limbs, eye stalks, etc., of crustacea. Nishisutsujii-Uwo and Pittendrigh (1968a,b), Page et al. (1977), and Page (1978) using cockroaches; Konopka (1972) using genetically mosaic flies; Engelmann and Mack (1978) using flies; and Koehler and Fleissner (1978) using beetles reported evidence of at least two major circadian oscillators, physically distinct, in the individual insect.

3. Jacklet and Geronimo (1971) and Benson and Jacklet (1977) showed that each eye of the mollusk *Aplysia* constitutes a self-sufficient circadian oscillation, as in fact do isolated pieces of the retina. Besides the two eyes, at least one other structure, the parietovisceral ganglion, has its own circadian clock (Strumwasser, 1974; Lickey et al., 1976).

4. For a long time it has been realized that the individual leaves of plants are capable of circadian rhythmicity independent of the rest of the plant; the same is true of flower petals (Bunning, 1973). Simon et al. (1976a, b), exploring the causes of the leaf moment rhythms typical of many plants, showed the oscillation to be self-sufficient in isolated pulvini of *Samanea*. Wilkins and Holowinsky (1965) reported circadian rhythms of respiration in cultured bits of callus tissues taken from the succulent plant *Bryophyllum*. Mergenhagen and Schweiger (1975a) found circadian rhythms of photosynthetic activity in fragments of a single cell of *Acetabularia*.

5. Circadian rhythmicity persists in pieces of diverse fungi.

That "The" Circadian Clock is Functionally Composite

Next, we examine the most convincing cases presently available that circadian clocks are *functionally* composite in multicellular organisms.

1.   In plants, as noted above, individual leaves and petals maintain rhythmicity on locally altered schedules, even when not physically separated from the rest of the plant. Mayer and Sadleder (1972) found the primary and secondary pulvini of legumes to function with different circadian periods.

2.   The several clocks of *Aplysia* show no indication of mutual coupling, so far.

In cases 3 through 7 to follow, only two components have been observed.

3.   In insects and in mammals, there are numerous indications that circadian clocks originate in nerve tissue. The bilateral symmetry of these organisms suggests two physical clocks. Can they function independently? The lasting alteration of future resetting behavior that accompanies phase resetting in flies' eclosion rhythms can be accounted for in quantitative detail in terms of:

a.   Two independent attractor-cycle oscillators, possibly located in the left and right sides of the brain (Winfree, unpublished computations, 1974) or

b.   A population of many clocks, in this case not necessarily with attracting cycles (Winfree, 1975b, 1976) or

c.   Two coupled attractor-cycle oscillators of the sort envisioned by Pavlidis (1976).

Koehler and Flesissner (1978) show functional dissociation of left-right clocks in insects. In bees, Medugorac and Lindauer (1967) and Beier et al. (1968) report experiments suggesting bimodal clock activity, possibly originating in experimental desynchronization of two clocks.

In cases 4 to 7 to follow, the two oscillators are not imagined to be equivalent as above but are thought to have quite different properties.

4.   Salisbury and Denney (1971), King (1975), and Bollig (1977) point to different clocks regulating leaf movement and flowering in three kinds of weed.

5.   Engelmann and Mack (1978) report that two clocks of 6% different period oscillate simultaneously in *Drosophila* pupae: One governs eclosion and the other governs the rhythm of adult activity.

6.   Pittendrigh (1960, Figure 12) was the first to report a peculiar splitting of daily activity rhythms in rodents, which has since become a fruitful subject of investigation not only in rodent species (Pittendrigh, 1967a, Figures 1–4; Daan et al., 1975; Pittendrigh and Daan, 1976c; Rusak, 1977) but also in other vertebrates (Hoffmann, 1969, Figure 3, 1971b, Figure 4 in the tree shrew; Gwinner 1974 in birds; Underwood, 1977 in lizards). In the latter case, the demonstration is particularly spectacular: Two components of the activity rhythm run through each other without synchronizing, just as they do in Koehler and Fleissner's beetle.

7. Wever (1975), Aschoff and Wever (1976), and Sulzman et al. (1977b) convincingly showed that there are two oscillators of quite different period in man and other primates, one predominantly affecting the activity rhythm and one predominantly affecting the body temperature rhythm. The two have been induced to scan through each other repeatedly:

> the human circadian system consists of a multiplicity of oscillators that are usually coupled to each other but [which] may change their phase relationship depending on conditions, and if they ever became desynchronized, free-running with different frequencies ... The rhythms of activity and of temperature are both influenced by two different classes of self-sustaining oscillators, but to different extents.
>
> Aschoff and Wever, 1976

Note that the experimental physiologists cited under 6 and 7 above all surmise that the "two" oscillators observed might each consist of many interacting oscillators, normally well-synchronized into clusters. (See suggestive rodent activity records in Pittendrigh and Bruce, 1959, their Figure 7; DeCoursey, 1961, her Figures 3, 9; Swade and Pittendrigh, 1967, their Figure 9; Morin et al., 1977a, her Figures 2 and 3.) Stimulated by such conjectures, I looked for similar behavior in a deliberately contrived population of electrical oscillators (Winfree, 1965, 1967a, and Chapter 11) and in computer simulations of better characterized oscillations. It turned out that both the coherence of each cluster and the splitting of a population into two clusters with somewhat different properties can be mathematically rationalized in terms of similar oscillators symmetrically coupled through a metabolic pool (Winfree, 1967a). Disaggregation into three or more clusters was also observed from time to time. It is also seen in rodent activity rhythms (e.g., Pittendrigh 1960, Figure 12 and Rusak, 1977, Figures 2 and 5). Pavlidis (1971, 1973) also constructed models of such phenomena.

The next section summarizes to the best of my knowledge what little information is presently available about interaction within aggregates of circadian clocks.

Interaction Among Circadian Oscillators

Cell Populations in a Shared Nutrient Medium. It is clear that in some cases circadian oscillators do interact to maintain mutual synchronization. For example, Page et al. (1977) found strong interactions between the left- and right-side clocks in the brain of the cockroach. Synchrony is normally maintained within the enormous cell of *Acetabularia* despite the fact that isolated pieces of the cell are capable of sustained circadian rhythmicity (Mergenhagen and Schweiger, 1975a). What about populations of separate individual unicellular organisms, such as *Acetabularia*, *Gonyaulax*, *Paramecium*, *Tetrahymena*, and *Euglena*? Rhythmicity in collections of *Acetabularia* plants promptly decays as individuals drift out of synchrony, but in the other four populations it persists without conspicuous damping for a long time in the absence of known external synchronizers. In *Euglena*, it persists for months (Brinkmann, 1966, 1967; Terry and Edmunds, 1970; Edmunds et al., 1971).

The latter cases could be accounted for in three ways:

1.   The rhythm is actually driven by a covert periodic influence; but this seems unlikely since the period seldom exactly matches the 24-hour period of the most plausible candidates.

2.   The members of the population all have very nearly the same period so that their initial coherence persists without need of mutually synchronizing interactions and without need of an external pacemaker. Njus (1975) estimates that in order to account for the continued narrowness of Gonyaulax's daily glow peak without invoking synchronizing interactions, one must assume a standard deviation of period not exceeding 1–4% (15 minutes to an hour). (The higher figure represents a standard deviation of the period within each cell of a population of statistically identical individuals; the lower figure represents the standard deviation of period between cells of a population of perfect oscillators.) This order of precision is not implausible. The circadian clocks of individual Drosophila pupae have periods which differ by only about this much (Winfree, 1973a, Appendix). But it is hard to account in this way for observations of undamped rhythmic settling or phototaxis in Euglena populations over the course of months (Brinkmann, 1971).

3.   The individual cells could interact in such a way as to maintain mutual synchronization. Suspensions of yeast cells maintain synchrony of glycolytic oscillations by metabolic interactions (Chapter 12). Southeast Asian fireflies maintain synchrony of flashing through the phase-shifting effect of a light flash on each individual's flashing rhythm (Buck and Buck, 1976; Hanson, 1978). Individual cells of the heart's pacemaker achieve the same electrically (Watanabe et al., 1967; Jongsma et al., 1975; Clay and DeHaan, 1979).

Empirically, one could test for such interactions among circadian clocks in three ways:

a.   As in Aldridge and Pavlidis (1976), using yeast cell suspensions, one could dilute the suspension in search of a threshold at which coherence is abruptly lost, i.e., the collective rhythm suddenly begins to decay (see Chapter 12). Such experiments have not been reported in circadian systems.

b.   As in Ghosh et al. (1971), using yeast cell suspensions once again, one could combine two populations of cells with different phases to see whether the combined result is simply a superposition of the two populations mixed or something different which might indicate interactions.

Nonsuperposition would prove interaction, though not necessarily of the kind that favors mutual synchronization. If the observed rhythms were conspicuously nonsinusoidal (e.g., a spike every 24 hours) then superposition might also prove noninteraction. But, with smooth sinusoidal observables, as in the cited cases, superposition is indistinguishable from a resetting of amplitude, an interpretation that can be excluded only by showing that the circadian clock in question adheres closely to a standard cycle. Such few circadian rhythms as have been examined in the required detail do not seem to adhere closely to a standard cycle (Drosophila, Chapter 20; and Kalanchoe, Chapter 21).

c. Alternatively, one could try phase-shifting a cell suspension by adding supernatant from a differently phased population. In this way an interaction might be discovered. It would remain to be proven that such interaction would result in stable mutual synchronization.

Experiments of types b and c have been reported: Mergenhagen and Schweiger (1974) using *Acetabularia*; Hastings and Sweeney (1958) using *Gonyaulax*; Brinkmann (1966, 1967) and Edmunds (1971) and Edmunds et al. (1971) using *Euglena*. In these trials exact superposition was not the usual result. But neither did the combined rhythm indicate immediate synchronization by taking up the large amplitude typical of a coherent population. Nor did addition of supernatant from another phase cause an obvious immediate phase shift. Gooch, Sulzman, and Hastings (1979 pers. comm.) using *Gonyaulax* did obtain mere superposition.

What these experiments show is that there are no violent and immediate mutually synchronizing interactions. More systematic and prolonged mixing experiments are required to ask about gentle and slow-acting interactions.

It may also be noteworthy that these tests have been applied not to the somatic cells of a multicellular organism, but to populations of anarchists, possibly not designed for cooperation. Moreover the populations studied all consist of green cells, necessarily exposed to the diurnal light-dark cycle in nature. Even if there were some advantage in synchrony, one would not necessarily expect selection for mutually synchronizing interactions, as there is no functional requirement for them. Support for this speculation comes from the leaves and petals of flowering plants, which *are* indeed capable of independently phased circadian rhythmicity: Whatever interactions may exist, they don't quickly induce synchrony. It might be that only the cells of nocturnal animals, cave fungi, the inner tissues of large animals etc. are sufficiently isolated from the environmental day-night cycle to experience some selective advantage in mutual synchronization. As noted above, Page et al. (1977) found strong synchronizing interactions between the two circadian pacemakers in the brain of a nocturnal animal.

Mutual Entrainment vs. Mutual Synchronization. I might note here that it seems to be widely assumed that if rhythmic interactions occur between circadian oscillators, then, supposing only that the coupling is strong enough relative to the dispersion of periods in the population, mutual synchronization is the natural outcome. It is crucial to distinguish here between mutual entrainment, which only implies phase locking at a common frequency, and mutual synchronization, which means that, additionally, the phases are all nearly the same. Mutual entrainment is commonly achieved between two similar oscillators, given strong enough coupling. Surprisingly, this was found not to be the case in a study of idealized oscillator populations (of more than two) by mathematical analysis and computer simulation (Winfree, 1967a). It was found that mutual interactions can either contribute to mutual synchronization or oppose it, depending on phase relations between each cell's rhythm of mutual influence and each cell's rhythm of sensitivity to that influence (see Chapter 4, Section B; Chapter 8, Section B; and Chapter 14,

Section B). The reason is basically that although mutual entrainment is commonly found between similar coupled oscillators, it does not usually consist of mutual synchronization. The mutually entrained oscillators may, e.g., be 180° out of phase. In a *population*, mutual entrainment is more difficult to arrange without mutual synchronization because the entraining rhythm felt by each oscillator, being a sum over all the others, is weaker to the extent that synchronization is imperfect (unless there is only one other). Thus, interacting mutants might be selected both in situations requiring mutual synchrony, and in situations such as Richter (1965) suggests might require collective arrhythmicity of individually rhythmic units.

20.   The Circadian Clocks of Insect Eclosion

> . . . indeed what reason may not go to Schoole to the wisdome of Bees, Aunts, and Spiders? what wise hand teacheth them to doe what reason cannot teach us? ruder heads stand amazed at those prodigious pieces of nature, Whales, Elephants, Dromidaries and Camels; these I confesse, are the Colossus and Majestick pieces of her hand; but in these narrow Engines there is more curious Mathematicks, and the civilitie of these little Citizens more neatly sets forth the wisedome of their Maker.
>
> Sir Thomas Browne, physician, regular correspondent of Henry Power

Understanding the circadian timing of eclosion in insects is a pretty big undertaking. A lot of technical detail is essential and a lot of close reasoning from meticulous experiments stands in the place of direct observations on the "clock's" unknown physiological mechanism. In that respect clockology has some of the intellectual delight of the earlier years of genetics. However, the whole argument has never been spelled out for publication in one place. This chapter once again attempts only an outline of the essentials. The story presented here seems to be generally valid for butterflies and moths, flies and mosquitos, wasps and bees (i.e., *Lepidoptera*, *Diptera*, and *Hymenoptera*), but I emphasize my own experimental beast, the fruitfly. For a review of insect eclosion systems from the viewpoint of physiology and ecology, see Remmert (1962).

A: Basics of Insect Eclosion Clocks

The Life Cycle

An insect starts as a fertilized egg. Within the egg, development culminates in the emergence of a larva (caterpillar, maggot, "worm"). The larva eats by burrowing through semisolid food until it attains suitable maturity. At this point, feeding stops and the animal sheds and hardens its outer cuticle, now called the pupal case. Then the next miracle begins: construction of the adult by metamorphosis. Prior to the pervasive transformations that metamorphosis comprises, the organs of an adult lie among the larva's organs as rudimentary bag-like sheets of un-differentiated cells. These are called imaginal disks. Most of the larva's organs dissolve during metamorphosis. The materials thus made available are used as nutrients by the growing imaginal disks. Some larval organs do survive into the adult though they undertake extensive remodeling. Examples include the brain and the insect's equivalent of a kidney, the Malphigian tubules. It is to these few that we must look for an understanding of the timer that schedules, well in advance of metamorphosis, the hour at which the winged adult will emerge from the former worm's coffin, the pupal case. This opening of the pupal case, called eclosion, happens all of a sudden at the appointed hour. This is a time of great vulnerability and conspicuousness for the young adult, so it is all gotten over with as quickly as possible. The wings are hardened, the cuticle tanned, and the adult is ready to go within the hour. Mating and fertilization normally follow an eight-hour fast. The life cycle is then complete.

The gating of pupal eclosion time has been used to experimentally study cir-cadian clock resetting in the cotton bollworm, Pectinophora (Pittendrigh and Minis, 1971); the silkworm moths Antheraea, Hyalophora, and Cecropia (Truman, 1971, 1972); the fruitflies Drosophila pseudoobscura and melanogaster (Skopik and Pittendrigh, 1967; and references throughout this chapter); the mosquito, Anopheles (Reiter and Jones, 1975); the midge, Clunio (Neumann, 1978); the Queensland fruitfly, Dacus (Bateman, 1955); and the fleshfly, Sarcophaga (Saunders, 1976b). For a review, see Saunders (1976). Most of what follows is about flies.

Clock Physiology and Development

Every line of evidence today suggests that the insect's eclosion clock is in its brain (Saunders, 1976):

1. The clock's photosensitivity, in flies and in moths, is at the head end, specifically in the brain.

2. Behavioral mutations affect the circadian rhythm of adult activity if and only if (in genetic mosaic flies) the brain tissue has the mutant genotype.

3. In moths, transplanting the brain is equivalent to transplanting the eclosion schedule.

The circadian clock that ultimately governs eclosion timing already exists when the egg hatches. In fact Minis and Pittendrigh (1968) in an elegant series of experiments showed that it exists in the egg prior to its acquiring the sensitivity to light by which its phase is normally determined. They achieved this in the moth Pectinophora gossypiella by using a single temperature pulse to induce rhythmicity in otherwise arrhythmic populations reared in constant conditions. But even so nonspecific a stimulus as a temperature pulse failed to "set the clock in motion" in eggs fertilized less than $5\frac{1}{2}$ days earlier: Evidently the clock had not yet formed. In the fruitfly Dacus, according to Bateman (1955), there is no arrhythmic stage: Larvae inherit the phase of their mother's clock. Brett (1955) and later Zimmerman (1969, 1971) showed that in Drosophila the clock persists in a regular cycle, from

the youngest larvae straight through metamorphosis, once it is started by terminating an exposure to visible light. Remarkable as it may seem, the circadian
clock appears to evade any substantial involvement in the developmental revolu-
tions taking place all around it, at least up to the last few days before the adult
emerges. It might have been expected that clock properties would change with
developmental age, but to an excellent first approximation this is not the case
during larval and pupal stages. It might also have been reasonably anticipated
(and it was: Harker, 1958; Skopik and Pittendrigh, 1967) that the clock that governs
the nominal termination of metamorphosis would also regulate the timing of
intermediate steps. It came as a disappointment (especially to me, having switched
universities in the middle of graduate study in order to explore developmental
regulation by circadian clocks) to find that this is not the case in *Drosophila*
(Pittendrigh and Skopik, 1970). This circadian clock seems less a synchronizer
of developmental and epigenetic dynamics than a behavioral device seated in the
nervous system.

Behavioral Output

Pupating insects are a great convenience for laboratory investigations. The
pupal case encloses a self-contained "life capsule" requiring only oxygen (and
not much of that) from outside. The animals are well protected in their life capsules
and can be handled almost as casually as so many rice grains. They do not move
around, so cages are dispensible. They don't need food and water, and accordingly
extrude no excreta. Best of all, the life capsule contains within it all the instrumen-
tation required to assay the phase of the circadian clock within, covertly marking
time in cycles of 24 hours.

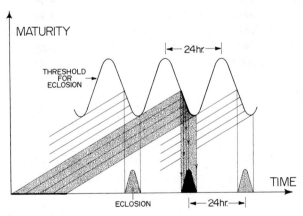

Figure 1. The timing of eclosion peaks in a mixed-age population betrays the phase of an oscillation
here imagined to determine a fluctuating threshold of developmental maturity at which the fly elects
to step forth and be counted. Pupae starting metamorphosis at various times during the darkened
interval at the left side on the time axis first encounter threshold along the darkened arc and so constitute the darkened eclosion peak. Their younger and older siblings fall into discretely separated peaks
unless the threshold rhythm's slope is everywhere shallower than the curve of maturation (here plotted
as a straight line without loss of generality).

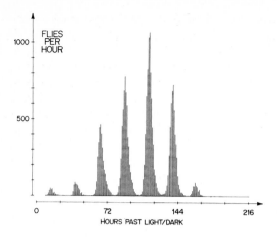

Figure 2. Summation of over 100 unperturbed control experiments collected from 1968 through 1976. Each recorded the number of flies emerging hourly after pupae were taken from prior continuous light into everlasting darkness. We are concerned with peak *timing* (horizontally in multiples of 24 hours), not with the number of flies emerging in each peak.

My belief is that it works like this. Each animal's developmental readiness to emerge increases continually with its age during metamorphosis. At some point, let's say when readiness reaches some threshold, the hormones are released which trigger eclosion. Suppose the threshold fluctuates diurnally. As Figure 1 suggests, individuals starting their climb to maturity as fertilized eggs at different times would then acquire threshold maturity in bunches 24 hours apart. These bunches constitute the train of discrete packets, or eclosion peaks, typically seen in populations of mixed age (Figure 2). Other things equal, the phase of this eclosion rhythm is taken to indicate the phase of the threshold rhythm, itself driven by the underlying photosensitive clock. This may be indirect, but there you have it.

In contrast, with a constant threshold, each individual would acquire threshold maturity a fixed time after fertilization, and the eclosion distribution would therefore exactly parallel the age distribution. This is in fact what happens in pupae whose clocks are suppressed by continuous light (Skopik and Pittendrigh, 1967) or otherwise induced to fluctuate much less conspicuously (see Box A).

There are two important facts to note about this diagram in present context:

A. The *size* of an eclosion peak is determined entirely by the age distribution. All pupae starting in the shaded range of the time axis emerge as adults in the shaded eclosion peak. The size of an eclosion peak has occasionally been confused as some measure of the "amplitude" of the circadian oscillation. But this is quite a different matter, as elaborated in Chapter 7. Only the *timing* of peaks, not their size, is determined by the phase and amplitude of the threshold rhythm.

B. The putative threshold rhythm must not be confused with "the waveform of the circadian oscillator", on at least two accounts:

1. The very notion of the waveform is ill defined and, I think, more nearly a troublesome fiction than helpful concept.

2. The eclosion peaks are not always regularly spaced along the time axis immediately after a phase-resetting stimulus. It would appear that eclosion is determined not directly and immediately by the reset clock, but by some physiological rhythm indirectly driven by a more nearly periodic clock, and that this

Box A: Rhythmic Gating and Arrhythmicity

Figure 1 presents one vision of how a smooth rhythm can determine the timing of a discrete event. In a population of organisms starting development at times strewn continuously along the time axis, this vision provides a way to think about "gating" of developmental events. If the rhythmic variation in the threshold has sufficient amplitude, then during part of each cycle the threshold is rising faster than developmental readiness

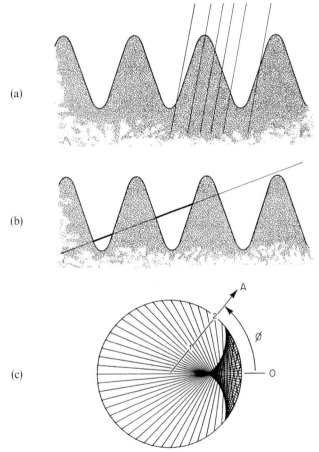

(a)

(b)

(c)

(a) With hastened maturation or attenuated circadian rhythmicity of threshold (relative to Figure 1), some pupae reach threshold at every hour. Eclosion peaks abruptly broaden and fuse at this point. (b) With development sufficiently slowed down, each pupa crosses repeatedly above threshold and back under. The darkened intervals might be detectable. (c) A polar coordinate contour map of age at eclosion as a function of the threshold rhythm's amplitude A (relative to rate of maturation) and its phase ϕ at any chosen reference age (or age at phase zero). Note the region around $\phi = 0$ at A-1 where the surface folds into three layers, representing the three possible intersections in Figure 1. The edges of this cusp are involutes of the unit circle. At any amplitude $A > 1$ the age at eclosion changes abruptly at this critical value of "age at phase zero", cleaving a mixed-age population into discrete peaks.

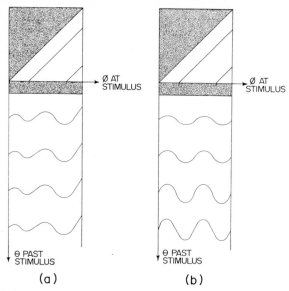

Figure 3. Two kinds of transients, portrayed in the format of Figures 31 through 33 of Chapter 1.
(a) Transients in the source of rhythmicity, e.g., gradual increase of amplitude after startup. (b) Transients in peripheral mechanisms causally downstream from the source of rhythmicity. Such rhythmically-driven mechanisms commonly require a few cycles to entrain stably to the recently started or reset source of rhythmicity.

Input to the Clock: Light and Temperature

So much for the output side of the clock. Since our experiments consist of varying the light intensity, it is also necessary to dwell briefly on the peculiarities of the input side.

The circadian clock in birds and mammals takes its cues in part from sophisticated sensory systems, e.g., hearing, vision, and social interactions. In lower organisms, we find only a very primitive sensory transduction. For example, the severe impact of sudden temperature changes on clock kinetics in the fruitfly *Drosophila* (Winfree, 1972a) is presumably through the temperature dependence of all reactions involved. There seems no reason to imagine a specialized thermoreceptor. Slower temperature changes are commonly buffered out (see Section A of Chapter 19).

Is the response to visible light this primitive? The visual nervous system is involved in some organisms, but this is not usual in the invertebrates. I know no persuasive evidence that clock photosensitivity in invertebrates attaches to any selectively specialized structures or pigments, nor that the absorbing substance is a part of the oscillating mechanism.[1] In fact, there are some reasons to think otherwise; e.g., some circadian clocks seem to lack photosensitivity altogether (Tweedy, 1970; Saunders, 1976a). Other hints in this direction come from detailed study of clock photosensitivity in the fruitfly *D. pseudoobscura*:

---

[1]   Though it might be in green plants (see Heide, 1977 about phytochrome).

1. It doesn't take much light to rephase. My standard light source provides 0.1 watt/m^2 near 450 nm. This is 4×10^{-7} moles of photons/m^2/s. Multiplying by the maximum conceivable absorption cross-section (10^5 l/mole-cm$^2 = 10^6$ m/ mole as in rhodopsin, chlorophyll, phytochrome) we find that 1% of the molecules, at the very most, absorb a photon each second. But in only several seconds, far less than the minimum expected waiting time for any given molecule to absorb a photon, the phase response saturates in dark-adapted pupae. Evidently, then, the phase response does not saturate for want of any surviving excitable or photo-labile molecules; the vast majority of them don't even know the light is on before the phase response is saturated, and continued exposure, though it must result in more absorptions, does little more to the clock so far as I am aware.

While this observation might be compatible with the notion that the photo-receptor is itself a rhythmically fluctuating component of the clock mechanism, it is not compatible with the implication sometimes invoked, i.e., that phase shifting is accomplished directly by substantially changing the amount of that substance. (However, many other circadian rhythms are far less photosensitive.)

2. In *D. pseudoobscura* (but not in *D. melanogaster*) prolonged illumination reduces the clock's photosensitivity during its next cycle by 10- to 20-fold. Sub-sequent dark adaptation takes days and is reversible by light that seems not to affect the clock. This suggests to me bleaching of a photoreceptor which is not a necessary part of an oscillating mechanism. I believe this possibility was first suggested in a mathematical model by Pavlidis (1967b) and in context of fruitfly experiments by Chandrashekaran and Loher (1969b). Similar "dark adaptation" has been noted in quite diverse species, e.g., in *Gonyaulax* (Christianson and Sweeney, 1973); in *Kalanchoe* (Engelmann et al., 1978); in *Chaoborus* (Bradshaw, 1972). However, in *Sarcophaga* (Saunders, 1978) and in *Pharbitis* (Bollig, 1975) photosensitivity *decreases* after transition from constant light into constant darkness. In this connection, note that Chandrashekaran and Engelmann (1973) suggest a correction on Winfree (1972b,e). They believe that dark adaptation occurs not smoothly but by stages at each successive T^* of the clock.

3. Even at the wavelength of greatest effectiveness (Frank and Zimmerman, 1969; Klemm and Ninnemann, 1976) 10 thousand J/m^2 is required for the "bleaching" in *D. pseudoobscura* (Winfree, 1972e). Given a photoreceptor with absorption cross-section comparable to rhodopsin or chlorophyll or phytochrome, that much light would hit each molecule 1,000 times on the average. It may be that only a rare hit permanently bleaches the molecule. But if bleaching is the usual concomitant of absorbing a single photon (as in rhodopsin, chlorophyll, and phytochrome), then the molecule's absorption cross-section or extinction coeffi-cient at the optimum wavelength falls below that of good biological photoreceptors by a factor of 1,000, discounted by whatever attenuation factor the overlying opaque tissues might provide. This is scarcely above the range typical of colorless organic molecules, a fact compatible with the failure of many attempts to see and isolate a clock pigment. This suggests to me that the extreme sensitivity of some circadian clocks to visible light is due less to efficient photon capture per mole of photoreceptor than to specific involvement of the excited molecule in clock chemistry.

4.   I know of no conclusive studies establishing whether a circadian clock has "a" photoreceptor pigment, i.e., is color blind, having a single color receptor. Alternatively (as in the case of temperature sensitivity), the oscillator might be affected in different ways through a variety of not-quite-colorless molecules. Bunning and Joerrens (1960) suggested as much for an insect. There are stronger indications of this in plants. The action spectrum for rephasing seems to depend on the phase at which it is measured, i.e., the resetting curve shape depends on the color of light used (Bunning and Moser, 1966; Halaban, 1969; Schrempf, 1975). In other words, the clock in plants has color vision. Frank and Zimmerman (1969) specifically tested for this possibility in *D. pseudoobscura* and found no indication of a phase-dependent action spectrum. However, their test was not delicate, comparing photoresponses at phases separated by only two hours. Moreover, their measurements (and those of Klemm and Ninneman, 1976) used pupae whose photosensitivity would have been an order of magnitude greater had they been exposed a day later; this increase might reflect the appearance of a second pigment. (In this connection see Winfree, 1975c.)

Do action spectra provide an efficient inroad to the clock's molecular mechanism if the photoreceptor only affects the clock while illuminated, but isn't part of the mechanism in the sense of affecting and being affected even in the dark? Or if the action spectrum is compounded of more than one absorption spectrum? I think not.

The circadian clock in most insects functions differently while exposed to sudden changes of temperature and/or to light, even quite dim light, of a suitable color. In no case has the photoreceptor been identified. In no case has its impact on the clock mechanism been described, except in terms of analogies to electronic oscillators, chemical effects on glycolytic oscillations, differential equations, and so on. Nonetheless, the clock does function differently while illuminated, so we can use a brief interlude of illumination to explore its state space, by starting from different phases in the usual cycle and continuing from there to various durations before reverting to darkness.

Under prolonged illumination, the clock mechanism approaches a time-independent state. (Pavlidis, 1978a, neutralizes a recent challenge to this inference.) The eclosion distribution is then no longer gated into discrete bursts. Prolonged illumination provides the standard initial condition for almost all our experiments. Upon transfer to darkness, oscillating kinetics is resumed from that initial state and the eclosion rhythm's phase is determined by the time of transfer. This trick was discovered by Bunning (1935).

Why Use Populations?

Having assembled a provisional picture of the individual physiological clock, in context of its sensory inputs and outputs to the timing of eclosion, we have finally to contend with the fact that eclosion measurements are typically conducted in populations. In principle, populations should not be necessary. Using cohorts of pupae collected as prepupae formed within a two-hour interval, Skopik and

Pittendrigh (1967) showed that eclosions occur in a peak less than three hours wide (measured as twice the RMS deviation of counts about the mean). They showed that the time of the peak reveals the phase of something (the clock) that varies periodically after transfer to darkness. In principle, phase shifts of the clock could be assayed with adequate precision by comparing the eclosion time of a single fly against Skopik and Pittendrigh's series of control experiments, in which the mean eclosion time was measured in cohorts after the clock was started (by transfer to darkness) at different ages. But a technical difficulty intervenes. The control series would have to be rerun in each experiment because as little as a $\frac{1}{2}$°C difference in mean temperature during metamorphosis is equivalent to a 5% change of developmental rate (and almost no change in clock period). In the 200 hours of *D. pseudoobscura* metamorphosis this amounts to a major fraction (10/24) of a 24-hour cycle. Nutritional differences caused by crowding presumably also affect the mean rate of development. Since eclosion time is jointly determined by clock phase and developmental readiness (Figure 1), any uncontrolled influence on developmental rate spoils the single-pupa phase assay. So the usual expedient is to use populations spanning a range of at least several days of age. This way, even if pupae achieve adequate maturity some hours or days earlier or later than usual, there are still some mature pupae available at any given time during the week when eclosion is monitored. Ideally, one would record the age distribution in a separate control experiment and correct eclosion peak times for the influence of departures from flatness using an algorithm suggested by Figure 1. It was my habit to do so, by routine computer processing of data, but it didn't make much difference in experiments that terminated in several discrete and narrow eclosion peaks. It suffices to ignore peak shape and accept any measure of central tendency as the phase of the clock, if one can live with the one- to two-hour irreproducibility of that measure.

Numerous warnings about interpretation of population data have been issued in the circadian rhythm literature since the mid-1950s (Harker, 1958; Wever, 1963, 1965b; Skopik and Pittendrigh, 1967; Ehret, 1971; Karlsson and Johnsson, 1972; Engelmann et al., 1973; Johnsson et al., 1973). Most of these prove irrelevant for insect eclosion systems producing discrete sharp peaks. However, population inhomogeneity does become a crucial factor for interpretation of experiments that produce vague or irregular rhythmicity, as discussed below.

Recapitulation

For present purposes, the central facts to bear in mind are:

1. The eclosion clock is started by transfer from constant illumination to everlasting darkness (perhaps punctuated by one or two light pulses well before eclosion time).

2. The position of the eclosion peak sequence along the time axis reveals the phase of the underlying clock if the eclosion peaks are sharp and uniformly spaced at 24-hour intervals.

3.   "Phase" means the fraction of a cycle elapsed, or 1 minus the fraction remaining. Thus two conventions must be specified.

a.   The time at which phase is evaluated: I evaluate old phase at the beginning of the first stimulus; I evaluate new phase (alias final phase) at the end of the last stimulus, which is the beginning of the final dark free-run to eclosion.

b.   A phase reference point to call phase zero on the eclosion assay: Let this be the center of mass (centroid) of the daily eclosion peak. Thus time $T = 0$ (the light-to-dark transition) is phase $1 - (17/24) = 0.2$ in my strain of $D.$ $pseudoobscura$, because eclosions follow the transition after 17 hours and thereafter at multiples of nearly 24 hours. The phase after some stimulus is $1 - (\theta/24)$, where θ is the number of hours (modulo 24) from the end of the stimulus to recurrent eclosions.

Much of the circadian rhythm literature uses a different convention, defining "subjective circadian time" (SCT) as hours elapsed (modulo the period) since the dark-to-light (not light-to-dark) transition in a 12 hour-light-12 hour-dark cycle. Since eclosion in $D.$ $pseudoobscura$ occurs at SCT 3 under these conditions, SCT is sometimes redefined as 3 plus the number of hours since eclosion, or 3 plus 24 minus the number of hours until next eclosion, assuming regular periodism.

B: Phase and Amplitude Resetting
in *Drosophila Pseudoobscura*

All this is a dream. Still, examine it by a few experiments. Nothing is too wonderful to be true if it be consistent with the laws of nature. And in such things as these experiment is the best test of such consistency.

Michael Faraday, 19 March 1849

Technology

In $D.$ $pseudoobscura$ the technology of eclosion monitoring centers around the fact that the freshly emerged fly is wet, white, weak, and wingless for about $\frac{1}{2}$ hour after eclosion. During this time, it stands motionless, clinging to the pupal case or some nearby object while wings inflate and other cuticle tans and hardens. So tenaciously do they cling that Pittendrigh found it necessary to use the auto-mated hammer blows of a heavy-duty solenoid at frequent intervals to shake them loose into a vial of soapy water. As a student in the lecture hall under Pittendrigh's laboratory at Princeton, I was accustomed to solving mathematics puzzles while "the Army of Science marched overhead", compelling the lecturer to briefly shuffle his papers in silence. Always alert to potential improvements of technology, Pittendrigh contrived a "superbang" machine in 1967. While optimizing the num-ber and magnitude of bangs, I noticed that "banging" is necessary only in the light. In the dark, flies fall spontaneously, and if mounted over a teflon-lined funnel (an innovation by W. Engelmann), they fall all the way to whatever fate one intends for them. Thus one can assemble a light-weight, compact fraction collector to handle as many as 100 separate pupal populations within the space of a desktop. Under an array of funnels slowly glides a lucite tray of chemically wetted water,

Figure 4. The time machine. Twelve brass cups house as many populations of pupae mounted to receive a measured exposure of blue light from a central rotating mirror when a timer activates the corresponding solenoid above each cup.

divided into compartments, each of which accumulates the flies emerging during one hour from one population. These hourly body counts underlie all the fly clock data plotted in this book. A ruby red safe-light, also introduced by Engelmann following Brett (1955), enormously facilitated manipulations originally conducted in absolute darkness (often with refugees from Pittendrigh's cockroach experiments crawling up my arm). The safe-light was eventually upgraded to the sodium doublet, improving visibility for flies and men both, without effect on the circadian clock (Winfree, 1975c). Blue light exposures were automated by a Rube Goldberg contraption of relays, mirrors, and solenoids called the "time machine" (Figure 4) for its responsibility to shift pupae into the past or into the future of their otherwise undisturbed cycles (see Materials and Methods, Winfree, 1973a). Data processing was automated first for an IBM 7094 at Princeton (1967), then for an IBM 360/50 at the University of Chicago (1970), and finally for a desk-top HP 9830 at Purdue (1973). During its 10 years of evolving operation, this system executed altogether 4,100 separate experiments (approximately one megafly), each experiment measuring the phase of one population's eclosion rhythm in darkness after suitably scheduled exposure to light by recording the eclosion time of each individual to within $\pm\frac{1}{2}$ hour. The majority were diverse control experiments, repeats on several different strains of *Drosophila* (all of which behaved in very nearly the same way), repeats or attempted repeats of other people's published experiments, and several unpublished expensive boondoggles.

But the remainder proved informative and enter significantly into our exploration of biological phase singularities:

1. The pinwheel experiment Chapter 2, Example 5; see next section, this chapter.

2. The time-machine experiment for plotting trajectories in Chapter 7.

3.   Experiments exploring conditions required to induce arrhythmicity, and the nature of that arrhythmicity (see below).

These experiments used about 1,200 populations of females of a sex-ratio strain that I assembled in 1967 from wild *D. pseudoobscura* kindly brought back from the Arizona desert by Ronald Quinn, following a suggestion of Pittendrigh, and cuticle marker mutants kindly provided by T. Dobzhansky. I allowed this strain to perish in 1975.

4.   Another 600 experiments went for the singularity trap experiment (Chapter 2, Example 6 and Box C) using males and females of R. Konopka's 19-hour clock mutant of *D. melanogaster* (see below).

The Timing of the Screw

This section describes an algebraic surface contrived to fit smoothly through the *D. pseudoobscura* data points obtained by hit-and-run experiments with a single pulse. The pulse of magnitude M is given T hours after the clock is started by transfering pupae from light into darkness. We then record the eclosion time θ, in hours measured from the stimulus or from stimulus anniversaries at multiples of 24 hours after the stimulus. The results were described in Figures 13 and 14 of Chapter 2, where a stereo plot of unadorned data was promised. Here it is (Figure 5). The coordinate axes are three:

1.   Horizontally to the right, the time or phase at which the light pulse was given, or the east-west location of a pupa on the imaginary desk top of Example 5, Chapter 2.

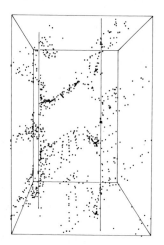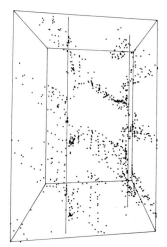

Figure 5.   Stereographic views of 879 measurements of eclosion time in *Drosophila pseudoobscura* following a brief light pulse. (For viewing see Chapter 12, Box A.) Coordinates and perspective as in Figure 14 of Chapter 2. The two interior uprights depict the presumed singularities, poles around which data points climb in corkscrew fashion. The second singularity is further in the foreground mainly on account of dark adaption in the clock's photoreceptor during the first 31 hours of darkness.

2.   Horizontally in depth, the duration or energy (the magnitude M) of the exposure. (In this range of durations, the clock is responding to energy, no matter how quickly or slowly it is delivered (Engelmann, 1969; Chandrashekaran and Engelmann, 1973). This is also the north-south location of a pupa on the imaginary desk top. These two coordinates constitute the stimulus plane: Each point on the desk top represents a unique combination of the two parameters that define a stimulus. These are the independent variables of the experimental question posed to the flies.

3.   Vertically downward, the times of eclosion. This is the flies' answer, the dependent variable, a series of times spaced 24 hours apart along the vertical axis above each stimulus point. In this plot time is measured from the stimulus downward. This is a simple, direct way to plot the rhythm's new reset phase upward. (See Chapter 1, Section D. Figures 31–33 in Chapter 1 are only two-dimensional because M takes a single fixed value in each. Here M varies in depth.)

It would be easier to grasp the dependence of the phase resetting on the stimulus parameters T and M if we could fit a smooth surface $\theta(T, M)$ to the complete cloud of 1,574 centroid data in the same way as we usually like to fit a smooth curve to our data on two-dimensional graphs. I did this by hand in Figure 14 of Chapter 2. I also contrived an algebraic expression to fit the data points almost as well as their reproducibility allow. The emergence times vary randomly with standard deviation $1\frac{1}{2}$ hours in unperturbed controls, whereas the reset peaks vary about this fitting function with standard deviation of only two hours. The surface is described by the following equation:

$$\tan\frac{2\pi}{24}(\theta + 1.1) = \frac{\sin(2\pi/24)(T - 7)}{1 - e^{-M}(1 + \cos(2\pi/24)(T - 7))},$$

where
$$M = \frac{\text{exposure duration}}{(6 + 140\exp(-T/8))}.$$

M is the "subjective" exposure magnitude, proportional to physical energy scaled by an exponential dark-adaption factor (Winfree, 1972b).[2] Note the singularity at $M = M^*$ ("M star") $= \log 2$, $T = T^*$ ("T star") $= 7 + 24n$ hours.

As noted above, the data stray from this surface by ± 2 hours, on the average. Some of this discrepancy is systematic due to my choice of a very simple fitting equation. But as the reproducibility of a phase measurement is only ± 1.5 hours anyway, it didn't seem worthwhile to tailor the fit more meticulously.

Plotted above the (T, M) plane, this $\theta(T, M)$ is a single surface wound around a series of screw axes, scarcely distinguishable from Figure 14 of Chapter 2. It resembles a vertical corkscrew or a spiral staircase linking together tilted planes. As it is periodic along both the T and θ coordinates, it resembles a two-dimensional crystal lattice. I call it a time crystal because its two periodic dimensions both represent time in units of one day.

---

[2]   Chandrashekaran and Engelmann (1973) suggested replacing this smooth function by a stepwise increase in photosensitivity with steps at each T^*. That would produce a tear in the resetting surface, an edge dislocation in the time crystal. I see no such tear in my Figure 5.

The surface in Figure 14 of Chapter 2 is in the same graphic format as developed for simple clocks in Chapter 3. Notice how different its shape is. Simple-clock resetting surfaces consisted of separate sheets, whereas here each unit cell of the crystal lattice contains one turn of a screw surface and these all fit together into a single periodic surface. Let's examine one of these unit cells in detail by serial sectioning, as though running it through a microtome for microscopy. Figure 13 of Chapter 2 shows serial sections through three consecutive unit cells of the idealized surface (the equation) at 12 levels of fixed eclosion time θ. These are successive positions of the eclosion wave on the stimulus plane (the desk top of the pinwheel experiment of Chapter 2). In other words, a pivoting wave, plotted in terms of wave arrival time above the plane on which it rotates, resembles a screw, a "helicoid". The axis of the screw stands vertically above the rotor's pivot at $T = T^*$, $M = M^*$. Thus a piecemeal resetting experiment, whose outcome resembles a crystal lattice of screw surfaces, is equivalent to a pinwheel experiment in which the graded stimulus evolves a rotating wave.

Figure 13 of Chapter 2 shows serial sections cut perpendicular to the θ axis. If we now instead cut perpendicular to the M axis, we have resetting maps, as described in Chapters 1, 3, 7, 14, and 19. As the composite in Figure 6 shows, they are all type 1 up to a critical stimulus magnitude M^*. At M^*, type 0 resetting abruptly emerges. The critical dose M^* turns out to be remarkably small by the standards we were accustomed to in the circadian business, viz less than a minute of dim blue light. Given at T^* to elicit singular behavior, M must be within 20% of the exact value M^*, and it has to be administered within $\frac{1}{2}$ hour of T^* (though T^* and M^* vary from one batch of pupae to the next). It is not hard to understand how this effect was overlooked until a theoretical model predicted the existence of a critical combination of time and dose and prescribed a recipe for finding it.

Figure 7 re-presents Figure 6 as phase response curves: Instead of plotting the new phase or eclosion time vertically, we plot the amount by which the stimulus changed it, the phase shift.

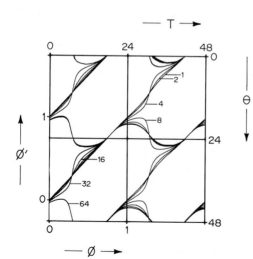

Figure 6.   Using the particular formula here adopted to approximate phase-resetting in *D. pseudoobscura* pupae, these are resetting curves for a series of stimulus magnitudes ranging from 0 to 64 seconds exposure to 0.1 W/m² light at about 410 nm. Each relates time T past LL/DD (or old phase ϕ) to time θ until eclosion (or new phase ϕ', neglecting multiples of 24 h).

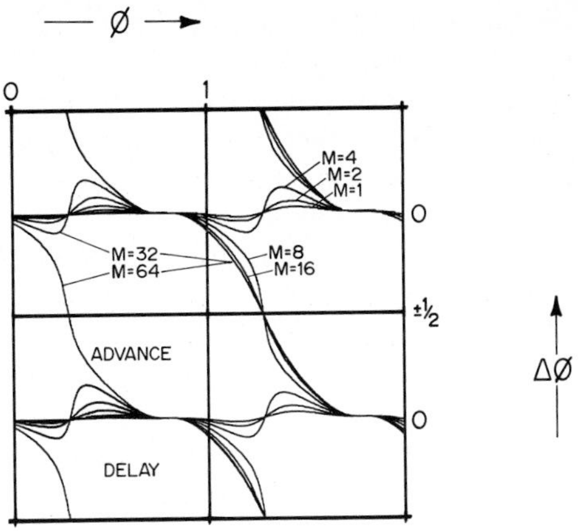

Figure 7. Phase shift (advances upward) vs. initial phase for a series of stimulus durations up to 64 seconds. This is Figure 6 sheared downward to the right by 45° without subtracting stimulus duration, since even 64 seconds is a negligible fraction of 24 hours.

In the next section we examine the singular (T^*, M^*) where no phase can be assigned to the clock.

Arrhythmicity in Fruitfly Populations

Circadian rhythms are usually monitored in populations of independently rhythmic clocks, be they individual organisms or single cells of a tissue. If they interact in unknown ways, then not much can be learned without monitoring each individual's rhythm, at least well enough to establish that the population is always coherent in phase on a common cycle. It commonly happens that populations created and reared in constant conditions are arrhythmic, raising the question as to whether collective arrhythmicity is just a consequence of incoherence among normally oscillating individuals. Brett's (1955) conclusion in respect to arrhythmic eclosion in populations of dark-reared *D. melanogaster* was "yes". The same conclusion was accepted for *D. pseudoobscura* (Pittendrigh, 1954, footnote 10; Pittendrigh and Bruce, 1957; 1958, pp. 251–252; 1960, p. 102; 1961, p. 116; Bunning, 1957). All the facts seemed consistent with this belief up through 1968, as Beck (1968, p. 62) summarized: "It is though that the light signal has the effect of synchronizing the rhythmic functions of all the members of the population." As Zimmerman (1969) noted, this belief meshes well with

> A. The general notion that the maintenance and entrainment of circadian oscillations is essential to the normal physiology and the development of eukaryotic organisms (Pittendrigh, 1960, 1961; Bunning, 1964) and
> B. The more specific hypothesis of Ehret and Trucco (1967) that the mechanism for circadian oscillations inheres in the physical organization—and therefore transcription—of the DNA is eukaryotic cells.

Zimmerman goes on to note that this belief had never been seriously tested in an animal, even after Bunning (1959, p. 522) had reported both cases in plants. Kalmus and Wigglesworth (1960) challenged that belief via a non-simple-clock model and Sweeney and Hastings (1960, p. 102) called for such a test, both at the first circadian clock symposium at Cold Spring Harbor.

In August of 1968 Zimmerman, at Amherst, submitted his paper critically testing and disproving this assumption by a trick using *D. pseudoobscura*'s type 1 resetting. In the same month I submitted a paper from Princeton achieving the same result by a different trick using type 0 resetting in the same organism (Winfree, 1968). Both unequivocally excluded the possibility of interpreting the arrhythmicity of dark-reared populations in the customary way. This outcome, in Zimmerman's words, "considered in conjunction with the fact that most organisms can reproduce and function normally in aperiodic environments, presents definite difficulties for . . ." notions A and B.

The question is important enough to lend interest to an examination of the evidence. What does it consist of, and how broad is its pertinence?

Zimmerman's Experiment

Zimmerman's experiment used type 1 resetting by a temperature pulse to implement the suggestion of Sweeney and Hastings (1960) that after a small phase shift, a population of clocks initially covering the cycle would still be spread completely around the cycle, so eclosion should still occur at all hours. As the experiment turned out, such a pulse applied to dark-reared arrhythmic pupae evokes sharp discrete daily bursts of eclosion. Therefore these dark-reared pupae could not have had normally running clocks. Zimmerman goes further, concluding that the clocks "were inherited at rest". This inference rests on a notion that has never been checked, i.e., that the clocks in dark-reared pupae are no more temperature-sensitive than the rhythmic populations on which the type 1 resetting was calibrated. Even the more conservative conclusion that the clocks could not have been running normally needs reinforcement at three points:

1. It implicitly supposes that pupae do not interact, e.g., to synchronize to the phase of a majority. They do in other species, e.g., in eclosing silkworm moths (Truman, 1972) and in another species of fly (Saunders, 1976a). Noninteraction in *Drosophila* had been explicitly supposed since Kalmus (1940), but the question had never been of real concern in *Drosophila* eclosion experiments because pupal rhythms had always been forced to synchrony anyway. So I checked it by three independent experimental tests (Winfree, 1970c appendix), none of which reveal any conspicuous or reproducible interactions. So this supposition seems safe.

2. My subsequent efforts to repeat Zimmerman's experiment failed because in my control experiments with rhythmic populations, the temperature pulse he used (28°C for 12 hours) did not give the same small phase shifts as in his controls. Even with much briefer incubations, I obtained large steady-state phase shifts (Winfree, 1972a), compatible with a type 0 curve and incompatible with Zimmerman's argument, which was based on the phase shifts being small.

3. Zimmerman's (1969) Figures 1 and 2 are incompatible in a way that suggests some confusion about the time scales. I have not been able to reproduce the reported timing of the eclosion rhythms according to either figure.

Supplementary Experiments

All this, together with widespread difficulties in working with temperature effects, leads me to doubt the proof while accepting the result. I accept the result because it also emerges from experiments using light instead of temperature:

1. After learning of Zimmerman's experiment, I repeated his format but used a very brief light pulse to evoke the required type 1 resetting. The outcome was qualitatively the same as his (Winfree, 1970c). It is also subject to the same quibble, that the light pulse might seem subjectively longer or brighter to dark-reared pupae than it does to the rhythmic pupae used to assay its potency. In fact, this now seems quite likely, in view of the subsequent discovery of 20-fold dark adaptation in *D. pseudoobscura* (Winfree, 1972b and e). A pulse that much brighter, like the more effective temperature pulse, would have qualitatively the same synchronizing effect as is observed even if the population were indeed composed of randomly phased, normally running clocks.

2. My 1968 experiment had employed a different trick, using the type 0 resetting evoked by a light pulse of saturating intensity and duration as in Pittendrigh and Bruce (1957). Thus questions of sensitivity are bypassed. The trouble with this experiment is that no one seems to understand it. So let me try to explain it in a different way this time, without belaboring quantitative details. Type 0 resetting necessarily has two ranges of old phase in which the new phase scarcely varies. Given an initially uniform distribution of old phase, the distribution of new phase should therefore exhibit two peaks, with lesser densities in between (Figure 8). Now there is a subtlety: We do not directly observe clock phase in an emergence distribution. Even a perfectly coherent population produces an eclosion peak six hours wide at two standard deviations, and wider at the base. (Skopik and Pittendrigh, 1967, showed that most of this breadth is due to the population's heterogeneity in regard to developmental age.) So what we really want to ask of the observed eclosion distribution is, "How can I represent these data as a superposition of 24 standard eclosion peaks phased around the clock, each with the right number of pupae in it to account for the observed distribution?" Resolving the observed eclosion distribution into standard peaks is valid if and only if:

a. Pupae at different phases do not interact. As noted above, this was verified.

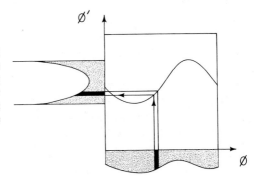

Figure 8. Any smooth distribution of old phase covering the cycle, transformed by type 0 resetting, becomes a distribution of new phases according to the conservation law $p(\phi)\,d\phi = p'(\phi')\,d\phi'$. Because $d\phi'/d\phi = 0$ at one or more pair of phases, the new phase distribution has one or more pair of horns.

b.   Even if clocks were not running normally prior to the type 0 light pulse, they are afterwards. There is no way to be sure of this but the result of the experiment turned out to be an eclosion rhythm indistinguishable from those obtained simply by moving pupae from the light into the dark. Also Honegger (1967) measured a resetting curve on such a population and this too looks the same as in pupae whose clocks were running normally after light-to-dark transition.

c.   It can be done without using negative densities at any phase, and in such a way that the results match observations within the expected statistical fluctuations associated with counting error.

A computer program was contrived to produce the best-fitting representations and chi-squared estimates of the match. It worked quite reliably in a variety of control experiments. For example, it resolved the phase distributions in populations artificially constructed by transferring batches of pupae from light to dark at different times of day, and then counting their eclosions together. It produced excellent matches to all the experimental data, and it failed on made-up data. Essentially the same trick, called "inverting the convolution integral", was used by Njus (1975) to extract the distribution of cell clock phases from the broadened glow rhythm of a *Gonyaulax* population. In the control experiments with pupal populations deliberately constructed to be arrhythmic (12 aliquots taken to dark at two-hour intervals around the clock, then exposed to the saturating light pulse) the result was as expected: a two-peaked distribution of phases within a few hours of the expected phase. In contrast, finally getting around to the results sought, dark-reared populations treated in the same way produced a phase distribution resembling a single narrow spike: a conclusively different result conclusively showing that dark-reared populations are homogeneous in their behavior.

Conclusion

In this case, it does seem legitimate to infer with Zimmerman that the pupae inherited their clocks at rest. At rest? Stuck at what phase? Dark-reared pupae react to a temperature pulse or to a light pulse much as do pupae made phaseless by a singular perturbation. This observation is awkward for any simple-clock model in which arrhythmicity is interpreted as arrest at some special phase on the cycle, as in experiments on the cell cycle. It suggests an alternative interpretation, i.e., that dark-reared clocks are created near a steady-state and linger there until kicked into oscillation about the steady-state.

The question as to whether ostensible arrhythmicity is only a matter of incoherence among individuals or cells of a population must be answered yes in order to sustain the view that lured me into this business, viz that the circadian clock plays a crucial organizing role in the machinery of life. But at the population level, we have just seen that *Drosophila* answers *no*. Then what at the cell level? If the eclosion rhythm can be regarded as an observation on the population of pupae then it can also be regarded as an observation on a larger population of cells. Does the inference made above carry through to the cell level? The answer is no because we would have to show that the cells do not interact. They probably *do* interact. The many cells of an individual pupa's brain somehow jointly determine the moment at which it takes a decision to open its pupal case. Here we know nothing. It could be that arrhythmicity in the individual pupa is only a manifestation of incoherence among its normally rhythmic cells. If so, then cytological incoherence is a major uncontrolled variable in clock experiments, and many years of experimentation may be ripe for a critical review of interpretations. If not then the belief that circadian rhythmicity is an essential principle of physiological organization must be abandoned, and so must be hypothetical mechanisms which have no place for a phase singularity. Since the consequences are interesting either way, I consider this high-priority business for the laboratory.

Other Modes of Arrhythmicity

There are at least four other contexts in which eclosion rhythmicity is lost. All four challenge the notion that circadian clocks are essential for temporal coordination:

1.   There are mutants in which chosen observables are arrhythmic. As in the case of rhythmic morphogenesis in fungi (Chapter 18), this may prove to be a matter of altered permissive conditions (some change of diet or of temperature might disinhibit the clock or restore coupling to the few aspects of physiology that we monitor) or it might be that an essential component of the mechanism is simply absent (Hastings and Schweiger, 1976, Part 5).

2.   Saunders (1976a) finds relatively arrhythmic eclosion in *Sarcophaga* following exposure to light-dark cycles in which $L + D = 12, 36, 60, \ldots$, hours. He thinks this may reflect incoherence among oscillators within each fly.

3.   The transition from type 1 to type 0 resetting is made through a phase singularity: An exposure of this size given at a critical phase induces lasting arrhythmicity in each pupa. There is some reason to conjecture that this state is the same as the dark-reared state (Winfree, 1968, 1970b) but proof is still lacking.

4.   Bright continuous light inhibits circadian rhythmicity in many organisms, including flies. This is what Zimmerman (1969) calls secondary arrhythmicity: imposed by continued exposure to nonpermissive conditions. Pittendrigh (1966) produced the first really convincing demonstrations that the clock actually stops in each pupa after 12 hours in the light. This notion is frequently challenged (e.g., Brady, 1974, p. 18), especially since the advent of experiments with dim continuous light (Chandrashekaran and Loher, 1969a; Winfree, 1974a; Pittendrigh, 1976). However, I know of no observations incompatible with suitable models in which sufficiently intense illumination does indeed stop a non-simple clock (Pavlidis, 1978a).

The Fruitfly Clock is Not a Simple Clock

The question is sometimes raised (I think first by Pittendrigh and Bruce, 1957, in context of experimental tests) as to whether the circadian clock is affected more by the abrupt transition from dark to light or light to dark ("differential action"[3]) than by the continual impact of photons on some ultimately photochemical mechanism ("continuous action"[3]). So long as these alternatives were only considered in connection with the widespread but implicit notion that the circadian clock is a simple clock (Chapter 3), it seemed most reasonable to adopt the differential action outlook (Pittendrigh, 1958 and especially 1960, p. 175). Yet it remained difficult to understand in these terms the lasting arrhythmicity observed forever after a transition from darkness to continuous light: Did the moment of transition (putatively the only stimulus perceived) switch the clock into some state of indeterminate phase? No such state exists in the cycle of a simple clock.

1. Pavlidis' Model.   Pavlidis (1967b) first resolved this difficulty explicitly in context of *D. pseudoobscura* data by postulating a disk of states inside the cycle of conventional phases.[4] In other words, Pavlidis adopted the alternative inference overlooked in Pittendrigh's experiment of 1960 cited above, i.e., that the circadian clock in *Drosophila* is not a simple clock. Moreover, *acceptance* of "continuous action" is the premise underlying Pavlidis' second model (1967b) and all the topological inferences in this volume. We owe to this paper the critical reestablishment of continuous action as a reasonable interpretation of *Drosophila* clock experiments, but now in the enlarged context of a clock whose state can vary in two ways, not only in phase.

On this basis Pavlidis (1968) first anticipated something that makes no sense in terms of simple-clock models: the possibility of stopping the *Drosophila* clock by a delicately timed light pulse. Pavlidis seems to have mistakenly thought his inference was in error, an exorciseable artifact of mathematical modeling:

> ... the model should be designed in such a way as to exclude the possibility that a small light stimulus would bring the system into R ... therefore *while a strong light stimulus would fail to damp out the oscillation, a weak one ... would damp it.* This is completely in contrast to the experimental findings and therefore we conclude that the critical point should be unstable ...
> [The citation is restricted to the italicized parts in Pavlidis, 1978a]

But the necessity of some such singular event could be inferred from fragments of published data on model-independent topological grounds (Winfree, 1967b). Shortly thereafter it was experimentally verified in *D. pseudoobscura* (Winfree, 1968) and then in other species (*Kalanchoe*: Johnsson et al., 1973; *Sarcophaga*: Saunders, 1976a; *D. melanogaster*: Winfree and Gordon, 1977).

---

[3]   In the more recent notation of Pittendrigh and Daans' five major papers of 1976, continuous action equals parametric and differential response equals nonparametric.

[4]   As with most other innovations, one could point to earlier independent discoverers: Kalmus and Wigglesworth (1960); Wever (1962–3–4); Moshkov et al. (1966) applied the same ideas in the same format to circadian rhythms in general, but they failed to pin this model explicitly to one data set. Strahm (1964) (cited only in Pavlidis, 1967a) had all the essentials, applied specifically to *D. pseudoobscura*; but he neglected to publish after submitting his thesis (see Chapter 7, Box A).

Acceptance of these inferences and of the experimental demonstration was retarded by the simple-clock concept, in ways typical of assumptions long and widely accepted, but never explicitly articulated. For example, the model-independent inference of a singularity rests on demonstration of type 0 resetting (Box B of Chapter 2 and Boxes B and C of Chapter 4) but up to that time it was common to stretch type 1 resetting curves through type 0 data, even if it was necessary to run the curve through more than 18 hours without a data point: only type 1 resetting was sensible in terms of the implicit simple-clock paradigm. Even Pavlidis once misrepresented in this way the type 0 resetting behavior of his potentially revolutionary model (1968, Figure 6b; Pavlidis and Kauzmann, 1969, Figure 7) bringing it into line with the published type 1 representations of type 0 experimental data. Such is the power of an idea.

As noted above, the *Drosophila* clock's peculiar behavior in dim continuous light (Winfree, 1974a; Pittendrigh, 1976) is also implicit in this class of clock models (Winfree, 1972e; Pavlidis, 1978a) but poses paradoxical difficulties in terms of the simple-clock assumption.

2. Amplitude Resetting Accompanies Phase Resetting.   Finally, there is the question of clock behavior after a prior rephasing stimulus. According to the simple clock idea, implicit in the definition of phase shift for one circadian clocks symposium (Aschoff, 1965b, p. xii: "Phase shift: a single displacement of an oscillation along the time axis") the clock's reaction to a second stimulus should exactly follow the same phase-specific schedule as before. This is the prediction of any model in which the circadian pacemaker, whatever its concrete nature, is limited to a cycle of fixed amplitude.

The assumption of a constant amplitude inherent in Pittendrigh's method of dealing with two successive phase-shifting stimuli, though reasonably successful with large stimuli eliciting small phase shifts, has been questioned and put to experimental tests by Johnsson et al. (1973). Their experiments using *Kalanchoe* (next chapter) did not give a conclusive result. However, experiments I carried out simultaneously using *D. pseudoobscura* did (Chapter 7). I found that the answer depends on the stimulus magnitude in the way expected of a non-simple-clock mechanism:

1.   Saturating stimuli take the oscillator from one phase to another on the cycle.

2.   More delicate perturbations reveal the presence of states "inside" the standard cycle of phases (Winfree, 1973a). These smaller stimuli, still causing phase shifts of all sizes but not saturating the response, reveal the continuum of alternative cycles that so distinguish the fruitfly's circadian clock from the earlier simple-clock models. Following such a stimulus, the eclosion rhythm looks entirely normal, only being displaced along the time axis as in the definition above. But its reaction to a stimulus offered up to two days later (i.e., the sensitivity rhythm) is systematically different in a way that makes sense in terms of a cycle of altered amplitude.

C: Other *Diptera*

How Different Is *D. Melanogaster*?

D. melanogaster, the geneticist's pet since Morgan adopted it around 1905, was introduced into the circadian rhythm business by Kalmus in 1935, largely for studies of the temperature dependence of the eclosion rhythm. From these experiments, Kalmus first introduced the idea of an endogenous oscillator as the cause of rhythmic behavior in this fly. Bunning (1935) discovered that although continuous light suppresses rhythmicity, rhythmicity can be reinitiated by simply turning off the lights, even after 14 arrhythmic generations have passed. Kalmus (1940) established the arrhythmicity of dark-reared cultures, and that a single one-minute light pulse suffices to elicit persistent 24-hour rhythmicity. Brett (1955) showed that all larval and pupal stages are susceptible to synchronization by light. Brett was also the first to use cohorts (populations of narrowly defined age) to investigate the development of clock properties.

Interest in *D. melanogaster* waned soon after Pittendrigh (1954) introduced *D. pseudoobscura*, which responds to light pulses with a much larger phase shift [though not because it is any more sensitive to light: The response saturates at about the same energy in both species (Zimmerman and Goldsmith, 1971; Winfree and Gordon, 1977)]. This diminished *melanogaster's* attractiveness during years of emphasis on phase control and entrainment. However, in 1971, during a renaissance of interest in clock genetics, Konopka and Benzer produced *melanogaster* mutants of long and short period. The "per^s" mutant has the advantage that its rephasing is "disinhibited" relative to the wild type and is in fact scarcely distinguishable from the rephasing pattern of *D. pseudoobscura* except that its period is compressed to 19 hours (see Chapter 2, Example 6, Box C, and Figure 18). Otherwise, the most conspicuous difference is a mere technical annoyance: freshly emerged flies, hanging by their feet in the dark, do not fall down before their wings inflate. Having built a light-weight plastic eclosion monitor without moving parts, I had no option to revert to banging to dislodge them. Fortunately, it turned out that a puff of carbon dioxide puts them to sleep without affecting the unenclosed pupae, so the eclosion monitor was "air-conditioned" with carbon dioxide, regulated for temperature and humidity, entering and being blown out at half-hour

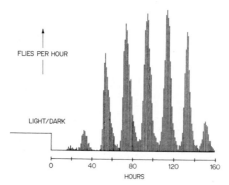

Figure 9. Eclosion rhythms in populations of *Drosophila melanogaster*'s 19-hour mutant, *per^s* resemble those of *Drosophila pseudoobscura* (Figure 2). This superficial similarity extends to the ways in which the *timing* of these rhythms depends on the timing of light pulses.

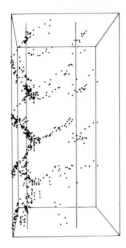 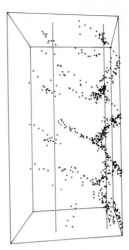

Figure 10. Stereographic views of 622 measurements of eclosion time in *Drosophila melanogaster* (19^h mutant) following a brief light pulse. (For viewing, see Chapter 12, Box A.) Coordinates and perspective as in Figure 18 of Chapter 2 but without reduplication of data points from $T < 19$ to $T + 19$. The two interior uprights depict the presumed singularities, poles around which data points climb in corkscrew fashion. (Data from Winfree and Gordon, 1977).

intervals. With this modification, the eclosion assay works almost indistinguishably for the two species (Figures 2 and 9). Phase-resetting results are also much the same, as Figure 18 of Chapter 2 showed. Figure 10 presents the same data in stereo for those who will take the trouble to locate a hand mirror. Note that no time-dependent distortion of the energy scale is conspicuous in this case, perhaps because *D. melanogaster's* clock dark-adapts very much more quickly or slowly than does *D. pseudoobscura's*.

How Different Is the Flesh Fly?

The principles of eclosion timing and of its regulation by a circadian clock are quite similar in the fly *Sarcophaga*, whose maggot begins life not in a potato meal mush but, as its name suggests, in decaying meat. Saunders first cultivated this fly as a host for the parasitic wasp *Nasonia* during his ingenious studies of the wasp's seasonal diapause (hibernation). It turned out that the flesh fly has a circadian clock of great interest in itself, quite similar to the eclosion clocks of fruitflies (see Chapter 2, Example 7). The salient differences reported by Saunders (1976, 1978) are:

1. Light pulses are administered only in the larval stage because the pupal stage is insensitive to light; even so, three orders of magnitude more light is required to elicit phase resetting of a magnitude similar to *Drosophila's*.

2. *Sarcophaga* ecloses three weeks after experimental interventions are completed and the pupae are put away to metamorphose. So transients are not observed and do not complicate the data analysis.

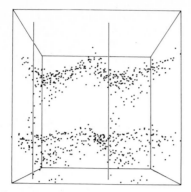 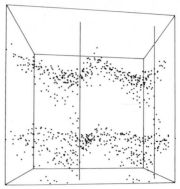

Figure 11.   Stereographic views of 188 measurements (courtesy of D. Saunders) of eclosion timing in *Sarcophaga* after a light pulse of some hours duration. (For viewing, see Chapter 12, Box A.) Coordinates and perspective as in Figure 19 of Chapter 2. The two interior uprights depict the presumed singularities, poles around which the data points climb in corkscrew fashion. All data are double-plotted both to the right and top-to-bottom (i.e., you see four identical copies of a unit cell).

3.   *Sarcophaga* pupae seem to interact to stimulate each other at eclosion time, as do silkworm pupae (Truman, 1972) and quail eggs (Vince et al., 1971). Saunders minimizes this effect by housing each animal separately in the eclosion monitor.

4.   Like the fruitfly's, the fleshfly's clock changes sensitivity, but apparently more with developmental age than with time in the dark. Figure 11 here corresponds to Figure 19 of Chapter 2. As with the other stereo pictures, the left eye's view is identical in perspective to the picture in Chapter 2.

21. The Flower of *Kalanchoe*

The tiny red flowers of *Kalanchoe blossfeldiana* open and close at 23-hour intervals. They do this for a week even when plucked from the plant and placed in a vial of sugar water under constant green light, at a constant temperature. Though blind to green, the flower's clock is sensitive to red light. By exposing the flower to red light of intensity several watts per square meter for minutes to hours, one disrupts the normal rhythmicity. In most cases, it recovers sufficiently within four days so that a phase shift can be measured.

A: Type 0 Resetting

The first systematic experiments of this sort were conducted by Zimmer (1962). Her results show what I take to be type 0, or "strong", resetting in response to a two-hour light pulse. The original publication draws a type 1 curve through the data (Figure 1) by inserting an 18-hour data-free discontinuity through the phase measurements. This seems to me an exaggerated concession to the theoretical prejudice that, in connection with phase shift plots, what goes up must come down. If the data must be construed as type 1 resetting, then there must be an extremely steep part of the curve and one would expect more variance of the measured phase shift near this point. As this is not observed, it seems to me that the phase jump probably doesn't exist and that the data fit better in the pattern of type 0 resetting.

Phase shifts can also be measured substantially in advance of the petal rhythm's settling down to a steady cycle (which, in fact, it never does in Figure 2). This is because of the fortunate circumstance that, unlike many circadian rhythms such as the eclosion rhythms of Chapter 20, *Kalanchoe*'s petal movement rhythm can be monitored almost continuously. Instead of having to wait for an event to signal passage of phase 0, at every moment we have at least two measurements in hand, namely, flower openness and the rate of change of that measurement. Let's call these position P and velocity V. If the mechanism of petal movement involves no more than two important variables then by measuring any two functions of state we should be able to distinguish the system's state from nearby other states. This should suffice to distinguish the latent phase (Chapter 6) of the rhythm while still far from an attracting cycle.

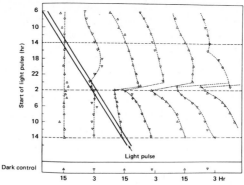

Figure 1.   An early resetting experiment using *Kalanchoe*'s petal movement rhythm. Alternating maxima and minima of openness are plotted to the right following a light pulse given at the time indicated along the diagonal slash. Prior to the stimulus (left), all flowers were synchronous. After stimulus and transients, these rhythms fall into the type 0 pattern, paralleling the stimulus rather than the controls. Zimmer's dotted curve, however, adheres to the type 1 pattern, progressively distorting a parallel to the controls (the prestimulus pattern). From Zimmer 1962, Abb. 2.

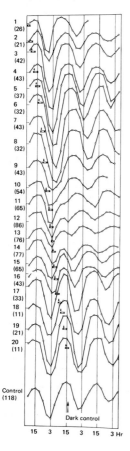

Figure 2.   The raw data from which the previous figures was derived (Zimmer, 1962, Abb. 1). The triangles and inverted triangles of Figure 1 mark the times of maxima and minima, respectively, on these curves. Note that the stimulus resets the amplitude of the rhythm along with its phase.

To test this notion, I plotted position vs. velocity for Zimmer's (1962) 13 cycles of unperturbed control rhythm (her curves 1-20 up to the light pulse, and Control curve, all in Figure 2). Much as in Chapter 7, each such plot constitutes a trajectory winding part way around the origin, or $2\frac{1}{2}$ times around in the case of Control. From the common start at "light out", I proceeded forward along each winding trajectory, writing down the present phase at three-hour intervals, using integers 1–8 repeatedly in a 24-hour cycle. (The rhythm's period is closer to 23^h, but 3×8 is close enough for present purposes.) All these numbers turn out to make a reasonably coherent pattern (Figure 3). For example, the region of the (position, velocity) plane extending upward in a narrow wedge from the origin turns out to be occupied by measurements taken from flowers at phase three-eighths of a cycle (plus whole cycles) after the beginning of darkness. These are the points at which the position is almost a local maximum (flower fully open) and its rate of change is near 0. The region extending to the left in the narrow wedge from the origin turns out to be occupied by measurements taken from flowers at phases five-eighths of a cycle (plus whole cycles) after the beginning of darkness. These are phases at which leaf position is neutral and its rate of decrease is almost a local maximum.

By plotting position vs. velocity at any moment, one can operationally define the phase of that flower in terms of unperturbed controls. If the (position, velocity)

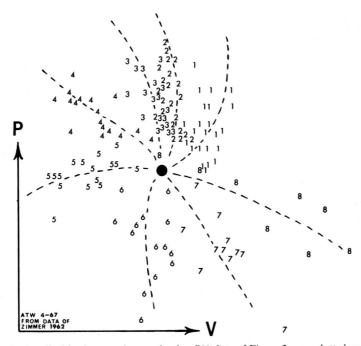

Figure 3. As described in the text, the prestimulus $P(t)$ data of Figure 2 were plotted against their own rates of change, $v(t) = P(t + 3 \text{ hours}) - P(t)$ at intervals of three hours. Successive (P, V) points were numbered 1, 2, 3, ... from the final dusk. The digits outline isochrons (dashed) by which a phase can be assigned to any (P, V) state without waiting for transients to subside.

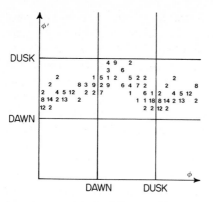

Figure 4. Each poststimulus datum in Figure 2 is plotted on Figure 3 to assign it a phase and thus to estimate the new phase some hours earlier when the stimulus ended. That new phase, ϕ', is here plotted (vertically) against the old phase, ϕ, of the prestimulus rhythm, when the stimulus began (horizontally). The digits record the number of such measurements giving each (new phase, old phase) pairing. Zeros are implicit where no digit appears. The measurements fall in a type 0 pattern.

plane is some kind of distorted image of the clock's state space as argued in Chapter 7, then the numbers 1–8 must correspond to equispaced latent phases. The clouds of numbers must be images of isochrons. So by plotting a perturbed rhythm in the same way, we should have a measure of latent phase from each successive data point. These numbers should increase consistently, i.e., by 1 every three hours after the stimulus. Zimmer's data permit a test of this method. She provides about 40 cycles of raw data from 20 resetting experiments, each following a two-hour red light pulse given at a different initial phase. These were digitized and individually used to evaluate latent phase according to the numbered regions in the (position, velocity) plane. The first three data (the first nine hours) after each light pulse gave inconsistent latent phase readings, suggesting that more than 2 degrees of freedom were involved in these early transients. The remaining measurements came out quite consistently along a type 0 curve (Figure 4). This is nothing exceptional. A type 0 curve is obtained by simply following the maxima on Figure 2, which run roughly parallel to the stimulus diagonal rather than running vertically, parallel to the control maxima.

However by using all the laboriously collected data instead of using only the times of maxima, it gives a clearer impression of the variability of measured phase. This makes it plainer that there is no phase of particular variability, as would be expected near a steep segment of the resetting curve. My inference is again that Zimmer's measurements outline type 0 resetting rather than type 1 resetting with a near-discontinuity in the curve, as it was depicted in Figure 1 and by subsequent authors (e.g., Aschoff, 1965, pp. 98, 109, and 279).

B: Resetting Data at Many Stimulus Magnitudes

The implication of type 0 resetting is that the complete resetting surface must contain a discontinuity. (Having exorcised one discontinuity, we have conjured forth another!) The most localized form it could take would be a phase singularity. *Kalanchoe*'s resetting surface was measured by Engelmann et al. (1973), repeating the format of the *Drosophila* experiments which has first shown the complete helicoidal resetting pattern of a circadian rhythm (Winfree, 1968). They published their data most completely in terms of phase shifts (their Figure 3); only a small subset of their data are given in the more direct stimulus reference format in their Figure 5. I replotted it all in the latter format for Figure 20 of Chapter 2. How were these figures obtained from those data points? The idea is to plot vertically the observed times of greatest flower openness, so the first step is to reconstruct the original observations from the reported phase shifts. The phase shift, by convention, is the difference in hours between the time of greatest opening in the experimental rhythm and in the nearest such time in an undisturbed control. Since the latter are not explicitly reported, it is in principle impossible to reconstruct observations from phase shifts. However, maxima normally occur at 23-hour intervals in the control rhythm, at about hours 20, 43, 66, etc., after transfer from light into the darkness. Thus a phase shift reported as a seven-hour delay indicates maximum openness at hours 27, 50, or 73 (etc.) after transfer to the dark, i.e., at hour $27 - (T + M)$, $50 - (T + M)$, $73 - (T + M)$ (etc.), after the end of a light pulse which began at hour T after transfer into the dark, and lasted M hours. But which one of those times is right? Their Figure 3 collects data taken in cycles 3 and 4 after the light pulse ends. Not knowing which is which, I plotted half the points, taken at random, in cycle 3, i.e., in hours 46 through $46 + 23 = 69$ after the light pulse ends. The remaining points were plotted in cycle 4, that is in hours 69 through 93. Engelmann also generously provided me with unpublished plots of data taken in cycle 1 and in cycle 2, also in phase shift format. These are accordingly used to reconstruct the original observations in the same way, namely: θ (hours after stimulus end) $= 20 +$ delay $(0–12$ hours$)$ or $-$ advance $(0–12$ hours$) - (T + M) + 23N$.

N is an integer chosen to reconstruct θ in the appropriate range 0–23, 23–46, 46–69, or 69–92 hours after the light pulse ends.

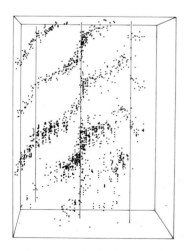 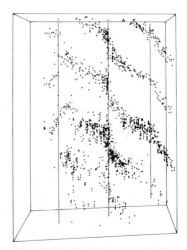

Figure 5. Stereographic views of 1,536 measurements of the timing of *Kalanchoe* flower opening after a light pulse. (For viewing, see Chapter 12, Box A.) Coordinates and perspective as in Figure 20 of Chapter 2. The two interior uprights, and one exterior on the left, depict the presumed singularities around which data climb in corkscrew fashion. θ values in the bottom two cycles are randomly altered by ± 23 hours relative to assignments in Figure 20 of Chapter 2.

These reconstructed data were plotted repeatedly, as seen from various per-spectives, until I could see the shape of a surface that approximates the 1,536 data points. They are shown in Figure 5 and in Figure 20 of Chapter 2 from the viewpoint of an observer at $T = 20$ hours, $M = -15$ hours, $\theta = 20$ hours. The box around the data consists of planes $T = 0$ ($=$ transition into darkness) and $T = 69$ hours ($=$ three cycles later), $\theta = 0$ (end of light pulse, beginning of un-disturbed free run), and $\theta = 92$ ($=$ four cycles later), $M = $ (undisturbed controls), and $M = 3$ hours.

Within this box the idealized surface was constructed as follows:

1.   At $M = 0$, $T + \theta = 20$ hours plus multiples of 23 hours. These are the undisturbed controls. These loci are drawn as diagonals on the front wall in Figure 20 of Chapter 2.

2.   At $T = 0$, the flower has been in the light for 12 hours and the phase of its rhythm in subsequent darkness is approximately set by the time of transfer from light to dark: $T + \theta = 20$ hours plus multiples of 23 hours *independent of M*. These loci are plotted as horizontals on the left wall.

3.   If the clock is really periodic, then item 2 above is repeated at 23-hour intervals of T. These loci are the horizontals in the remaining unit cells of this 3×4 lattice. [Actually, *Kalanchoe*'s clock is not quite periodic: Like *Drosophila* p.'s, it grows more photosensitive with time in the dark. But that doesn't matter for locating a phase at which phase (if not necessarily amplitude) is utterly insen-sitive to light.]

4.   The data appear to wind around two vertical screw axes. If this plant's clock behaves like the only other circadian clocks for which we presently have such complete resetting data (the three kinds of fly, Chapter 20), then we might expect:

A.   Near the screw axis $T = T^*$, $M = M^*$, this surface resembles a spiral staircase.

B. T^* is about one-quarter cycle and M^* just measures the organism's sen-sitivity to the light used, which might increase as the organism dark adapts. In *Kalanchoe* using red light of 2.3 watts/m², Engelmann et al. find $T^* = 7$ hours plus multiples of 23 hours. At three successive T^*'s, they find $M^* = 4$ hours, 2 hours, and $1\frac{1}{2}$ hours.

C.   At T^*, a light of duration $M < M^*$ has little effect on phase and a light of duration $M > M^*$ hours inverts phase. This situates the spiral staircase verti-cally: Radii extend at fixed T and θ to the $M = 0$ controls on the front surface and to the large-M resetting curve on the back surface.

5.   These features can be visualized in terms of the contour map of θ above the (T, M) plane (Figure 6). It seems that features A, B, C can be linked smoothly by joining the contour lines as shown in dashes. The linking surfaces are then sketched on the three-dimensional plot. This part was done on tracing paper over the data, trying both to follow the data and to make the surface as smooth as pos-sible and as nearly as possible the same in each vertical repeat.

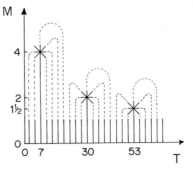

Figure 6.   A sketch of text items 4A, B, C in terms of the level contours of θ (or of ϕ') on the stimulus plane. The dashed segments constitute one possible way to smoothly join the described features.

C: A Phase Singularity

Figure 5 and Figure 20 of Chapter 2 show these results plotted in the three-dimensional time crystal format used elsewhere in this book. The data cloud looks like a screw with a phase singularity near two hours exposure, beginning about $1\frac{1}{4}$ cycles after transfer of the flowers to constant darkness and again at 23-hour intervals later (Engelmann et al., 1978). As in *Drosophila*'s circadian clock, photosensitivity increases in the dark: The helicoid in cycle 1 has its singularity one-quarter cycle after the light-to-dark transition, but at twice the exposure duration needed in cycle 2. There is little further change in cycle 3.

Engelmann et al. (1978) show that rhythmicity of petal movement is commonly lost after near-singular exposure. This exposure inflicts a discontinuity on the phase response possibly because it extinguishes the oscillation until such time as a second stimulus reawakens it.

In some of these experiments there is a substantial dispersion of phase among the 16 individual flowers collectively monitored for rhythmicity. Engelmann et al. (1973) note that the singular stimulus magnitude M^* might be underestimated from such population data. In fact the apparent type 0 resetting, and so the whole helicoid, could be artifacts of phase scatter within a small population of flowers each of which exhibits only type 1 resetting (Johnsson et al., 1973, p. 158; Winfree, 1976). However, Engelmann et al. (1978) verify type 0 resetting on individual flowers, and M^* in individual flowers turns out to be nearly the same as in populations.

Engelmann et al. (1974) obtained quite similar results in less extensive experiments replacing the red light by exposure to 30°C. The point here, I think, is that these phenomena reflect the basic dynamic structure of the circadian clock and have little to do with the particular mode by which its functioning is disturbed.

D: Arrhythmicity Not an Artifact of Populations

What happens at the screw axis? This singular stimulus is found to annihilate rhythmicity in each *Kalanchoe* flower (Englemann et al., 1978). Does it do so by arresting rhythmicity in each cell of the flower or by merely inducing incoherent rhythmicity in the population of cells whose collective behavior comprises the gross movement of the petals? This latter possibility is suggested by the unlimited steepness of the resetting surface at its singularity. The slightest variation of initial phase or of photosensitivity would seem sufficient to cause large variations in the new phase of a cellular clock.

In the format Zimmerman and I used for fruitfly pupae (Chapter 20, Section B), Engelmann et al. (1978) pose the question of incoherence to the cells of a single flower in which the petal-opening rhythm had been annihilated by singular perturbation. A type 1 resetting stimulus, S*/4, was then given, which should have only slightly synchronized cell clocks if each were independently rhythmic on the normal cycle. As in the *Drosophila* experiments, high-amplitude rhythmicity was immediately evoked. If the rhythm in the whole flower is a mere superposition of cell rhythms of turgor, then this result is more readily interpreted as displacement from a homogeneous equilibrium state than as resynchronization from scattered phases on the usual cycle. Bunning (1959, p. 523) was the first to discover (in a plant) that the rhythm evoked by a single stimulus derives not from synchronization of preexisting rhythms in single cells, but from initiating rhythmicity in each cell; the same seems to apply to the *Kalanchoe* flower returned by a prior singular stimulus to an arrhythmic condition.

E: Amplitude Resetting

The range of petal movement in *Kalanchoe* flowers provides a more direct measure of amplitude than the laboriously acquired measure based on resetting curves, applied using *Drosophila* in Chapter 7. This range measure can be used to plot contours of reset amplitude just as we have often plotted contours of reset phase on the (T, M) stimulus plane. Engelmann et al. (1978) show that amplitude falls off by concentric rings around the singularity. In this respect also, *Kalanchoe*'s pattern resembles *Drosophila*'s.

22.   The Cell Mitotic Cycle

Rules of Reasoning in Philosophy. Rule II: . . . We are certainly not to relinquish the evidence of experiments for the sake of dreams and vain fictions of our own devising; nor are we to recede from the analogy of Nature, which is wont to be simple, and always consonant to itself.

Isaac Newton, *The System of the World*

In no case is the process well understood whereby the growing cell "decides" to replicate its genome, segregate its chromosomes into two nuclei, and wall them off from each other by cell fission. Such a fundamental biological process presumably has some universal aspects. Its appeal as an object of investigation is further enhanced by the seductive belief that the mechanism of replication is constrained to some kind of simplicity by these facts:

1.   The end state is close to the initial state (namely, a freshly divided cell); and

2.   The whole process repeats at fixed intervals of time, at least in certain kinds of cells in optimal growth conditions.

A brief foray into this intriguing puzzle is included here for three reasons:

A.   At least three influential notions about clock-like dynamics arose first in this context and later found application in studies of circadian rhythms (Chapters 19–21).

B.   There is a close but poorly understood causal link between circadian rhythms and the mitotic cycle.

C.   Techniques of analysis evolved in circadian experiments later stimulated parallel enquiries into the cell cycle, (mostly by Kauffman et al.: see below), with results that bear upon our exploration of phase singularities (Chapters 2 and 10).

I devote a few pages to each of these three topics. Section D goes into a little more detail about one specific experimental system, the true slime mold *Physarum*. Like the three other members of kingdom fungi celebrated in this Bestiary (yeasts in Chapter 12, cellular slime molds in Chapter 15, colonial ascomycetes in Chapter 18), the true slime mold is conspicuously periodic in its biochemical habits. In this case a clock-like regularity of mitosis is most prominently featured. With this periodism comes the by now familiar crisis of circular logic. It is resolved in this case not by a phase singularity but by a discontinuity that appears to reveal an honest physiological cataclysm.

A: Three Basic Concepts and Some Models

A Sequence of States

The earliest inquiries into the cell cycle provide a paradigm that seems to pervade much of the circadian literature, though only in recent years has it become explicit and articulate. This paradigm is best summarized graphically in the diagrams used by people professionally involved in unraveling the mechanisms of rhythmic mitosis (Figures 1 and 2). As generally envisioned, the cell cycle is a series of discrete steps or states that a cell must execute in sequence if it is to replicate normally. Pictured abstractly, these steps constitute a one-way cycle, a ring device, though possibly adorned with baroque appendages. There may be loops in the cycle, or alternate pathways here and there. There may be stopping places at which the cell is unable to go on to the next step until some permissive signal is

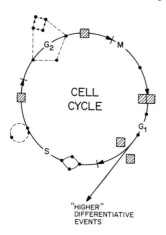

Figure 1.   Diagram of a generalized cell cycle. The four classical phases are shown: G_1 (gap 1), S (synthetic), G_2 (gap 2), and M (mitotic). In addition, a number of arbitrarily chosen points are designated by dots in order to illustrate possible alternative pathways, branched network, loops, and blockage points. Any precise resemblance to a particular cell is purely coincidental. (From Edmunds, 1974, Figure 1 with permission.)

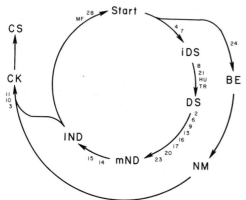

Figure 2.   The circuitry of the yeast cell cycle. Events connected by an arrow are proposed to be related such that the distal event is dependent for its occurrence upon the prior completion of the proximal event. Numbers refer to *cdc* genes that are required for progress from one event to the next; HU and TR refer to the DNA synthesis inhibitors hydroxyurea and trenimon, respectively; MF refers to the mating factor, α factor. (From Hartwell et al, 1974, copyright 1974 by the American Association for the Advancement of Science.

given. But the central point for my purpose here is the intrinsically one-dimensional character of the process. The dynamics more resemble driving a car on roadways than piloting a sailboat or airplane in continua with more degrees of freedom.

It is this vision of the cell cycle that gave rise to the concept of a simple clock (Campbell, 1964, and Chapter 3), and, by analogy, to much thinking about circadian clocks, and to my experimental efforts to ask whether a circadian clock is a simple clock.

Phase-Specific Arrest Within the Cycle

If the cell cycle consists of a sequence of processes, some of which are necessary and sufficient causes of others, then the first job in preparing a "trouble shooter's manual of the cell cycle" is to construct a flow chart of these cause-and-effect connections, similar to a flow chart of metabolic pathways (Figures 1 and 2). With the normal cycle thus verbally divided up into stages, it would be illuminating to seek mutants incapable of each process named, just as biochemical geneticists unraveled the intricate contingencies that constitute the manufacture of small molecules in intermediary metabolism by finding out which subsequent events are also arrested in such a mutant.

Hartwell et al. (1974) discovered in this way that the cell cycle in yeast consists of at least two parallel sequences of events, one of which can proceed, even repetitiously, while the other is blocked. One involves DNA synthesis and nuclear division, and the other involves budding and cytoplasmic division.

Rephasing as a Quantitative Test

Systematic phase-resetting experiments to probe the mechanism underlying a physiological rhythm made their first appearance in the cell cycle literature. The protocol was typically as follows:

a. Synchronize mitoses by selecting cells that are a certain size or sticky in a certain range of phases, or by synchronous release of a population from arrest by an inhibitor, or by fusion of randomly phased plasmodia.

b. At some initial phase ϕ (measured as hours elapsed since mitosis), interfere with normal operation of the cell cycle by changing the temperature, introducing a poison, or whatever.

c. Continue the stimulus for a time M (for magnitude of disturbance) and then return the cells to standard culture conditions.

d. Observe the setback or excess delay of mitosis time beyond what would be expected if the cell cycle were simply paralysed during treatment time M. Excess delay D is defined as the actual period P (measured in units of one normal cycle) from the last mitosis before perturbation to the first mitosis after perturbation minus the usual period between mitoses (which we take for our unit of time) minus stimulus duration M:

$$D = P - 1 - M = P - M \text{ (modulo 1 period)}.$$

perpendicular one-half cycle discontinuities as in Chapter 1, Figure 24. The data are sometimes irregular but there is no clear hint of a phase singularity (1c) penetrating the $v = \frac{1}{2}$ plane.

More recently Kauffman et al. (see below) reexamined the question with particular alertness to evidence of attractor cycle kinetics in the mitotic cycle. They pursued the two main classes of experiment that have become traditional in this field:

1. Two plasmodia at different phases are fused, resulting in a phase compromise as their two mitotic rhythms synchronize to one. The results are examined for evidence of critical phases, at which either cell's physiology abruptly changes, or for a combination of phases such that the compromise phase is unpredictable in the ways anticipated from the attractor cycle conjecture (Chapter 6) (Kauffman, 1974; Tyson and Kauffman, 1975; Kauffman and Wille, 1975).

2. A physiological shock, such as incubation at 37°C, retards or advances the next mitosis. Clues are sought in the amount and direction of phase shift as a function of the initial phase when the shock is administered (Wille et al., 1977).

Both experiments are potentially capable of revealing a phase singularity if one exists, and both are capable of demonstrating type 0 resetting.

However, at this writing neither phenomenon has emerged from *Physarum*. [I here spare the reader by deleting most of what I prepared for this chapter, viz an extensive examination of the data obtained in those experiments, and the interpretations given them. A sufficient part of this has appeared independently in Tyson and Sachsenmaier (1978), who came to the same conclusion.] In fact, those of the fusion data from Kauffman and Wille (1975) in which thorough mixing and synchrony are reported do fill out the 1972 picture more thoroughly and quite consistently. Kauffman and Wille (1975) conclude that mitosis in *Physarum* behaves more like a relaxation oscillator than does their model.

In this circumstance I feel compelled for the present to abide by the first of Newton's two exhortations at the opening of this chapter, despite the attractiveness of theoretical schemes involving phase singularities. At the same time it must be remembered that *Physarum* fusion experiments can never resolve events more closely than the 45 minutes required for plasmodia to fuse. Any number of singularities might lurk unresolved in the hour of ostensible discontinuity preceding mitosis.

Who Cares?

Knowledge of cell dynamics has obvious practical importance and obviously involves considerations of circular logic akin to those outlined in Chapter 1. I remember with pleasure recurrent discussions on this fact with Graeme Mitchison in 1971, continuing hours after tables had been cleared in the Cambridge MRC Lab cafeteria. We wondered whether phase singularities could reasonably be pursued in this matter, and decided against it. But Stuart Kauffman decided on a more aggressive approach to the question, taking it to the master, *Physarum*. If

Kauffman and friends have accurately interpreted *Physarum*'s answer, then cell division is regulated by a continuous attractor cycle oscillation which operates independently of nuclear division itself. Such a discovery would belong in a conspicuous place in the main text of this volume as perhaps the main fruit of the search for phase singularities. I have chosen not to put it that much in the spotlight because I am apprehensive that *Physarum*'s behavior may be very much more complicated than *any* of the simplistic models here entertained. But more data may come and I do recommend the matter to your attention, as the outcome may be important.

23. The Female Cycle

Depend upon it: there is nothing so unnatural as the commonplace.

Sherlock Holmes,
The Case of Identity

A: Women, Hormones, and Eggs

Monthly bleeding may have been commonplace among nuns, spinsters, and the infertile centuries ago, but it could hardly have been common among the women to whose uteri we all owe our existence. When they were not pregnant, their breast feeding encouraged lactation, which suppressed ovarian cycling. Short (1976) estimates that it may have been uncommon to experience three consecutive menstrual cycles in a lifetime under these conditions. Accordingly, the female endocrine system's menstrual cycle has not been subjected to selection pressure for its clock-like attributes. In fact, there are diverse clues that some fraction of women are reflex ovulators, not spontaneous cyclers at all (Clark and Zarrow, 1971).[1] In a reflex ovulator, mature follicles await rupture by a surge of hormone which is elicited only by sexual stimulation. The ovum then starts its journey down the fallopian tube, and pregnancy (or, less likely, recycling) ensues.

One of the stranger byproducts of improved survival of infants and of western sexual mores is that lactation is discouraged in industrialized societies. Consequently, the female endocrine system nowadays commonly functions in a periodic mode. It cycles only because after an ovum works its way into the well-prepared uterus without finding a sperm, there is little alternative but to start over with the next follicle.

The one baffling mystery in this glib story surrounds the fact that the next-ripest follicles are so immature that they will take weeks to ripen, thus ensuring a definite minimum interval between ovulations. No one seems to have a clue as to the mechanism that selects a single ovum out of a teeming ovary and encourages its growth for a full month. In fact, a multiplicity of candidate follicles are initially

---

[1] Rats too (Taleisnik et al., 1966; Moss et al., 1977).

Box A: The Mechanisms of the Normal Human Female Cycle

The chief actors in this drama are two glands, the ovary and the anterior pituitary. Neither alone is intrinsically cyclic in its activity. Corresponding to the two glands are two pairs of hormones.

First, there are the two glycoprotein gonadotropins, so called because they affect the gonad (the ovary). These are luternizing hormone (LH) and follicle-stimulating hormone (FSH). These are secreted by specific cells in the anterior pituitary in response to hormonal releasing factors from the hypothalamus.

Then there are the two classes of steroid sex hormones, aimed at the anterior pituitary (by way of the hypothalamus) and at the uterus (see below). These are estrogenic steroids (E) and progestational steroids (P), all synthesized in the gonad (the ovary) and secreted.

The principle gland of the ovary is the follicle, of which there are many, all being matured under the stimulation of gonadotropic hormones to secrete estrogen. Some mature as far as to release an egg then secrete mixed estrogen and progesterone. The first step of the cycle, when the follicles are only making estrogen, is called the *follicular* phase. After one of them ovulates, it turns into a yellowish body called the corpus luteum which, in addition to estrogen, synthesizes and releases progesterone. This postovulatory half of the cycle is therefore called the *luteal* phase. These ovarian steroid secretions in turn control the pattern of pituitary gonadotropin release both directly and by way of the hypothalamus, thus closing the causal cycle. In humans, the corpus luteum has a built-in lifetime of about 14 days, toward the end of which its progesterone and estrogen outputs decline. This disinhibits pituitary gonadotropin release with the result that a new follicle begins its sequence of estrogen production, rupture, and combined estrogen and progesterone secretion, while its predecessor degenerates.

Not linked in this closed chain of cause and effect, the lining of the uterus proliferates under the sequential stimulation first of estrogens and then of progesterone. The luteal progesterone says, "The follicle has ruptured, prepare for implantation". When the corpus luteum begins to fade after two weeks, it no longer provides enough progesterone to support the uterine decidum, which thus earns its name by sloughing off. This marker event is the onset of monthly flow, conventionally indicating day one of the cycle. Because it is affected by, but does not affect the hormonal interactions, it is not properly a part of the clock-like mechanism. (According to another convention of labeling, menses start not on day 1 but on day 14, ovulation taking the preferred position of day 1 in the cycle.)

Women differ from animals with a breeding season in that gonadotropins are secreted all year round in amounts adequate to encourage follicles to develop. In seasonal breeders, gonadotropin levels are kept below a minimal level, thus inhibiting cycling, except when the photoperiod is right (and the photoperiod is monitored by a circadian clock). Most women also differ from reflex ovulators in that gonadotropin secretion reaches a peak when the follicle is mature, causing it to release its ovum. In a reflex ovulator, in contrast, adequate LH output and rupture of the follicle occurs only upon command to the hypothalamus by stimuli associated with mating. In rodents that cycle spontaneously this stimulus to the hypothalamus and thence to the anterior pituitary may be provided daily by an internal circadian clock in or mediated by the suprachiasmatic nucleus. In humans the hypothalamic stimulus seems independent of circadian factors and independent of sexual stimulation.

> The sketch provided above is obviously misleading in its oversimplification. The female cycle, like many biological regulatory mechanisms, is characterized by redundancy, by multiple fail-safe devices. This makes it hard to unravel any one "mechanism". It also allows great evolutionary diversification, since many parts of the interacting machinery can mutate independently without imperiling the main function they collectively subserve. Thus, every published experiment on female cycle mechanisms must necessarily be flagged with a label: "guinea pig", "hamster", "rat", "marsupial", "ungulate", "primate", or "human".

recruited but most of them eventually wither, yielding to the one chosen. This competition may contribute to the notorious variability of the female cycle, most of which is confined to the preovulatory, follicular half of the cycle. It is not even certain that release of the egg is an essential part of the causal loop that constitutes the female cycle at all. According to one view, it is only a peripheral process governed by, but not importantly affecting, the female cycle. The evidence is basically that anovulatory cycles take about the same length of time whether ovulation occurs or somehow fails to occur. To me, such evidence speaks rather for two parallel alternative pathways of cycle completion, one with a follicular estrogen spurt and subsequent progesterone manufacture by the corpus luteum, and one without. The hormonal receptors of the female endocrine system might be indifferent to these alternatives only in the limited sense that both paths might happen to take about two weeks. Fuller understanding awaits clarification of the mechanisms of competition and selection among follicles in the ovary.

Though ovulation may or may not be a peripheral, gated process, monthly sloughing of uterine decidua definitely is. The human female cycle proceeds normally in the absence of the uterus and there is no known hormonal or neural feedback from the uterus to the endocrine glands (see Box A, outlining the normal endocrine cycle).

B: Statistics ("Am I Overdue?!")

Oblivious to our uncertainties about mechanism, the human ovary in nonpregnant, nonlactating women releases an ovum every four weeks or so. The consequence in modern society is monthly flow. So widespread and long established is this phenomenon that women *complain* of menstrual irregularities and even prefer birth control pills that artificially induce periodic bleeding (perhaps as reassurance that pregnancy has in fact been forestalled for another 28 days). How regular *is* the female cycle? It appears that successive intervals between recurrences of menstrual onset (the most obvious phase marker) are nearly independent samples from a probability distribution. The mean interval depends in a regular way on the individual age (Treloar et al., 1967). For women aged 20 to 40 the standard deviation is typically a few days (that is, about two-thirds of intervals fall within a few days of the mean and about nineteen-twentieths fall within twice that range).

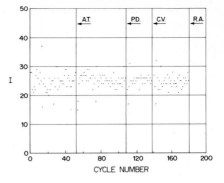

Figure 1. Vertically, intervals between 189 successive menstrual onsets, in days, recorded by DSW. Note the four long intervals caused by pregnancies. Ovulation was not regulated by "the pill" or other artificial synchronizers.

In the example shown in Figure 1 (data collected by my mother from age 18 to age 33), 189 consecutive cycles (interrupted by three pregnancies and terminated after a fourth) have a mean of 25.0 days with a standard deviation of 3.2 days, most of which is presumably (according to Presser, 1972) in the preovulatory half of each cycle. These statistics do not vary markedly between the four segments punctuated by pregnancies (Figure 2). Figure 3 plots each interval against the next. If flow onset occurred sometimes early or sometimes late in a more regular underlying cycle, or if the calendar dates were inaccurately recorded, then this cloud of dots would be elongated along a $-45°$ slope, since errors that lengthen one interval would shorten the next by the same amount. If there were long term variations of period (for example, longer periods in winter or when older), the

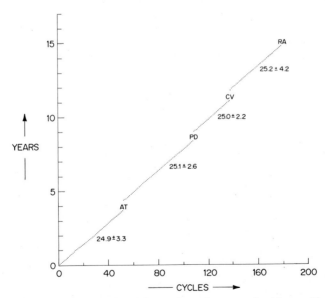

Figure 2. As in Figure 1 but the vertical scale is *cumulative* days, spanning 15 years. The mean interval and standard deviation are indicated before each pregnancy terminates that record. The four means are about one standard deviation shorter than Presser's (1972, p. 150) population mean, and exhibit about the same variability as he found (op. cit., p. 154).

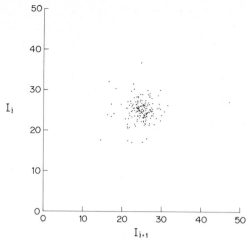

Figure 3. The successive intervals of Figure 1 are plotted each against the next (omitting the four pregnancies). The symmetry of the scatter indicates no positive or negative serial correlation of intervals.

cloud would be elongated along the $+45°$ slope, since long intervals would tend to follow long intervals, etc. Neither effect stands out conspicuously in these data. Gunn (1937) reported little correlation between consecutive cycles on the basis of much larger samples and more thorough statistical analysis.

If a hormonal disturbance caused a sudden phase shift, how noticeable would it be? How many cycles N should we wait for passage of any transients before observing the asymptotic phase shift inflicted on the central timing mechanism? Using the same data base, Figure 4 portrays the unperturbed rhythms' probability distribution of times of next flow onset, of the second flow onset, and of the third, all measured from a given onset at $T = 0$. If there were no correlations between successive intervals, a phase shift of less than about twice the standard deviation

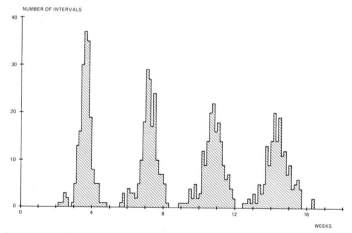

Figure 4. On a horizontal scale of weeks, the number of intervals of each duration is plotted: Peak 1, from one onset to the next, i.e., from i to $i + 1$. Peak 2, from i to $i + 2$. Peak 3, from i to $i + 3$. Peak 4, from i to $i + 4$. (Data from Figure 1.)

times the square root of N would be lost in the intrinsic variability of the period. With a standard deviation of 3.2 days and allowing two cycles in case the first period is abnormal, the minimum reliably measurable phase shift would be $(\pm 2 \times 3.2 \times \sqrt{2}) = 9$ days out of 25. Figure 4 indicates that this is indeed approximately the observed variability in the second cycle, and it gets worse after three or four cycles. This makes it difficult to work quantitatively with the female cycle's resetting curves.

C: Rephasing Schedules

Despite quantitative variability, at least one qualitatively essential feature should be discernible. We should be able to determine the female cycle's topological types of resetting. Does it achieve type 0, given a sufficiently strong perturbation? If so, then some interesting discontinuities are implicit at lower doses. The possibility even exists that the female cycle has a phaseless manifold (see Chapter 6), from which it never comes out again spontaneously, once perturbed into that range of states. Accidents of this sort could conceivably underlie some varieties of amenorrhea. Lasting amenorrhea, induced deliberately by a single fleeting perturbation, might have some advantages as a contraceptive technique. Preliminary evidence comes from limited clinical experiments, from computer simulations hopefully summarizing the best contemporary information on the hormonal mechanisms of the human menstrual cycle, and from more extensive experiments using the estrous cycles and menstrual cycles of large mammals.

First, the clinical results: Hormonal phase shifting is possible and the first full cycle after perturbation seems normal. The phase shift is a sufficiently reproducible function of phase to permit reliable entrainment to perturbations administered on a regular schedule at periods close to the normal cycle duration (Cseffalvay, 1966; Boutselis et al., 1971, 1972; Yen and Tsai, 1972; Arrata and Chatterton, 1974; Abraham et al., 1974; Dhont et al., 1974; and Shaikh and Klaiber, 1974). There are some reports that visible light also serves this purpose (Lacey, 1975; Dewan, et al., 1978). Large phase shifts are possible, as is required for type 0 resetting. For example, the large dose of estrogen commonly administered in cases of rape promptly dumps the uterine decidua and therefore presumably resets the female cycle to a standard phase; however, I have not seen data demonstrating that subsequent cycles are normal nor that the reset is nearly independent of the initial phase in the follicular and hormonal cycle.

Second, the results from reasoning through the detailed mechanism of menstrual rhythmicity: A variety of numerical models of the endocrine interactions thought to underlie the female cycle have recently appeared in the technical literature. The deficiencies of some of these models have been conspicuous and their evolution has been prompt. It is by no means clear that reality is in hand at present. However no type 0 resetting curves have yet been produced from these models. Using the model of Shack et al. (1971) and Stetz (1971), I tried progesterone doses of various sizes applied at various times throughout the cycle and obtained only small phase resets (1972, unpublished). Using the much more sophisticated model of Bogumil et al. (1972), my student Eric Best and I tested the effect of estrogen injections at

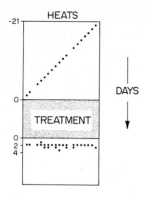

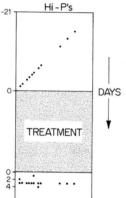

Figure 5.   Data from Cooper and Rowson, 1975, Figure 1, replotted in the format of Figures 31 through 33 of Chapter 1. Dots indicate onset of behavioral heat in heifers. The normal period length is 21 days; these experiments thus scan one cycle. Onsets occur two to three days after the standard stimulus regardless of when in the cycle it was administered.

Figure 6.   As in Figure 5 but data are from Palmer and Jousset, 1975a, Tables 2 and 3. Dots represent the beginning of high progesterone phase in mares, which also have a 21-day cycle. In both Figures 5 and 6 many of the single dots represent several duplicate experiments using different individual animals.

various times in the cycle, looking for a permanent rephasing of the subsequent endocrine rhythmicity. The data obtained (Best, 1975) showed conspicuous discontinuities as a function of the initial phase and in many cases lapsed into strange cycles that seemed to raise doubts about the realism of our implementation of the model of Bogumil et al. (1972). Subsequent simulations by Bogumil examined the effect of estrogen infusions administered at various times in the cycle. The resulting resetting curve showed three glaring discontinuities in the follicular half of the cycle and none at ovulation or in the luteal phase (Bogumil, 1976, personal communication).[2]

Third, the results from animal experiments: The best animal data come from veterinary endocrinologists trying to optimize the efficiency of artificial insemination in cattle breeding (Cooper and Rowson, 1975; Palmer and Jousset, 1975a,b). If discontinuities be disallowed a priori, then the problem of synchronizing a randomly phased herd amounts to the problem of obtaining a very flat type 0

---

[2]   Note that real discontinuities in the rephasing response do not necessarily reveal discontinuities in the oscillator mechanism. If they did, one might have to look askance on Bogumil's simulation because its discontinuities correspond to no conspicuous physiological event. For example, a smooth kinetic scheme whose attracting cycle is knotted has a phaseless set of peculiar topology in the pertinent state space. This results in suprising discontinuities of new phase over a whole range of (old phase, stimulus magnitude) combinations.

resetting curve. Resetting curve data obtained by these investigators in response to schedules of hormonal perturbation applied at various phases in the cycle are replotted in Figures 5 and 6. Close synchronization is obtained. The data look superficially like type 0 curves. However, it must be remembered that these observations terminated with a demonstration of excellent synchrony soon after the last of a sequence of injections. Confirmation of type 0 topology awaits a repeat with verification that subsequent rhythmicity is normal following injection at each phase. (It should also be noted that the estrous cycle of ungulates differs in mechanism from the human menstrual cycle.)

D: The Question of Smoothness

Another issue involved here is even more fundamental than ascertaining the reality of the type 0 resetting which is so important in theoretical constructions: Can the female cycle be thought of realistically in terms of smooth dynamics on a time scale of days, or at worst, hours? Most thinking on the subject of the female cycle employs approximate notions which though they may be appropriate for pioneering investigations of very complicated machinery, neglect considerations of continuity. Like the cell cycle (Chapter 22), the female cycle is commonly described much as people describe the operation of an internal combustion engine: as a sequence of discrete steps linked by thresholds and logical switches, all proceeding in a closed loop (e.g., Danziger and Elmergreen, 1957). In this view, those parts of state space not visited along the attracting cycle are ignored as though they did not exist. But, if they do exist, they are important when the cycle doesn't pursue its normal course. Contemporary fertility technology ensures that states off the normal cycle will indeed be traversed, just as cancer chemotherapy ensures it for the cell cycle. Nonetheless, it is possible that *this language* of discrete state transitions is not only a convenient artifice during an early stage of investigation, but also an accurate reflection of the dynamics of the female cycle. If the female cycle embodies processes with radically different time scales then intervals of continuous dynamics may be joined in a switch-like way by means of thresholds, with the result that on a time scale of days and hours we face more discontinuities than the minimum assortment required topologically. In that case many of the methods exploited in this book would be superfluous because they only serve to identify a discontinuity implicit in smooth kinetics.

E: Circadian Control of Ovulation

In contrast to the human menstrual cycle, the four- to five-day estrous cycle of some rodents has a short enough period to succumb to regulation by circadian rhythms. This may even have selective value in terms of optimal phasing for coordination of sexual activity in the day-night cycle. The principle that mammalian physiology is organized on a 24-hour basis manifests itself in this situation as a regular time of day for ovulation in such rodents as the laboratory rat, the laboratory mouse and the golden hamster. The signal that ruptures the Graffian follicle is an luteinizing hormone (LH) pulse from the pituitary, which is in turn responding to a daily signal from the suprachiasmatic nuclei (SCN) (Fitzgerald and Zucker, 1976; Moore and Eichler 1976; Stetson and Watson-Whitmyre 1976; Morin et al. 1977a; Rusak 1977). The SCN, located just in front of the crossing of left and right optic nerves, are in turn governed by the external light-dark cycle by way of the retinohypothalamic projection. But if that nerve cable is cut, the SCN still exhibit circadian periodicity, now free-running at the individual's own period near 24 hours. The estrous cycle remains entrained four-to-one or five-to-one by a circadian LH pulse: only every fourth or fifth day does the daily pulse find a follicle mature enough to rupture and produce a corpus luteum, whose secretory activity ultimately starts the next cycle.

In this situation, the female cycle's phase shifts in a predictable way. In response to hormonal perturbation, it shifts by an integer number of circadian cycles depending on when in the four- or five-day cycle the hormone is administered. Note that because the phase shift is thus quantized, the resetting "curve" is discontinuous. Topological arguments thus have no pertinence. To make any finer adjustment of phase than this, it is necessary to rephase the SCN clock, e.g., by delaying or advancing the light-dark cycle. In constant darkness, resetting by a single light pulse would also work, but results are more variable (Daan and Pittendrigh, 1976b).

Not only does the circadian clock regulate the female cycle, but also conversely, estradiol feeds back to the circadian clock to shorten its period (in hamsters) and also to alter the daily timing of its activity rhythm (Morin et al., 1977b). In birds, testosterone has recently been shown to "split" the circadian rhythm and to alter its period (Gwinner, 1974). The 1970s continue to treat us to a very exciting chapter of physiological investigation as the neural, neurosecretory and humoral interactions among the pineal gland, the SCN, and the sex glands are progressively revealed.

In contrast, no substantial role of circadian fluctuations has been established in human or other primate female cycles, despite several deliberate inquiries.

The circadian clock plays a dominant role also in the seasonal suppression and release of ovarian activity, not only in rodents but also in many other varieties of mammal, bird, and reptile as well as in the invertebrate animals.

References

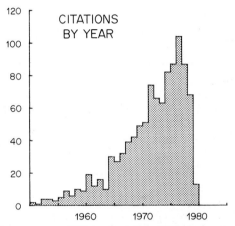

CITATIONS BY YEAR

This histogram shows the most recent 1024 citations of the bibliography, classified by year of publication. The upward trend of the graph through 1976 represents essentially complete coverage of the literature most germane to my topics, I think. How many publications of similar quality might I have overlooked during preparation of this book in 1977–1979? If the graph bears extrapolating, there might be 25 dated 1977, another 60 dated 1978, and another 140 dated 1979. You might find them by using Science Citations Index with the key references of 1975 and 1976.

Abraham, G. E., Chakmakjian, Z. H., Buster, J. E., and Marshall, J. R. (1974) The effect of exogenous conjugated estrogen on plasma gonadotropins and ovarian steroids during the menstrual cycle. *Obstet. Gynec.* **43**, 676–684.

Adkisson, P.L., and Walcott, C. (1965) Physiology of insect diapause. XV. The transmission of photoperiod signals to the brain of the oak silkworm, *Antheraea pernyi. Biol. Bull.* **128**, 497–507.

Adler, I. (1974) A model of contact pressure in phyllotaxis. *J. Theor. Biol.* **45**, 1–49.

Aizawa, Y. (1975) Perturbational approach to the phenomena of mode-locking in nonlinear systems. *Phys. Lett.* **54A**, 485–486.

Aizawa, Y. (1976) Synergetic approach to the phenomena of mode-locking in nonlinear systems. *Prog. Theor. Phys.* **56**, 703–716.

Alcantara, F., and Monk, M. (1974) Signal propagation during aggregation in the slime mould *Dictyostelium discoideum. J. Gen. Microbiol.* **85**, 321–334.

Aldridge, J. (1976) Short-range Intercellular communication, biochemical oscillations, and circadian rhythms. In *Handbook of Engineering in Medicine and Biology*, D. Fleming, ed. CRC Press, Cleveland, Ohio, pp. 55–147.

Aldridge, J., and Pavlidis, T. (1976) Clocklike behavior of biological clocks. *Nature* **259**, 343–344.

Aldridge, J., and Pye, E. K. (1976) Cell density dependence of oscillatory metabolism. *Nature* **259**, 670–671.

Aldridge J., and Pye, E. K. (1979a) Divalent cation effects on oscillatory metabolism in yeast. Unpublished.

Aldridge, J., and Pye, E. K. (1979b) Amplitude resetting in populations of strongly coupled oscillators. Unpublished.

Alexopoulos, C. J., (1952) *Introductory Mycology*, Wiley, New York.

Allessie, M. A., (1977) Circulating Excitation in the heart, Thesis, Chapter 1. Rijksuniversiteit Limburg, the Netherlands.

Allessie, M. A., Bonke, F. I. M. and Schopman, F. J. G. (1973) Circus movement in rabbit atrial muscle as a mechanism of tachycardia. *Circul. Res.* **33**, 54–62.

Allessie, M. A., Bonke, F. I. M., Schopman, F. J. G. (1976) Circus movement in rabbit atrial muscle as a mechanism of tachycardia. II. The role of non-uniform recovery of excitability of the occurrence of uni-directional block, as studied with multiple microelectrodes. *Cir. Res.* **39**, 168–177.

Allessie, M. A., Bonke, F. I. M., and Schopman, F. J. G. (1977) Circus movement in rabbit atrial muscle as a mechanism of tachycardia: III. The "leading circle" concept: a new model of circus movement in cardiac tissue without the involvement of an anatomical obstacle. *Circul. Res.* **41**, 9–18.

Alleva, J. J., Waleski, M. V., and Alleva, F. R. (1971) A biological clock controlling the estrous cycle of the hamster. *Endocrinology* **88**, 1368–1379.

Anderson, L. C., Brumberger, H., and Marchessault, R. H. (1967) Morphology of poly-L-alanine spherulites. *Nature* **216**, 52–54.

Andrews, R., and Folk, G. (1964) Circadian metabolic patterns in cultured hamster adrenal. *Comp. Biochem. Physiol.* **11**, 393–400.

Andrews, R. (1971) Circadian rhythms in adrenal organ cultures. *Gegenbaurs Morph. Jahrb. Leipzig* **117**, 89–98.

Andronov, A. A., Vitt, A. A., and Khaikin, S. E. (1966) *Theory of Oscillators*, Pergamon, Oxford.

Arechiga, H. (1977) Circadian rhythmicity in the nervous system of crustaceans. *Fed. Proc.* **36**, 2036–2041.

Aris, R. (1975) The mathematical theory of diffusion and reaction in permeable catalysts. vols. I and II. Oxford University Press (Clarendon), London and New York.

Arnold, V. I., and Avez, A. (1968) *Ergodic Problems of Classical Mechanics*, W. A. Benjamin, New York.

Aronson, D. G. and Weinberger, H. F. (1975) Nonlinear diffusion in population genetics, combustion and nerve propagation, in *Proceedings of the Tulane Program in P.D.E. and Related Topics*, Lecture Notes in Mathematics, 446, 5–49; ed. J. Goldstein, Springer-Verlag, New York, 1975.

Arrata, W. S. M., and Chatterton, R. T. (1974) Effect of prostaglandin F-2 alpha on the luteal phase of the cycle in non-pregnant women. *Amer. J. Obstet. Gynec.* **120**, 954–959.

Arshavskii, Y. I., Berkinblit, M. V., and Dunin-Barkovskii, V. L. (1965) Propagation of pulses in a ring of excitable tissues. *Biofizika* **10**, 1048–1054.

Arshavskii, Y. I., Berkinblit, M. B., Kovalev, S. A., and Chailakhyan, L. M. (1964) Periodic transformation of rhythm in a nerve fiber with gradually changing properties. *Biofizika* **9**, 365–371.

Arvanitaki, A., and Chalazonitis, N. (1968) Electrical properties and temporal organization in oscillatory neurons. in *Neurobiology of Invertebrates*. J. Salanki, ed. Plenum, New York, pp. 169–174.

Aschoff, J. (1965a) Circadian rhythms in man. *Science* **148**, 1427–1432.

Aschoff, J. (1965b) *Circadian Clocks*. North-Holland, Amsterdam.

Aschoff, J. (1969) Desynchronization and resynchronization of human circadian rhythms. *Aerospace Med*. **40**, 844–849.

Aschoff, J. (1976) Circadian systems in man and their implications. *Hosp. Prac*. **11**, 51–57.

Aschoff, J., Gerecke, U., and Wever, R. (1967) Desynchronization of human circadian rhythms. *Japan, J. Physiol*, **17**, 450–457.

Aschoff, J., Klotter, K., and Wever, R. (1965) Circadian clocks. In *Circadian Clocks*, J. Aschoff, ed. North-Holland, Amsterdam, pp. 10–19.

Aschoff, J., and Wever, R. (1976) Human circadian rhythms: a multi-oscillatory system. *Fed. Proc*. **35**, 2326–2332.

Ashkenazi, I. E., Hartman, Hr, Strulovitz, B., and Dar, O. (1975) Activity rhythms of enzymes in human red blood cell suspensions. *J. Interdiscipl. Cycle Res*. **6**, 291–301.

Ashkenazi, M., and Othmer, H. G. (1978) Spatial patterns in coupled biochemical oscillators. *J. Math. Biol*. **5**, 305–350.

Ayers, J. L., and Selverston, A. I. (1977) Synaptic control of an endogenous pacemaker network. *J. Physiol. (Paris)* **73**, 453–461.

Ayers, J. L. and Selverston, A. I. (1979) Monosynaptic entrainment of an endogenous pacemaker network: a cellular mechanism for von Holst's magnet effect. *J. Comp. Physiol*. **129**, 5–17.

Bailey, J., and Marshall, J. (1970) The relationship of the post-ovulatory phase of the menstrual cycle to the total cycle length. *J. Biosoc. Sci*. **2**, 123–132.

Balakhovskii, I. S. (1965) Several modes of excitation movement in ideal excitable tissue. *Biophysics* **10**, 1175–1179.

Barker, R. J., Cohen, C. F., and Mayer, A. (1964) Photoflashes: a potential new tool for control of insect populations. *Science* **145**, 1195–1197.

Barlow, J. S. (1962) A phase-comparator model for the diurnal rhythm of emergence of *Drosophila. Ann. NY Acad. Sci*. **98**, 788–805.

Barnett, A, (1966) A circadian rhythm of mating type reversals in *Paramecium multimicronucleatum* Syngen 2, and its genetic control, *J. Cell Physiol*. **67**, 239–270.

Barneveld, H. H. (1971) Statistical analysis of repetitive neuronal activity: clock-spikes in *Calliphora erythrocephala*. Rijksuniversiteit te Groningen thesis.

Bateman, M. A. (1955) The effect of light and temperature on the rhythm of pupal ecdysis in the Queensland fruitfly, *Dacus tryoni. Aust. J. Zool*. **3**, 22–33.

Bateson, W. (1894) *Materials for the Study of Variation*, MacMillan, London.

Beck, M. T., and Varadi, Z. B. (1971) Periodic homogeneous chemical reactions in space and time. *Magyar Kemiai Folyoirat* **77**, 167–168.

Beck, M. T., and Varadi, Z. B. (1972) One, two and three-dimensional spatially periodic chemical reactions. *Nature Physical Sci*. **235**, 15–16.

Beck, M. T., Varadi, Z. B., and Hauck, K. (1976) Quantitative description of one-dimensional spatially-periodic phenomena in the potassium bromate-malonic acid-catalyst-sulphuric acid system. *Acta Chimica Academiae Scientiarum Hungaricae* **91**, 13–26.

Beck, S. D. (1968) *Insect Photoperiodism*, Academic, New York.

Beier, W., Medugorac, I., and Lindauer, M. (1968) Synchronization et dissociation de "l'horloge interne" des abilles par des facteurs externes *Ann. Epiphyties* **19**, 133–144.

Belousov, B. P. (1958) collection of abstracts of radiation medicine *Sb. Ref. Radiats. Med.* Medgiz, Moscow, p. 145.

Benham, C. W. (1894) The artificial spectrum top. *Nature* **51**, 200.

Benson, J. A., and Jacklet, J. W. (1977) Circadian rhythm of output from neurones in the eye of *Aplysia* (I, II, III, IV) *J. Exp. Biol.* **70**, 151–211.

Berg, H. (1974) Dynamic properties of bacterial flagellar motors. *Nature* **249**, 77–79.

Berkinblit, M. B., Kalinin, D. I., Kovalev, S. A., and Chailakhyan, L. M. (1975) Study with the Noble model of synchronization of the spontaneously active myocardial cells bound by a highly permeable contact. *Biofizika* **20**, 121–125.

Berliner, M. D., and Neurath, P. W. (1965) The rhythms of three clock mutants of *Ascobolus immersus*. *Mycologia* **57**, 809–817.

Berliner, M. D., and Neurath, P. W. (1966) Control of rhythmic growth of a *Neurospora* 'clock' mutant by sugars. *Can. J. Micro.* **12**, 1068–1070.

Bessey, E. A. (1950) *Morphology and Taxonomy of Fungi*, Blakiston, Philadelphia.

Best, E. N. (1975) Exploration of a menstrual cycle model. *Simulation Today* **25**, 117–120.

Best, E. N. (1976) Null space and phrase resetting curves for the Hodgkin-Huxley equations. Ph.D. thesis, Purdue University.

Best, E. N. (1979) Null space in the Hodgkin-Huxley equations: a critical test. Biophys. J. **27**, 87–104.

Betz, A. (1966) Metabolic flux in yeast cells with oscillatory controlled glycolysis. *Physiol. Plant.* **19**, 1049–1054.

Betz, A., and Becker, J.-U. (1975a) Metabolite concentrations and phase relations in D-fructose induced glycolytic oscillations. *Physiol. Plant.* **33**, 285–289.

Betz, A., and Becker, J. U. (1975b) Phase dependent phase shifts induced by pyruvate and acetaldehyde in oscillating NADH of yeast cells. *J. Interdiscipl. Cycle Res.* **6**, 167–173.

Betz, A., and Chance, B. (1965a) Influence of inhibitors and temperature from the oscillation of reduced pyridine nucleotides in yeast cells. *Arch. Biochem. Biophys.* **109**, 579–584.

Betz, A., and Chance, B. (1965b) Phase relationship of glycolytic intermediates in yeast cells with oscillatory metabolic control. *Arch. Biochem. Biophys.* **109**, 585–594.

Bhéreur, P., Roberge, F. A., and Nadeau, R. A. (1971) A cardiac pacemaker model. *Med. and Biol. Eng.* **9**, 3–12.

Birukow, G. (1960) Innate types of chronometry in insect orientation. *Cold Spring Harbor Symposia on Quantitative Biology*, vol. 25, 403–412.

Blandamer, M. J. and Roberts, D. L. (1978) Response of the Belousov-Zhabotinskii oscillating reaction to addition of either bromide or cerium (IV) ions. *J. Chem. Soc. Far. Trans. I*, **74**, 1636–1646.

Bogumil, R. J., Ferin, M., Rootenberg, J., Speroff, L., and Vande Wiele, R. L. (1972) Mathematical studies of the human menstrual cycle. I. Formulation of a mathematical model. *J. Clin. Endo. Metab.* **35**, 126–143. II. Simulation performance of a model of the human menstrual cycle. Ibid. **35**, 144–156.

Boiteux, A., Goldbeter, A., and Hess, B. (1975) Control of oscillating glycolysis of yeast by stochastic, periodic, and steady source of substrate: a model and experimental study. *Proc. Nat. Acad. Sci.* **72**, 3829–3833.

Bolker, D. (1973) The spinor spanner. *Amer. Math. Mo.* **80**, 977–984.

Bollig, I. (1975) Photoperiodic time measurement and circadian leaf movement in *Pharbitis nil* controlled by the same clock? *Z. Pflanzenphysiol.* **77**, 54–69.

Bollig, I. (1977) Different circadian rhythms regulate photoperiodic flowering response and leaf movement in *Pharbitis nil* (L.) Choisy. *Planta* **135**, 137–142.

Bonhoeffer, K. F. (1948) Activation of passive iron as a model for excitation of nerve. *J. Gen. Physiol.* **32**, 69–91.

Bonhoeffer, K. F. (1953) Modelle der Nervenerregung. *Naturwissenschaften* **40**, 301–311.

Bonhoeffer, K. F., and Langhammer, G. (1948) Über periodische Reactionen. IV. *Zeit Elektrochem.* **52**, 67–72.

Bonner, J. T. (1967) *The Cellular Slime Molds*, Princeton University Press.

Boon, P. N. M., and Strackee, J. (1976) A population of coupled oscillators. *J. Theor. Biol.* **57**, 491–500.

Bortoff, A. (1969) Intestinal motility. *N. Eng. J. Med.* **280**, 1335–1337.

Both, R., Finger, W., and Chaplain, R. A. (1976) Model predictions of the ionic mechanisms underlying the beating and bursting pacemaker characteristics of molluscan neurons. *Biol. Cybern.* **23**, 1–11.

Bouligand, Y. (1965) Sur une architecture torsadée repandué dans de nombreuses cuticles d'arthropodes, *C. R. Acad. Sc. Paris* **261**, 3665–3668.

Bouligand, Y. (1969) Sur l'existence de "pseudomorphoses cholesteriques" chez divers organismes vivants. *J. Physique* (coll. C4), suppl. au 11–12, **30**, 90–103.

Bouligand, Y. (1971) Les orientations fibrillaires dans le squelette des arthropodes. *J. Microscopie* **11**, 441–472.

Bouligand, Y. (1972a) Twisted fibrous arrangements in biological materials and cholesteric mesophases. *Tissue and Cell* **4**, 187–217.

Bouligand, Y. (1972b) Recherches sur les textures des etats mesomorphes 1. les arrangements focaux dans les smectiques: rappels et considerations theoriques. *Journal De Physique* **33**, 525–547.

Bouligand, Y. (1972c) Recherches sur les textures des etats mesomorphes 2. les champs polygonaux dans les cholesteriques. *Journal De Physique* **33**, 715–736.

Bouligand, Y. (1973a) Recherches sur les textures des etats mesomorphes. 3. les plages a evantails dans les cholesteriques. *Journal De Physique* **34**, 603–614.

Bouligand, Y. (1973b) Recherches sur les textures des etats mesompophes 4. la texture a plans et al morphogenese des principales textures dans les cholesteriques. *Journal De Physique* **34**, 1011–1020.

Bouligand, Y. (1974a) Recherches sur les textures des etats mesomorphes. 5. noyaux, fils et rubans de moebius dans les nematiques et les cholesteriques peu torsades. *Journal De Physique* **35**, 215–236.

Bouligand, Y. (1974b) Recherches sur les textures des etats mesomorphes. 6. dislocations coins et signification des cloisons de granjean-cano dans les cholesteriques. *Journal De Physique* **35**, 959–981.

Bouligand, Y. (1978) Cholesteric order in biopolymers, Ch. 15, pp. 237–247 of *Mesomorphic Order in Polymers*, A.C.S. Symposium Series 74, ed. by A. Blumstein, Amer. Chem. Soc., Washington D.C.

Bourret, A., Lincoln, R. G., Carpenter, B. H. (1969) Fungal endogenous rhythms expressed by spiral figures. *Science* **166**, 763–764.

Bourret, J. A. (1971) Modification of the period of a non-circadian rhythm in *Nectria cinnabarina*. *Plant Physiol.* **47**, 682–684.

Boutselis, J. G., Dickey, R. P., and Vorys, N. (1972) Control of ovulation time with steroid and non-steroid compounds. *Amer. J. Obstet. Gynec.* **112**, 171–177.

Boutselis, J. G., Vorys, N., and Dickey, R. (1971) Control of ovulation time with low-dose estrogens. *Obstet. Gynec.* **38**, 863–868.

Bradshaw, W. E. (1972) Action spectra for photoperiodic response in a diapausing mosquito. *Science* **175**, 1361–1363.

Brady, J. (1975) The physiology of insect circadian rhythms. *Adv. Ins. Physiol.* **10**, 1–99.

Braemer, W. (1960) A critical review of the sun-azimuth hypothesis. *Cold Spring Harbor Symposia on Quantitative Biology*, vol. 25, 413–427.

Brayer, F. T., Chiacze, L., and Duffy, B. J. (1969) Calendar rhythm and menstrual cycle change. *Fertil. Steril.* **20**, 279–288.

Brett, W. J. (1955) Persistent diurnal rhythmicity in *Drosophila* emergence. *Ann. Ent. Soc. Amer.* **48**, 119–131.

Brink, R., Bronk, D., and Larrabee, M. (1946) Chemical excitation of nerve. *Ann. N.Y. Acad. Sci.* **47**, 457–486.

Brinkmann, K. (1966) Temperatureinflüsse auf die circadiane Rhythmik von *Euglena gracilis* bei Mixotrophie und Autotrophie. *Planta* **70**, 344–389.

Brinkmann, K. (1967) Biologische Rhythmik, der Einfluss von Populationseffekten auf die circadiane Rhythmik von *Euglena gracilis*. *Nachrichten Akad Wissenschaft, Göttingen.* II. **10**, 138–140.

Brinkmann, K. (1971) Metabolic control of temperature compensation in the circadian rhythm of *Euglena gracilis*. pp 567–593. In *Biochronometry*, M. Menaker, ed. Nat. Acad. Sci., Washington, D.C.

Brody, S., and Harris, S. (1973) Circadian rhythms in *Neurospora*: spatial differences in pyridine nucleotide levels. *Science* **180**, 498–500.

Brody, S. and Martins, S. A. (1979) Circadian rhythms in *Neurospora crassa*; effects of unsaturated fatty acids. *J. Bacteriology* **137**, 912–915.

Brooks, C., and Lu, H. (1972) *The Sino-atrial Pacemaker of the Heart*, C. C. Thomas, Springfield, Ill.

Brooks, R. F. (1975) The kinetics of serum induced initiation of DNA synthesis in BHK-21-C-13 cells and the influence of exogenous adenosine. *J. Cell. Physiol.* **86**, 369–377.

Brooks, R. F. (1976) Regulation of the fibroblast cell cycle by serum. *Nature* **260**, 248–250.

Brown, B. H., Duthie, H. L., Horn, A. R. and Smallwood, R. H. (1975) A linked oscillator model of electrical activity of human small intestine. *Amer. J. Physiol.* **229**, 384–388.

Brown, F. A., and Webb, H. M. (1948) Temperature relations of an endogenous daily rhythmicity in the fiddler crab, *Uca. Physiol. Zool.* **21**, 371–381.

Bruce, V. G. (1965) Cell division rhythm and the circadian clock. In *Circadian Clocks*, J. Aschoff, ed. North-Holland, Amsterdam, pp. 125–138.

Bruce, V. G., and Minis, D. H. (1969) Circadian clock action spectrum in a photoperiodic moth. *Science* **163**, 583–585.

Bruce, V. G., and Pittendrigh, C. S. (1956) Temperature independence in a unicellular "clock". *Proc. Nat. Acad. Sci.* **42**, 676–682.

Bruce, V. G., Wright, F., and Pittendrigh, C. S. (1960) Resetting the sporulation rhythm in *Pilobolus* with short light flashes of high intensity. *Science.* **131**, 728–730.

Brumberger, H. (1970) Rhythmic crystallization of poly-L-alanine. *Nature* **227**, 490–491.

Bryant, P. J., Bryant, S. V., and French, V. (1977) Biological regeneration and pattern formation. *Sci. Amer.* **237** (1), 66–81.

Bryant, S. V. (1976) Regenerative failure of double half limbs in *Notophthalmus viridescens*. *Nature* **263**, 676–679.

Buck, J. (1938) Synchronous rhythmic flashing of fireflies. *Quart. Rev. Biol.* **13**, 301–314.

Buck, J., and Buck, E. (1968) Mechanism of rhythmic synchronous flashing of fireflies. *Science* **159**, 1319–1327.

Buck, J., and Buck, E. (1976) Synchronous fireflies. *Sci. Amer.* **234** (5), 74–85.

Buerle, R. L. (1956). Properties of a mass of cells capable of regenerating pulses. *Phil. Trans. Roy. Soc.* B **240**, 55–94.

Bullock, T. H. (1955) Compensation for temperature in the metabolism and activity of poikilotherms. *Biol. Rev.* **30**, 311–342.

Bullough, W., and Laurence, E. (1964) Mitotic control by internal secretion: the role of the chalone-adrenalin complex. *Exp. Cell. Res.* **33**, 176–194.

Bunning, E. (1935) Zur Kenntnis der endogenen Tagesrhythmik bei Insekten und Pflanzen. *Ber. dtsch. Bot. Ges.* **53**, 594–623.

Bunning, E. (1956a) Die physiologische Uhr. *Naturw. Rdsch., Stuttg.* **9**, 354–356.

Bunning, E. (1956b) Versuche zur Beeinflussung der endogenen Tagesrhythmik durch chemische Faktoren. *Z. Bot.* **44**, 515–529.

Bunning, E. (1957) Endogenous diurnal cycles of activity in plants. In *Rhythmic and Synthetic Processes in Growth*, D. Rudnick, ed. p. 111. Princeton Univ. Press, Princeton.

Bunning, E. (1959) Physiological mechanism and biological importance of the endogenous circadian periodicity in plants and animals. Publ. #55 of AAAS., from *Photoperiodism and Related Phenomena in Plants and Animals*, R. B. Withrow, ed. p. 523.

Bunning, E. (1960) Opening address: biological clocks. In *Cold Spring Harbor Symposium on Quantitative Biology*, vol. 25, 1–10.

Bunning, E. (1964, 1973) *The Physiological Clock*, Springer-Verlag, New York.

Bunning, E., and Joerrens, G. (1960) Tagesperiodische antagonistische Schwankungen der Blauviolett und Gelbrot-Empfindlichkeit als Grundlage der photoperiodischen Diapauseinduktion bei *Pieris brassicae. Z. Naturf.* **15b**, 205–213.

Bunning, E. and Moser, I. (1966) Response-Kurven bei der circadianen Rhythmik von *Phaseolus*. Planta **69**, 101–110.

Bunow B., J. Kernevez, G. Joly, and D. Thomas (1980) "Pattern formation by immobilized enzymes: application to *Drosophila* morphogenesis" submitted to J. Theor. Biol.

Bures, J. (1959) Reversible decortication and behavior. In *The Central Nervous System and Behavior*, M. A. B. Brazier, ed. Josiah Macy Jr. Foundation, New York. p. 207.

Bures, J., Buresova, O., and Krivanek, J. (1974) *The Mechanism and Application of Leao's Spreading Depression of Electroencephalographic Activity*. Academic, New York.

Burnett, J. H. (1968) *Fundamentals of Mycology*, St. Martins, New York, p. 68.

Burns, F. J., and Tannock, I. F. (1970) On the existence of a G phase in the cell cycle. *Cell Tissue Kinet.* **3**, 321–334.

Burton, A. C. (1971) Cellular communication, contact inhibition, cell clocks, and cancer: the impact of the work and ideas of W. R. Loewenstein. *Perspectives in Biology and Medicine* **14**, 301–318.

Burton, A. C., and Canham, P. B. (1973) The behavior of coupled biochemical oscillators as a a model of contact inhibition of cellular division. *J. Theor. Biol.* **39**, 555–580.

Busse, H. G. (1969) A spatial periodic homogeneous chemical reaction. *J. Phys. Chem.* **73**, 750.

Busse, H. G., and Hess, B. (1973) Information transmission in a diffusion-coupled oscillatory chemical system. *Nature* **244**, 203–205.

Butterfield, J. F. (1968) Subjective (induced) color television. *Journal of the Society for Motion Picture and Television Engineers* **77**, 1025–1028.

Caldarola, P. C., and Pittendrigh, C. S. (1974) A test of the hypothesis that D_2O affects circadian oscillations by diminishing the apparent temperature. *Proc. Nat. Acad. Sci.* **71**, 4386–4388.

Cameron, R. G. (1969) Effect of flying on the menstrual function of air hostesses. *Aerospace Medicine* **40**, 1020–1023.

Campbell, A. (1964) The theoretical basis of synchronization by shifts in environmental conditions. In *Synchrony in Cell Division and Growth*, E. Zeuthen, ed. Wiley, New York, pp. 469–484.

Campenhausen, C. von (1969) The colors of Benham's top under metameric illuminations. *Vision Res.* **9**, 677–682.

Carpenter, G. (1979) Bursting phenomena in excitable membranes. *SIAM J. Appl. Math.* **36**, 334–372.

Carroll, L. 1971. *The Rectory Umbrella and Mischmasch*, Dover, New York, pp. 31–33.

Chance, B., Estabrook, R. W., and Ghosh, A. (1964) Damped sinusoidal oscillations of cytoplasmic reduced pyridine nucleotide in yeast cells. *Proc. Nat. Acad. Sci.* **51**, 1244–1251.

Chance, B., Pye, K., and Higgins, J. (1967) Waveform generation by enzymatic oscillators. *IEEE Spectrum* **4**, 79–86.

Chance, B., Schoener, B., and Elsaesser, S. (1965a) Metabolic control phenomena involved in damped sinusoidal oscillations of reduced diphosphopyridine nucleotide in a cell-free extract of *Saccharomyces carlsbergensis*. *J. Biol. Chem.* **240**, 3170–3181.

Chance, B., Williams, J. R., and Schoener, B. (1965b) Properties and kinetics of reduced pyridine nucleotide fluorescence of the isolated and in vivo rat heart. *Biochem. Z.* **341**, 357–377.

Chance, B., Williamson, G., Lee, I., Mela, L., DeVault, D., Ghosh, A., and Pye, E. K. (1968) Synchronization phenomena in oscillations of yeast cells and isolated mitochondria. *Biological and Biochemical Oscillators*, B. Chance et al., eds. (1973) Academic, New York, pp. 285–300.

Chance, B., and Yoshioka, T. (1966) Sustained oscillations of ionic constituents of mitochondria. *Arch. Biochem. Biophys.* **117**, 451–465.

Chandrashekaran, M. K. (1967) Studies on phase-shifts in endogenous rhythms. *Z. vergl. Physiol.* **56**, 154–170.

Chandrashekaran, M. K. (1974) Phase shifts in the *Drosophila pseudoobscura* circadian rhythm evoked by temperature pulses of varying durations. *J. Interdiscipl. Cycle Res.* **5**, 371–380.

Chandrashekaran, M. K., and Engelmann, W. (1973) Early and late subjective night phases of the *Drosophila pseudoobscura* circadian rhythm require different energies of blue light for phase shifting. *Z. Naturforsch.* **28**, 750–753.

Chandrashekaran, M. K., and Engelmann, W. (1976) Amplitude attenuation of the circadian rhythm in *Drosophila* with light pulses of varying irradiance and duration. *Int. J. of Chronobiol.* **3**, 231–240.

Chandrashekaran, M. K., Johnsson, A., and Engelmann, W. (1973) Possible "dawn" and "dusk" roles of light pulses shifting the phase of a circadian rhythm. *J. Comp. Physiol.* **82**, 347–356.

Chandrashekaran, M. K., and Loher, W. (1969a) The effect of light intensity on the circadian rhythms of eclosion in *Drosophila pseudoobscura*. *Z. Vergl. Physiol.* **62**, 337–347.

Chandrashekaran, M. K., and Loher, W. (1969b) The relationship between the intensity of light pulses and the extent of phase shifts of the circadian rhythm in the eclosion rate of *Drosophila pseudoobscura*. *J. Exp. Zool.* **172**, 147–152.

Chaplain, R. A. (1976) Metabolic regulations of the rhythmic activity in pacemaker neurons II: metabolically induced conversions of beating to bursting pacemaker activity in isolated *Aplysia* neurons. *Brain Res.* **106**, 307–319.

Chapman, R. A. (1967) Dependence on temperature of the conduction velocity of the action potential of the squid giant axon. *Nature* **213**, 1143–1144.

Chevaugeon, J., and Van Huong, N. (1969) Internal determinism of hyphal growth rhythms. *Trans. Br. Mycol. Soc.* **53**, 1–14.

Child, C. M. (1941) *Patterns and Problems of Development*, University of Chicago Press.

Chin, B., Friedrich, P. D. and Bernstein, I. A. (1972) Stimulation of mitosis following fusion of plasmodia in the myxomycete *Physarum polycephalum*. *J. Gen. Microbiol.* **71**, 93–101.

Christianson, R., and Sweeney, B. M. (1972) Sensitivity to stimulation, a component of the circadian rhythm in luminescence in *Gonyaulax*. *Plant Physiol.* **49**, 994–997.

Christianson, R., and Sweeney, B. M. (1973) The dependence of the phase response curve for the luminescence rhythm in *Gonyaulax* on the irradiance in constant conditions. *Int. J. Chronobiol.* **1**, 95–100.

Christie, G. A., Moore-Robinson, M., Gullet, C. C., and Bergin, K. G. (1970) Project Pegasus: some physiological effects of travel across time zones. *Clin. Trials J.*, **7** 45–53.

Chu, H., Norris, D. M., and Carlson, S. D. (1975) Ultrastructure of the compound eye of the diploid female Beetle, *Xyleborus ferrugineus*. *Cell Tissue Res.* **165**, 23–26.

Clark, J. H., and Zarrow, M. X. (1971) Influence of copulation on time of ovulation in women. *Amer. J. Obstet. Gynec.* **109**, 1083–1085.

Clark, R. L. and Steck, T. L. (1979) Morphogenesis in *Dictyostelium*: an orbital hypothesis. *Science*, **204**, 1163–1168.

Clay, J. R., and DeHaan, R. L. (1979) Fluctuations in interbeat interval in rhythmic heart-cell clusters. *Biophys. J.* **28**, 377–390.

Cloudsley-Thompson, J. L. (1961) *Rhythmic activity in animal Physiology and Behavior*, NY Acad. Sci. Press, New York, p. 236.

Coddington, E. A., and Levinson, N. (1955) *Theory of Ordinary Differential Equations*, Maple, York, Penn., p. 323, Theorem 2.2.

Code, F., and Szurszewski, H. (1970) The effect of duodenal and mid-small-bowel transection on the frequency gradient of the pacesetter potential in the canine small intestine. *J. Physiol.* **207**, 281–289.

Cohen, D. S., Hoppensteadt, F. C., and Miura, R. M. (1977) Slowly modulated oscillations in nonlinear diffusion processes. *SIAM J. Appl. Math.* **33**, 217–229.

Cohen, D. S., Neu, J. C. and Rosales, R. R. (1978) Rotating spiral wave solutions of reaction-diffusion equations. *SIAM J. Appl. Math.* **35**, 536–547.

Cohen, H. (1971) *Nonlinear Diffusion Problems*, vol. 7, Studies in Applied Mathematics. A. H. Taub., ed. University of California, Berkeley, pp. 27–64.

Cohen, M. H. (1971) Models for the control of development. *Symp. Exper. Biol.* **25**, 455–476.

Cohen, M. H. (1972) Models of clocks and maps in developing organisms. *Some Mathematical Questions in Biology*, Jack Cowan, ed. **2**, 3–32. Amer. Math. Soc., Providence, R. I.

Cohen, M. H., and Robertson, A. (1971a) Wave propagation in the early stages of aggregation of cellular slime molds. *J. Theor. Biol.* **31**, 101–118.

Cohen, M. H., and Robertson, A. (1971b) Chemotaxis and the early stages of aggregation in cellular slime mold. *J. Theor. Biol.* **31**, 119–130.

Cohen, M. S. (1977) The cyclic AMP control system in the development of *Dictyostelium discoideum* I. cellular dynamics. *J. Theor. Biol.* **69**, 57–85.

Cohen, M. S. (1978) The cyclic AMP control system in the development of *Dictyostelium discoideum* II. an allosteric model. *J. Theor. Biol.* **72**, 231–255.

Cole, K. S. (1968) *Membranes, Ions, and Impulses*. University of California Press, Berkeley.

Collins, M. A., and Ross, J. (1978) Chemical relaxation pulses and waves. Analysis of lowest order multiple time scale expansion. *J. Chem. Phys.* **68**, 3774–3784.

Colquhoun, W. P. (ed.) (1971) *Biological Rhythms and Human Performance*, Academic, New York.

Connor, J. A. (1978) Slow repetitive activity from fast conductance changes in neurons. *Fed. Proc.* **37**, 2139–2145.

Connor, J. A., Prosser, C. L., and Weems, W. A. (1974) A study of pacemaker activity in intestinal smooth muscle. *J. Physiol.* **240**, 671–701.

Connor, J. A., and Stevens, C. F. (1971a) Inward and delayed outward membrane currents in isolated neural somata under voltage clamp. *J. Physiol.* **213**, 1–19.

Connor, J. A., and Stevens, C. F. (1971b) Voltage clamp studies of a transient outward membrane current in gastropod neural somata. *J. Physiol.* **213**, 21–30.

Connor, J. A., and Stevens, C. F. (1971c) Prediction of repetitive firing behaviour from voltage clamp data on an isolated neurone soma. *J. Physiol.* **213**, 31–53.

Conroy, R. T. W. L., and Mills, J. M. (1970) *Human Circadian Rhythms*, J. N. A. Churchill, London.

Conway, E., Hoff, D., and Smoller, J. (1978) Large time behavior of solutions of systems of nonlinear reaction-diffusion equations. SIAM *J. Appl. Math.* **35**, 1–16.

Cooke, J., and Zeeman, E. C. (1976) A clock and wave front model for control of the number of repeated structures during animal morphogenesis. *J. Theor. Biol.* **58**, 455–476.

Cooper, M. J., and Rowson, L. E. A. (1975) Control of the oestrous cycle in Friesian heifers with ICI 80996. *Ann. Biol. Anim. Biochem. Biophys.* **15**, 427–436.

Cottrell, A. (1978) in *The Encyclopedia of Ignorance* eds. R. Duncan and M. Weston-Smith (p. 133). Pergamon, N.Y.

Crick, F. H. C. (1970) Diffusion in embryogenesis. *Nature* **225**, 420–422.

Crick, F. H. C. (1971) The scale of pattern formation. *Symp. Soc. Exp. Biol.* **25**, 429–438.

Croll, N. A. (1966) Activity and the orthokinetic response of larval trichoneme to light. *Parasitology* **56**, 307–312.

Cseffalvay, V. T. (1966) Vom Zyklus abhängige Leistungsfähigkeit der Frau und die Transponierung der Menstruation. *Medizin und Sport* **1**, 1–5.

Cumming, B. (1972) The role of circadian rhythms in photoperiodic induction in plants. pp 33–85. In: *Proc. Int. Symp. Circadian Rhythmicity*, J. F. Bierhuizen, ed. Centre for Agri. Publ. and Doc., Waginengen.

Cumming, B., and Wagner, E. (1968) Rhythmic processes in plants. *Ann. Rev. Pl. Physiol.* **19**, 381–416.

Cummings, F., and Prothero, J. W. (1978) A model of pattern formation in multicellular organisms. *Collective Phenomena* **3**, 41–53.

Cummings, F. (1975) A biochemical model of the circadian clock. *J. Theor. Biol.* **55**, 455–470.

Daan, S., and Berde, C. (1978) Two coupled oscillators: simulations of the circadian pacemaker in mammalian activity rhythms. *J. Theor. Biol.* **70**, 297–314.

Daan, S., Damassa, D., Pittendrigh, C. S., and Smith, E. R. (1975) An effect of castration and testosterone replacement on a circadian pacemaker in mice (*Mus musculus*). *Proc. Nat. Acad. Sci.* **72**, 3744–3747.

Daan, S., and Pittendrigh, C. (1976a) A functional analysis of circadian pacemakers in nocturnal rodents. II. The variability of phase response curves. *J. Comp. Physiol.* **106**, 253–266.

Daan, S., and Pittendrigh, C. (1976b) A functional analysis of circadian pacemakers in nocturnal rodents. III. Heavy water and constant light: homeostasis of frequency? *J. Comp. Physiol.* **106**, 267–290.

Danielevsky, A. S., Goryshin, N. I., and Tyschchenko, V. P. (1970) Biological rhythms in arthropods. *Ann. Rev. Entomol.* **15**, 201–244, R. F. Smith, ed.

Danziger, L. and Elmergreen, G. L. (1956) The thyroid-pituitary homeostatic mechanism. *Bull. Math. Biophys.* **18**, 1–13.

Danziger, L., and Elmergreen, G. L. (1957) Mathematical models of endocrine systems. *Bull. Math. Biophys.* **19**, 9–18.

Darmon, M., Brachet, P., and Pereira De Silva, L. (1975) Chemotactic signals induce cell differentation in *Dictyostelium discoideum. Proc. Nat. Acad. Sci.* **72**, 3163–3166.

Davenport, H. W. (1977) Motility of the small intestine. Chapter 4 in *The Digestive Tract*, Yearbook Medical Pub., Chicago, pp. 58–71.

DeCoursey, P. (1959) Daily activity rhythms in the flying squirrel, *Glaucomys volans*. Ph.D. thesis, University of Wisconsin.

DeCoursey, P. (1961) Effect of light on the circadian activity rhythm of the flying squirrel, *Glaucomys volans. Z. vergl. Physiol.* **44**, 331–354.

deDuve, C. (1969) In *Lysosomes in Biology and Pathology*, J. T. Dingle and Honor B. Fell eds. North-Holland, Amsterdam, London.

Defant, A. (1961) *Physical Oceanography*, vol. 2, Pergamon, New York.

Defay, R., Prigogine, I., and Sanfeld, A. (1977) Surface thermodynamics. *J. Coll. Int. Sci.* **58**, 498–510.

Degn, H. (1969) Compound III kinetics and chemiluminescence in oscillatory oxidation reactions catalyzed by horseradish peroxidase. *Biochim. Biophys. Acta* **180**, 271–290.

DeHaan, R. L., and Fozzard, H. A. (1975) Membrane response to current pulses in spheroidal aggregates of embryonic heart cells. *J. Gen. Physiol.* **65**, 207–222.

DeHaan, R. L., and Sachs, H. G. (1972) Cell coupling in developing systems: the heart-cell paradigm. In *Current Topics in Dev. Biol.* **7**, 193–228.

Delbruck, M., and Reichardt, W. (1956) I. Systems analysis for the light growth reactions of *Phycomyces*. In *Cellular Mechanisms in Differentation and Growth*, D. Rudnick, ed. Princeton University Press, pp. 3–43.

DeSimone, J. A., Beil, D. L. and Scriven, L. E. (1973) Ferroin-collodion membranes: dynamic concentration patterns in planar membranes. *Science* **180**, 946–948.

Devi, V. R., Guttes, E., and Gettes, S. (1968) Effects of ultraviolet light on mitosis in *Physarum polycephalum*. *Exp. Cell Res.* **50**, 589–598.

Dewan, E. M. (1967) On the possibility of a perfect rhythm method of birth control by periodic light stimulation. *Amer. J. Obstet. Gynec.* **99**, 1016–1019.

Dewan, E. M., Menkin, M. F., and Rock, J. (1978) Effect of photic stimulation on the human menstrual cycle *Photochem. Photobiol.* **27**, 581–585.

Dewey, W., Miller, H., and Nagasawa, H. (1973) Interactions between S and Gl cells. *Exp. Cell Res.* **77**, 73–78.

Dharmananda, S., and Feldman, J. F. (1979) Assay of spatial distribution of circadian clock phase in aging cultures of *Neurospora crassa*. *Plant Physiol.* **63**, 1049–1054.

Dhont, M., Vandekerskhove, D., and Vermeulin, A. (1974) The effect of ethinyl estradiol administration during the follicular phase of the menstrual cycle. *J. Clin. Endo. Metab.* **39**, 465–470.

Diamant, N. E., and Bortoff, A. (1969) Nature of the intestinal slow wave frequency gradient. *Amer. J. Physiol.* **216**, 734–743.

Diamant, N. E., Rose, T. K., and Davison, E. J. (1970) Computer simulation of intestinal slow-wave frequency gradient. *Am. J. Physiol.* **219**, 1684–1690.

Diamant, N. E., Wong, J., and Chen, L. (1973) Effects of transection on small intestinal slow-wave propagation velocity. *Amer. J. Physiol.* **225**, 1497–1500.

Dodson, K. S., Macnaughton, M. C., and Coutis, J. R. T. (1975) Infertility in women with apparently ovulatory cycles. *Br. J. Obstet. Gynaec.* **82**, 615–624.

Donachie, W., and Masters, M. (1969) Temporal control of gene expression in bacteria. In *The Cell Cycle: Gene-Enzyme Interactions*, G. Padilla, G. Whitson, and I. Cameron, eds. Academic, New York, pp. 37–76.

Duranceau, C. M. (1977) Control of growth and pattern formation in the imaginal wing disc of *Drosophila melanogaster*. University of California, Irvine, Ph.D. thesis.

Durham, A. C. H. and E. B. Ridgway (1976) Control of chemotaxis in *Physarum polycephalum J. Cell Biol.* **69**, 218–223.

Durston, A. J. (1973) *Dictyostelium discoideum* aggregation fields as excitable media. *J. Theor. Biol.* **42**, 483–504.

Durston, A. J. (1974a) Pacemaker activity during aggregation in *Dictyostelium discoideum*. *Dev. Biol.* **37**, 225–235.

Durston, A. J. (1974b) Pacemaker mutants of *Dictyostelium discoideum*. *Dev. Biol.* **38**, 308–319.

Duysens, L. N., and Amesz, J. (1957) Fluorescence spectophotometry of reduced phosphopyridine nucleotide in intact cells in the near-ultraviolet and visible region. *Biochim. Biophys. Acta* **24**, 19–26.

Dynnik, V. V., and Sel'kov, E. E. (1973) On the possibility of self-oscillations in the lower part of the glycolytic system. *FEBS Lett.* **37**, 342–346.

Dynnik, V. V., and Sel'kov, E. E. (1975a) Generator of oscillations in the lower part of glycolytic system. *Biofizika* **20**, 288–292.

Dynnik, V. V., and Sel'kov, E. E. (1975b) Double frequency oscillations in the glycolytic system. Mathematical model. *Biofizika* **20**, 293–297.

Dynnik, V. V., Sel'kov, E. E., and Ovtchinnikov, I. A. (1977) Self-oscillations in the lower part of the glycolytic system, two alternative mechanisms. *Studia Biophysica* **63**, 9–23.

Dynnik, V., Sel'kov, E. E., and Semashko, L. R. (1973) Analysis of the adenine nucleotide effect on the oscillatory mechanism in glycolysis. *Studia Biophysica* **41**, 193–214.

Dzuik, P. J., Hines, F. C., Mansfield, M. E., and Baker, R. D. (1964) Follicle growth and control of ovulation in the ewe following treatment with 6-methyl-17-acetoxyprogesterone. *J. Anim. Sci.* **23**, 787–790.

Edelstein, B. B. (1970) Instabilities associated with dissipative structure. *J. Theor. Biol.* **26**, 227–241.

Edmunds, L. N. (1971) Persistent circadian rhythm of cell division in *Euglena*: some theoretical considerations and the problem of intercellular communication. *Biochronometry*, M. Menaker, ed. Nat. Acad. Sci., Washington, D.C.

Edmunds, L. N. (1974a) Circadian clock control of the cell developmental cycle in synchronized cultures of *Euglena. Mechanisms of Regulation of Plant Growth*, R. L. Bieleski, A. R. Ferguson, M. M. Cresswell, eds. Bulletin 12, The Royal Society of New Zealand, Wellington, pp. 287–297.

Edmunds, L. N. (1974b) Phasing effects of light on cell division in exponentially increasing cultures of *Tetrahymena* grown at low temperatures, *Exp. Cell. Res.* **83**, 367–379.

Edmunds, L. N. (1975) Temporal differentiation in *Euglena*: circadian phenomena in non-dividing populations and in synchronously dividing cells. *Colloques Internationaux du C.N.R.S. les Cycles Cellulaires et Leur Blochee Chez Pouesieurs Protistes* **240**, 53–67.

Edmunds, L. N. (1977) Clocked cell cycle clocks. *Waking and Sleeping* **1**, 227–252.

Edmunds, L. N., Chuang, L., Jarrett, R. M., and Terry, O. W. (1971) Long-term persistence of free-running circadian rhythms of cell division in *Euglena* and the implication of autosynchrony. *J. Interdiscipl. Cycle Res.* **2**, 121–132.

Edmunds, L. N., Sulzman, F. M., and Walther, W. G. (1974) Circadian oscillations in enzyme activity in *Euglena* and their relation to the circadian rhythm of cell division. *Chronobiology*, 61–66.

Ehret, C. F. (1971) Discussion In *Biochronometry*, M. Menaker, ed. Nat. Acad. Sci., Washington, D.C., pp. 108–109.

Ehret, C. F. (1974) The sense of time: evidence for its molecular basis in the eukaryotic gene-action system. In *Adv. Biol. Med. Physics*, 15, J. H. Lawrence and J. W. Gofman, eds. Academic, New York, pp. 47–77.

Ehret, C. F., and Barlow, J. S. (1960) Towards realistic model of a biological period-measuring mechanism. In *Cold Spring Harbor Symposium* for Quantitative Biology, **25**, 217–220.

Ehret, C. F., Potter, V. R., and Dobra, K. (1975) Chronotypic action of theophylline and of pentobarbital as circadian zeitgebers in the rat. *Science* **188**, 1212–1215.

Ehret, C. F., and Trucco, E. (1967) Molecular models for the circadian clock. I. The chronon concept. *J. Theor. Biol.* **15**, 240–262.

Ehret, C. F., and Wille, J. J. (1970) The photobiology of circadian rhythms in protozoa and other eucaryotic micro-organisms. Chapter 13 of *Photobiology of Micro-organisms*, Per Halldal, ed. Wiley, New York, pp. 370–416.

Elliott, A. L., Mills, J. N., Minors, D. S., and Waterhouse, J. M. (1972) The effect of real and simulated time-zone shifts upon circadian rhythms of body temperature, plasma 11-

hydroxycorticosteroid and renal excretion in human subjects. *J. Physiol.* (London) **221**, 227–257.

El-Sharkaway, T. Y., and Daniel, E. E. (1975) Electrical activity of small intestinal smooth muscle and its temperature dependence. *Amer. J. Physiol.* **229**, 1268–1276.

Engelberg, J. (1968) On determinstic origins of mitotic variability. *J. Theor. Biol.* **20**, 249–259.

Engelmann, W. (1969) Phase shifting of eclosion in *Drosophila pseudoobscura* as a function of the energy of the light pulse. *Z. vergl. Physiol.* **64**, 111–117.

Engelmann, W., Eger, I., Johnsson, A., and Karlsson, H. G. (1974) Effect of temperature pulses on the petal rhythm of *Kalanchoe*: an experimental and theoretical study. *Int. J. Chronobiol.* **2**, 347–358.

Engelmann, W., and Honegger, H. W. (1966) Tagesperiodische Schlupfrhythmik einer augenlosen *Drosophila melanogaster*-Mutante. *Naturwissenschaften* **22**, 1–2.

Engelmann, W., and Honegger, H. W. (1967) Versuche zur Phasenverschiebung endogener Rhythmen: Blutenblattbewegung von *Kalanchoe blossflediana*. *Z. Naturforsch.* **22b**, 200–204.

Engelmann, W., Karlsson, H. G., and Johnsson, A. (1973) Phase shifts in the *Kalanchoe* petal rhythm caused by light pulses of different durations. *Int. J. Chronobiol.* **1**, 147–156.

Engelmann, W., Johnsson, A., Kobler, H. G., and Schimmel, M. L. (1978) Attenuation of *Kalanchoe's* petal movement rhythm with light pulses. *Physiol. Plant,* **43**, 68–76.

Engelmann, W., and Mack, J. (1978) Different oscillators control the circadian rhythm of eclosion and activity in *Drosophila*. *J. Comp. Physiol.* **127**, 229–237.

Enright, J. T. (1970) Ecological aspects endogenous rhythmicity. *Amer. Rev. Ecol. Syst.* **1**, 221–237.

Enright, J. T. (1972) A virtuoso isopod. Circa-lunar rhythms and their tidal fine structure. *J. Comp. Physiol.* **77**, 141–162.

Enright, J. T. (1975) The circadian tape recorder and its entrainment. in *Physiological Adaptation to the Environment: Proceedings of a Symposium*, F. J. Vernberg, ed. N.Y., Intext Educ. Publ. pp. 465–476.

Enright, J. T., and Hamner, W. M. (1967) Vertical diurnal migration and endogenous rhythmicity. *Science* **157**, 937–941.

Erneux, T., and Herschkowitz-Kaufman, M. (1975) Dissipative structures in two dimensions. *Biophys. Chem.* **3**, 345–354.

Erneux T., and Herschkowitz-Kaufman, M. (1977) Rotating waves as asymptotic solutions of a model chemical reaction. *J. Chem. Phys.* **66**, 248–250.

Esch, H. (1976) Body temperature and flight performance of honey bees in a servo-mechanically controlled wind tunnel. *J. Comp. Physiol.,* **109**, 265–277.

Eskin, A. (1971) Some properties of the system controlling the circadian activity rhythm in sparrows. In *Biochronometry*, M. Menaker, ed. Nat. Acad. Sci. Washington, D.C., pp. 55–80.

Eyring, H., and Henderson, D. editors (1978) *Theoretical Chemistry* 4, Periodicities in chemistry and biology. Academic, N.Y.

Fahrenbach, W. H. (1969) The morphology of the eyes of *Limulus*. II. Ommatidia of the compound eye. *Z. Zellforsch.* **93**, 451–483.

Fantes, P. A., Grant, W. D., Pritchard, R. H., Sudbery, P. E., and Wheales, A. E. (1975) The regulation of cell size and the control of mitosis. *J. Theor Biol.* **50**, 213–244.

Farley, B. G. (1964) Ojai Symposium on *Neural Theory and Modeling*, R. F. Reiss, ed. Stanford University Press, California, p. 43.

Farley, B. G. (1965) A neural network model and the "slow potentials" of electrophysiology. In *Computers in Biomedical Res.*, **1**, 265–294, ed. R. W. Stacey.

Farley, B. G., and Clark, W. A. (1961) Activity in networks of neuron-like elements. In *Information Theory*, 4th London Symposium, C. Cherry, ed. Butterworth's London.

Feinn, D., and Ortoleva, P. (1977) Catastrophe and propagation in chemical reactions. *J. Chem. Phys.* **67**, 2119–2131.

Feldman, J. (1975) Circadian periodicity in *Neurospora*: alteration by inhibitors of cyclic AMP phosphodiesterase. *Science* **190**, 789–790.

Feng, L. J., Rodbard, D., Rebar, R., and Ross, G. T. (1977) Computer simulation of the human pituitary-ovarian cycle: studies of follicular phase estradiol infusions and the midcycle peak. *J. Clin. Endo*, **45**, 775–787.

Fenichel, N., (1974) Asymptotic stability with rate conditions. *Ind. Univ. Math. J.* **23**, 1109–1137.

Fenichel, N., (1977) Asymptotic stability with rate conditions II. *Ind. Univ. Math. J.* **26**, 81–93.

Festinger, L., Allyn, M. R. and White, C. W. (1971) The perception of color with achromatic stimulation. *Vision Res.* **11**, 591–612.

Feynman, R. P., Leighton, R. B., and Sands, M. (1964) *Lectures on Physics*, Addison-Wesley, Reading, Mass., **2**, 9–3.

Field, R. J. (1972) A reaction periodic in space and time. *J. Chem. Ed.* **49**, 308–311.

Field, R. J. (1975) Limit cycle oscillations in the reversible oregonator. *J. Chem. Phys.* **63**, 2289–2296.

Field, R. J., Koros, E., and Noyes, R. M. (1972) Oscillations in chemical systems. II. Thorough analysis of temporal oscillation in the bromate-cerium-malonic acid system. *J. Amer. Chem. Soc.* **94**, 8649–8664.

Field, R. J., and Noyes, R. M. (1972) Explanation of spatial band propagation in the Belousov reaction. *Nature* **237**, 390–392.

Field, R. J., and Noyes, R. M. (1974a) Oscillations in chemical systems. IV. Limit cycle behavior in a model of a real chemical reaction. *J. Chem. Physics* **60**, 1877–1884.

Field, R. J., and Noyes, R. M. (1974b) Oscillations in chemical systems. V. Quantitative explanation of band migration in the Belousov-Zhabotinskii reaction. *J. Chem. Soc.* **96**, 2001–2006.

Field, R. J., and Noyes, R. M. (1975) A model illustrating amplification of perturbations in an excitable medium. In *Symposia of the Faraday Society*. **9**, *Physical Chemistry of Oscillatory Phenomena*, pp. 21–27.

Field, R. J., and Troy, W. C. (1979) The existence of solitary travelling wave solutions of a model of the Belousov-Zhabotinskii reaction. *SIAM J. Appl. M.* **37**, 561–587.

Fife, P. C. (1976a) Singular perturbation and wave front techniques in reaction-diffusion problems. *SIAM-AMS Proceedings* **10**, 23–49.

Fife, P. C. (1976) Pattern formation in reacting and diffusing systems. *J. Chem. Phys.* **64**, 554–564.

Fife, P. C. (1977) Asymptotic analysis of reaction-diffusion wave fronts. *Rocky Mtn. J. Math.* **7**, 389–415.

Fife, P. C. (1978) Asymptotic states for equations of reaction and diffusion. *B. Amer. Math. Soc.* **84**, 693–726.

Fife, P.C. (1979) Mathematical aspects of reacting and diffusing systems. *Lecture Notes in Biomathematics* **28**, pp. 181, ed. S. Levin, Springer Verlag, N.Y.

Firth, D. R. (1966) Interspike interval fluctuations in the crayfish stretch receptor. *Biophys. J.* **6**, 201–215.

Fitzgerald, M., and Zucker, I. (1976) Circadian organization of the estrous cycle of the golden hamster. *Proc. Nat. Acad. Sci.* **73**, 2923–2927.

FitzHugh, R. (1960) Thresholds and plateaus in the Hodgkin-Huxley nerve equations. *J. Gen. Physiol.* **43**, 867–896.

FitzHugh, R. (1961) Impulses and physiological states in theoretical models of nerve membrane. *Biophys. J.* **1**, 445–466.

Flink, E. B., and Doe, R. P. (1959) Effect of sudden time displacement by air travel on synchronization of adrenal function. *Proc. Soc. Exp. Biol. Med.* **100**, 498–501.

Fohlmeister, J., Poppele, R., and Purple, R. (1974) Repetitive firing: dynamic behavior of sensory neurons reconciled with a quantitative model. *J. Neurophysiol.* **37**, 1213–1227.

Franck, U. F. (1956) Models for biological excitatory processes. *Progress in Biophysics and Biophysical Chem.* **6**, 171–206.

Franck, U. F. (1968) Oscillatory behavior, excitability, and propagation behavior on membranes and membrane-like interfaces, pp. 7–30. In: *Biological and Biochemical Oscillators*, B. Chance et al. eds. Academic, London (1973).

Franck, U. F., and Mennier, L. (1953) Gekoppelte periodische Electrodenvor-gänge. *Z. Naturforsch.* **8b/8**, 396–424.

Frank, K. D., and Zimmerman, W. F. (1969) Action spectra for phase shifts of a circadian rhythm in *Drosophila*. *Science* **163**, 688–689.

Frankel, J. (1962) The effect of heat, cold and p-fluorophenylalanine on morphogenesis in synchronized *Tetrahymena pyriformis*. *CR. Trav. Lab. Carlsberg* **33**, 1–52.

Frankel, J. (1969) The relationship of protein synthesis to cell division and oral development in synchronized *Tetrahymena pyriformis* GL-C: an analysis employing cycloheximide. *J. Cell Physiol.* **74**, 135–148.

French, A. R. (1977) Periodicity of recurrent hypothermia during hibernation in the pocket mouse, *Perognathus longimembris*. *J. Comp. Physiol.* **115**, 87–100.

French, V., Bryant, P., and Bryant, S. V. (1976) Pattern regulation in epimorphic fields. *Science* **193**, 969–981.

French V. (1978) Intercalary regeneration around the circumference of the cockroach leg. *J. Embryol. Exp. Morph.* **47**, 53–84.

Frenkel, R. (1966) Reduced diphosphopyridine nucleotide oscillations in cell-free extracts from beef heart. *Arch. Biochem. Biophys.* **115**, 112–121.

Frisch, H. L., and Wasserman, E. (1961) Chemical topology. *J. Amer. Chem. Soc.* **83**, 3789–3795.

Fujii, H., and Sawada, Y. (1978) Phase-difference locking of coupled oscillating chemical systems. *J. Chem. Phys.* **69**, 3830–3832.

Fuller, C. A., Sulzman, F. M., and Moore-ede, M. C. (1978) Thermoregulation is impaired in an environment without circadian time cues. *Science* **199**, 794–796.

Galilei, G. (1623) *Il Saggiatore*. Academia dei Lincei, Rome, cited in *Gravitation* by C. Misner, K. Thorne, and J. Wheeler, 1973, Freeman, San Francisco, p. 304.

Garfinkel, D., Frenkel, R. and Garfinkel, L. (1968) Simulation of the detailed regulation of glycolysis in a heart supernatant preparation. *Comp. Biomed. Res.* **2**, 68–91.

Garrey, W. E. (1914) Nature of fibrillary contraction in the heart. *Amer. J. Physiol.* **33**, 397–414.

Gaze, R. M. (1970) *The Formation of Nerve Connections*, Academic, New York. Acad. Press, N.Y.

Geller, J. S., and Brenner, M. (1978) Measurements of metabolites during c*AMP* oscillations of *Dictyostelium discoideum*. *J. Cell. Phys.* **97**, 413–419.

Gerisch, G. (1961) Zellfunktionen und Zellfunktionswechsel in der Entwicklung von *Dictyostelium discoideum*. II. Aggregation homogener Zellpopulationen und Zentrenbildung. *Dev. Biol.* **3**, 685–700.

Gerisch, G. (1965) Stadienspezifische Aggregationsmuster bei *Dictyostelium discoideum*. *Wilhelm Roux Archiv Entwicklungsmech. Organismen* **156**, 127–144.

Gerisch, G. (1968) Cell aggregation and differentiation in *Dictyostelium*. In *Current Topics in Development Biology*, vol. 3 pp. 157–197, A. Moscona and A. Monroy, eds. Academic, New York.

Gerisch, G. (1971) Periodische signale steuern die musterbildung in zellverbanden. *Naturwissenschaften* **58**, 430–438.

Gerisch, G. (1978) Cell interactions by cyclic AMP in *Dictyostelium*. *Biologie Cellulaire*, **32**, 61–68.

Gerisch, G., Fromm, H., Huesgen, A., and Wick, U. (1975) Control of cell-contact sites by cyclic AMP pulses in differentiating *Dictyostelium* cells. *Nature* **255**, 547–549.

Gerisch, G., and Hess, B. (1974) Cyclic AMP-controlled oscillations in suspended *Dictyostelium* cells: their relation to morphogenetic cell interactions. *Proc. Nat. Acad. Sci.* **71**, 2118–2122.

Gerisch, G., Hulser, D., Malchow, D., and Wick, U. (1975) Cell communication by periodic cyclic-AMP pulses. *Phil. Trans. Roy. Soc. Lond. B.* **272**, 181–192.

Gerisch, G., and Wick, U. (1975) Intracellular oscillation and release of cyclic AMP from *Dictyostelium* cells. *Biochem. Biophys. Res. Comm*, **65**, 364–370.

Gerola, H. and Seiden, P. E. (1978) Stochastic star formation and spiral structure of galaxies. *Astrophys. J.* **223**, 129–139.

Gersbach, A. (1922) Über die Wendigkeit der Kolonieauslaufer des *Bacillus mycoides*. *Centr. F. Bakt.* **88**, 97–103.

Ghosh, A. and Chance B. (1964) Oscillations of glycolytic intermediates in yeast cells. *Biochem. Biophys. Res. Comm.* **16**, 174–181.

Ghosh, A. K., Chance, B. and Pye, E. K. (1971) Metabolic coupling and synchronization of NADH oscillations in yeast cell populations. *Arch. Biochem. Biophys.* **145**, 319–331.

Gibbs, J. W. (1881) in Letter of acceptance of the Rumford medal, 10 Jan 1881; quoted by L. P. Wheeler, *Josiah Willard Gibbs—The History of a Great Mind*, Yale University Press, New Haven, 1962. (From Larry Greller.)

Gierer, A., and Meinhardt, H. (1972) A theory of biological pattern formation. *Kybernetik* **12**, 30–39.

Gierer, A., Meinhardt, H. (1974) *Biological Pattern Formation Involving Lateral Inhibition*, Lectures on Mathematics in the Life Sciences, vol. 7. Amer. Math. Soc., Providence, R.I., pp. 163–183.

Giersch, C., Betz, A., and Richter, O. (1975) A new method for determining the individual reaction rates of the glycolytic system. *Biosystems* **7**, 147–153.

Gilbert, D. A. (1974) Nature of the cell cycle and the control of cell proliferation. *Biosystems* **5**, 197–206.

Gilkey, J. C., Jaffe, L. F., Ridgway, E. B., and Reynolds, G. T. (1978) A free calcium wave traverses the activating egg of the medaka, *Oryzias latipes*. *J. Cell Biol.* **76**, 448–466.

Glansdorff, P., and Prigogine, I. (1971) *Thermodynamic Theory of Structure, Stability, and Fluctuations*. Wiley, London.

Glass, L. (1977) Patterns of supernumerary limb regeneration. *Science* **198**, 321–322.

Glass, L. and Mackey, M. C. (1979) A simple model for phase locking of biological oscillators. Submitted.

Gmitro, J. I., and Scriven, L. E. (1966) A physicochemical basis for pattern and rhythm. In *Intracellular Transport*, K. B. Warren, ed. Academic, New York, pp. 221–255.

Gola, M. (1976) Electrical properties of bursting pacemaker neurons. In *Neurobiology of Invertebrates: Gastropoda Brain*. Hungarian Acad. of Sciences, Budapest, pp. 381–423.

Goldbeter, A. (1973) Patterns of spatiotemporal organization in an allosteric enzyme model. *Proc. Nat. Acad. Sci.* **70**, 3255–3259.

Goldbeter, A. (1975) Mechanism for oscillatory synthesis of *cAMP* in *Dictyostelium discoideum*. *Nature* **253**, 540–542.

Goldbeter, A., and Caplan, S. R. (1976) Oscillatory enzymes. *Ann. Rev. Biophys. and Bioenerg.* **5**, 449–476.

Goldbeter A., and Erneux, T. (1978) Oscillations entretenues et excitabilite dans la reaction de la phosphofructokinase *C.R. Acad. Sc. Paris* **286**, 63–66.

Goldbeter, A., Erneux, T., and Segel, L. A. (1978) Excitability in the adenylate cyclase reaction in *Dictyostelium discoideum*. *FEBS Lett.* **89**, 237–241.

Goldbeter, A. and Lefever, R. (1972) Dissipative structures for an allosteric model, application to glycolytic oscillations. *Biophys, J.* **12**, 1302–1315.

Goldbeter, A., and Nicolis, G. (1976) An allosteric enzyme model with positive feedback applied to glycolytic oscillations. In *Progress in Theoretical Biology*, vol. 4, R. Rosen, ed. Academic, New York, pp. 65–160.

Goldbeter, A., and Segel, L. A. (1977) Unified mechanism for relay and oscillation of cyclic AMP in *Dictyostelium discoideum*. *Proc. Nat. Acad. Sci.* **74**, 1302–1315.

Goldstein, S., and Rall, W. (1974) Changes of action potential shape and velocity for changing core conductor geometry. *Biophys. J.* **14**, 731–757.

Golenhofen, K. (1970) Slow rhythms in smooth muscle. Chap. 10 of *Smooth Muscle*, E. Bulbring, A. Branding, A. Jones, T. Tomita, eds. Williams and Wilkins, Baltimore, pp. 316–342.

Gollub, J., Bruner, T., and Danly, B. (1978) Periodicity and chaos in coupled nonlinear oscillators. *Science* **200**, 48–50.

Gooch, V. D., and Packer, L. (1971) Adenine nucleotide control of heart mitochondrial oscillations. *Biochim. Biophys. Acta* **245**, 17–20.

Gooch, V. D., and Packer, L. (1974a) Oscillatory systems in mitochondria. *Biochim. Biophys. Acta* **346**, 245–260.

Gooch, V. D., and Packer, L. (1974b) Oscillatory states of mitochondria studies on the oscillatory mechanism of liver and heart mitochondria. *Arch. Biochem. Biophys.* **163**, 759–768.

Goodwin, B. C. (1963) *Temporal Organization in Cells: A Dynamic Theory of Cellular Control Processes*, Academic, New York.

Goodwin, B. C. (1964) *A Statistical Mechanics of Temporal Organization in Cells*. Symposia of the Society for Experimental Biology **18**, 301–326.

Goodwin, B. C. (1966) An entrainment model for timed enzyme synthesis in bacteria. *Nature* **209**, 477–481.

Goodwin, B. C. (1967) Biological control processes and time. *Ann. N.Y. Acad. Sci.* **138**, 748–758.

Goodwin, B. C. (1969) Synchronization of *Escherichia coli* in a chemostat by periodic phosphate feeding. *Eur. J. Biochem.* **10**, 511–514.

Goodwin, B. C. (1970) Temporal order as the origin of spatial order in embryos. *Studium Generale* **23**, 273–282.

Goodwin, B. C. (1973) Theoretical physiology. In *Selected Topics in Physics, Astrophysics, and Biophysics*, E. A. deLaredo and N. K. Jurisic, eds., Reidel, Boston, pp. 408–419.

Goodwin, B. C. (1974) Excitability and spatial order in membranes of developing systems. *Farad. Symp. Chem. Soc.* **9**, 226–232.

Goodwin, B. C. (1975) A membrane model for polar ordering and gradient formation. *Adv. Chem. Phys.* **29**, 269–280.

Goodwin, B. C. (1976) *Analytical Physiology of Cells and Developing Organisms.* Academic, New York.

Goodwin, B. C., and Cohen, M. H. (1969) A phase-shift model for the spatial and temporal organization of developing systems. *J. Theor. Biol.* **25**, 49–107.

Gordon, H. (1976) Corneal geometry: an alternative explanation of the effect of 6-diazo-5-oxo-L-norleucine on the development of chick cornea. *Dev. Biol.* **53**, 303–305.

Gordon, H., and Winfree, A. T. (1978) A single spiral artefact in arthropod cuticle. *Tissue and Cell* **10**, 39–50.

Gosling, J. T., and Hundhausen, A. J. (1977) Waves in the solar wind. *Sci. Amer.* **236**(3), 36–43.

Gouras, P. (1958) Spreading depression of activity in amphibian retina. *Amer. J. Physiol.* **195**, 28–32.

Grant, R. P. (1956) Mechanism of A-V arrhythmias, with an electronic analogue of the human A-V node. *Amer. J. Med.* **20**, 334–344.

Grasman J. and M. J. W. Jansen (1979) Mutually synchronized relaxation oscillators as prototypes of oscillating systems in biology. *J. Math. Biol.* **7**, 171–197.

Grattarola, M., and Torre, V. (1977) Necessary and sufficient conditions for synchronization of nonlinear oscillators with a given class of coupling. *IEEE Transactions on Circuits and Systems* **CAS-24**, 209–215.

Green, M. S., and Sengers, J. V. (eds.) (1966) Critical phenomena. *Nat. Bur. Std. Misc. Publ. #273.* (Introduction, p. X).

Greenberg, J. M. (1976) Periodic solutions to reaction-diffusion equations. *J. Appl. Math.* **30**, 199–205.

Greenberg, J. M. (1978) Axi-symmetric, time-periodic solutions of reaction-diffusion equations. *SIAM J. Appl. Math.* **34**, 391–397.

Greenberg, J. M., Hassard, B. D., and Hastings, S. P. (1978) Pattern formation and periodic structures in systems modeled by reaction-diffusion equations. *Bull. Amer. Math. Soc.* **84**, (6). 1296–1327.

Greenberg, J. M., and Hastings, S. P. (1978) Spatial patterns for discrete models of diffusion in excitable media. *SIAM J. Appl. Math.* **34**, 515–523.

Greenwood, J. A., and Durand, D. (1954) The distribution of length and components of the sum of *n* random unit vectors. *Ann. Math. Stat.* **26**, 233–246.

Gregory, J. C. (1931) *A Short History of Atomism.* A. and C. Black, London, pp. 159–162.

Greller, L. D. (1977) Glycolytic oscillations and dynamics on the cusp catastrophe manifold, University of Penn. thesis.

Gross, J. D. (1975) Periodic cyclic AMP signals and cell differentiation. *Nature* **255**, 522–523.

Gross, J. D., Peacey, M. J., and Trevan, D. J. (1976) Signal emission and signal propagation during early aggregation in *Dictyostelium discoideum. J. Cell. Sci.* **22**, 645–656.

Guckenheimer, J. (1975) Isochrons and phaseless sets. *J. Math. Biol.* **1**, 259–273.

Guckenheimer, J. (1976) Constant velocity waves in oscillating chemical reactions. In *Lecture Notes on Mathematics,* P. Hilton, ed. **525**, 99–103.

Guckenheimer, J., Oster, G., and Ipaktchi, A. (1977) The dynamics of density dependent population models. *J. Math. Biol.* **4**, 101–147.

Gul'ko, F. B., and Petrov, A. A. (1970) Mathematical model of the processes of excitation in Purkinje fibres. *Biofizika* **15**, 513–520.

Gul'ko, F. B., and Petrov, A. A. (1972) Mechanism of formation of closed pathways of conduction in excitable media. *Biofizika* **17**, 261–270.

Gulrajani, R. M., and Roberge, F. A. (1978) Possible mechanisms underlying bursting pacemaker discharges in invertebrate neurons. *Fed. Proc.* **37**, 2146–2152.

Gumbel, E. J., Greenwood, J. A., and Durand, D. (1953) The circular normal distribution: theory and tables. *Amer. Stat. Assn. J.* **48**, 131–152.

Gunn, D. L. (1937) Menstrual periodicity: statistical observations on a large number of normal cases. *J. Gyn. Brit. Empire.* **44**, 839–877.

Gurel, O. (1972) Bifurcation models of mitosis. *Physiol. Chem. Physics* **4**, 139–152.

Guttes, E., Devi, V. R., and Guttes, S. (1969) Synchronization of mitosis in *Physarum polycephalum* by coalescence of postmitotic and premitotic plasmodial fragments. *Experientia* **25**, 615–616.

Guttman, R., Lewis, S. and Rinzel, J. (1980) Control of repetitive firing in squid axon membrane as a model for a neuron oscillator, *J. Physiol.* (Lond.), **305**, 377–395.

Gwinner, E. (1974) Testosterone induces "splitting" of circadian locomotor activity rhythms in birds. *Science* **185**, 72–74.

Hahn, H. S., Nitzan, A., Ortoleva, P., and Ross, J. (1974) Threshold excitations, relaxation oscillations, and effect of noise in an enzyme reaction. *Proc. Nat. Acad. Sci.* **71**, 4067–4071.

Halaban, R. (1968) The circadian rhythm of leaf movement of *Coleus blumei x C. frederici*, a short day plant. II. The effects of light and temperature signals. *Plant Physiol.* **43**, 1887–1893.

Halaban, R. (1969) Effects of light quality on the circadian rhythm of leaf movement of a short-day plant. *Plant Physiol.* **44**, 973–977.

Halberg, F. (1974) Protection by timing treatment according to bodily rhythms—an analogy to protection by scrubbing before surgery. *Chronobiologia* **1**, 27–68.

Halberg, F. (1977) Implications of biologic rhythms for clinical practice. *Hosp. Pract.* **12**, 139–149.

Halberg, F., and Connor, R. L. (1961) Circadian organization and microbiology: variance spectra and a periodogram on behavior of *Escherichia coli* growing in a fluid culture. *Proceedings of the Minnesota Academy of Science* **29**, 227–239.

Halberg, F., Johnson, E. A., Brown, B. W., and Bittner, J. J. (1960) Susceptibility rhythm to *E. coli* endotoxin and bioassay. *Proc. Soc. Exp. Biol. Med.* **103**, 142–144.

Hale, J. K. (1963) *Oscillations in Nonlinear Systems.* McGraw-Hill, New York, p. 94, Theorem 10-1.

Hale, J. K. (1969) *Ordinary Differential Equations.* Wiley-Interscience, New York, p. 217, Theorem 2.1.

Hale, J. K. (1977) *Theory of Functional Differential Equations.* Applied Math. Sciences, vol. 3, Springer-Verlag, New York and Heidelberg.

Halvorson, H., Carter, B., and Tauro, P. (1971) Synthesis of enzymes during the cell cycle. In *Advances in Microbial Physiology* **6**, 47–106.

Halvorson, H., Gorman, J., Tauro, P., Epstein, R., and LaBerge, M. (1964) Control of enzyme synthesis in synchronous cultures of yeasts. *Fed. Proc.* **23**, 1002–1008.

Hamburger, K. (1962) Division delays induced by metabolic inhibitors in synchronized cells of *Tetrahymena pyriformis. CR Trav. Lab. Carlsberg.* **32**, 359–370.

Hamner, K. C., Finn, J. C., Sirohi, G. S., Hoshizaki. T., and Carpenter, B. H. (1962) The biological clock at the South Pole. *Nature* **195**, 476–480.

Hamner, K. C., and Hoshizaki, T. (1974) Photoperiodism and circadian rhythms: an hypothesis. *Bioscience* **24**, 407–414.

Hanson, F. E. (1978) Comparative studies of firefly pacemakers *Fed. Proc.* **37**, 2158–2164.

Hanson, F. E., Case, F., Buck, E., and Buck, J. (1971) Synchrony and flash entrainment in a New Guinea firefly. *Science* **174**, 161–164.

Harcombe, E. S., and Wyman, R. J. (1977) Output pattern generation by *Drosophila* flight motoneurons. *J. Neurophys.* **40**, 1066–1078.

Hardeland, R. (1972) Circadian rhythmicity of tyrosine amino-transferase in suspension of isolated rat liver cells. *J. Interdiscipl. Cycle Res.* **3**, 109–114.

Hardeland, R. (1973) Circadian rhythmicity in cultured liver cells. *Int. J. Biochem.* **4**, 581–595.

Hardt, S., Naparstek. A., Segel, L. A., and Caplan, S. R. (1976) Spatiotemporal structure formation and signal propagation in a homogeneous enzymatic membrane. In *Analysis and Control of Immobilized Enzyme Systems*, D. Thomas and J. P. Kernevez, eds. North-Holland, Amsterdam.

Harker, E. J. (1958) Diurnal rhythms in the animal kingdom. *Camb. Phil. Soc. Biol. Rev.* **33**, 1–52.

Harmon, L. D., and Lewis, E. R. (1966) Neural modelling. *Physiol. Rev.* **46**, 513–591.

Harris, G. W., and Naftolin, F. (1970) The hypothalamus and control of ovulation. *Br. Med. Bull.* **26**, 3–9.

Harris, P. J. C. and Wilkins, M. B. (1978) Circadian rhythm in *Bryophyllum* leaves: phase control by radiant energy. *Planta* **143**, 323–328.

Hartline, D. K. (1976) Simulation of phase-dependent pattern changes to perturbations of regular firing in crayfish stretch receptor. *Brain Res.* **110**, 245–257.

Hartline, D. K., Gassie, D. V., and Sirchia, C. D. (1980) Burst reset properties in an endogenously bursting network-driver cell. *Brain Res.* Withdrawn in exasperation.

Hartwell, L., Culotti, J., Pringle, J., and Reid, B. (1974) Genetic control of the cell division cycle in yeast. *Science* **183**, 46–51.

Hastings, J. W. (1964) The role of light in persistent daily rhythms. In *Photophysiology.* vol. 1, A. C. Giese, ed., Academic, New York.

Hastings, J. W., and Sweeney, B. M. (1957) On the mechanism of temperature independence in a biological clock. *Proc. Nat. Acad. Sci.* **43**, 804–811.

Hastings, J. W., and Sweeney, B. M. (1958) A persistent diurnal rhythm of luminescence in *Gonyaulax polyedra. Biol. Bull.* **115**, 440–458.

Hastings, J. W., and Sweeney, B. M. (1964) Phased cell division in marine dino-flagellates. Chap. 10., In *Synchrony in Cell Division and Growth*, Erik Zeuthen, ed., Interscience, New York, pp. 307–321.

Hastings, J. W., and Schweiger, H. G. (1976) *The Molecular Basis of Circadian Rhythms*, Abakon, Berlin.

Hastings, S. P. (1975) Some mathematical problems from neurobiology. *Amer. Math. Monthly* **82**, 881–895.

Hastings, S. P. (1976) Periodic plane waves for the oregonator. *Stud. Appl. Math.* **55**, 293–299.

Hastings, S. P., and Murray, J. D. (1975) The existence of oscillatory solutions in the Field-Noyes model for the Belousov-Zhabotinskii reaction. *SIAM J. Appl. Math.* **28**, 678–688.

Haus, E., Halberg, F., Kuhl, J. F. W., and Lakatua, D. J. (1974) Chronopharmacology in animals. *Chronobiologia* **1**, 122–156.

Haus, E., Halberg, F., Scheving, L. E., Pauly, J. E., Cardoso, S., Kuhl, J. F. W., Sothern, R. B., Shiotsuka, R. N., Whang, D. S. (1972) Increased tolerance of leukemic mice to arabinosyl cytosine with schedule adjusted to circadian system. *Science* **177**, 80–82.

Hauty, G. T., and Adams, T. (1965) Phase shifting of the human circadian system. In *Circadian Clocks*, J. Aschoff, ed. North-Holland, Amsterdam, pp. 413–425.

Hayashi, C. (1964) *Nonlinear oscillations in physical systems*, Electrical and Electronic Engineering Series. McGraw-Hill. New York.

Hayes, D. K., Sullivan, W. N., Oliver, M. Z., and Schecter, M. S. (1970) Photoperiod manipulation of insect diapause: a method of pest control? *Science* **169**, 382–383.

Heide, O. M. (1977) Photoperiodism in higher plants: an interaction of phytochrome and circadian rhythms. *Physiol. Plant.* **39**, 25–32.

Hein, I. (1930) Liesegang phenomena in fungi. *Amer. J. Bot.* **17**, 143–151.

Heinrich, B., and Bartholomew, G. (1972) Temperature control in flying moths. *Sci. Amer.* **226**, (6) 70–77.

Heinrich, B., and Casey, T. M. (1973) Metabolic rate and endothermy in sphinx moths. *J. Comp. Physiol.* **82**, 195–206.

Hejnowicz, Z. (1970) Propagated disturbances of transverse potential gradient in intracellular fibrils as the source of motive forces for longitudinal transport in cells. *Protoplasma* **71**, 343–364.

Hellbrugge, T. (1960) The development of circadian rhythms in infants. *Cold Spring Harbor Symposia on Quantitative Biology* **25**, 311–323.

Hellwell E. J. and I. R. Epstein (1979) Chemical oscillation and "chaos" in a single system. *J. Phys. Chem.* **83**, 1359–1361.

Helmholtz, H. (1867) On integrals of the hydrodynamic equations, which express vortex motion. *Phil. Mag. Suppl.* **33**, 485–510.

Hengstenberg, R. (1971) Das Augenmuskelsystem der Stubenfliege *Musca domestica*. I. Analyse der "clock-spikes" und ihrer Quellen, *Kybernetik*, **2**, 56–77.

Herschkowitz-Kaufman, M. (1970) Structures dissipatives dans une reaction chemique homogene. *CR Acad. Sc. Paris* **270**, 1049–1052.

Herschkowitz-Kaufman, M., and Nicolis, G. (1972) Localized spatial structures and nonlinear chemical waves in dissipative systems. *J. Chem. Phys.* **56**, 1890–1895.

Hess, B., and Boiteux, A. (1968) *Control of Glycolysis in Regulatory Functions of Biological Membranes*, J. Jarnefelt, ed. BBA Library, vol. 11. Elsevier, Amsterdam.

Hess, B., and Boiteux, A. (1971) Oscillatory phenomena in biochemistry. *Ann. R. Biochem.* **40**, 237–258.

Hess, B., Boiteux, A., and Kruger, J. (1969) Cooperation of glycolytic enzymes, in *Enzyme Regul.* **7**, 149–167, Pergamon Press, London.

Higgins, J. (1964) A chemical mechanism for oscillation of glycolytic intermediates in yeast cells. *Proc. Nat. Acad. Sci.* **51**, 989–994.

Higgins, J. (1967) The theory of oscillating reactions. *Ind. Eng. Chem.* **59**, 18–62.

Higgins, J., Frendel, R., Hulme, E., Lucas, A., and Rangazas, G. (1968) The control theoretic approach to the analysis of glycolytic oscillators. In *Biological and Biochemical Oscillators*, B. Chance et al., eds. Academic, New York, 1973, pp. 127–175.

Hill, A. V. (1933) Wave transmission as the basis of nerve activity. *Cold Spring Harbor Symposia on Quantitative Biology* **1**, 146–151.

Hoagland, H. (1933) The physiological control of judgement of duration: evidence for a chemical clock. *J. Gen. Psychol.* **9**, 267–287.

Hodgkin, A. L., and Huxley, A. F. (1952) A quantitative description of membrane current and its application to conduction and excitation in nerve. *J. Physiol.* **117**, 500–544.

Hoffmann, B. F., and Cranefield, P. F. (1960) *The Electrophysiology of the Heart*, McGraw-Hill, New York.

Hoffmann, K. (1960) Experimental manipulation of the orientational clock in birds. In: *Cold Spring Harbor Symposium on Quantitative Biology* **25**, 379–387.

Hoffmann, K. (1969a) Circadiane periodik bei tupajas in konstanten bedingungen. *Zool. Anz. Suppl. Verh. Zool. Ges.* **33**, 171–177.

Hoffmann, K. (1969b) Zum Einfluss der Zeitgeberstärke auf die Phasenlage der synchronisierten circadianen Periodik. *Z. vergl. Physiol.* **62**, 93–110.

Hoffmann, K. (1976) *The Adaptive Significance of Biological Rhythms Corresponding to Geophysical Cycles*, Dahlem Workshop on *The Molecular Basis of Circadian Rhythms*, pp. 63–76 J. W. Hastings and H. G. Schweiger, eds. Abakon, Berlin.

Hoffmann, K. (1971a) Biological clocks in animal orientation and in other functions. *Proc. Int.*

Symp. Circadian Rhythmicity, J. F. Bierhuizen, ed. Centre for Agri. Publ. and Doc., pp. 175–205 Wageningen.

Hoffmann, K. (1971b) Splitting of the circadian rhythm as a function of light intensity. In *Biochronometry*, M. Menaker, ed. Nat. Acad. Sci., Washington, D.C., pp. 134–151.

Holaday, D. A., Volk, H., and Mandell, J. (1958) Electrical activity of the small intestine with special reference to the origin of rhythmicity. *Amer. J. Physiol.* **195**, 505–515.

Honegger, H. W. (1967) Zur Analyse der Wirkung von Lichtpulsen auf das Schlüpfen von *Drosophila pseudoobscura*. *Z. vergl. Physiol.* **57**, 244–262.

Horridge, G. A. (1967) The eyes of the firefly *Photuris*. *Proc. Roy. Soc. B.* **171**, 445–463.

Horrobin, D. F. (1969) The female sex cycle. *J. Theor. Biol.* **22**, 80–88.

Howard, F. (1932) Nuclear division in plasmodia of *Physarum*. *Ann. Bot.* **46**, 451–458.

Howard, L. N., and Kopell, N. (1974) Wave trains, shock fronts, and transition layers in reaction-diffusion equations. *SIAM AMS Proceedings* **8**, 1–12. (Ed. D. Cohen) Amer. Math. Soc., Providence, R. I.

Howard, L. N., and Kopell, N. (1977) Slowly varying waves and shock structures in reaction-diffusion equations. *Stud. Appl. Math.* **56**, 95–145.

Hunter, J. E. (1974) A model of activity and sleep. *J. Theor. Biol.* **46**, 481–499.

Huygens, C. (1673) *Horologium Oscillatorium*, F. Muguet, Paris. pp. 117.

Ingold, C. T., and Cox, V. J. (1955) Periodicity of spore discharge in *Daldinia*. *Ann. Bot. N.S.* **19**, 201–209.

Inskeep, E. K., Howland, B. E., Pope, A. L., and Casida, L. E. (1964) Some effects of progesterone on experimentally induced corpora lutea in ewes. *J. Anim. Sci.* **23**, 790–794.

Jack, J. J. B., Noble, D., and Tsien, R. W. (1975) *Electric Current Flow in Excitable Cells*, Clarendon, Oxford.

Jacklet, J. W. (1977) Neuronal circadian rhythm: phase shifting by a protein synthesis inhibitor. *Science* **198**, 69–71.

Jacklet, J. W., and Geronimo, J. (1971) Circadian rhythms: population of interacting neurons. *Science* **174**, 299–302.

Jacobson, M. (1970) *Developmental Neurobiology*, Holt, Rinehart and Winston, New York.

Jalifé, J. and Antzelevitch, C. (1979) Phase resetting and annihilation of pacemaker activity in cardiac tissues. *Science* **206**, 695–697.

Jalifé, J., Hamilton, A. J., Lamanna, V. R., and Moe, G. K. (1980) Effects of current flow on pacemaker activity of the isolated kitten SA node. *Am. J. Physiol.* **238**, 307–316.

Jalifé J., and Moe, G. K. (1976) Effect of electrotonic potentials on pacemaker activity of canine purkinje fibers in relation to parasystole, *Circ. Res.* **39**, 801–808.

Jalifé J., and Moe, G. K. (1979) Phasic effects of vagal stimulation on pacemaker activity of the isolated sinus node of the young cat. *Circul. Res.* **45**, 595–607.

Jenerick, H. (1963) Phase plane trajectories of the muscle spike potential. *Biophys. J.* **3**, 363–377.

Jensen, D. (1966) The hagfish. *Sci. Amer.* **214**, (2), 82–90.

Jerebzoff, S. (1965) Growth rhythms. in *The Fungi*, vol. 1, G. C. Ainswoth, ed., Academic Press, N.Y., pp. 625–645.

Johnson, M. (1939) Effect of continuous light on periodic spontaneous activity of white-footed mice (*Peromyscus*). *J. Exp. Zool.* **82**, 315–328.

Johnson, A. (1973) Oscillatory transpiration and water uptake of *Avena* plants. I. Preliminary observations. *Physiol. Plant* **28**, 40–50.

Johnsson, A. (1976) Oscillatory water regulation in plants. *Bull. Inst. Math. Appl.* **12**, 22–26.

Johnsson, A., Brogårdh, T., and Holje, φ. (1979) Oscillatory transpiration of *Avena* plants: perturbation experiments provide evidence for a stable point of singularity. *Physiol. Plant.* **45**, 393–398.

Johnsson, A., and Israelsson, D. (1969) "Phase-shift in geotropical oscillations—a theoretical and experimental study. *Physiol. Plant.* **22**, 1226–1237.

Johnsson, A., and Karlsson, H. G. (1971) Biological rhythms: singularities in phase-shift experiments as predicted from a feedback model. In *Proc. of First European Biophysical Congress*, E. Broda and A. Locker, eds. Vienna Acad. Med. Press, pp. 263–267.

Johnsson, A., and Karlsson, H. G. (1972) A feedback model for biological rhythms. I. Mathematical description and basic properties of the model. *J. Theor. Biol.* **36**, 153–174.

Johnsson, A., Karlsson, H. G., and Engelmann, W. (1973) Phase shift effects in the *Kalanchoe* petal rhythm due to two or more light pulses. *Physiol. Plant.* **28**, 134–142.

Jongsma, H. J., Masson-Pevet, M., Hollander, C. C., and J. de Bruyne. (1975) Synchronization of the beating frequency of cultured rat heart cells. In *Development and Physiological Correlates of Cardiac Muscle*. M. Leiberman and T. Sano, eds. Raven, New York, pp. 185–196.

Jorne, J. (1977) The diffusive Lotka-Volterra oscillating system. *J. Theor. Biol.* **65**, 133–139.

Jouvet, M., Mouret, J., Chouvet, G., and Siffre, M. (1974) Toward a 48 hour day: experimental bicircadian rhythum in man. Chapter 41 in the *Neurosciences. 3rd Study Program* F. O. Schmidt and F. G. Worden, eds. M.I.T., Cambridge, pp. 491–497.

Joyner, R. W., Ramon, F., and Moore, J. W. (1975) Simulation of action potential propagation in an inhomogeneous sheet of coupled excitable cells. *Circ. Res.* **36**, 654–661.

Judd, H. L., and Eichelberger, M. (1967) The effects of anoxia on the circadian rhythm of eclosion in *Drosophila pseudoobscura*. Bachelor's thesis, Princeton University.

Jung, C., Rothstein, A. (1967) Cation metabolism in relation to cell size in synchronously grown tissue culture cells. *J. Gen. Physiol.* **50**, 917–932.

Junge, D., and Stephens, C. L. (1973) Cyclic variation of potassium conductance of a burst-generating neurone of *Aplysia*. *J. Physiol.* (London) **235**, 155–181.

Kaempfer, E. (1727) *The History of Japan* (*with a description of the Kingdom of Siam*). Transl. J. G. Scheuchzer. Hans Sloane, London. Two vols. in one. See vol. 1, p. 45 or pp. 78–79 of vol. 1 of 1906 reprint by J. McLehose and Sons, Glasgow.

Kalmus, H. (1935) Periodizitat und Autochronie (Ideochronie) als zeitregelnde Eigenschaften der Organismen. *Biol. Generalis* **11**, 93–114.

Kalmus, H. (1940) Diurnal rhythms in the axolotl larva and in *Drosophila*. *Nature* **145**, 72–73.

Kalmus, H., and Wigglesworth, L. A. (1960) Shock excited systems as models for biological rhythms. In *Cold Spring Harbor Symposium on Quantitative Biology*, **25**, 211–216.

Karakashian, M., and Schweiger, H. G. (1976) Circadian properties of the rhythmic system in individual nucleated and enucleated cells of *Acetabularia mediterranea*. *Exp. Cell Res.* **97**, 366–377.

Karfunkel, H. R. (1975) Zur Theorie der Anregbarkeit und Ausbreitung von Erregungswellen in chemischen Reaktionssystemen. Dissertation, Universitat zu Tubingen.

Karfunkel, H. R., and Kahlert, C. (1977) Excitable chemical reaction systems. II. Several Pulses on the ring fiber. *J. Math. Biol.* **4**, 183–185.

Karfunkel, H. R., and Seelig, F. F. (1972) Reversal of inhibition of enzymes and the model of a spike oscillator. *J. Theor. Biol.* **36**, 237–253.

Karfunkel, H. R., and Seelig, F. F. (1975) Excitable chemical reaction systems. I. Definition of excitability and simulation of model systems. *J. Math. Biol.* **2**, 123–132.

Karlsson, H. G., and Johnsson, A., (1972) A feedback model for biological Rhythms. II. Comparisons with experimental results, especially on the petal rhythm of *Kalanchoe*. *J. Theor. Biol.* **36**, 175–194.

Karve, A. D., and Salanki, A. S. (1964) The phenomenon of phase shift in geotropically induced feedback oscillations in *Carthamus tinctorius L.* seedlings. *Z. Bot.* **52**, 113–117.

Kuramoto, Y. (1975) Self-entrainment of a population of coupled nonlinear oscillators. In *Int. Symp. on Math. Prob. in Theor. Physics*, H. Araki, ed., Springer-Verlag New York, Lecture Notes in Physics **39**, 420–422.

Kuramoto, Y. (1978) Diffusion-induced chaos in reaction system. *Prog. Theor. Phys.* **64**, (supplement). 346–367.

Kuramoto, Y., and Tsuzuki, T. (1975) On the formation of dissipative structures in reaction-diffusion systems. *Prog. Theor. Phys.* **54**, 687–699.

Kuramoto, Y., and Tsuzuki, T. (1976) Persistent propagation of concentration waves in dissipative media far from thermal equilibrium. *Prog. Theor. Phys.* **55**, 356–369.

Kuramoto, Y., and Yamada, T. (1975) Pattern formation in chemical reactions. *Prog. Theor. Phys.* **54**, 1582–1583.

Kuramoto, Y., and Yamada, T. (1976) Pattern formation in oscillatory chemical reactions. *Prog. Theor. Phys.* **56**, 724–740.

Kuriyama, H., and Suzuki, H. (1976) Changes in electrical properties of rat myometrium during gestation and following hormonal treatments. *J. Physiol.* **260**, 315–333.

Lacey, L. (1975) *Lunaception: A Feminine Odyssey into Fertility and Contraception*, Coward.

LaFontaine, E. J., Lavernhe, J., Courillon, J., Medvedeff, M., and Ghata, J. (1967) Influence of air travel east-west and vice versa on circadian rhythms of urinary elimination of potassium and 17-hydroxycorticosteroids. *Aerospace Med.* **38**, 944–947.

Lamport, H. (1940) Periodic changes in blood estrogen. *Endocrinology* **27**, 673–679.

Land, E. H. (1977) The retinex theory of color vision. *Sci. Amer.* **237**, (6), 108–128.

Lanzavecchia, G. (1977) Morphological modulations in helical muscles (Aschelminthes and Annelida). *Int. Rev. Cytol.* **51**, 133–186.

Lavenda, B., Nicolis, G., and Herschkowitz-Kaufman, M. (1971) Chemical instabilities and relaxation oscillations. *J. Theor. Biol.* **32**, 283–292.

Lawrence, P. A. (1970) Polarity and patterns in postembryonic development of insects. *Adv. Ins. Physiol.* **7**, 197–265.

Lawrence, P. A. (1971) The organization of the insect segment. In *Symp. Soc. Exp. Biol. 25 Control Mechanisms of Growth and Differentiation*, 379–390.

Lawrence, P. A., Crick, F. H. C., and Munro, M. (1972) A gradient of positional information in an insect, *Rhodnius*. *J. Cell. Sci.* **11**, 815–853.

Lawton, I. E., and Schwartz, N. B. (1967) Pituitary-ovarian function in rats exposed to constant light: a chronological study. *Endocrinology*. **81**, 497–508.

Leao, A. A. P. (1944) Spreading depression of activity in the cerebral cortex. *J. Neurophysiol.* **7**, 359–390.

Lees, A. D. (1972) The role of circadian rhythmicity in photoperiodic induction in animals. In *Circadian Rhythmicity*. J. F. Bierhuizen et al., eds. Proc. Intern. Symp. Circ. Rhythmicity, Wageningen, 87–110.

Leigh, E. G. (1965) On the relation between the productivity, diversity, and stability of a community. *Proc. Nat. Acad Sci.* **53**, 777–781.

Leutscher-Hazelhoff, J. T., and Kuiper, J. W. (1966) Clock spikes in the *Calliphora* optic lobe. Inter. Symposium on the Functional Organization of the Compound Eye, C. G. Bernhard, ed. Pergamon, New York, pp. 483–492.

Levin, S. A., ed. (1978) *Studies in Mathematical Biology* (Studies in mathematics 15) Mathe. Assn. Amer., Washington.

Lewis, T. (1911) *The Mechanism of the Heartbeat*. Shaw and Sons, London.

Li, Tien-Yien., and Yorke, J. A. (1975) Period three implies chaos. *Amer. Math. Mo.* **82**, 985–992.

Lickey, M. E., Block, G. D., and Hudson, D. J. (1976) Circadian oscillators and photoreceptors in the Gastropod *Aplysia*. *Photochem. Photobiol.* **23**, 253–273.

Ličko, V., and Landahl, H. D. (1971) Analog simulation of A-V conduction block and Wenckebach phenomenon. *Comput. Biol. Med.* **1**, 185–192.

Liesegang, R. E. (1939) Spiralenbildung bei Niederschlägen in Gallerten. *Koll. Z.* **87**, 57–58.

Linkens, D. A. (1974) Analytical solution of large numbers of mutually coupled nearly sinusoidal oscillators. *IEEE Trans. Circ. and Sys.* **CAS-21**, 294–300.

Linkens, D. A. (1976) Stability of entrainment conditions for a particular form of mutually coupled van der Pol oscillators. *IEEE Trans. Circ. and Sys.* **CAS-23**, 113–121.

Linkens, D. A. (1977) The stability of entrainment conditions for RLC coupled van der Pol oscillators used as a model for intestinal electrical rhythms. *Bull. Math. Biol.* **39**, 359–372.

Linkens, D. A., and Datardina, S. (1977) Frequency entrainment of coupled Hodgkin-Huxley-type oscillators for modeling gastro-intestinal electrical activity. *IEEE Trans. Biomed. Eng.* **BME-24**, 362–365.

Linkens, D. A., Duthie, H. L., and Brown, B. H. (1973) Dynamical model of the human small intestine. In *Regulation and Control in Physiological Systems*, A. S. Iberall, A. C. Guyton, eds. Instr. Soc. of Amer., Pittsburgh, pp. 190–142.

Linkens, D. A., Rimmer, S. J., and Datardina, S. P. (1978) Spectral analysis of coupled non-linear oscillators under modulation conditions with reference to intestinal modelling. *Comput. Biol. Med.* **8**, 125–137.

Linkens, D. A., Taylor, I., and Duthie, H. L. (1976) Mathematical modeling of the colorectal myoelectrical activity in humans. *IEEE Trans. Biomed. Eng.* **BME-23**, 101–110.

Livolant, F., Giraud, M., and Bouligand, Y. (1978) A goniometric effect observed in sections of twisted fibrous materials. *Biol. Cellulaire* **31**, 159–168.

Llanos, J. M. E., and Nash, R. E. (1970) Mitotic circadian rhythm in a fast-growing and a slow-growing hepatoma: mitotic rhythm in hepatomas. *J. Nat. Canc. Inst.* **44**, 581–585.

Locke, M. (1959) Skin grafting in insects. *New Scientist* **5**, 1337–1339.

Locke, M. (1960) The cuticular pattern in an insect: the intersegmental membranes. *J. Exp. Biol.* **37**, 398–406.

Locke, M. (1961) Pore canals and related structures in insect cuticle. *J. Biophys. Biochem. Cytol.* **10**, 589–618.

Locke, M. (1967) The development of patterns in the integument of insects. *Advances in Morphogenesis* **6**, 33–88.

Loftus-Hills, J. J. (1974) Analysis of an acoustic pacemaker in Strecker's chorus frog, *Pseudacris streckeri* (Anura: Hylidae). *J. Comp. Physiol.* **90**, 75–87.

Loher, W. (1972) Circadian control of stridulation in the cricket *Teleogryllus commodus* Walker. *J. Comp. Physiol.* **79**, 173–190.

Lotka, A. J. (1920) Undamped oscillations derived from the law of mass action. *J. Amer. Chem. Soc.* **42**, 1595–1598.

Luce, G. G. (1971) *Biological Rhythms in human and Animal Physiology*, Dover, New York. (This is the same as *Biological Rhythms in Psychiatry and Medicine*. Public Health Service Publication 2088 (1970), Superintendent of Documents, Washington, D.C.).

Lukat, R. (1978) Circadian growth layers in the cuticle of behaviourally arrhythmic cockroaches (*Blaberus fuscus*, Ins., Blattoidea). *Experientia* **34**, 477.

MacGregor, R. J., and Lewis, E. R. (1977) *Neural Modelling*, Plenum, New York, pp. 153–159.

Mackay, S. A. (1978) Computer simulation of aggregation in *Dictyostelium discoideum*. *J. Cell. Sci.* **33**, 1–16.

Maden, M. (1977) The regeneration of positional information in the amphibian limb. *J. Theor. Biol.* **69**, 735–753.

Maden, M. and Turner, R. N. (1978) Supernumerary limbs in the axolotl. *Nature* **273**, 232–235.

Maier, R. W. (1973) Phase-shifting of the circadian rhythm of eclosion in *Drosophila pseudoobscura* with temperature-pulses. *J. Interdiscipl. Cycle Res.* **4**, 125–135.

Malchow, D., Nanjundiah, V. and Gerisch G. (1978) pH-oscillations in cell suspensions of *Dictyostelium discoideum*: their relation to cyclic-AMP signals. *J. Cell. Sci.* **30**, 319–330.

Mano, Y. (1975) Systems constituting the metabolic sequence in the cell cycle. *Biosystems* **7**, 51–65.

Marek, M., and Stuchl, I. (1975) Synchronization in two interacting oscillatory systems. *Biophys. Chem.* **4**, 241–248

Martins-Ferreira, H., de Oliveiro Castro, G., Struchinea, C. J., and Rodrigues, P. S. (1974) Circling spreading depression in isolated chick retina. *J. Neurophys.* **37**, 773–784.

Mathieu, P. A., and Roberge, F. A. (1971) Characteristics of pacemaker oscillations in *Aplysia* neurones. *Can. J. Physiol. Pharmacol.* **49**, 787–795.

May, R. M. (1972) Limit cycles in predator-prey communities. *Science* **177**, 900–902.

May, R. M. (1973) *Stability and Complexity in Model Ecosystems*, Princeton University Press, Princeton.

May, R. M. (1976) Simple mathematical models with very complicated dynamics. *Nature* **261**, 459–467.

Mayer, A. G. (1908) Rhythmical pulsation in scyphomedusae. *Papers of the Tortugas Lab. of the Carnegie Inst. of Wash*, **1**, 115–131.

Mayer, A. G. (1914) The relation between degree of concentration of one electrolytes of seawater and rate of nerve conduction in *Cassiope*. *Papers of the Tortugas Lab. of the Carnegie Inst. of Wash.*, **6**, 25–54.

Mayer W. and Sadleder, D. (1972) Different light intensity dependence of the free-running periods as the cause of internal desynchronization of circadian rhythms in *Phaseolus coccineus*. *Planta* (Berl) **108**, 173–178.

Mayeri, E. (1973) A relaxation oscillator description of the burst-generating mechanism in the cardiac ganglion of the lobster. *J. Gen. Physiol.* **62**, 473–488.

McAllister, R. E., Noble, D., and Tsien, R. W. (1975) Reconstruction of the electrical activity of cardiac Purkinje fibres. *J. Physiol.* (London) **251**, 1–59.

McClintock, M. K. (1971) Menstrual synchrony and suppression. *Nature* **229**, 244–245.

McCormick, J. M., and Salvadori, M. (1964) *Numerical Methods in Fortran*, Prentice-Hall, N.J.

McCoy, E. J., and Baker, R. D. (1969) Intestinal slow waves: decrease in propagation velocity along upper small intestine. *Amer. J. Digestive Diseases* **14**, 9–13.

McDonald, T. F., and Sachs, H. G. (1975a) Electrical activity in embryonic heart cell aggregates: developmental aspects. *Pflügers Arch.* **354**, 151–164.

McDonald, T. F., and Sachs, H. G. (1975b) Electrical activity in embryonic heart cell aggregates: pacemaker oscillations. *Pflügers Arch.* **354**, 165–176.

McGovern, J. P., Smolensky, M. H., and Reinberg, A. (1977) *Chronobiology in Allergy and Immunology*, Charles C. Thomas, Springfield, Ill.

McKean, H. P., Jr. (1970) Nagumo's equation. *Adv. in Math.* **4**, 209–223.

McMurry, L., and Hastings, J. W. (1972) No desynchronization among four circadian rhythms in the unicellular alga, *Gonyaulax polyedra*. *Science* **175**, 1137–1139.

McWilliam, J. A., (1887) Fibrillar contraction of the heart. *J. Physiol.* **8**, 296–310.

Medugorac, I., and Lindauer, M. (1967) Das Zeitgedächtnis der Bienen unte dem Einfluss von Narkose und von sozialen Zeitgebern. *Z. vergl. Physiol.* **55**, 450–474.

Meinhardt, H., and Gierer, A. (1974) Applications of a theory of biological pattern formation based on lateral inhibition. *J. Cell Sci.* **15**, 321–346.

Menaker, M. (1974) Aspects of the physiology of circadian rhythmicity in the vertebrate central nervous system. *The Neurosciences: Third Study Program* F. O. Schmidt and F. G. Worden, eds. M.I.T. Press. Cambridge, **3**, 479–489.

Menaker, M., Takahashi, J. S., and Eskin, A. (1978) The physiology of circadian pacemakers. *Ann. Rev. Physiol.* **40**, 501–526.

Menaker, M., and Zimmerman, N. (1976) Role of the pineal in the circadian system of birds. *Amer. Zool.* **16**, 45–55.

Menaker, W., and Menaker, A. (1959) Lunar periodicity as a unit of time in human reproduction. *Amer. J. Obstet. and Gynec.* **77**, 905–914.

Menzel, R., Moch, K., Wladarz, G., and Lindauer, M. (1969) Tagesperiodische Ablagerungen in der Endokutikula der Honigbiene. *Biol. Zbl.* **88**, 61–67.

Mercer, D. M. A. (1965) The behavior of multiple oscillator systems. In *Circadian Clocks* ed. J. Aschoff, North Holland, Amsterdam pp. 64–73.

Mergenhagen, D., and Schweiger, H. G. (1974) Circadian rhythmicity: does intercellular synchronization occur in *Acetabularia*? *Plant Sci. Lett.* **3**, 387–389.

Mergenhagen, D., and Schweiger, H. G. (1975a) Circadian rhythm of oxygen evolution in cell fragments of *Acetabularia mediterranea*. *Exp. Cell Res.* **92**, 127–130.

Mergenhagen, D., and Schweiger, H. G. (1975b) The effect of different inhibitors of transcription and translation of the expression and control of circadian rhythm in individual cells of *Acetabularia*. *Exp. Cell Res.* **94**, 321–326.

Michelson, A. A. (1911) On the metallic coloring of birds and insects *Phil. Mag.* **21**, 554–567.

Michelton, W. E. K. (1952) Vision through the atmosphere. University of Toronto Press, Toronto.

Miles, L. E. M., Raynal, D. M., and Wilson, M. A. (1977) Blind man living in normal society has circadian rhythms of 24.9 hours. *Science* **198**, 421–423.

Mills, J. N. (1973) (ed.) *Biological Aspects of Circadian Rhythms*. Plenum, New York.

Mills, J. N., Minors, D. S., and Waterhouse, J. M. (1977) The physiological rhythms of subjects living on a day of abnormal length. *J. Physiol.* (London) **268**, 803–826.

Milnor, J. (1965) *Topology From the Differentiable Viewpoint*. Univ. of Va. Press, Charlottesville.

Mines, G. R. (1913) On dynamic equilibrium in the heart. *J. Physiol.* (London) **46**, 349–383.

Mines, G. R. (1914) On circulating excitations on heart muscles and their possible relation to tachycardia and fibrillation. *Trans. R. Soc. Can.* **4**, 43–53.

Minis, D. H. (1965) Parallel peculiarities in the entrainment of a circadian rhythm and photoperiodic induction in the pink boll worm (*Pectinophora gossypiella*). In *Circadian Clocks*. J. Aschoff, ed., North-Holland, Amsterdam, pp. 333–343.

Minis, D., and Pittendrigh, C. (1968) Circadian oscillation controlling hatching: its ontogeny during embryogenesis of a moth. *Science* **159**, 534–536.

Minor, P. D., and Smith, J. A. (1974) Explanation of degree of correlation of sibling generation times in animal cells. *Nature* **248**, 241–243.

Minorsky, N. (1962) *Nonlinear Oscillations*, Van Nostrand, Princeton, N.J.

Misner, C. W., Thorne, K. S., and Wheeler, J. A. (1973) *Gravitation*. Freeman, San Francisco.

Mitchison, G. J. (1977) Phyllotaxis and the fibonacci series. *Science* **196**, 270–275.

Mitchison, J. M. (1971) *The Biology of the Cell Cycle*, Cambridge University Press.

The Mobile Medical Team of the Wuhan Fourth Hospital (1976) Autotransplantation of severed limbs. *Chinese Medical Journal* **2**, 417–422.

Mochan, E., and Pye, E. K. (1973) Respiratory oscillations in adapting yeast cultures. *Nature N.B.* **242**, 177–179.

Moe, G. K. (1962) On the multiple wavelet hypothesis of atrial fibrillation. *Arch. Int. Pharmacodyn.* **140**, 183–188.

Moe, G. K., and Abildskov, J. A. (1959) Atrial fibrillation as a self-sustaining arrhythmia independent of focal discharge. *Amer. Heart J.* **58**, 59–70.

Moe, G. K., Rheinboldt, W. C., and Abildskov, J. A. (1964) A computer model of atrial fibrillation. *Amer. Heart J.* **67**, 200–220.

Moghissi, K. S., Syner, F. N., and Evans, T. N. (1972) A composite picture of the menstrual cycle. *Amer. J. Obstet. Gynec.* **114**, 405–418.

Moller, U., Larsen, J. K., and Faber, M. (1974) The influence of injected tritiated thymidine on the mitotic circadian rhythm in the epithelium of the hamster cheek pouch. *Cell. Tiss. Kinet.* **7**, 231–239.

Moore, G. P., Segundo, J. P., and Perkel, D. H. (1963) Stability patterns in interneuronal pacemaker regulation. *Proc. of the San Diego Symposium for Biomedical Engr.*, pp. 184–193.

Moore, R. Y. (1978) Central neural control of circadian rhythms. *Frontiers in Neuroendocrinology*, **5**, 185–206.

Moore, R. Y., and Eichler, V. B. (1976) Central neural mechanisms in diurnal rhythm regulation and neurendocrine responses to light. *Psychoneurendocrinology* **1**, 265–279.

Moore-Ede, M. C., Kass, D. A., and Herd, J. A. (1977) Transient circadian internal desynchronization after light-dark phase shift in monkeys. *Amer. J. of Physiol.* **232**, R31–R37.

Moore-Ede, M. C., Meguid, M. M., Fitzpatrick, G. F., Boyden, C. M., and Ball, M. R. (1978) Circadian variation in response to potassium infusion. *Clin. Pharm. Ther.* **23**, 218–227.

Moore-Ede, M. C., Schelzer, W. S., Kass, D. A., and Herd, J. A. (1976) Internal organization of the circadian timing system in multicellular animals. *Fed. Proc.* **35**, 2333–2338.

Mori, S. (1947) A concept on mechanisms of the endogenous daily rhythmic activity. *Mem. Coll. Sci.*, Univ. Kyoto, B, **19**, 1–4.

Morin, L. P., Fitzgerald, K. M., Rusak, B., and Tucker, I. (1977a) Circadian organization and neural reduction of hamster reproductive rhythms. *Psychoneuroendo.* **2**, 73–98.

Morin, L. P., Fitzgerald, K. M., and Tucker, I. (1977b) Estradiol shortens the period of hamster circadian rhythms. *Science* **196**, 305–307.

Morse, P. M., and Feshbach, H. (1953) *Methods of Theoretical Physics* McGraw-Hill, New York Part 1, p. 20.

Moshkov, B. S., Fukshanskii, L. Ya., and Yuzefovich, G. I. (1966) A mathematical model of a biological clock. *Dokl. Akad. Nauk. SSSR* **167**, 440–443 (in Russian).

Moss, R. L., Dudley, C. A., and Schwartz, N. B. (1977) Coitus-induced release of luteinizing hormone in the proestrus rat: fantasy or fact? *Endocrinology* **100**, 394–397.

Mrosovsky, N. (1970) Mechanism of hiberation cycles in ground squirrels: circannian rhythm or sequence of stages. *Pennsylvania Academy of Science* **44**, 172–175.

Mueller, M. W. and Arnett, W. D. (1976) Propagating star formation and irregular structure in spiral galaxies. *Astrophys. J.* **210**, 670–678.

Murray, J. D. (1975) Non-existence of wave solutions for the class of reaction-diffusion equations given by the Volterra interacting-population equations with diffusion. *J. Theor. Biol.* **52**, 459–469.

Murray, J. D. (1976a) On travelling wave solutions in a model for the Belousov-Zhabotinskii reaction. *J. Theor. Biol.* **56**, 329–353.

Murray, J. D. (1976b) Spatial structures in predator-prey communities: a nonlinear time delay diffusional model. *Math. Bios.* **30**, 73–85.

Murray, J. D. (1977) *Lectures on Non-Linear-Differential-Equation Models in Biology*, Clarendon, Oxford.

Nadeau, R. A., and Roberge, F. A. (1969) Mechanism of A-V arrhythmias. In *Electrical Activity of the Heart*, G. W. Manning and S. P. Ahuja, eds. Charles C. Thomas, Springfield, Ill., pp. 117–128.

Nagumo, J., Arimoto, S., and Yoshizawa, S. (1962) An active pulse transmission line simulating nerve axon. *Proc. IRE* **50**, 2061–2070.

Nanjundiah, V. (1973) Chemotaxis, signal relaying and aggregation morphology. *J. Theor. Biol.* **42**, 63–105.

Nanjundiah, V. (1976) Signal relay by single cells during wave propagation in a cellular slime mold. *J. Theor. Biol.* **56**, 275–282.

Narlikar, J. V. (1970) The interaction of laboratory science and astronomy *Amer. Sci.* **58**, 290–297.

Nayar, J. K. (1968) The pupation rhythm in *Aedes taeiorhynchus.* IV. Further studies of the endogenous diurnal (circadian) rhythm of pupation. *Ann. Entomol. Soc. Amer.* **61**, 1408–1417.

Nazarea, A. D. (1974) Critical length of the transport-dominated region for oscillating non-linear reactive processes. *Proc. Nat. Acad. Sci.* **71**, 3751–3753.

Neill, D., and Day, B. N. (1964) Relationship of developmental stage two regression of the corpus luteum in swine. *Endocrinology* **74**, 355–360.

Nelsen, T. S., and Becker, J. C. (1968) Simulation of the electrical and mechanical gradient of the small intestine. *Amer. J. Physiol.* **214**, 749–757.

Neu, J. C. (1979a) Chemical waves and the diffusive coupling of limit cycle oscillators. *SIAM.* In press.

Neu, J. C. (1979b) Coupled chemical oscillators. *SIAM J. Appl. Math.* **37**, 307–315.

Neu, J. C. (1979c) Large populations of coupled chemical oscillators. *SIAM.* In press.

Neumann, D. (1978) Entrainment of a semilunar rhythm by simulated tidal cycles of mechanical disturbance. *J. Exp. Mar. Biol. Ecol.* **35**, 73–85.

Neville, A. C. (1975) *Biology of Arthropod Cuticle*, Springer-Verlag, New York.

Newell, R. C. (1969) Effect of fluctuations of temperature on the metabolism of intertidal invertebrates. *Amer. Zool.* **9**, 293–307.

Newman, J. R. (1956) *The World of Mathematics*, Simon and Schuster, New York. Arthropod.

Newton, Isaac (1730) *Opticks*, 4th edition, reprinted by Dover, New York (1952) p. 401.

Nickerson, M. (1944) An experimental analysis of barred pattern formation in feathers. *J. Exp. Zool.* **95**, 361–397.

Nicolis, G., and Portnow, J. (1973) Chemical oscillations. *Chem. Rev.* **73**, 365–366.

Nicolis, G., and Prigogine, I. (1977) *Self-Organization in Nonequilibrium Systems*, Wiley, New York.

Nishiitsutsuji-Uwo, J., and Pittendrigh, C. S. (1968a) Central nervous system control of circadian rhythmicity in the cockroach. *Z. vergl. Physiologie* **58**, 1–46.

Nishiitsutsuji-Uwo, J., and Pittendrigh, C. S. (1968b) The neuroendocrine basis of midgut tumour induction in cockroaches. *J. Insect Physiol.* **13**, 851–859.

Nitzan, A., Ortoleva, R., and Ross, J. (1974) Nucleation in systems with multiple stationary states. *Farad. Symp. Chem. Soc.* **9**, 241–253.

Njus, D. (1975) Thesis, Harvard University.

Njus, D., Sulzman, F. M., and Hastings, J. W. (1974) Membrane model for the circadian clock. *Nature* **248**, 116–120.

Noble, J. V. (1974) Geographic and temporal development of plagues. *Nature* **250**, 726–729.

Novak, B., and Seelig, F. F. (1976) Phase-shift model for the aggregation of amoebae: a computer study. *J. Theor. Biol.* **56**, 301–327.

Nudelman, H. B., and Glantz, R. M. (1977) Sustained oscillations, entrainment, and lateral inhibition in the crayfish visual system. *Fed. Proc.* **36**, 2042–2044.

Oatley, J., and Goodwin, B. C. (1971) In *Biological Rhythms and Human Performance*, W. P. Colquhoun, ed. Academic, London–New York.

Offner, F., Weinberg, A., and Young, G. (1940) Nerve conduction theory: some mathematical consequences of Bernstein's model. *Bull. Math. Biophys.* **2**, 89–103.

Olsen, L. F. and Degn, H. (1977) Chaos in an enzyme reaction. *Nature* **267**, 177–178.

Orban, M., and Körös, E. (1978) Chemical oscillations during the uncatalyzed reaction of aromatic compounds with bromate. 1. Search for chemical oscillators. *J. Phys. Chem.* **82**, 1672–1674.

Ortoleva, P. (1976) Local phase and renormalized frequency in inhomogeneous chemioscillations. *J. Chem. Phys.* **64**, 1395–1406.

Ortoleva, P., and Ross, J. (1973) Phase waves in oscillatory chemical reactions. *J. Chem. Phys.* **58**, 5673–5680.

Ortoleva, P., and Ross, J. (1974) On a variety of wave phenomena in chemical reactions. *J. Chem. Phys.* **60**, 5090–5107.

Ortoleva, P., and Ross, J. (1975) Theory of propagation of discontinuities in kinetic systems with multiple time scales: fronts, front multiplicity, and pulses. *J. Chem. Phys.* **63**, 3398–3408.

Osterhout, W. J. W., and Hill, S. E. (1938) Pacemakers *in Nitella*. II. arrhythmia and block. *J. Gen. Physiol.* **22**, 115–130.

Ostriker, J. P. (1971) The nature of pulsars. *Sci. Amer.* **224**, (1), 48–59.

Ostwald, W. (1900) Periodische Erscheinungen bei der Auflösung des Chrom. in Säuren. *Zeit Phys. Chem.* **35**, 33–76 and 204–256.

Othmer, H. G. (1975) On the temporal characteristics of a model for the Zhabotinskii-Belousov reaction. *Mathe. Biosci.* **24**, 205–238.

Othmer, H. G. (1976a) Temporal oscillations in chemically-reacting systems. *Proc. IEEE Int. Conf. on Systems, Man and Cybernetics.* pp. 314–320.

Othmer, H. G. (1976b) On the significance of finite propagation speeds in multicomponent reacting systems. *J. Chem. Phys.* **64**, 460–470.

Othmer, H. G. (1977) Current problems in pattern formation. In *Lectures on Mathematics in the Life Sciences* vol. 9, S. A. Levin, ed. *Amer. Math. Soc.*, Providence, R. I., pp. 57–85.

Othmer, H. G., and Aldridge, J. A. (1978) The effects of cell density and metabolite flux on cellular dynamics. *J. Math. Biol.* **5**, 169–200.

Ottesen, E. (1965) Analytical studies on a model for the entrainment of circadian systems. unpublished B. A. thesis, Biology dept., Princeton University.

Page, T. L. (1978) Interactions between bilaterally paired components of the cockroach circadian system. *J. Comp. Physiol.* **124**, 225–236.

Page, T. L., Caldarola, P. C., and Pittendrigh, C. S. (1977) Mutual entrainment of bilaterally distributed circadian pacemakers. *Proc. Nat. Acad. Sci.* **74**, 1277–1281.

Palmer, E., and Jousset, B. (1975a) Synchronization of oestrus and ovulation in the mare with a two PG-HCG sequence treatment. *Ann. Biol. Anim. Biochem. Biophys.* **15**, 471–480.

Palmer, E., and Jousset, B. (1975b) Synchronization of oestrus in mares with a prostaglandin analog and HCG. *J. Reprod. Fert. Suppl.* **23**, 269–274.

Palmer, J. D. (1976) *Introduction to Biological Rhythms* Academic, New York.

Palmer, J. D. (1977) Human rhythms. *Bioscience* **27**, 93–99.

Parameswaran, N. (1975) Zur Wandstruktur von Sklereiden in einigen Baumrinden. *Protoplasma* **85**, 305–314.

Parker-Rhodes, A. F. (1955) Fairy ring kinetics. *Trans. Br. Mycol. Soc.* **38**, 59–72.

Pastelin, G., Mendez, R., and Moe, G. K. (1978) Participation of atrial specialized conduction pathways in atrial flutter. *Circ. Res.* **42**, 386–393.

Patton, R. J., and Linkens, D. A. (1978) Hodgkin-Huxley type electronic modelling of gastrointestinal electrical activity. *Med. Biol. Eng. Comput.* **16**, 195–202.

Pavlidis, T. (1967a) A mathematical model for the light affected system in the *Drosophila* eclosion rhythm. *Bull. Math. Biophys.* **29**, 291–310.

Pavlidis, T. (1967b) A model for circadian clocks. *Bull. Math. Biophys.* **29**, 781–791.

Pavlidis, T. (1968) Studies on biological clocks: a model for the circadian rhythms of nocturnal organisms. In *Lectures on Mathematics in the Life Sciences*, vol. 1, M. Gerstenhaber, ed., Amer. Math. Soc., Providence, R. I., pp. 88–112.

Pavlidis, T. (1969) Populations of interacting oscillators and circadian rhythms. *J. Theor. Biol.* **22**, 418–436.

Pavlidis, T. (1971) Populations of biochemical oscillators as circadian clocks. *J. Theor. Biol.* **33**, 319–338.

Pavlidis, T. (1973) *Biological Oscillators: Their Mathematical Analysis*, Academic, New York,

Pavlidis, T. (1975) Spatial organization of chemical oscillators via an averaging operator. *J. Chem. Phys.* **63**, 5269–5273.

Pavlidis, T. (1976) Spatial and temporal organization of populations of interacting oscillators. In Dahlem Workshop on *The Molecular Basis of Circadian Rhythms*, J. W. Hastings, H. G. Schweiger, eds. Abakon, Berlin, pp. 131–148.

Pavlidis, T., (1978a) What do mathematical models tell us about circadian clocks? *B. Math. Biol.* **40**, 625–635.

Pavlidis, T. (1978b) Qualitative similarities between the behavior of coupled oscillators and circadian rhythms. *Bull. Mathe. Biol.* **40**, 675–692.

Pavlidis, T., and Kauzmann, W. (1969) Toward a quantitative biochemical model for circadian oscillators. *Arch. Biochem. Biophys.* **132**, 338–348.

Pavlidis, T., and Pinsker, H. M. (1977) Oscillator theory and neurophysiology. *Fed. Proc.* **36**, 2033–2059. (Workshop with several papers.)

Pavlidis, T., Zimmerman, W. F., and Osborn, J. (1968) A mathematical model for the temperature effects on circadian rhythms. *J. Theor. Biol.* **18**, 210–221.

Pearson, M. J., and McLaren, D. I. (1977) A criticism of catastrophe modelling of the differentiative process in amphibian development. *J. Theor. Biol.* **69**, 721–734.

Perkel, D. H., Schulman, J. H., Bullock, T. H., Moore, C. P., and Segundo, J. P. (1964) Pacemaker neurons: effects of regularly spaced synaptic input. *Science* **145**, 61–63.

Peskin, C. S. (1975) *Mathematical aspects of Heart Physiology*, Courant Inst. of Math. Sci. Publication, New York.

Peterson, E. L. and Jones, M. D. R. (1979) Do circadian oscillators ever stop in constant light? *Nature* **280**, 677–679.

Petsche, H., Prohaska, O., Rappelsberger, P., Vollmer, R., and Kaiser, A. (1974) Cortical seizure patterns in multidimensional view: the information content of equipotential maps. *Epilepsia* **15**, 439–463.

Petsche, H., and Rappelsberger, P. (1970) Influence of cortical incisions on synchronization pattern and traveling waves. *Electroenceph. Clin. Neurophysiol.* **28**, 592–600.

Petsche, H., Rappelsberger, P., and Trappl, R. (1970) Properties of cortical seizure potential fields. *Electroenceph. Clin. Neurophysiol.* **29**, 567–578.

Petsche, H., and Sterc, J. (1968) The significance of the cortex for the travelling phenomenon of brain waves. *Electroenceph. Clin. Neurophysiol.* **25**, 11–22.

Pinsker, H. M. (1977) *Aplysia* bursting neurons as endogenous oscillators. I: Phase response curves for pulsed inhibitory synaptic input. and II. Synchronization and entrainment by pulsed inhibitory synaptic input. *J. Neurophysiol.* **40**, 527–543 and 544–556.

Pinsker, H. M., and Kandel, E. R. (1977) Short-term modulation of endogenous bursting rhythms by monosynaptic inhibition in *Aplysia* neurons: effects of contigent stimulation. *Brain Res.* **125**, 51–64.

Pinsker, H. M. (1979) Phase plane analysis of endogenous neuronal oscillators in *Aplysia*. Biol. Cybern., in press.

Pittendrigh, C. S. (1954) On temperature independence in the clock system controlling emergence time in *Drosophila*. *Proc. Nat. Acad. Sci.* **40**, 1018–1029.

Pittendrigh, C. S. (1958) Perspectives in the study of biological clocks. In Symposium on *Perspectives in Marine Biology*, University of California Press, Berkeley, pp. 239–268.

Pittendrigh, C. S. (1960) Circadian rhythms and the circadian organization of living systems. In *Cold Spring Harbor Symposium on Quantitative Biology* **25**, 159–184.

Pittendrigh, C. S. (1961) On temporal organization in living systems. *Harvey Lect.* **56**, 93–125.

Pittendrigh, C. S. (1965) On the mechanism of the entrainment of a circadian rhythm by light cycles. In *Circadian Clocks*, J. Aschoff, J. ed. North-Holland, Amsterdam, pp. 277–297.

Pittendrigh, C. S. (1966) The circadian oscillation in *Drosophila pseudoobscura* pupae: a model for the photoperiodic clock. *Z. fur Pflanzenphysiol.* **54**, 275–307.

Pittendrigh, C. S. (1967a) Circadian systems. I. The driving oscillation and its assay in *Drosophila pseudoobscura*. *Proc. Nat. Acad. Sci.* **58**, 1762–1767.

Pittendrigh, C. S. (1967b) Circadian rhythms, space research, and manned space flight. In *Life Sciences and Space Research* **5**, 122–134. North-Holland, Amsterdam.

Pittendrigh, C. S. (1972) Circadian surfaces and the diversity of possible roles of circadian organization in photoperiodic induction. *Proc. Nat. Acad. Sci.* **69**, 2734–2737.

Pittendrigh, C. S. (1974) Circadian oscillations in cells and the circadian organization of multicellular systems. In *The Neurosciences: Third Study Program*, F. O. Schmitt and F. G. Worden, eds. M.I.T. Press, Cambridge, pp. 437–458.

Pittendrigh, C. S. (1976) *Circadian Clocks: What are They?* In Dahlem Workshop on *The Molecular Basis of Circadian Rhythms*. J. W. Hastings and H. G. Schweiger, eds. Abakon Berlin.

Pittendrigh, C. S., and Bruce, V. (1957) An oscillator model for biological clocks. In *Rhythmic and Synthetic Processes in Growth.*, D. Rudnick, ed. Princeton University Press, p. 75–109.

Pittendrigh, C. S., and Bruce, V. G. (1959) Daily rhythms as coupled oscillator systems; and their relation to thermo- and photo-periodism. In *Photoperiodism and Related Phenomena in Plants and Animals*, R. B. Withrow, ed., AAAS #55, Washington, D.C., pp. 475–505.

Pittendrigh, C. S., Bruce, V., and Kaus, P. (1958) On the significance of transients in daily rhythms *Proc. Nat. Acad. Sci.* **44**, 965–973.

Pittendrigh, C., and Daan, S. (1976a) A functional analysis of circadian pacemakers in nocturnal rodents. I. The stability and lability of spontaneous frequency. *J. Comp. Physiol.* **106**, 223–252.

Pittendrigh, C. S., and Daan, S. (1976b) A functional analysis of circadian pacemakers in nocturnal rodents. IV. Entrainment: pacemaker as a clock. *J. Comp. Physiol.* **106**, 291–331.

Pittendrigh, C., and Daan, S. (1976c) A functional analysis of circadian pacemakers in nocturnal rodents. V. pacemaker structure: A clock for all seasons. *J. Comp. Physiol.* **106**, 333–355.

Pittendrigh, C. S., and Minis, D. H. (1964) The entrainment of circadian oscillations by light and their role as photoperiodic clocks. *Amer. Nat.* **98**, 261–294.

Pittendrigh, C. S., and Minis, D. H. (1971) The photoperiodic time measurement in *Pectinophora gossypiella* and its relation to the circadian system in that species. In *Biochronometry*, M. Menaker, ed. Nat. Acad. Sci., Washington, D.C.

Pittendrigh, C. S., and Skopik, S. D. (1970) Circadian systems. V. The driving oscillation and the temporal sequence of development. *Proc. Nat. Acad. Sci.* **65**, 500–507.

Pizarello, D. J., Isaak, D., Chua, K. E., and Rhyne, A. L. (1964) Circadian rhythmicity in the sensitivity of two strains of mice to whole-body radiation. *Science* **145**, 286–291.

Plant, R. E. (1976) A simple model for the crustacean cardiac pacemaker mechanism. *Math. Biosci.* **32**, 275–290.

Plant, R. E. (1977) Crustacean cardiac pacemaker model—an analysis of the singular approximation. *Math. Biosci.* **36**, 149–171.

Plant, R. F., and Kim, M. (1975) On the mechanism underlying bursting in the *Aplysia* abdominal ganglion R15 cell. *Math. Biosci.* **26**, 357–375.

Plant, R. E., and Kim, M. (1976) Mathematical description of a bursting pacemaker neuron by a modification of the Hodgkin-Huxley equations. *Biophys. J.* **16**, 227–244.

Plant, R. E. (1979) Temporal pattern generation in coupled oscillators—model for the *Tritonia* trigger group. *Mathe. Biosci.* **43**, 239–263.

Platzman, G. W. (1972) North Atlantic Ocean: preliminary description of normal modes. *Science* **178**, 156–157.

Pochobradsky, J. (1972) Change in the menstrual period through reproductive life. *J. Interdiscipl. Cycle Res.* **3**, 245–249.

Pol, van der, B. (1926) On relaxation oscillations. *Phil. Mag.* **2**, 978–992.

Pol, van der, B. (1940) Biological rhythms considered as relaxation oscillations. *Acta Med. Scand. Suppl.* **108**, 76–87.

Pol, van der, B., van der Mark, J. (1928) The heartbeat considered as a relaxation oscillation and as electrical model of the heart. *Phil. Mag. Suppl.* **6**, 763–765.

Pollack, G. H. (1977) Cardiac pacemaking: an obligatory role of catecholamines? *Science* **196**, 731–738.

Polya, G. (1954) *Induction and Analogy in Mathematics*, Princeton Univ. Press, Princeton, N. J. p. 168.

Polya, G. (1957) *How to Solve it*. Doubleday and Co., Garden City N.Y. p. 65.

Presser, H. B. (1972) Temporal data relating to the human menstrual cycle. In *Biorhythms and Human Reproduction*, (M. Ferin, ed.) Chap. 10, Wiley, New York, pp. 145–160.

Pritchard, R. H., Barth, P. T., and Collins, J. (1969) Control of DNA synthesis in bacteria. *Symp. Soc. Gen. Microbiol.* **19**, 263–297.

Probine, M. C., and Barber, N. F. (1966) The structure and plastic properties of the cell wall of *Nitella* in relation to extension growth. *Aust. J. Biol. Sci.* **19**, 439–457.

Pye, E. K. (1964) Dissertation, University of Manchester.

Pye, E. K. (1969) Biochemical mechanisms underlying the metabolic oscillation of yeast. *Can. J. Bot.* **47**, 271–285.

Pye, E. K., Chance, B. (1966) Sustained sinusoidal oscillations of reduced pyridine nucleotide in a cell-free extract of *Saccharomyces carlsbergensis*. *Proc. Nat. Acad. Sci. U.S.A.* **55**, 888–894.

Pye, E. K. (1968) Glycolytic oscillations in cells and extracts of yeast—some unsolved problems. In *Biological and Biochemical Oscillators*, B. Chance et al. eds. Academic, New York, 1973, pp. 269–284.

Pye, E. K. (1971) Periodicities in intermediary metabolism. In *Biochronometry*, M. Menaker. ed. Nat. Acad. Sci., Washington, D. C. pp. 623–636.

Rankin, J. S. (1969) Hormonal patterns and interactions in the normal menstrual cycle. *Clin. Obstet. Gynec.* **21**, 741–754.

Rao, P. N., and Johnson, R. (1970) Mammalian cell fusion: studies on the regulation of DNA synthesis and mitosis. *Nature* **225**, 159–164.

Rao, P. N., Hittelman, W. N., and Wilson, B. A. (1974) Mammalian cell fusion. Part 6. Regulation of mitosis in binucleate *hela* cells. *Exp. Cell Res.* **90**, 40–46.

Rapoport, A. (1952) Periodicities of open linear systems with positive steady states. *Bull. Math. Biophys.* **14**, 171–183.

Rapp, P. E. (1979) Bifurcation theory, control theory and metabolic regulation: a mathematical investigation of biological and biochemical oscillators. *Biological Systems, Modelling and Control.* D. A. Linkens ed., Peregrinus, London.

Rapp, P. E., and Berridge, M. J. (1977) Oscillations in calcium-cyclic AMP control loops from the basis of pacemaker activity in other high frequency biological rhythms. *J. Theor. Biol.* **66**, 497–525.

Rasmussen, L., and Zeuthen, E. (1962) Cell division and protein synthesis in *Tetrahymena*, as studied with p-fluorophenylalanine. *CR Trav. Lab. Carlsberg.* **32**, 333–358.

Rastogi, R. P., Singh, K., Rastogi, P., and Rai, R. B. (1977) Nucleation phenomenon in oscillatory reactions. *Indian Journal of Chemistry* **15A**, 295–297.

Rawson, K. S. (1956) Homing behavior and endogenous activity rhythms. Ph.D. thesis, Harvard University.

Rebbi, G. (1979) Solitons. *Sci. Amer.* **235** (1), 92–116.

Reinberg, A., and Halberg, F. (1971) Circadian chronopharmacology. *N. Rev. Pharm.* **11**, 455–492.

Reiter, P., and Jones, M. D. R. (1975) An eclosion timing mechanism in the mosquito. *J. Ent.* (A) **50**, 161–168.

Remmert, H. (1962) *Der Schüpfrhythmus der Insekten*, Franz Steiner, Wiesbaden.

Remy, U. (1968) Thesis. Mediz. Fak., Ruprecht-Karl-Universität, Heidelberg.

Rensing, R. (1971) Hormonal control of circadian rhythms in *Drosophila*. In *Biochronometry*, M. Menaker, ed. Nat. Acad. Sci., Washington, D.C., pp. 527–539.

Rescigno, A., Stein, R. B., Purple, R. L., and Poppele, R. E. (1970) A neuronal model for the discharge patterns produced by cyclic inputs. *Bull. Math. Biophys.* **32**, 337–353.

Reshodko, L. V. (1974) Computer models and machine experiments in biological research systems, for example smooth muscle. *J. Gen. Biol.* **35**, 80–87. (in Russian)

Reshodko, L. V., and Bures, J. (1975) Computer simulation of reverberating spreading depression in a network of cell automata. *Biol. Cyb.* **18**, 181–189.

Reusser, E. J., and Field, R. J. (1979) The transition from phase waves to trigger waves in a model of the Zhabotinskii reaction. *J. Amer. Chem. Soc.* **101**, 1063–1071.

Richards, W. (1971) The fortification illusions of migraines. *Sci. Amer.* **224**, 5, 88–96.

Richter, C. (1960) Biological clocks in medicine and psychiatry; shock-phase hypothesis. *Proc. Nat. Acad. Sci.* **46**, 1506–1530.

Richter, C. P. (1965) *Biological Clocks in Medicine and Psychiatry*, Springfield, C. C. Thomas.

Richter, O., Betz, A., and Giersch, C. (1975) The response of oscillating glycolysis to perturbations in the NADH/NAD system: a comparison between experiments and a computer model. *Biosystems* **7**, 137–146.

Rinzel, J., and Keller, J. (1973) Traveling wave solutions of a nerve conduction equation. *Biophys. J.* **13**, 1313–1337.

Ritzema Bos, J. (1901) Heksenringen *Tijdschr. Plantenziekten* **7**, 97–126.

Robertson, A. (1972) Quantitative analysis of the development of cellular slime molds. From *Some mathematical Questions in Biology*, **3**, 48–73. Amer. Math. Soc., Providence, R. I.

Robertson, A., Drage, D. J., and Cohen, M. H. (1972) Control of aggregation in *Dictyostelium discoideum* by an external periodic pulse of cyclic adenosine monophosphate. *Science* **175**, 333–335.

Robinson, C. (1966) The cholesteric phase in polypeptide solutions and biological structures. *Mol. Cryst.* **1**, 467–494.

Rogers, L. A., and Greenbank, G. R. (1930) The intermittent growth of bacterial cultures. *J. Bacteriol.* **19**, 181–190.

Roos, W., and Gerisch, G. (1976) Receptor-mediated adenylate cyclase activation in *Dictyostelium discoideum*. *FEBS Lett.* **68**, 170–172.

Roos, W., Scheidegger, C., and Gerisch, G. (1977) Adenylate cyclase activity oscillations as signals for cell aggregation in *Dictyostelium discoideum*. *Nature* **266**, 259–261.

Rosen, R. (1971) *Dynamical system theory in biology*, vol. 1, Wiley-Interscience, New York.

Rosenfeld, A. (1969) *Picture Processing by Computer*, Academic, New York.

Ross, J. (1976) Temporal and spatial structure in chemical instabilities. *Berichte Der Bunsengesellschaft F. Physikalische Chemie* **80**, 1112–1125.

Rössler, O. E. (1972a) Basic circuits of fluid automata and relaxation systems. *Z. Naturforsch.* **27b**, 333–343 (in German).

Rössler, O. E. (1972b) A principle for chemical multivibration. *J. Theor. Biol.* **36**, 413–417.

Rössler, O. E. (1972c) Design for automonous chemical growth under different environmental constraints. *Progress in Theoretical Biology* **12**, 167–211.

Rössler, O. E. (1974a) Chemical automata in homogeneous and reaction-diffusion kinetics. *Lecture Notes in Biomathematics* **4**, 399–418.

Rössler, O. E. (1974b) A synthetic approach to exotic kinetics (with examples). *Lecture Notes in Biomathematics* **4**, 546–582.

Rössler, O. E. (1975) Steps toward a temperature compensated homogeneous chemical clock pp. 99–105, San Diego 1975 Biomedical Symposium. I. Martin, ed. San Diego Biomedical Symposium Press, San Diego.

Rössler, O. E. (1976a) An equation for continuous chaos. *Phys. Lett.* **57A**, 397–398.

Rössler, O. E. (1976b) Chaos in abstract kinetics: two prototypes. *Bull. Math. Biol.* **39**, 275–289.

Rössler, O. E. (1976c) Chaotic behavior in simple reaction systems. *Z. Naturforsch.* **31a**, 259–264.

Rössler, O. E. (1976d) Chemical turbulence: chaos in a simple reaction-diffusion system. *Z. Naturforsch.* **31a**, 1168–1172.

Rössler, O. E. (1978) Chemical turbulence—a synopsis. In *Synergetics*, H. Haken, ed. Lecture Notes in Physics, Springer-Verlag, Berlin.

Rössler, O. E., and Hoffman, D. (1972) Repetitive hard bifurcation in a homogeneous reactions system. In *Analysis and Simulation*, North-Holland, Amsterdam, pp. 91–101.

Rössler, O. E. and Kahlert, C. (1979) Winfree meandering in a 2-dimensional 2-variable excitable medium. *Z. Naturf* **34a**, 565–570.

Rössler, O. E., and Ortoleva, P. J. (1978) Strange attractors in 3-variable reaction systems, in *Theoretical Approaches to Complex Systems*, Springer-Verlag, New York (Lecture Notes in Biomathematics 21) 67–74.

Rössler, O. E., and Wegmann, E. (1978) Chaos in Zhabotinskii reaction. *Nature* **271**, 89–90.

Rossomando, E. F., and Sussman, M. (1973) A 5'-adenosine monophosphate-dependent adenylate cyclase and an adenosine 3':5'-cyclic monophosphate-dependent adenosine triphosphate pyrophosphohydrolase in *Dictyostelium discoideum*. *Proc. Nat. Acad. Sci.* **70**, 1254–1257.

Rothmund, V. (1907) *Löslichkeit und Löslichkeitsbeeinflussung*, Leipzig.

Rozenshtraukh, L. V., Kholopov, A. V., and Yushmanova, A. V. (1970) Vagus inhibition—cause of formation of closed pathways of conduction of excitation in the auricles. *Biofizika* **15**, 690–700.

Rubin, J., and Robertson, A. (1975) The tip of the *Dictyostelium discoideum* pseudoplasmodium as an organizer. *J. Embryol. Exp. Morph.* **33**, 227–241.

Ruch, T. C., and Patton, H. D. (1973) *Physiology and Biophysics: Digestion, Metabolism, Endocrine Function, and Reproduction*, W. B. Saunders, Philadelphia, pp. 24–31.

Rusak, B. (1977) Role of suprachiasmatic nuclei in generation of circadian rhythms in golden-hamster, *Mesocricetus auratus*. *J. Comp. Phys.* **118**, 145–164.

Rusch, H. P., Sachsenmaier, W., Beherns, K., and Gruter, V. (1966) Synchronization of mitosis by the fusion of the plasmodia of *Physarum polycephalum*. *J. Cell Biol.* **31**, 204–209.

Rustad, R. C. (1970) Variations in the sensitivity to X-ray induced mitotic delay during the cell division cycle of the sea urchin egg. *Radiat. Res.* **42**, 498–512.

Sachs, H. G., and DeHaan, R. L. (1973) Embryonic myocardial cell aggregates: volume and pulsation rate. *Dev. Biol.* **30**, 233–240.

Sachsenmaier, W. (1976) Control of synchronous nuclear mitosis in *Physarum polycephalum*. In Dahlem Workshop on *The Molecular Basis of Circadian Rhythms*, J. W. Hastings and H. G. Schweiger, eds. Dahlem Konferenzen, Berlin, pp. 409–420.

Sachsenmaier, W., Donges, K. H., Rupff, H., and Czihak, G. (1970) Advanced initiation of synchronized mitosis in *Physarum polycephalum* following UV-irradiation. *Z. Naturforsch.* **256**, 866–871.

Sachsenmaier, W., Remy, U., and Plattner-Schobel, R. (1972) Initiation of synchronous mitosis in *Physarum polycephalum*. *Exp. Cell Res.* **73**, 41–48.

Salisbury, F. B., and Denney, A. (1971) Separate clocks for leaf movement and photoperiodic flowering in *Xanthium strumarium* L. (cocklebur). In *Biochronometry*, M. Menaker, ed. Nat. Acad. Sci., Washington, D.C., pp. 292–311.

Sander, G., and Pardee, A. (1972) Transport changes in synchronously growing CHO and L cells. *J. Cell Physiol.* **80**, 267–272.

Sanglier, M., and Nicolis, G. (1976) Sustained oscillations and threshold phenomena in an operon control circuit. *Biophys. Chem.* **4**, 113–121.

Sano, T., Sawanobori, T., and Adaniya, H. (1978) Mechanism of rhythm determination among pacemaker cells of the mammalian sinus node. *Am. J. Physiol.* **235**, H379–H384.

Sargent, M. L., and Briggs, W. R. (1967) The effects of light on a circadian rhythm of conidiation in *Neurospora*. *Plant Physiol.* **42**, 1504–1510.

Sargent, M. L., and Woodward, D. O. (1969) Genetic determinants of circadian rhythmicity in *Neurospora*. *J. Bacteriol.* **97**: 861–866.

Sarna, S. K. and Bowes, K. (1976) Gastric pacemakers. *Gastroenterology* **70**, 226–231.

Sarna, S. K. and Daniel, E. E. (1973) Electrical stimulation of gastric electrical control activity. *Am. J. Physiol.* **225**, 125–131.

Sarna, S. K., and Daniel, E. E. (1974) Threshold curves and refractoriness properties of gastric relaxation oscillators. *Amer. J. Physiol.* **226**, 749–755.

Sarna, S. K., Daniel, E. E., and Kingma, Y. J. (1971) Simulation of slow-wave electrical activity of small intestine. *Amer. J. Physiol.* **221**, 166–175.

Sarna, S. K., Daniel, E. E., and Kingma, Y. J. (1972a) Effect of partial cuts on gastric electrical control activity and its computer model. *Amer. J. Physiol.* **223**, 332–340.

Sarna, S. K., Daniel, E. E., and Kingma, Y. J. (1972b) Simulation of the electric-control activity of the stomach by an array of relaxation oscillators. *Digestive Diseases.* **17**, 299–310.

Saunders, D. S. (1974) Evidence for "dawn" and "dusk" oscillators in the *Nasonia* photoperiodic clock. *J. Insect Physiol.* **20**, 77–88.

Saunders, D. S. (1976a) The circadian eclosion rhythm in *Sarcophaga argyrostoma*: comparisons with the photoperiodic clock. *J. Comp. Physiol.* **110**, 111–133.

Saunders, D. S. (1976b) *Insect Clocks*, Pergamon, Oxford.

Saunders, D. S., and Thomson, E. J. (1977) Strong phase response curve for the circadian rhythm of locomotor activity in a cockroach (*Nauphoeta cinerea*). *Nature* **270**, 241–243.

Saunders, D. S. (1978) An experimental and theoretical analysis of photoperiodic induction in the flesh-fly, *Sarcophaga argyrostoma. J. Comp. Physiol.* **124**, 75–96.

Scheffey, C., and Wille, J. J. (1978) Cycloheximide-induced mitotic delay in *Physarum polycephalum. Exp. Cell Res.* **113**, 259–262.

Scheving, J. E., and Pauly, J. E. (1973) Cellular mechanisms involving biorhythms with emphasis on those rhythms associated with the S and M stages of the cell cycle. *Int. J. Chron.* **1**, 269–286.

Scheving, L. E., and Pauly, J. E. (1974) Circadian rhythms: some examples and comments on clinical application. *Chronobiologia* **1**, 3–21.

Scheving, L. E., Vedral, D. F., and Pauly, J. E. (1968) Circadian susceptibility rhythm in rats to pentobarbitol sodium. *Anat. Rec.* **160**, 741–760.

Schmitt, O. H. (1960) Biophysical and mathematical models of circadian rhythms. *Cold Spring Harbor Symposium* for quantitative Biology, **25**, 207–210.

Schmitz, R. A., Renola, G. T. and Garrigan, P.C. (1979) Observations of complex dynamic behavior in the H_2—O_2 reaction on nickel. *Ann. N. Y. Acad. Sci.* **316**. 638–651.

Schmitz, R. A., Graziani, K. R., and Hudson, J. L. (1977) Experimental evidence of chaotic states in Belousov-Zhabotinskii reaction. *J. Chem. Phys.* **67**, 3040–3044.

Schrempf, M. (1975) Eigenschaften und Lokalisation des Photorezeptors für Phasenverschiebendes Störlicht bei der Blütenblattbewegung von *Kalanchoe blossfeldiana* (v. Poelln.). Thesis, Tübingen.

Schulman, J. (1969) Information transfer across an inhibition to pacemaker synapse at the crayfish stretch receptor. Ph.D. thesis, University of Calif. at Los Angeles, Dept. Zoology.

Scott, A. C. (1970a) Distributed multimode oscillators of one and two spatial dimensions. *IEEE Trans. Circuit Theory* **17**, 55–60.

Scott, A. C. (1970b) *Active and Nonlinear Propagation in Electronics.* Wiley-Interscience, N.Y.

Scott, S. W. (1979) Stimulation Simulations of young yet cultured beating hearts. Ph.D. thesis, State Univ. of N.Y. at Buffalo.

Segel, L. A., and Jackson, J. L. (1972) Dissipative structure: an explanation and an ecological example. *J. Theor. Biol.* **37**, 545–559.

Segal, S. J. (1974) The physiology of human reproduction. *Sci. Amer.* **231**, (9), 52–62.

Selfridge, O. (1948) Studies of flutter and fibrillation. V. *Arch. Inst. Cardiologie de Mexico* **18**, 177–187.

Sel'kov, E. E. (1968a) Self-oscillations in glycolysis. A simple kinetic model. *Eur. J. Biochem.* **4**, 79–80.

Sel'kov, E. E. (1968b) Self-oscillations in glycolysis. Simple single-frequency model. *Mol. Biol.* **2**, 252–266.

Sel'kov, E. E. (1970) Two alternative auto-oscillatory steady states in the metabolism of the thiols; Two alternative types of cell multiplication: normal and malignant. *Biofizika* **15**, 1065–1073.

Sel'kov, E. E. (1972) Nonlinear theory of regulation of the key step of glycolysis. *Studia Biophysica* **33**, 167–176.

Sel'kov, E. E. (1973) Analysis of temporal organization of multienzyme systems in Proc. IV Int. Biophysics Cong., pp. 437–475. Acad Sci. USSR. Puschino (in Russian).

Sel'kov, E. E. (1975) Stabilization of energy charge, generation of oscillations and multiple steady states in energy metabolism as a result of purely stoichiometric regulation. *Eur. J. Biochem.* **59**, 151–157.

Sel'kov, E. E., and Betz, A. (1968) On the mechanism of single-frequency glycolytic oscillations. In *Biological and Biochemical Oscillators*, B. Chance et al. ed. Academic, New York, 1973, pp. 197–220.

Sel'kov, E. E., and Dynnik, S. N. (1976) Hysteresis, alternative stationary states, and auto-oscillations in an open futile cycle, containing a substrate-inhibited reaction. *Biochem. (USSR)* **41**, 1365–1370.

Sel'kov, E. E., and Sozinov, L. A. (1970) Stable circadian rhythms as a property of cell populations. *Life Sciences and Space Res.* **8**, 157–167.

Shack, W. J., Tam, P. Y., and Lardner, T. J. (1971) A mathematical model of the human menstrual cycle. *Biophys. J.* **11**, 835–848.

Shackelford, P. G., and Feigin, R. D. (1973) Periodicity of susceptibility to pneumococcal infection: influence of light and adrenocortical secretions. *Science* **182**, 285–287.

Shaffer, B. M. (1957) Variability of behaviour of aggregating cellular slime molds. *Quar. J. Micro. Sci.* **98**, 393–405.

Shaffer, B. M. (1962) The acrasina. *Adv. Morphogenesis.* **2**, 109–182.

Shaikh, A. A., and Klaiber, E. L. (1974) Effects of sequential treatment with estradiol and PGF 2 alpha on the length of the primate menstrual cycle. *Prostaglandins* **6**, 253–263.

Shcherbunov, A. I., Kukushkin, N. I. and Saxon, M. E. (1973) Reverberator in a system of interrelated fibers described by the Noble equation. *Biofizika* **18**: 519–525.

Sheppard, J. J. (1968) *Human Color Perception*, Chap. 8: Temporal phenomena, Elsevier, New York.

Shibata, M., and Bures, J. (1972) Reverberation of cortical spreading depression along closed-loop pathways in rat cerebral cortex. *J. Neurophys.* **35**, 381–388.

Shibata, M., and Bures, J. (1974) Optimum topographical conditions for reverberating cortical spreading depression in rats. *J. Neurobiol.* **5**, 107–118.

Shields, R. (1976) New view of the cell cycle. *Nature* **260**, 193–194.

Shields, R. (1977) Transition probability and the origin of variability in the cell cycle. *Nature* **267**, 704–707.

Shilo, B., Shilo, V., and Simchen, G. (1976) Cell cycle initiation in yeast follows first-order kinetics. *Nature* **264**, 767–770; and Matters Arising. *Nature* **267**, 647–649 (1977).

Shinbrot, M. (1966). Fixed point theorems. *Sci. Amer.* **214**, (1), 105–110.

Shipp, E., and Gunning, R. V. (1975) Endogenous rhythm of nerve activity in the housefly eye. *Nature* **258**, 520–521.

Short, R. V. (1976) Definition of the problem. The evolution of human reproduction. *Proc. Roy. Soc. Lond. B* **195**, 3–24.

Simmons, D. J., Sherman, N. E., and Lesker, P. A. (1974) Allograft induced osteoinduction in rats. *Clin. Ortho.* **103**, 252–261.

Simon, E., Satter, R. L., and Galston, A. W. (1976a) Circadian rhythmicity in excised *Samanea* pulvini. I. Sucrose-white light interactions. *Plant Physiol.* **58**, 417–420.

Simon, E., Satter, R. L., and Galston, A. W. (1976b) Circadian rhythmicity in excised *Samanea* pulvini. II. Resetting the clock by phytochrome conversion. *Plant Physiol.* **58**, 421–425.

Sipski, M. L., and Wagner, T. E. (1977) The total structure and organization of chromosomal fibers in eutherian sperm nuclei. *Biology of Reproduction* **16**, 428–440.

Skopik, S. D., and Pittendrigh, C. S. (1967) Circadian systems. II. The oscillation in the individual *Drosophila* pupa; its independence of developmental stage. *Proc. Nat. Acad. Sci.* **58**, 1862–1869.

Skrabal, A. (1915) Vorlesungsversuch zur Demonstration eines Falles der Abnahme der Reaktionsgeschwindigkeit mit der Temperatur. *Zeitschrift für Elektrochemie* **21**, 461–463.

Slack, J. M. W. and Savage, S. (1978) Regeneration of reduplicated limbs in contravention of the complete circle rule. *Nature* **271**, 760–761.

Slonczewski, J. C., and Malozemoff, A. P. (1978) Physics of domain walls in magnetic garnet films in *Physics of Magnetic Garnets*, Soc. Ital. di Fisica, Bologna, Italy pp. 134–195.

Smith, E. E. and Guyton, A. C. (1961) An iron heart model for study of cardiac impulse transmission *Physiologist* **4**, 112.

Smith, J. A., and Martin, L. (1973) Do cells cycle? *Proc. Nat. Acad. Sci.* **70**, 1263–1267.

Smoes, M.-L. (1976) Toward a mathematical description of phase waves. *J. Theor. Biol.* **58**, 1–14.

Smoes, M.-L. (1979) Zhabotinskii system—three different meanings of phase waves (abstract). *B. Am. Phys. Soc.* **24**, 477.

Smoes, M.-L., and Dreitlein, J. (1973) Dissipative structures in chemical oscillations with concentration-dependent frequency. *J. Chem. Phys.* **59**, 6277–6285.

Sollberger, A. (1976) Rhythmic changes in clinical laboratory values. *C. R. C. Crit. Rev. of Clin. Lab. Sci.* **6**, 247–285.

Somero, G. N. and Hochachka, P. W. (1969) Isozymes and short-term temperature compensation in poikilotherms. *Nature* **223**, 194–195.

Sommerville, D. (1958) *An Introduction to Geometry of N Dimensions*, Dover, New York.

Sondhi, K. C. (1963) The biological foundations of animal patterns. *Quart. Rev. Biol.* **38**, 289–327.

Spangler, R. A., and Snell, F. M. (1961) Sustained oscillations in a catalytic chemical system. *Nature* **191**, 457–458.

Spanier, E. H. (1966) *Algebraic Topology*, McGraw-Hill, N.Y.

Specht, P. C., and Bórtoff, A. (1972) Propagation and electrical entrainment of intestinal slow waves. *Digestive Diseases* **17**, 311–316.

Speroff, L., and Vande Wiele, R. L. (1971) Regulation of the human menstrual cycle. *Amer. J. Obstet. Gynec.* **109**, 234–247.

Stahl, F. W. (1967) Circular genetic maps. *J. Cell. Physiol.* (Suppl.) **70**, 1–12.

Stanley, L. (1874) *The First Voyage Round the World, by Magellan*, The Hakluyt Society, London, vol. 52.

Stanshine, J. A. (1976) Asymptotic solutions of the Field-Noyes model for the Belousov reaction. II. plane waves. *Stud. Appl. Math.* **55**, 327–349.

Stein, P. S. G. (1974) Neural control of interappendage phase during locomotion. *Amer. Zool.* **14**, 1003–1016.

Stern, C. D., and Goodwin, B. C. (1977) Waves and periodic events during primitive streak formation in the chick. *J. Embryol. Exp. Morph.* **41**, 15–22.

Stern, J. (1967) *Bibliography of Liesegang Rings* (2nd ed.), National Bureau of Standards Miscellaneous Publication 292, Washington, D.C.

Stetson, M. H., and Watson-Whitmyre, M. (1976) Nucleus suprachiasmaticus: the biological clock in the hamster? *Science* **191**, 197–199.

Stetz, C. W. (1971) A mathematical model of the human menstrual cycle. B.S. thesis, M.I.T., Cambridge.

Stibitz, G. R., and Rytand, D. A. (1968) On the path of the excitation wave in atrial flutter. *Circulation* **37**, 75–81.

Strahm, N. D. (1964) Mathematical models for circadian rhythms. M.S. thesis, M.I.T., Cambridge.

Stratonovich, R. L. (1967) *Topics in the Theory of Random Noise* Gordon and Breach, New York, vol. II, Chap. 9.

Strumwasser, F. (1968) Membrane and intra-cellular mechanism governing endogenous activity in neutrons. *Physiological and Biochemical Aspects of Nervous Integration.* VIII. F. D. Carlson, ed. Prentice-Hall, Englewood Cliffs, N.J., pp. 329–341.

Strumwasser, F. (1971) The cellular basis of behavior in *Aplysia. J. Psychiatr. Res.* **8**, 237–257.

Strumwasser, F. (1974) Neuronal principles organizing periodic behavior. In *The Neurosciences Third Study Program*, F. O. Schmitt and F. G. Worden, eds. M.I.T., Cambridge, pp. 459–478.

Stupfel, M. (1975) Biorhythms in toxicology and pharmacology. *Biomedicine* **22**, 18–24.

Sturtevant, R. P. (1973) Circadian variability in *Klebsiella* demonstrated by cosinor analysis. *Int. J. Chronobiol.* **1**, 141–146.

Sudbery, P. E. and Grant, W. D. (1976) The control of mitosis in *Physarum polycephalum*. *J. Cell. Sci.* **22**, 59–65.

Sulzman, F. M., Fuller, C. A., and Moore-Ede, M. C. (1977a) Feeding time synchronizes primate circadian rhythms. *Physiol. Behav.* **18**, 775–779.

Sulzman, F. M., Fuller, C. A., and Moore-Ede, M. C. (1977b) Spontaneous internal desynchronization of circadian rhythms in the squirrel monkey. *Comp. Biochem. Physiol.* **58A**, 63–67.

Sulzman, F. M., Fuller, C. A. and Moore-Ede, M. C. (1978) Extent of circadian synchronization by cortisol in the squirrel monkey. *Comp. Biochem. Physiol.* **592**, 279–283.

Suzuki, R. (1976) Electrochemical neuron model. *Adv. Biophys.* **9**, 115–156.

Suzuki, R., Sato, S., and Nagumo, J. (1963) *Electrochemical Active Network*, Notes of Professional Group on Nonlinear Theory of IECE (Japan), Feb. 26, 1963.

Swade, R. H. (1969) Circadian rhythms in fluctuating light cycles: toward a new model of entrainment. *J. Theor. Biol.* **24**, 227–239.

Swade, R. H., and Pittendrigh, C. S. (1967) Circadian locomotor rhythms of rodents in the arctic. *Amer. Naturalist* **101**, 431–466.

Sweeney, B. M. (1960) The photosynthetic rhythm in single cells of *Gonyaulax polyedra*. In *Cold Spring Harbor Symposium for Quantitative Biology* **25**, 145–147.

Sweeney, B. M. (1969) *Rhythmic Phenomena in Plants*, Academic, New York.

Sweeney, B. M. (1974) A physiological model for circadian rhythms derived from the *Acetabularia* rhythm paradoxes. *Int. J. Chronobiol.* **2**, 25–33.

Sweeney, B. M., and Hastings, J. (1958) Rhythmic cell division in populations of *Gonyaulax polyedra*. *J. Protozool.* **5**, 217–224.

Sweeney, B. M. and Hastings, J. W. (1960) Effects of temperature upon diurnal rhythms. *Cold Spring Harbor Symposium on Quantitative Biology* **25**, 87–103.

Tachibana, A. and Fukui, K. (1978) Differential geometry of chemically reacting systems. *Theoret. Chim. Acta. (Berl.)* **49**, 321–347.

Taddei-Ferretti, C., and Cordella, L. (1976) Modulation of *Hydra attenuata* rhythmic activity: phase response curve. *J. Exp. Biol.* **65**, 737–751.

Taguchi, Y. H., and Tabachnick, J. (1974) The effect of clipping guinea-pig hair and chronic radio dermatitis on diurnal circadian rhythms in epidermal labeling and mitotic indices. *Arch. Dermatol. Forsch.* **249**, 167–177.

Takimoto, A., and Hamner, K. C. (1964) Effect of temperature and preconditioning on photoperiodic response of *Pharbitis nil*. *Plant Physiol.* **39**, 1024–1030.

Taleisnik, S., Caligaris, L., and Astrada, J. J. (1966) Effect of copulation on the release of pituitary gonadotropins in male and female rats. *Endocrinology* **79**, 49–54.

Tatterson, D. F., and Hudson, J. L. (1973) An experimental study of chemical wave propagation. *Chem. Eng. Commun.* **1**, 3–11.

Teorell, T. (1971) A biophysical analysis of mechano-electrical transduction in *Handbook of Sensory Physiology. I. Principles of Receptor Physiology*, ed. Loewenstein, W. R. Springer-Verlag, Heidelberg.

Terry, O. W., and Edmunds, L. N. (1970) Rhythmic settling induced by temperature cycles in continuously stirred autotrophic cultures of *Euglena gracilis*. *Planta* **93**, 128–142.

Tharp, G., and Folk, G. (1965) Rhythmic changes in rate of mammalian heart and heart cells during prolonged isolation. *Comp. Biochem. Physiol.* **14**, 255–273.

Thoenes, D. (1973) Spatial oscillations in the Zhabotinsky reaction. *Nature Phys. Sci.* **243**, 18–21.

Thom, R. (1975) *Structural Stability and Morphogenesis*, W. A. Benjamin, Reading, Mass.

Thomas, C. A. (1967) The rule of the ring. *J. Cell. Physiol.* (Suppl.) **70**, 13–34.

Thomson, W. I. (1867) On vortex atoms. *Phil. Mag.* **34**, 15–24.

Thompson, D. (1961) *On Growth and Form*, abridged ed., J. T. Bonner, ed. Cambridge University Press. Cambridge, Engl.

Thompson, H. E., Horgan, J. D., and Delfs, E. (1969) A simplified mathematical model and simulations of the hypophysis-ovarian endocrine control system. *Biophys. J.* **9**, 278–291.

Thompson, J. M. T. (1975) Experiments in catastrophe. *Nature* **254**, 392–395.

Thompson, J. M. T. and Hunt, G. W. (1977) The instability of evolving systems. *Interadisc. Sci Rev.* **2**. 240–262.

Thormar, H. (1959) Delayed division in *Tetrahymena pyriformis* induced by temperature changes. *CR Trav. Lab. Carlsberg* **31**, 207–225.

Tilden, J. (1974) The velocity of spatial wave propagation in the Belousov reaction. *J. Chem. Phys.* **60**, 3349–3350.

Torre, V. (1976) A theory of synchronization of two heart pace-maker cells. *J. Theor. Biol.* **61**, 55–71.

Treloar, A. E., Boynton, R. E., Behn, B. G., and Brown, B. W. (1967) Variation of the human menstrual cycle through reproductive life. *Int. J. Fertility* **12**, 77–126.

Trelstad, R., and Coulombre, A. (1971) Morphogenesis of the collagenous stroma in the chick cornea. *J. Cell Biol.* **50**, 840–858.

Troy, W. C. (1977a) The disappearance of solitary travelling wave solutions of a model of the Belousov-Zhabotinskii reaction. *Rock. Mtn. J. Math.* **7**, 467–478.

Troy, W. C. (1977b) Threshold phenomena in the Field-Noyes model of the Belousov-Zhabotinskii reaction. *J. Math. Anal and Appl.* **58**, 233–248.

Troy, W. C. (1978) Mathematical modeling of excitable media in neurobiology and chemistry. *Theoretical Chemistry* **4**, 133–157. H. Eyring and D. Henderson, eds. Academic, New York.

Troy, W. C., and Field, R. J. (1977) The amplification before decay of perturbations around stable states in a model of the Zhabotinskii reaction. *SIAM J. Appl. Math.* **32**, 306–322.

Truman, J. W. (1971) Hour-glass behavior of the circadian clock controlling eclosion of the silkmoth *Antheraea pernyi*. *Proc. Nat. Acad. Sci.* **68**, 595–599.

Truman, J. W. (1972) Physiology of insect rhythms. II. The silkmoth brain as the location of the biological clock controlling eclosion. *J. Comp. Physiol.* **81**, 99–114.

Tsien, R. W., and Carpenter, D. O. (1978) Ionic mechanisms of pacemaker activity in cardiac Purkinje fibers. *Fed. Proc.* **37**, 2127–2131.

Tucker, A. W., and Bailey, H. S., Jr. (1950) Topology. *Sci. Amer.* **182**, (1), 18–24.

Tuckwell, H. C., and Miura, R. M. (1978) A mathematical model for spreading cortical depression. *Biophys. J.*, **23**, 257–276.

Turek, F. W., McMillan, J. P., and Menaker, M. (1976) Melatonin: effects on the circadian locomotor rhythm of sparrows. *Science* **194**, 1441–1443.

Turing, A. M. (1952) The chemical basis of morphogenesis. *Phil. Trans. Roy. Soc. B***237**, 37–72.

Tuttle, J. W. and Doughery, C. R. eds. (1963) *G.E. Glow Lamp Manual*, General Electric, Cleveland.

Tweedy, D. G., and Stephen, W. P. (1970) Light refractive emergence rhythm in the leafcutter bee, *Megachile rotundata* (F.) (Hymenoptera: Apoidea). *Experientia* **26**, 377–379.

Tyshchenko, V. P. (1966) Two-oscillatory model of the physiological mechanism of insect photoperiodic reaction. *Zh. Obshch. Biol.* **27**, 209–222 (in Russian).

Tyson, J. J. et al. (1976) Mathematical background: group report. Dahlem Workshop on *The Molecular Basis of Circadian Rhythms*, J. W. Hastings and H. G. Schweiger, eds. Abakon, Berlin. pp. 85–108.

Tyson, J. J. (1976) *The Belousov-Zhabotinskii Reaction*, Lecture Notes in Biomathematics, vol. 10, S. Levin, ed. Springer-Verlag, Berlin.

Tyson, J. J. (1977a) Analytic representation of oscillations, excitability, and traveling waves in a realistic model of the Belousov-Zhabotinskii reaction. *J. Chem. Phys.* **66**, 905–915.

Tyson, J. J. (1977b) Multiple stationary states in oregonator (letter). *J. Chem. Phys.* **67**, 4297–4298.

Tyson, J. J. (1978) Appearance of chaos in a model of Belousov reaction. *J. Math. Biol.* **5**, 351–362.

Tyson, J. J. (1979) Periodic enzyme synthesis: reconsideration of the theory of oscillatory Repression. *J. Theor. Biol.* **80**, 27–38.

Tyson, J. J., and Kauffman, S. (1975) Control of mitosis by a continuous biochemical oscillation:synchronization; spatially inhomogeneous oscillations. *J. Math. Biol.* **1**, 289–310.

Tyson, J. J., and Sachsenmaier, W. (1978) Is nuclear division in *Physarum* controlled by a continuous limit cycle oscillator? *J. Theor. Biol.* **73**, 723–737.

Underwood, H. (1977) Circadian organization in lizards: the role of the pineal organ. *Science* **195**, 587–589.

Urquhart, J., ed. (1973) *Temporal Aspects of Therapeutics*, Plenum, New York.

Ursprung, H. (1966) The formation of patterns in development. In *Major Problems of Development*, M. Locke, ed., Academic, New York, pp. 177–216.

Utsumi, K., and Packer, L. (1967) Oscillatory states of mitochondria. II. Factors controlling period and amplitude. *Arch. Biochem. Biophys.* **120**, 404–412.

Vanden Driessche, T. (1971) Structural and functional rhythms in the chloroplasts of *Acetabularia*. In *Biochronometry*, M. Menaker, ed. Nat. Acad. Sci., Washington, D.C., pp. 612–621.

Vande Wiele, R. L., Bogumil, J., Dyrenfurth, I., Ferin, M., Jewelewicz, R., Warren, M., Rizkallah, T., and Mikhial, G. (1970) Mechanisms regulating the menstrual cycle in women. *Recent Progress in Hormone Research* **26**, 63–102.

Varadi, Z. B., and Beck, M. T. (1975) One-dimensional periodic structures in space and time. *Biosystems* **7**, 77–82.

Vavilin, V. A., Zhabotinskii, A. M., and Zaikin, A. N. (1968) Effect of ultraviolet radiation on the oscillating oxidation reaction of malonic acid derivatives. *Russian J. Physical Chemistry* **42**, 1649–1651.

Verma, A. R., and Krishna, P. (1966) *Polymorphism and Polytypism in Crystals*, Wiley, New York.

Verzeano, M. (1963) The synchronization of brain waves. *Acta. Neurol. Latinoamer.* **9**, 297–301.

Vince, M. A., Green, J., and Chinn, S. (1971) Changes in the timing of lung ventilation and late foetal development in quail. *Comp. Biochem. Physiol.* **39A**, 769–783.

Volterra, V. (1926) Variazioni e fluttuazioni del numero d'individui in specie animali conviventi. *Mem. Acad. Lincei.*, **2**, 31–113. (Translation in an appendix to Chapman's *Animal ecology*, New York, 1931.)

Wagner, E., Deitzer, G. F., Fischer, S., Frosch, S., Kempf, O., and Stroebele, L. (1975) Endogenous oscillations in pathways of energy transduction as related to circadian rhythmicity and photoperiodic control. *Biosystems* **7**, 68–76.

Wagner, E. (1976) Endogenous rhythmicity in energy metabolism: basis for timer-photoreceptor interactions in photoperiodic control. In Dahlem Workshop on *The Molecular Basis of Circadian Rhythms*, J. W. Hastings and H. G. Schweiger, eds., Abakon, Berlin, pp. 215–238.

Wagner, E., and Cumming, B. G. (1970) Betacyanin accumulation, chlorophyll content, and flower initiation in *Chenopodium rubrum* as related to endogenous rhythmicity and phytochrome action. *Can. J. Bot.* **48**, 1–18.

Wahl. O. (1932) Neue Untersuchungen über das Zeitgedächtnis der Bienen. *Z. vergl. Physiol.* **16**, 529–589.

Walker, T. J. (1969) Acoustic synchrony: two mechanisms in the snowy tree cricket. *Science* **166**, 891–894.

Watanabe, A., Obara, S., and Akiyama, T. (1967) Pacemaker potentials for the periodic burst discharge in the heart ganglion of a stomatopod *Squilla oratoria. J. Gen. Physiol.* **50**, 839–862.

Watson, J. (1976) *Molecular Biology of the Gene*, 3rd ed., Benjamin, Menlo Park, p. 510.

Watson-Whitmyre, M. M., and Stetson, M. H. (1977) Circadian organization in regulation of reproduction-identification of a circadian pacemaker in hypothalamus of hamster. *J. Interd. Cy. Res.* **8**, 360–367.

Wegmann, K., and Rössler, O. E. (1978) Different kinds of chaotic oscillations in the Belousov-Zhabotinskii reaction. *Z. Naturforsch.* **33a**, 1179–1183.

Wehner, R., Herrling, P., Brunnert, A., and Klein, R. (1973) Periphere Adaptation und zentralnervöse Umstimmung in optisches Systems von *Cataglyphis bicolor. Review Suisse Zool.* **79**, (suppl. au Mai), 197–228.

Wei, J. (1962) Axiomatic treatment of chemical reaction systems. *J. Chem. Phys.* **36**, 1578–1584.

Weidemann, S. (1951) Effect of current flow on the membrane potential of cardiac muscle. *J. Physiol.* **115**, 227–236.

Weitzman, E. D. (1976) Biologic rhythms and hormone secretion patterns. *Hosp. Prac.* **11**, 8, 79–86.

Werner, G. (1970) The topology of the body representation in the somatic afferent pathway. Chapter 54 in *The Neurosciences Second Study Program*, F. O. Schmidt, ed. pp. 605–617. Rockefeller U. P., N. Y.

West, T. C., and Landa, J. F. (1962) Minimal mass required for induction of a sustained arrhythmia in isolated atrial segments. *Amer. J. Physiol.* **202**, 232–236.

Wever, R. (1962) Zum Mechanismus der biologischen 24-Stunden-Periodik. *Kybernetik* **1**, 139–154.

Wever, R. (1963) Zum Mechanismus der biologischen 24-Stunden-Periodik. II. Der Einfluss des Gleichwertes auf die Eigenschaften selbsterregter Schwingungen. *Kybernetik* **1**, 213–231.

Wever, R. (1964) Zum Mechanismus der biologischen 24-Stunden-Periodik. III. Anwendung der Modell-Gleichung. *Kybernetik* **2**, 127–144.

Wever, R. (1965a) A mathematical model for circadian rhythms. *Circadian Clocks*, J. Aschoff, ed. North-Holland, Amsterdam, pp. 47–63.

Wever, R. (1965b) Einzelorganismen und Populationen im circadianen Experiment: eine methodische Analyse. *Z. vergl. Physiol.* **51**, 1–24.

Wever, R. (1973) Internal phase angle differences in human circadian rhythms: causes for changes and problems of determinations. *Int. J. Chronobiol.* **1**, 371–390.

Wever, R. (1975) The circadian multi-oscillator system of man. *Inst. J. Chronobiol.* **3**, 19–55.

Wiedenmann, G. (1977) Weak and strong phase shifting in the activity rhythm of *Leucophaea maderae* (Blaberidae) after light pulses of high·intensity. *Z. Naturforsch.* **32c**, 464–465.

Wiener, N., and Rosenblueth, A. (1946) The mathematical formulation of the problem of conduction of impulses in a network of connected excitable elements, specifically in cardiac muscle. *Arch. Inst. Cardiologia de Mexico* **16**, 205–265.

Wigner, E. (1960) The unreasonable effectiveness of mathematics in the natural sciences. *Comm. in Pure and Applied Mathematics* **13**, 222–237.

Wilhelm, H. E., and van der Werff, T. J. (1977) Nonlinear wave equations for chemical reactions with diffusion in multicomponent systems. *Chem. Phys.* **67**, 3382–3387.

Wilhelm, R. H. (1966) Parametric pumping—A model for active transport. In *Intracellular Transport*, J. Danielli, ed. Academic, New York.

Wilhelm, R. H., Rice, A. W., Rolke, R. W., and Sweed, N. H. (1968) Parametric pumping: dynamic principle for separating fluid mixtures. *Ind. Eng. Chem. Funda.* **7**, 337–349.

Wilkins, M. B. (1960) The effect of light upon plant rhythms. In *Cold Spring Harbor Symposium on Quantitative Biology* **25**, 115–129.

Wilkins, M. B. (1962) Effects of temperature changes on phase and period of the rhythm. *Proc. Roy. Soc. B* **156**, 220–241.

Wilkins, M. B. (1965) The influence of temperature and temperature changes on biological clocks. In *Circadian Clocks*, J. Aschoff, ed. North-Holland, Amsterdam, pp. 146–163.

Wilkins, M. B. (1973) An endogenous circadian rhythm in the rate of carbon dioxide ouput of *Bryophyllum*. *J. Exper. Bot.* **24**, 488–496.

Wilkins, M. B., and Holowinsky, A. T. (1965) The occurrence of an endogenous circadian rhythm in a plant tissue culture. *Plant Physiol.* **40**, 907–909.

Wilkins, M. H. F. (1963) X-ray diffraction studies of the molecular configuration of nucleic acids. In *Aspects of Protein Structure*, G. N. Ramachandran, ed. Academic, London, pp. 23–27

Wille, J. J. (1979) Biological rhythms in protozoa in *Biochemistry and Phyisology of Protozoa* **2**, 67–149. Ed. M. Levandowsky, Academic, N.Y.

Wille, J. J., Scheffey, C., and Kauffman, S. A. (1977) Novel behavior of mitotic clock in *Physarum*. *J. Cell Sci.* **27**, 91–104.

Wilson, B. J. (1969) *Famous Clocks*. Book 46, Bancroft Tiddlers, London.

Wilson, H. R. (1973) Cooperative phenomena in a homogeneous cortical tissue model. In *Synergetics*, H. Haken, ed. Teubner, Stuttgart.

Wilson, H. R., and Cowan, J. D. (1972) Exitatory and inhibitory interactions in localized populations of model neurons. *Biophys. J.* **12**, 1–24.

Wilson, H. R., and Cowan, J. D. (1973) A mathematical theory of the functional dynamics of cortical and thalamic nervous tissue. *Kybernetik* **13**, 55–80.

Wilson, W. A., and Wachtel, H. (1974) Negative resistance characteristic essential for maintenance of slow oscillations in bursting neurons. *Science* **186**, 932–934.

Winfree, A. T. (1965) An experimental study, theoretically of biological rhythms. Bachelor's thesis, Dept. of Engineering Physics, Cornell University.

Winfree, A. T. (1966) A progress report. Published by Thomas C. Jenkins, Dept. of Biophysics, John Hopkins University.

Winfree, A. T. (1967a) Biological rhythms and the behavior of populations of coupled oscillators. *J. Theor. Biol.* **16**, 15–42.

Winfree, A. T. (1967b) Puzzles and paradoxes. (Mimeo circulated in limited number.)

Winfree, A. T. (1968) The investigation of oscillatory processes by perturbation experiments. In *Biological and Biochemical Oscillators*, B. Chance et al., eds. Academic, London (1973), pp. 461–501.

Winfree, A. T. (1970a) Oscillatory control of cell differentiation in *Nectria*? *Proc. IEEE Symp. Adapt. Process* **9**, pp. 23.4.1–23.4.7.

Winfree, A. T. (1970b) The temporal morphology of a biological clock. In *Lectures on Mathematics in the Life Sciences*. vol. 2, M. Gerstenhaber, ed. Amer. Math. Soc., Providence, R.I., pp. 109–150.

Winfree, A. T. (1970c) An integrated view of the resetting of a circadian clock. *J. Theor. Biol.* **28**, 327–374.

Winfree, A. T. (1971a) Corkscrews and singularities in fruitflies: resetting behavior of the circadian eclosion rhythm. In *Biochronometry*, M. Menaker, ed. Nat. Acad. Sci., Washington, D.C., pp. 81–109.

Winfree, A. T. (1971b) Comment. In *Biochronometry*, M. Menaker, ed. Nat. Acad. Sci., Washington, D.C., pp. 150–151.

Winfree, A. T. (1972a) Acute temperature sensitivity of the circadian rhythm in *Drosophila*. *J. Insect. Physiol.* **18**, 181–185.

Winfree, A. T. (1972b) Slow dark-adaptation in *Drosophila's* circadian clock. *J. Comp. Physiol.* **77**, 418–434.

Winfree, A. T. (1972c) Spiral waves of chemical activity. *Science* **175**, 634–636.

Winfree, A. T. (1972d) Oscillatory glycolysis in yeast: the pattern of phase resetting by oxygen. *Arch. Biochem. Biophys.* **149**, 388–401.

Winfree, A. T. (1972e) On the photosensitivity of the circadian time sense in *Drosophila pseudoobscura*. *J. Theor. Biol.* **35**, 159–189.

Winfree, A. T. (1973a) Resetting the amplitude of *Drosophila's* circadian chronometer. *J. Comp. Physiol.* **85**, 105–140.

Winfree, A. T. (1973b) Scroll-shaped waves of chemical activity in three dimensions. *Science* **181**, 937–939.

Winfree, A. T. (1973c) Spatial and temporal organization in the Zhabotinsky reaction. In *Advances in Biological and Medical Physics*, **16**, 115–136, ed J. H. Lawrence et al. (1977) Academic, New York.

Winfree, A. T. (1973d) Polymorphic pattern formation in the fungus *Nectria*. *J. Theor. Biol.* **38**, 363–382.

Winfree, A. T. (1973e) Time and timelessness in biological clocks. In *Temporal Aspects of Therapeutics* J. Urquardt and F. E. Yates, eds. Plenum, New York, pp. 35–57.

Winfree, A. T. (1974a) Suppressing *Drosophila's* circadian rhythm with dim light. *Science* **183**, 970–972.

Winfree, A. T. (1974b) Rotating chemical reactions. *Sci. Amer.* **230**, (6), 82–95.

Winfree, A. T. (1974c) Patterns of phase compromise in biological cycles. *J. Math. Biol.* **1**, 73–95.

Winfree, A. T. (1974d) Two kinds of wave in an oscillating chemical solution. *Farad. Symp. Chem. Soc.* **9**, 38–46.

Winfree, A. T. (1974e) Wavelike activity in biological and biochemical media. In *Lecture Notes in Biomathematics* **2**, pp. 243–260. P. van den Driessche, ed. Springer-Verlag, Berlin.

Winfree, A. T. (1974f) Rotating solutions to reaction/diffusion equations. *SIAM-AMS Proceedings* **8**, 13–31. (Ed. D. Cohen). Amer. Math. Soc., Providence R.I.

Winfree, A. T. (1975a) Resetting biological clocks. *Physics Today* **28**, 34–39.

Winfree, A. T. (1975b) Unclocklike behavior of a biological clock. *Nature* **253**, 315–319.

Winfree, A. T. (1975c) *Drosophila's* clock photoreceptor is still not rhodopsin. *Biophys. J.* **15**, 177a.

Winfree, A. T. (1976) On phase resetting in multicellular clockshops. In Dahlem Workshop on *The Molecular Basis of Circadian Rhythms*, J. W. Hastings and H. G. Schweiger, eds. Abakon, Berlin. pp. 109–129.

Winfree, A. T. (1977) Phase control of neural pacemakers. *Science* **197**, 761–762.

Winfree, A. T. (1978a) Stably rotating patterns of reaction and diffusion, in *Theoretical Chemistry* **4**, 1–51, ed. H. Eyring and D. Henderson; Academic, N.Y.

Winfree, A. T. (1978b) Patterns of phase compromise in biological cycles in *Studies in Mathematics* **15**, *Studies in mathematical Biology*, pp. 266–294, ed. S. A. Levin, Math. Assoc. of Amer. Providence, R.I.

Winfree, A. T. (1978c) Chemical clocks: a clue to biological rhythms. *New Scientist* **80**, 10–13.

Winfree, A. T. (1979) Twenty-four hard problems about the mathematics of 24-hour rhythms, in Lectures in Applied Mathematics **17**, *Nonlinear Oscillations in Biology* ed. F. C. Hoppensteadt. Amer. Math. Soc., Providence R.I.

Winfree, A. T., and Gordon, H. (1977) The photosensitivity of a mutant circadian clock. *J. Comp. Physiol.* **122**, 87–109.

Wisnieski, B. J., and Fox, C. F. (1976) Correlations between physical state and physiological activities in eukaryotic membranes, especially in response to temperature. Dahlem Workshop on *The Molecular Basis of Circadian Rhythms*, J. W. Hastings and H. G. Schweiger eds. Abakon, Berlin. pp. 247–266.

Wojtowicz, J. (1972) Oscillatory behavior in electrochemical systems. *Modern Aspects of Electrochemistry* **8**, J. D. Bockrist and B. E. Conway, eds. pp. 47–102.

Wolken, J. (1968) Cover photograph of *Science* **161**, no. 3845.

Wolken, J. J. (1971) *Invertebrate Photoreceptors, A Comparative Analysis*, Academic, New York.

Wolpert, L. (1969) Positional information and the spatial pattern of cellular differentiation. *J. Theor. Biol.* **25**, 1–47.

Wongwiwat, M., Sukapanit, S., Triyanond, C., and Sawyer, W. D. (1972) Circadian rhythm of the resistance of mice to acute pneumococcal infection. *Infection and Immunity* **5**, 442–448.

Woody, C. O., and Pierce, R. A. (1974) Influence of day of estorus cycle-like treatment and response to estrous cycle regulation by norethandrolone implants and estradiol valerate injections. *J. Animal Sci.* **39**, 902–906.

Wurster, B. (1976) Temperature dependence of biochemical oscillations in cell suspensions of *Dictyostelium discoideum*. *Nature* **260**, 703–704.

Yakhno, V. G. (1975) A model of the leading centre. *Biofizika* **20**, 669–674.

Yamada, T., and Kuramoto, Y. (1976) Spiral waves in a nonlinear dissipative system. *Prog. Theor. Phys.* **55**, 2035–2036.

Yamanishi, J., Kawato, M., and Suzuki, R. (1979) Studies on human finger tapping neural networks by phase transition curves. *Biol. Cybern.* **33**, 199–208.

Yamazaki, H., Oono, Y., and Hirakawa, K. (1978) Experimental study on chemical turbulence. *J. Phys. Soc. Japan Physics* **44**, 335–336, and II. **46**, 721–728.

Yates, F. E., Russell, S. M., and Maran, J. W. (1971) Brain-adenohypophysial communication in mammals. *Ann. Rev. Physiol.* **33**, 393–444.

Yen, S. S. C., and Tsai, C. C. (1972) Acute gonadotropin release induced by exogenous estradiol during the mid-follicular phase of the menstrual cycle. *J. Clin. Endocrinol. Metab.* **34**, 298–305.

Yoshimoto, Y. and Kamiya, N. (1978) Studies on contraction of the plasmodial strand. *Protoplasma* **95**, 89–122.

Yoshizawa, S., Amari, S., and Nagumo, J. (1971) Reverberatory activity of excitation on nerve networks. *Denshi Tsushin Gakki Shi.* (*Journal of Electronics-Communication Society, Japan*) **54**, 1364–1373 (in Japanese).

Young, R. A. (1977) Some observations on temporal coding of color vision: psychophysical results. *Vision Res.* **17**, 957–965.

Young, R. W. (1978) Visual cells, daily rhythms, and vision research. *Vision Research* **18**, 573–578.

Zahler, S., and Sussmann, J. (1977) Claims and accomplishments of applied catastrophe theory. *Nature* **269**, 759–763.

Zaikin, A. N. (1975) Instability and spread of excitation in a model of the catalytic reaction. II. Model of a distributed system. *Biofizika* **20**, 772–777.

Zaikin, A. N., and Kawczynski, A. L. (1977) Spatial effects in active chemical systems. 1. Model of leading center. *J. Non-Equilib. Thermodyn.* **2**, 39–48.

Zaikin, A. N., and Zhabotinsky, A. M. (1970) Concentration wave propagation in two-dimensional liquid-phase self-oscillating systems. *Nature* **225**, 535–537.

Zeeman, E. C. (1972) Differential equations for the heartbeat and nerve impulse. *Towards a Theoretical Biology* **4**, C. H. Waddington, ed. Aldine, Chicago, 8–67.

Zeeman, E. C. (1977) *Catastrophe Theory*. Addison Wesley, Reading, Mass.

Zelazny, B., and Neville, A. C. (1972) Endocuticle layer formation controlled by non-circadian clocks in beetles. *J. Insect Physiol.* **18**, 1967–1979.

Zeuthen, E., and Williams, N. E. (1969) Division-limiting morphogenetic processes in *Tetrahymena*. In *Nucleic Acid Metabolism, Cell Differentiation and Cancer Growth*, E. V. Cowdry and S. Seno, eds. Pergamon Press, Oxford, pp. 203–216.

Zhabotinsky, A. M. (1968) A study of self-oscillatory chemical reaction III: space behavior. In *Biochemical Oscillators, Proceedings of the 1968 Prague Symposium*, B. Chance, et al., eds. Academic, New York (1973).

Zhabotinsky, A. M. (1970) Investigations of homogeneous chemical auto-oscillating systems (in Russian). Thesis, Puschino (Biophysics 03091).

Zhabotinsky, A. M. (1974) *Spontaneously Oscillating Concentrations*, Science Publishers, Moscow (In Russian.)

Zhabotinsky, A. M., and Zaikin, A. N. (1973) Autowave processes in a distributed chemical system. *J. Theor. Biol.* **40**, 45–61.

Zimmer, R. (1962) Störlichwirken auf tagesperiodische Blütenblattbewegungen. *Planta* **58**, 283–300.

Zimmerman, N. H., and Menaker, M. (1975) Neural connections of sparrow pineal role in circadian control of activity. *Science* **190**, 477–479.

Zimmerman, W. F. (1966) The effects of temperature on the circadian rhythm of eclosion in *Drosophila pseudoobscura*. Ph.D. thesis, Princeton University.

Zimmerman, W. F. (1969) On the absence of circadian rhythmicity in *Drosophila pseudoobscura* pupae. *Biol. Bull.* **136**, 494–500.

Zimmerman, W. F. (1971) Some photophysiological aspects of circadian rhythmicity in *Drosophila*. In *Biochronometry*, M. Menaker, ed. Nat. Acad. Sci., Washington, D.C., pp. 381–391.

Zimmerman, W. F., and Goldsmith, T. H. (1971) Photosensitivity of the circadian rhythm and of visual receptors in carotenoid-depleted *Drosophila*. *Science* **171**, 1167–1169.

Zimmerman, W. F., Pittendrigh, C. S., and Pavlidis, T. (1968) Temperature compensation of the circadian oscillation in *Drosophila pseudoobscura* and its entrainment by temperature cycles. *J. Insect. Physiol.* **14**, 669–684.

Zucker, I. (1976) Light, behavior, and biologic rhythms. *Hosp. Prac.* **11**, 10, 83–91.

Zwanzig, R. (1976) Interactions of limit-cycle oscillators. In *Topics in Statistical Mechanics and Biophysics*, R. A. Piccirelli, ed. Amer. Inst. of Physics, New York.

Zweig, S. (1938) *Conqueror of the Seas, The Story of Magellan*, The Literary Guild of America, New York, pp. 287–293.

Index of First Names

Abraham et al. 1974 455
Aizawa 1976 85,211
Alcantara and Monk 1974 338,339,342
Aldridge and Pavlidis 1976 204,212,398
Aldridge and Pye 1976 293,297
Aldridge and Pye 1979a 60,109,111,156,204,211,265,296
Aldridge and Pye 1979b 60,109,156,204,211,263,265,296,297
Aldridge 1976 211,212,288,295
Allessie et al. 1973 67,150,240,270,332,333,334,336
Allessie et al. 1976 171,336
Allessie et al. 1977 67,171,336
Anderson et al. 1967 13
Andrews 1971 395
Andronov et al. 1966 133,163,169,233,234
Arechiga 1977 395
Aris 1975 229
Aristotle 44
Arndt 1937 338
Arnold and Avez 1968 208
Arrata and Chatterton 1974 455
Arshavskii et al. 1964 327,328
Arshavskii et al. 1965 330
Arvanitaki and Chalazonitis 1968 319
Aschoff and Wever 1976 376,397
Aschoff et al. 1967 376
Aschoff 1965a 376
Aschoff 1965b 423,430
Aschoff 1969 376
Aschoff 1976 376
Ashkenazi and Othmer 1978 208,211,227,230
Ashkenazi et al. 1975 395
Augustine, Saint 40
Ayers and Selverston 1977 318
Ayers and Selverston 1979 172
Bacon, Francis 247
Balakhovskii 1965 240,249,284,331,339
Barnett 1966 442
Barneveld 1971 317
Bateman 1955 402
Bateson 1894 358
Beck and Varadi 1971 124,213,220,306
Beck and Varadi 1972 124,125,308
Beck 1968 417
Beck, Varadi and Hauck 1976 124,220
Beier et al. 1968 396
Belousov 1958 300,302
Benham 1894 16
Benson and Jacklet 1977 395
Berg 1974 78
Bernal, J. D. 231
Berkinblit et al. 1975 208
Bessey 1950 370
Best 1975 456

Best 1976 156,171,172
Best 1979 156,274,318
Best, Eric 455
Betz and Becker 1975a 176
Betz and Becker 1975b 296,297
Betz and Chance 1965b 176,298
Betz 1966 176,297
Betz, Augustin 287
Bhereur et al. 1971 209
Bierce, Ambrose 133
Birukow 1960 21
Blandamer and Roberts 1978 176
Bogumil et al. 1972 175,455,456
Bogumil, R. J. 175,456
Boiteux et al. 1975 289
Bollig 1975 409
Bollig 1977 396
Bonhoeffer 1948 153
Bonner 1967 337
Boon and Strackee 1976 208
Bortoff 1969 319
Bortoff, A. 327
Both et al. 1976 319,320
Bouligand 1965 365
Bouligand 1969-1974 365
Bouligand 1972a 150,365
Bouligand 1978 365
Bourret et al. 1969 247,266,369,370
Bourret 1971 246
Boutselis et al. 1971 455
Boutselis et al. 1972 455
Bradshaw 1972 409
Brady 1974 81,362,421
Braemer 1960 20
Brett 1955 92,380,402,413,417,424
Brewer and Rusch 1968 439
Brink et al. 1946 319
Brinkmann 1966 380,397,399
Brinkmann 1967 83,397,399
Brinkmann 1971 398
Brody and Martins 1979 379
Brooks 1975, 1976 443
Brown and Webb 1948 379
Brown et al. 1975 229,304,326
Brown, Frank 93
Bruce et al. 1960 110,369
Bruce 1965 442
Bruce, Victor 278
Bryant et al. 1977 65,345
Bryant 1976 358
Buck and Buck 1968 325
Buck and Buck 1976 323,398
Buerle 1956 240
Bullock 1955 381
Bullough and Lawrence 1964 441
Bunning and Joerrens 1960 410
Bunning and Moser 1966 410
Bunning 1935 49,410,424
Bunning 1956a 93
Bunning 1956b 378
Bunning 1957 417
Bunning 1959 203,417,434
Bunning 1960 172,385
Bunning 1964 172,417
Bunning 1973 391,395
Bunow et al. 1980 356
Bures et al. 1974 246
Bures 1959 336
Burns and Tannock 1970 443,444
Burton and Canham 1973 173,441,443
Burton 1971 441,443
Busse and Hess 1973 213,217,229,312
Busse 1969 220,300,304,306
Butterfield 1968 17
Caldarola and Pittendrigh 1974 ...382

Campbell 1964 77,93,137,164,437
Carpenter 1979 320
Carroll, Lewis 11,361
Casten, Richard 126,129
Cavendish, Henry 392
Chance and Yoshioka 1966 256
Chance et al. 1965a 111,290
Chance et al. 1967 209,290
Chance et al. 1968 297,298
Chance 1965 289,291,292
Chance, Britton 287,289
Chandrashekaran and Engelmann 1976 ...196
Chandrashekaran and Engelmann 1973 ...409,415
Chandrashekaran and Loher 1969b 49,406,409
Chandrashekaran and Loher 1969a 421
Chandrashekaran et al. 1973 209
Chandrashekaran 1974 60,380
Chaplain 1976 292,320
Chapman 1967 246
Chevaugeon and van Huong 1969 368
Child 1941 267
Chin et al. 1972 446
Christianson and Sweeney 1973 111,385,386,409
Christie et al. 1970 376
Clark and Steck 1979 67,172,252
Clark and Zarrow 1971 450
Clay and Dehaan 1979 324,398
Coddington and Levinson 1955 148
Cohen and Robertson 1971a,b 339,340
Cohen 1971 233
Cohen 1972 443
Cohen 1977 233
Cohen 1978 233,341
Cohen, Hoppensteadt, and Miura 1977 ..230
Cohen, Neu, and Rosales 1978 227,230,247
Cole 1968 316
Collins and Ross 1978 234
Colquhoun 1971 376
Connor and Stevens 1971a,b,c 320
Connor 1978 320
Conroy and Mills 1970 376
Conway et al. 1978 227,306
Cooke and Zeeman 1976 267
Cooper and Rowson 1975 456
Cottrell 1978 308
Cseffalvay 1966 455
Creator, The 29
Cumming 1972 110
Cummings and Prothero 1978 250,267,346,359
Cummings 1975 263,379
Daan and Berde 1978 117,119,209,282
Daan and Pittendrigh 1976a 147,422
Daan and Pittendrigh 1976b 81,82,86,147,382,422,458
Daan et al. 1975 209,396
Dante Allegheri 91
Danziger and Elmergreen 1956 137,199,200,379
Danziger and Elmergreen 1957 175,262,457
Darmon et al. 1975 338
Darwin, Charles 355
Davenport 1977 326,327
De Luve 1969 74
Decoursey 1959 93
Decoursey 1961 397
Defant 1961 46-48
Defay et al. 1977 308,344
Degn 1969 176
Dehaan and Sachs 1972 324
Dehaan, Robert 324
Delbruck and Reichardt 1956 82
Delbruck, Max 393
Desimone et al. 1973 229,301,344
Devi et al. 1968 439
Dewan et al. 1978 455
Dewey et al. 1973 212

Dharmananda and Feldman 1979 123,224,228,266,368,370
Dhont et al. 1974 455
Diamant and Bortoff 1969 327
Diamant et al. 1970 229,304,326
Diamant et al. 1973 329
Dicke, Richard 392
Dobzhansky, T. 414
Duranceau 1977 356
Durham and Ridgeway 1976 229
Durston 1973 281,333
Duysens and Amesz 1957 286,288
Dynnik and Sel'Kov 1973, 1975a,b 209,290
Dynnik et al. 1977 209
Edmunds et al. 1971 397,399
Edmunds 1971 399
Edmunds 1974a 441
Edmunds 1974b 79,442
Edmunds 1975 91,436,441,442
Edmunds 1977 442
Ehret and Barlow 1960 381
Ehret and Trucco 1967 92,262,381,417,441
Ehret and Wille 1970 393,442
Ehret et al. 1975 376
Ehret 1971 411
Ehret 1974 442
Eigen, Manfred 273
El-Sharkaway and Daniel 1975 319
Elliott et al. 1972 376
Engelmann and Honegger 1967 110
Engelmann and Mack 1978 395,396
Engelmann et al. 1964 380
Engelmann et al. 1973 59,111,197,262,411,430,431,433
Engelmann et al. 1974 60,111,262,433
Engelmann et al. 1978 109,111,198,262,264,409,433,434
Engelmann 1969 49,406,415
Engelmann, Wolfgang 264,265,369,412,413,431
Engelberg 1968 440
Enright and Hamner 1967 375
Enright 1970 390
Enright 1975 93
Erneux and Herschkowitz-Kaufman 1975 .229
Erneux and Herschkowitz-Kaufman 1977 .227,229,246
Esch 1976 380
Fahrenbach 1969 363
Fantes et al. 1975 173,443
Faraday, Michael 392,412
Farley and Clark 1961 240,331
Farley 1964 331
Farley 1965 284
Feinn and Ortoleva 1977 234
Fenichel 1974, 1977 148
Festinger et al. 1971 17
Feynman et al. 1964 45,132
Field and Noyes 1972 219,229,302
Field and Noyes 1974b 229,302
Field and Troy 1979 302
Field et al. 1972 170
Fife 1979 126,234
Firth 1966 317
Fitzgerald and Zucker 1976 458
FitzHugh 1960 137,169,200,316
FitzHugh 1961 137,153,169,200,316
Flink and Doe 1959 376
Fohlmeister et al. 1974 171
Ford, Joseph 230
Franck and Mennier 1953 304
Franck 1968 169
Frank and Zimmerman 1969 49,384,409,410
Frankel 1962 440
Frankel 1969 440
French et al. 1976 65,250,266-268,273,345-350,361,363
French 1978 346
Fujii and Sawada 1978 83,85,114,208
Fuller et al. 1978 376

Galilei, Galileo 4,19
Garfinkel et al. 1968 290
Garrey 1914 330
Gaze 1970 8
Geller and Brenner 1978 342
Gerisch and Hess 1974 82,297,338,342
Gerisch et al. 1975 212,338
Gerisch 1961 339,340
Gerisch 1965 66,240,284,331,338,342
Gerisch 1968 337
Gerisch 1971 339,340
Gerola and Seiden 1978 242
Ghosh and Chance 1964 176
Ghosh et al. 1971 24,61-63,118,208,297,299,398
Gibbs, Josiah Willard 345,358
Gilbert 1974 173,441
Gilkey et al. 1978 246
Glass and Mackey 1979 171
Glass 1977 13,65,250,266,273,346,354,359,361
Gmitro and Scriven 1966 229
Gola 1976 177,320
Goldbeter and Caplan 1976 220
Goldbeter and Erneux 1978 233,286,295,307,341
Goldbeter and Lefever 1972 289,299
Goldbeter and Nicolis 1976 286,289
Goldbeter and Segel 1977 233,341
Goldbeter et al. 1978 233,234,307,341
Goldbeter 1975 341,379
Goldstein ,Byron 379
Golenhofen 1970 325,381
Gollub et al. 1978 208
Gooch and Packer 1971 212
Gooch, Sulzman, and Hastings 1979 399
Goodwin and Cohen 1969 127,250,266,373
Goodwin 1963 287,381,391,440
Goodwin 1966 209,391,440
Goodwin 1973 121
Goodwin 1974 121
Goodwin 1975 121
Goodwin 1969 79
Goodwin 1976 93,121,147,266,268,345,355
Gordon and Winfree 1978 269,366
Gordon 1976 365
Gordon, Herman 2,266,363,365
Gosling and Hundhausen 1977 242
Gouras 1958 336
Grant 1956 209
Grasman and Jansen 1979 210,230
Grattarola and Torre 1977 115,208,210
Greenberg and Hastings 1978 240,246
Greenberg et al. 1978 240
Greenberg 1976 230
Greenberg 1978 230,312
Greenberg, James 310
Greenwood and Durand 1954 9
Greller 1977 60,109,111,198,265,295,296,298
Guckenheimer 1975 148,150,222,261,294
Guckenheimer 1976 230,246,309
Gulko and Petrov 1972 68,240,246,332-335,341
Gulrajani and Roberge 1978 320
Gumbel et al. 1953 9
Gunn 1937 454
Guttes et al. 1969 444
Guttman et al. 1979 79,156,177,274,318
Gwinner 1974 396,458
Hahn et al. 1974 233
Halaban 1968 110
Halaban 1969 410
Halberg and Connor 1961 377,393
Halberg et al. 1960 376
Halberg 1977 376,442
Hale 1963, 1969, 1977 148
Halvorson et al. 1971 91,212,297,441
Hamburger 1962 440

Hamner and Hoshizaki 1974 209
Hamner et al. 1962 45
Hamner, Karl 63
Hanson et al. 1971 323
Hanson 1978 111,119,120,323,398
Harcombe and Wyman 1977 323
Hardeland 1972, 1973 395
Harker 1958 403,411
Harmon and Lewis 1966 316
Harmon, Leon 283
Harris and Wilkins 1978 111
Harrison, John 19
Hartline et al. 1979 34,111,318
Hartline 1976 34,171,317,318,321
Hartwell et al. 1974 91,436,437,440
Hastings and Murray 1975 170
Hastings and Schweiger 1976 421
Hastings and Sweeney 1957 21,381,382
Hastings and Sweeney 1958 32,110,385,387,399
Hastings 1976 229,230,234
Hastings, Stuart 309
Haus et al. 1972 376
Haus et al. 1974 376
Heide 1977 408
Heinrich and Bartholomew 1972 380
Heinrich and Casey 1973 380
Hejnowicz 1970 121
Hellbrugge 1960 376
Hellwell and Epstein 1979 303
Helmholtz 1867 256
Helmholtz, Hermann 41
Hengstenberg 1971 317
Herschkowitz-Kaufman 1970 220
Hess and Boiteux 1968 289,290
Hess and Boiteux 1971 220,286
Hess et al. 1969 288
Hess, Benno 287
Hesse, Hermann 276,288
Higgins et al. 1968 290,299
Higgins 1964 289,299,379
Higgins 1967 220,261,287,299,378
Higgins, Joseph 287,289
Hill 1933 170,385
Hoagland 1933 93,381
Hobbit, The 275
Hodgkin and Huxley 1952 234,315
Hoffmann 1960 20
Hoffmann 1969a 396
Hoffmann 1969b 380
Hoffmann 1970 396
Hoffmann 1971b 396
Hoffmann 1971a 390
Hoffmann 1976 390
Holaday et al. 1958 326
Holmes, Sherlock 73,107,178,344,450
Honegger 1967 110,420
Howard and Kopell 1974 229
Howard and Kopell 1977 229
Howard 1932 444
Hoyle, Fred 72
Hunter 1974 200
Huygens 1673 150
Ingold and Cox 1955 369
Jack et al. 1975 233,316
Jacklet and Geronimo 1971 395
Jacklet 1977 111
Jacobson 1970 8
Jalife and Antzelevitch 1979 172,274,318
Jalife and Moe 1976 111,119,172,318,324,325
Jalife and Moe 1979 172,322
Jalife et al. 1980 168
Jalife, Jose 119,318
Jenerick 1963 177
Jensen 1966 324

Johnson 1939 93
Johnsson and Israelsson 1969 111,188
Johnsson and Karlsson 1971 164,262
Johnsson and Karlsson 1972 262
Johnsson et al. 1973 198,411,422,423,433
Johnsson et al. 1979 156,198,265,275
Johnsson 1976 109,111,156,265,275
Jongsma et al. 1975 324,398
Jorne 1977 229
Jouvet et al. 1974 376
Jung and Rothstein 1967 441
Junge and Stevens 1973 319
Jurgen 375
Kaempfer 1727 277
Kalmus and Wigglesworth 1960 137,156,174,200,263,417,422
Kalmus 1935 49,379
Kalmus 1940 92,379,418,424
Karakashian and Schweiger 1976 111
Karfunkel and Seelig 1975 233,234
Karfunkel 1975 240,246,341
Karlsson and Johnsson 1972 411
Karve and Salanki 1964 ....111, 385
Katchalsky and Spangler 1968 169,233
Katholi et al. 1977 209
Katz and Pick 1962 333
Kauffman and Wille 1975 34,173,261,274,319,439,440,448
Kauffman et al. 1978 267,356
Kauffman 1969 173,441
Kauffman 1974 150,173,443,440,448
Kauffman, Stuart 247,448
Kaus 1976 97,190,407
Kaus, Peter 201
Kawato and Suzuki 1978 148,150
Kawato et al. 1979 208
Keller and Segel 1970 343
Keller and Segel 1971a,b 344
Keller 1958 251
Keller 1960 378
Kepler, Johannes 1,11,95
King and Cumming 1972 110
King 1975 396
Kingma and Min 1975 329
Kinosita 1937 330
Klein and Wegmann 1975 376
Klein 1976 341
Klemm and Ninnemann 1976 409,410
Klevecz et al. 1978 440
Knight 1972 171,318
Koehler and Fleissner 1978 202,208,395,396
Kolata 1977 276
Konijn et al. 1967 337
Konopka and Benzer 1971 424
Konopka 1972 111,395
Kopell and Howard 1973a 124,125,213,220,229,305
Kopell and Howard 1973b 210,227,229,306
Kopell and Howard 1974 229
Koros and Orban 1978 300
Kraepelin and Franck 1973 370
Kraft et al. 1970 376
Kramm 1973 147
Krasnow, Richard 2,168
Krinskii and Kholopov 1967a 270,329
Krinskii and Kholopov 1967b 270
Krinskii et al. 1967 270
Krinskii et al. 1972 208
Krinskii 1978 340
Krinskii 1966 249,270,331
Krinskii 1968 171,240,242,270
Krinskii 1973 340
Kuffler and Nicholls 1976 316
Kuhnert and Linde 1977 300
Kuramoto and Tsuzuki 1976 230
Kuramoto and Yamada 1975 229
Kuramoto and Yamada 1976 83,85,129,229

Kuramoto 1975 ....115,118,120,147,210,**229**
Kuramoto 1978 228,230,271
Kuriyama and Suzuki 1976 319
Lacey 1975 455
Lafontaine et al. 1967 376
Lanzavecchia 1977 269
Laszlo, Andre 248
Lavenda et al. 1971 169
Lawrence et al. 1972 249,267
Lawrence 1970 249,267
Lawrence 1971 249,267
Leao 1944 335
Lederberg, Joshua 5
Lees 1972 389
Leigh 1965 443
Leutscher-Hazelhoff and Kuiper 1966 ..317
Levin 1978 150
Lickey et al. 1976 395
Licko and Landahl 1971 209
Linkens and Datardina 1977 208
Linkens et al. 1976 169,170,229
Linkens 1974 211
Linkens 1976 114,208
Linkens 1977 114,208
Livolant et al. 1978 365
Llanos and Nash 1970 442
Locke 1960 249
Lotka 1920 135
Luce 1971 376
Lukat 1978 362
MacGregor and Lewis 1977 316,331
MacKay 1978 339,340
Maden 1977 356
Magellan, Ferdinand 11
Maier 1973 380
Malchow et al. 1978 111,119,198,212,282,297,343
Mano 1975 381,440
Marek and Stuchl 1975 208
Martins-Ferreira et al. 1974 240,336
Mathieu and Roberge 1971 319
Maxwell, James Clerk 41,71,131,145,359
May 1973 287,391
Mayer and Sadleder 1972 396
Mayer 1908, 1914 330
Mayeri 1973 318
McAllister et al. 1975 171
McDonald and Sachs 1975 319
McKean 1970 129,234
McWilliam 1887 67
Medugorac and Lindauer 1967 396
Menaker and Zimmerman 1976 379
Menaker et al. 1978 377
Mercer 1965 97,137
Mergenhagen and Schweiger 1974 399
Mergenhagen and Schweiger 1975a 395,397
Michelson 1911 364,365
Miles et al. 1977 376
Mills et al. 1977 376
Mills 1973 376
Milnor 1965 28
Mines 1913 330
Mines 1914 271,330
Minis and Pittendrigh 1968 402
Minor and Smith 1974 443
Minorsky 1962 133,169,208,234
Misner et al. 1973 261
Mitchison 1971 439,440
Mitchison, Graeme 28,448
Mittenthal, Jay 126,211,360
Mobile medical team 1976...........348
Moe and Abildskov 1959 270
Moe et al. 1964 240,270,331,335
Moe 1962 270
Moe, Gordon 271,272,332

Moller et al. 1974 441
Moore and Eichler 1976 458
Moore et al. 1963 172
Moore-Ede et al. 1976 212,376,394
Moore-Ede et al. 1977 376
Moore-Ede et al. 1978 376
Morgan, Thomas Hunt 424
Morin et al. 1977a 397,458
Morin et al. 1977b 458
Morse and Feshbach 1953 71
Moshkov et al. 1966 .... 137, 200, 422
Moss et al. 1977 450
Mrosovsky 1970 91
Mueller and Arnett 1976 242
Murray 1975 229
Murray 1976a 219
Murray 1976b 229
Murray 1977 126,229
Nadeau and Roberge 1969 209
Nagumo et al. 1963 240,331
Nanjundiah 1973 344
Narlikar 1970 71
Nayar 1968 110
Nazarea 1974 229
Nelsen and Becker 1968 170,208,229
Neu 1979a 127,147,208,230
Neu 1979b 114,147,230
Neu 1979c 115,118,147,210,230
Neumann 1966 402
Neville 1975 365
Newman 1956 20,71
Newton, Isaac 31,412
Nickerson 1944 368
Nicolis and Portnow 1973 220,286
Nicolis and Prigogine 1977 200,211,228,230
Nishisutsujii and Pittendrigh 1968....395
Nitzan et al. 1974 129,210
Njus et al. 1974 380
Njus 1975 398,420
Novak and Seelig 1976 229
Nudelman and Glantz 1977 317
Oatley and Goodwin 1971 391
Offner et al. 1940 233,234,316
Olsen and Degn 1977 303
Orban and Koros 1978 300
Ortoleva and Ross 1973 127,229,312
Ortoleva and Ross 1974 229
Ortoleva and Ross 1975 234
Ortoleva 1976 230
Ostriker 1971 242
Ostwald 1900 331
Othmer and Aldridge 1978 297
Othmer 1975 233,306
Othmer 1976b 226
Othmer 1977 227
Ottesen 1965 34,118
Page et al. 1977 202,208,395,397,399
Page 1978 208,395
Palmer and Jousset 1975a,b 456
Palmer 1976 45,93
Palmer 1977 376
Parker-Rhodes 1955 243
Pastelin et al. 1978 330
Patton and Linkens 1978 229,304,326
Pavlidis and Kauzmann 1969 156,379-381,423
Pavlidis 1967a 137,172,188,199,200,262,379,422
Pavlidis 1967b 137,188,199,200,261,262,409,422
Pavlidis 1968 262,422,423
Pavlidis 1969 211,378
Pavlidis 1971 117,281,387,397
Pavlidis 1973 34,117,132,150,199,211,281,397
Pavlidis 1975 230
Pavlidis 1976 208,262,263,396
Pavlidis 1978a 262,410,421-423

Pavlidis 1978b 208
Pavlidis, Theodosios 211
Pearson and McLaren 1977 276
Perkel et al. 1964 .... 111,118,164,170,172,317,318,322,**323**
Peron, Madame 17
Peskin 1975 119,171,208,318
Peterson and Jones 1979 161
Petsche and Rappelsberger 1970 213
Petsche and Sterc 1968 335
Petsche et al. 1970 213,335
Petsche et al. 1974 335
Pigafetta (navigator) 11
Pinsker 1977 34,111,150,171,172,318,319,322
Pinsker 1979 177
Pittendrigh and Bruce 1957 32,93,94,172,209,378,380
                                     387,406,417,419,422
Pittendrigh and Bruce 1959 49,94,110,384,387,397
Pittendrigh and Daan 1976a 147,390,422
Pittendrigh and Daan 1976b 147,390,422
Pittendrigh and Daan 1976c 147,261,262,390,396,422
Pittendrigh and Minis 1964 118,164,383,406
Pittendrigh and Minis 1971 402
Pittendrigh and Skopik 1970 403
Pittendrigh et al. 1958 407
Pittendrigh 1954 380,417,424
Pittendrigh 1958 417,422
Pittendrigh 1960 83,110,396,397,417,422
Pittendrigh 1961 417
Pittendrigh 1966 92,421
Pittendrigh 1967a 396
Pittendrigh 1967b 388
Pittendrigh 1974 209,282
Pittendrigh 1976 421,423
Pittendrigh, Colin 48,50,93,188,375,412-414
Pizarello et al. 1964 376
Plant and Kim 1975,1976 320
Plant 1976 320
Plant 1977 320
Plant 1979 209
Plato 231,258
Platzman 1972 46
Poincare, Henri 146,208
Pollack 1977 320
Polya 1954 91,276
Polya 1957 367
Power 1663 285,401
Presser 1972 453
Proclus 91
Pritchard et al. 1969 443
Pye 1969 289,291,297,378
Pye 1971 286
Pye, E. Kendall 291,303
Quinn, Ronald 414
Rao and Johnson 1970 444
Rao et al. 1974 444
Rapp and Berridge 1977 320
Rasmussen and Zeuthen 1962 173,385,440,442
Rastogi et al. 1977 313
Rawson 1956 93
Rebbi 1979 72
Reinberg and Halberg 1971 376
Reiter and Jones 1975 402
Remmert 1962 390,401
Remy 1968 446
Rescigno et al. 1970 170,318
Reshodko and Bures 1975 240,336
Reshodko 1974 330
Reusser and Field 1979 229,306
Richards 1971 336
Richter 1965 118,400
Rinzel and Keller 1973 233,234
Ritzema Bos 1901 243
Roberts 1962 172
Robertson et al. 1972 266,344

Robinson 1966 13
Rogers and Greenbank 1930 377,393
Roos and Gerisch 1976 340,341
Roos et al. 1977 342
Ross 1976 230,230,344
Rossler and Hoffmann 1972 307
Rossler and Kahlert 1979 228,335
Rossler and Wegmann 1978 303
Rossler 1972b 233
Rossler 1972c 233
Rossler 1974a 233
Rossler 1974b 233
Rossler 1975 173,381
Rossler 1976 208
Rossler 1976-1979 158,444
Rossler 1978 228,234,271,335,444
Rossler, Otto 174,358
Rossomando and Sussman 1973 341
Rozenshtraukh et al. 1970 331
Rusak 1977 396,397,458
Rusch et al. 1966 173,442
Rustad 1970 440
Sachsenmaier et al. 1970 439
Sachsenmaier et al. 1972 173,443,445,446
Sachsenmaier 1976 445
Sachsenmaier, W. 446
Salisbury and Denney 1971 396
Sander and Pardee 1972 441
Sanglier and Nicolis 1976 233
Sano et al. 1978 318,324,325
Sarna and Bowes 1976 330
Sarna and Daniel 1973 329,330
Sarna and Daniel 1974 329
Sarna et al. 1971 229,304,326,329
Sarna et al. 1972a,b 229,304,329
Saunders and Thomson 1977 111
Saunders 1974 209
Saunders 1976a 34,59,111,386,408,418,421,422,425,426
Saunders 1976b 389,402
Saunders 1978 34,58,79,264,379,409,425
Scheving and Pauly 1973 442
Scheving et al. 1968 376
Scheffey and Wille 1978 438
Scheffey, Carl 114,208
Schmitt 1960 378
Schmitz et al. 1977 234
Schmitz et al. 1979 303
Schulman 1969 34,172,317,318,323
Scott 1970 229
Scott 1979 172,318,322,324,325
Skrabal 1915 381
Selfridge 1948 240,249
Sel'Kov and Betz 1968 290
Sel'Kov and Sozinov 1970 442
Sel'Kov 1968a,b 289,290
Sel'Kov 1970 173,442
Sel'Kov 1972 299
Sel'Kov 1973 381,440,442
Shack et al. 1971 455
Shackelford and Feigin 1973 376
Shaffer 1957 337
Shaffer 1962 66
Shaikh and Klaiber 1974 455
Shaw, Napier 290
Shcherbunov et al. 1973 314,332
Sheppard 1968 17,41
Shibata and Bures 1972, 1974 240,336
Shields 1976, 1977 443
Shilo et al. 1976 443
Shinbrot 1966 7
Short 1976 450
Simmons et al. 1974 376
Simon et al. 1976a 395
Simon et al. 1976b 111,395

Sipski and Wagner 1977 365
Skopik and Pittendrigh 1967 402-404,410,411,419
Slack and Savage 1978 358
Slonczewski and Malozemoff 1978 72,365
Smith and Guyton 1961 331
Smith and Martin 1973 443,444
Smoes and Dreitlein 1973 219
Smoes 1976 219
Smoes 1979 305
Sollberger 1976 376
Spangler and Snell 1961 381
Spanier 1966 28
Specht and Bortoff 1972 229,329,330
Stahl 1967 6
Stanshine 1976 229
Stein 1974 24,322
Stern and Goodwin 1977 267
Stern 1967 368
Stetson and Watson-Whitmyre 1976 458
Stetz 1971 455
Stibitz and Rytand 1968 330
Strahm 1964 137,199,200,262,422
Stratonovich 1967 120
Strumwasser 1968 319
Strumwasser 1971 319
Strumwasser 1974 395
Stupfel 1975 376
Sturtevant 1973 377,393
Sudbery and Grant 1976 443
Sulzman et al. 1977a 376
Sulzman et al. 1977b 376,397
Sulzman et al. 1978 212
Suzuki 1976 240,331
Swade and Pittendrigh 1967 397
Swade 1969 81,83,93,147,262
Sweeney and Hastings 1958 442
Sweeney and Hastings 1960 418
Sweeney 1969 110
Sweeney 1974 385
Taddei-ferretti and Cordolla 1976 ...111,318,322
Taguchi and Tabachnick 1974 441
Takimoto and Hamner 1964 209
Taleisnik et al. 1966 450
Tatterson and Hudson 1973 229
Teorell 1971 156,274,318
Terry and Edmunds 1970 397
Tharpe and Folk 1965 395
Thoenes 1973 124,125,129,213,219,220
Thom 1975 261
Thomas 1967 6
Thompson and Hunt 1977 73,344
Thompson 1973 73
Thompson, Francis 74
Thomson 1867 256
Thormar 1959 438
Thurber, James 274
Torre 1976 208
Treloar et al. 1967 452
Trelstad and Coulombre 1971 365
Troy and Field 1977 307
Troy 1977a 229,234
Troy 1977b 229,234,329
Troy 1978 307,329
Truman 1971 402
Truman 1972 402,426
Tuckwell and Miura 1978 335
Turek et al. 1976 379
Turing 1952 228
Twain, Mark 176
Tweedy 1970 408
Tyschenko 1966 209
Tyson and Kauffman 1975 92,173,208,440,447,448
Tyson and Sachsenmaier 1978 173,208,445,446,448
Tyson et al. 1976 264,378

Tyson 1976 219,234,300,302,310
Tyson 1977a 229,234,307
Tyson 1978 303
Underwood 1977 396
Urquhardt 1973 376
Van der Pol and van der Mark 1928 169,208
Van der Pol 1926 153,170
Varadi and Beck 1975 124,213,220
Vavilin et al. 1968 170,213
Verzeano 1963 335
Vince et al. 1971 426
Volterra 1926 135,137
Von Campenhausen 1969 17
Wagner and Cumming 1970 385
Wagner et al. 1975 286,385
Wagner 1976 286,385
Wahl 1932 379
Walker 1969 119,318,322,323
Watanabe et al. 1967 325,398
Watson 1976 92
Wegmann and Rossler 1978 176,303
Wei 1962 166
Weitzman 1976 376
Wever 1962 201,422
Wever 1963 84,201,411,422
Wever 1964 83,188,201,422
Wever 1965a 84,188,201
Wever 1973 376
Wever 1975 376,397
Whitehead, Alfred North 358
Wiedenmann 1977 111,275,362,383
Wiener and Rosenblueth 1946 130,171,242,245,246,249,333,336
Wigner, Eugene 308
Wilhelm and van der Werff 1977 229
Wilhelm 1966 121
Wilhelm et al. 1968 121
Wilkins and Holowinsky 1965 395
Wilkins 1963 13
Wilkins 1965 204
Wille et al. 1977 448
Wille 1979 442
Wilson and Wachtel 1974 320
Winfree 1942 453
Winfree 1965 117,397
Winfree 1967a 81,93,115,117,118,147,281,397,399
Winfree 1967b 422
Winfree 1968 34,201,261,421,430
Winfree 1970a 215,230,266
Winfree 1970b 49,64,123,264,421
Winfree 1970c 49,64,418,419
Winfree 1971a 64
Winfree 1971b 283
Winfree 1972a 408
Winfree 1972b 201,415,419
Winfree 1972c 213,245,305,307,329
Winfree 1972d 109,265
Winfree 1972e 200,201,419,423
Winfree 1973a 108,111,177,180,201,423
Winfree 1973c 171,228,234,327
Winfree 1973d 215,230,266,367
Winfree 1974a 197,201,421,423
Winfree 1974b 225,233,234,246,308,327,341
Winfree 1974c 150,208,299,443
Winfree 1974d 236,254,305,307
Winfree 1974e 307,327
Winfree 1974f 225,233,234,246,308,341
Winfree 1975c 410,413
Winfree 1976 208,433
Winfree 1977 111,274,323
Winfree 1978a 225,246,273,302
Winfree 1978b 150,208,299,443
Winfree and Gordon 1977 111,422,425
Winfree, Dorothy 453,454
Wisnieski and Fox 1976 380
Wojtowicz 1972 288

Wolken 1971 362,364
Wolpert 1969 346
Wongiwat et al. 1972 376
Yakhno 1975 234,314
Yamada and Kuramoto 1976 227,230,246
Yamakazi et al. 1978 303
Yamakazi et al. 1979 303
Yamanishi et al. 1979 111,318
Yen and Tsai 1972 455
Yoshimoto and Kamiya 1978 229
Yoshizawa et al. 1971 240
Young 1977 17
Young 1978 303,376
Young, Thomas 41
Zahler and Sussmann 1977 276
Zaikin and Kawczynski 1977 234,314
Zaikin and Zhabotinsky 1970 220,229,248,284,300,304
Zaikin 1975 233,234
Zeeman 1972 171,234,306
Zeeman 1977 406,444
Zelasny and Neville 1972 362,364
Zeuthen and Williams 1969 443
Zhabotinsky and Zaikin 1973 234,240,314
Zhabotinsky 1968 220,271,305
Zhabotinsky 1970 240
Zhabotinsky 1974 300
Zhabotinsky, A. M. 332
Zimmer 1962 110,384,387,427-430
Zimmerman and Goldsmith 1971 424
Zimmerman and Menaker 1975 379
Zimmerman et al. 1968 380
Zimmerman 1969 64,201,402,417-421,434
Zimmerman 1971 402
Zimmerman, William 380,386,418
Zwanzig 1976 115,208,230,234
Zweig 1938 11

Index of Subjects

Acetabularia, 111, 395, 397, 399
Acetaldehyde, 60, 118, 265, 288, 295-298
Acrasiales. See Dictyostelium
Action potential, 70, 110, 315, 316
Active scattering. See Mutual repulsion
Adenosine phosphates, 212, 256, 285-299 (Chapter 12), 341, 342. See also
    Cyclic AMP
Advance vs. delay ambiguity, 385-387
Advance vs. advancing, 49, 386
Aether, 256, 359
Aggregate: amplitude, 98, 206; phase, 96, 98, 206; rhythmicity, 84, 96, 97,
    206, 323, 398; rhythmicity, defined, 95-97. See also Mutual entrainment
Alternating flows, method of, 137, 164
Amoeba, social. See Dictyostelium
Amphidromic point, defined, 46, 260
Amplitude of fundamental, zero, 98, 102, 106, 133, 260
Amplitude resetting: in Avena, 198; in clock models, 83, 84, 95-106, 112,
    131-144 (Chapter 5), 262, 263, 398, 403-406; in Dictyostelium, 198, 343;
    in Drosophila, 176-204, 208, 262, 263, 406, 423, 434; in Kalanchoe, 198,
    423, 427, 428, 433, 434; in yeast, 198, 293
Angular velocity: in oscillator models, 79-94, 147; modulation of, 81, 85,
    143, 163; negative, 81, 82, 87, 99
Animal navigation, 20
Antiphase, 114, 116, 208, 281, 282
Aplysia, 292, 318, 395, 396
Archimedes' spiral, 64, 244, 245, 369
Ascobolus immersus, 369
Ascomycete fungi (also see by species name), 246-250, 266, 367-374 (Chapter
    18)
Assay pulse, defined, 179
Attracting cycle: deficiencies of models, 174; defined, 146; knotted, 456;
    mutual entrainment, 207-212; none observed in Drosophila clock, 189
Attracting steady-state of oscillator, 156, 172, 174, 198, 263, 265, 274,
    275, 296, 455
Attractor basin, 146, 157, 275, 296, 455
Attractor-cycle oscillator. See Attracting cycle
Attractor-repellor pair, 80, 81, 87, 99, 100, 233
Autophasing, 93
Avena sativa, 109, 156, 198, 265, 275
Azimuth, defined (fungus), 123
Basidiomycete fungi, 243
Belousov-Zhabotinsky reagent. See Malonic acid reagent
Benham's top, 16
Bilateral symmetry. See Symmetry
Black hole, 72, 157, 261
Blastema, 267, 348, 353, 354
Bloch points, 72
Brain: electrical waves in, 213, 240, 256, 270-272, 335, 336; insect, 202,
    208, 402, 403
Breakpoint. See Phase jump, unreal
Butterfield color encoder, 17, 433

Calcium, 60, 79, 295, 296, 316, 320
Cap (bounded by cycle), 68-70, 157, 158, 161, 162, 225, 235-242, 267, 273
Cell division. See Mitotic cycle
Cell wall architecture, 365
Central disk. See Rotor's core
Centroid, 183, 412
Chaetomium robusta, 64, 222, 224, 370
Chaos, 158, 174, 208, 228, 271, 272, 303, 443, 444
Chemical oscillator. See Dictyostelium, Malonic acid, Glycolysis
Chemical waves. See Waves
Chitin fibrils, 269, 361-366 (Chapter 17)
Chloride ions, in Merck ferroin, 301
Chronon, 92, 262, 441, 442
Circadian clock: and estrous cycle, 451, 457, 458; and mitosis, 441-443; as
    relaxation oscillator, 172, 173, 283, 385, 389; at the South Pole, 45;
    dark adaptation, 188, 201, 403-406; evolution, 387-392, 399; general
    characteristics, 377, 378; in development, 64, 196, 264, 402-407, 411,
    420, 421, 426; in flies. See Drosophila, Sarcophaga; in fungi. See
    Daldinia, Neurospora, Pilobolus; in Gonyaulax, 21, 204, 385, 398, 409,
    420; in insect epidermis, 361, 362, 397; in medicine, 376, 442; in plants.
    See Gonyaulax, Kalanchoe; in procaryotes, 377, 390, 393; in rodents, 86,
    93, 117, 209, 282, 396, 397; inhibition by constant light, 197, 378,
    404-406, 421; initiation, 64, 196, 200, 201, 264, 265, 434; mechanism
    (conjectures), 378, 379, 392, 393, 417; models and dynamics, 172-174,
    199-204, 208, 209, 261-264, 277-284 (Chapter 11); mutual entrainment,
    263, 396-400; not compounded of quicker clocks, 378; period unaffected
    by amplitude, 201; photoreceptor, 113, 188, 201, 384, 408-410;
    physiologically essential?, 417-421; probable diversity of mechanisms,
    379, 392, 393; splitting of the rhythm. See Splitting; temperature
    effects on period, 172, 173, 377-382, 424; temperature effects on phase,
    60, 408, 418, 433; trajectories, 176-204 (Chapter 7) transient after
    inhibition, 186-189, 201; two-oscillator models, 94, 117, 202, 208, 209,
    282, 396, 397, 406; unstable-amplitude model, 200, 201
Circadian, defined, 375
Circular logic, defined, 25
Circulating wave. See Circus wave, Waves on rings
Circus wave, 67, 240, 270, 271, 333-336
Clockface model, 65, 250, 266-268, 345-360 (Chapter 16)
Clockshop model, 84, 174, 203, 263, 281-284, 394-400. See also Incoherence
Collective. See Aggregate
Color vision: in circadian clock, 196, 410; in humans, 10, 15, 16, 40-42,
    259
Color wheel, 42, 259
Complete circle rule, 347
Composite clock. See Clockshop
Composition vs. concentration, 248, 249
Continuity, defined, 7
Continuous action, 422
Contour line, defined, 8
Convolution, inverting, 420
Cophase, defined, 34, 141, 149, 322, 438
Cotidal contour, 46-48, 260
Crickets, 119, 318, 322, 323
Critical point, 115
Crystal, fibrous, 13, 14, 25, 269, 364-366
Cs, defined, 18, 138
Cuticle of arthropods, 65, 268, 269, 361-366 (Chapter 17)
Cyclic AMP (not in Dictyostelium), 290, 320; See also Dictyostelium
Daldinia concentrica, 369
Dark adaptation in circadian clocks, 188, 201, 403-406, 409, 415
Dark-reared insects. See Naive
Degrees of freedom, more than two, 132, 133, 157-159, 171, 201, 234, 260-
    263, 297, 430
Delay, defined, 386, 387
Delay vs. delaying, 49, 386
Developmental: fields, 65, 345-360 (Chapter 16); readiness, 200, 403-406
Dictyostelium discoideum, 66, 172, 233, 240, 246, 252, 269, 270, 281, 332,
    337-344 (Chapter 15); mutual entrainment, 119, 343; type 0 resetting,
    172, 274, 343
Differential action, 422
Diffusion, molecular, coupling by, 121, 205-257 (Chapters 8 and 9), 297-
    299, 302, 306

Diffusive instability, 211, 220, 228, 306, 308, 343, 344
Dimension. See Degrees of freedom
Direction maps, topological type, 20
Directional derivative, 113
Discrete state device. See Logical automata
Dispersion. See Incoherence
Dissipative structure. See Diffusive instability
Distal transformation, 347
Division protein. See Mitogen
DNA, circular, 6, 92, 262
DNA, helicoidally packed, 365
Dot convention for d/dt, 31, 76, 98
Double periodicity, 36-38
Dove prism, 22, 23
Drosophila melanogaster: clock ,54-58, 264, 379, 380, 424, 425; imaginal
    disks, 345-360 (Chapter 16), 402
Drosophila pseudoobscura clock: amplitude resetting, 176-204, 208, 262,
    263, 406, 423, 434; arrhythmicity, 261-264, 417-421; dark adaptation,
    188, 201, 408-410; isochrons, 191-195, 416; gating, 403-406; phase
    resetting curves, 108, 176-204 (Chapter 7), 383, 416; pinwheel
    experiment, 108, 180-196, 383, 416, 419; temperature effects, 60, 380,
    408, 418; time crystal, 51, 109, 414-416; time machine experiments and
    trajectories, 176-204 (Chapter 7); transients (central), 185-189, 201;
    transients (peripheral), 94, 97, 406-408; two-oscillator
    interpretations, 94, 202, 208, 209, 396, 406; variance, 113, 398
Dynamics, defined, 76
D2, defined, 5
Eclosion, defined, 402
Ecological cycles, 135, 229
Eikonal equation, 251
Elasticity of image (symmetric diffusion), 127, 225-227
Emergence (eclosion), 178-180, 401-406
Entrainment: defined, 114, 399, 400; mutual. See Mutual entrainment;
    mutual, active resistance to, 116-118, 281, 324, 399, 400; not mutual,
    93, 143, 164, 201, 318, 322, 323, 455
Enzyme kinetics, 287-290, 381
Epilepsy, 171, 336
Epimorphosis, defined, 355
Equations: gating, 406; image elasticity, 226; involute, 245; isochrons,
    150-157; oscillator resetting, 137-140· simple clock, 85; resetting
    surface, 415; wave broom, 121; waves on rings, 127-129
Equilibrium. See Steady-state
Estrous cycle: 209, 450-458 (Chapter 23); and circadian rhythms, 451, 457,
    458; no type 0 resetting, 70, 110, 455-457
Evolution: of circadian clocks, 387-392, 399; of female cycles, 452
Excess delay, 437-439
Excitability, 172, 231-257 (Chapter 9)
Excitable media: continuous vs. cellular, 339, 340; models, 130, 231-257
    (Chapter 9), 277-284, 295. See also Brain, Dictyostelium, Glycolysis, Heart,
    Malonic acid reageent, Retina
Excitable membranes, 68, 110, 153, 171, 208, 270, 271, 274, 307, 308, 315-
    336 (Chapter 14), 340
Excitable oscillator, 231-233, 246, 247, 326-329
Explanations as mappings, 258
Eye. See Insect or Retina
Fairy ring, 243
Fate map, 351
Female cycle. See Estrous or Menstrual
Fibrillation, 67, 240, 270-272, 330-335
Final phase, defined, 87, 438
Firefly: cuticle, 362-364; flashing, 119, 120, 277, 279, 323, 325, 398
Flutter, 67, 240, 270-272, 330-335
Frontier, defined, 122
Fruitfly. See Drosophila
Fundamental sine wave, 96, 97, 206, 260, 272
Gating, 403-406
Genome, 6, 91, 92, 262, 241
Geometrical optics, 251
Geophysical rhythms, 93
Germination of spores, 222, 224
Glycolysis (anaerobic oscillations): amplitude resetting, 196, 293;
    excitability, 233, 295; mutual entrainment, 60-63, 212, 297-299, 398;

oscillator models, 156, 208, 209, 288-290; phase compromise, 24, 60-63, 298; phase resetting, 60-63, 109, 291-297; steady-state repellor, 265, 295-297; suppressed by oxygen, 161, 288; time machine experiment, 198

Glycolysis: and neural pacemakers, 320; defined and diagrammed, 285-287; regulation, 288-290

Gonyaulax polyedra, 21, 204, 385, 398, 409, 420

Gradients: circular, 250, 346-348, 362; concentration, crossed, 231, 235-237, 267, 273, 310; morphogen. See Morphogenetic; parameter, 213, 220, 228, 305, 325-330; phase. See Phase gradients; stimulus magnitude, 64, 109, 217, 235, 239, 294; vector cross-product of, 273

Heart muscle: models. See Hodgkin-Huxley; mutual coupling and entrainment, 119, 208, 313, 324, 325, 398; rotor and singularity, 67, 240, 246, 271, 318, 330-335

Helianthus annuus, 188, 385

Helical border, 54, 291, 296

Helicoid. See Screw-like surface

Heterogeneous nuclei. See Pacemaker

Hit and run. See Phase resetting

Hodgkin-Huxley equations, 68, 110, 153, 171, 208, 274, 307, 308, 315-336 (Chapter 14)

Homeostasis: instability , 287, 391, 392, 440; period, 86, 382

Hourglass: 74-94, 389, 443; defined, 78

Hue, 10, 15-17, 26, 42, 259

Hue, combination, 25

Huygens' principle, 245, 251, 308, 336

Hysteresis, 211

Image: of diffusion-coupled media, 124, 125, 127-129, 215, 225-227, 236-241, 248; of embryonic tissues, 215, 351-360; of stimulus plane, 101, 104, 160, 161, 167, 168, 218

Image, elasticity of, 127, 225, 226

Imaginal disks (fruitfly), 345-360 (Chapter 16), 402

Impermeable barrier, 272, 305, 312, 315, 373

Incoherence: circadian clocks, 84, 263, 281, 282, 396-400; computation of, 113; Drosophila, 199-204, 410, 411, 417-421; Kalanchoe, 433, 434; neural tissues, 332-335; simple clocks, 104, 112; yeast cells, 296-298

Influence rhythm: 117, 147, 399; defined, 117

Inhomogeneities, 68, 270, 271, 332, 333, 336, 340, 411

Insect: cuticle, 250; epidermis, 345-360 (Chapter 16), 361-363; eye, 362-364; limbs, 345-360 (Chapter 16)

Instabilities, 207, 287, 320, 323, 356, 391, 440. See also Diffusive

Integrate-and-fire model, 170, 171, 208

Intercalation, rule of, 347

International Date Line, 10-13, 32, 45, 347

Intestine, mammal, 208, 229, 325-329, 381

Involute description, deficiencies, 245, 246, 309, 310

Involute spiral, 242, 245, 246, 251, 308, 405

Iridescence of cuticle, 365

Iron wire, 153, 240, 331

Isochrons: 64, 100, 102, 109, 145-175 (Chapter 9), 218, 219; closed-ring, 168; defined, 13, 98, 148-150, 190, 250; in a mycelium, 215; in malonic acid reagent, 250-252; measured (Drosophila), 191-195, 416; measured (Kalanchoe), 177, 191, 429, 430; measured (yeast), 294; radial vs. spiral vs. ornate, 152-155

I1, I2, I3 defined, 5

Kalanchoe blossfeldiana: 105, 380, 384, 423, 427-434 (Chapter 21); clock transients, 188, 428-432; isochrons measured in, 177, 191, 429, 430; time crystal and singularity, 59, 60, 109, 264, 422, 431-434; type 0 resetting, 427-430, 433

Kinematic wave, 213, 220, 305

Lambda-omega oscillator, 147, 210, 233

Latent phase: 34, 141, 145-175 (Chapter 6), 215, 221, 407, 427, 438; defined, 146-150; list of properties, 158, 159

Left and right brain clocks, 117, 202, 208, 209, 282, 396, 397

Lens (exocone) of arthropod eye, 65, 268, 269, 361-366 (Chapter 17)

Liesegang rings, 249, 250, 368

Life cycle, 275

Limbs, supernumerary, 347-349, 353-355, 359

Limit cycle. See Attracting cycle

Linear filter, 97, 190

Liquid crystal, 269, 364-366

Logical automata, 70, 168, 171, 173, 240, 242, 270, 330, 331, 336, 339, 340, 441, 457

Longitude, 18-20, 43
Malonic acid reagent: and intestine, 329; as relaxation oscillator, 170;
    bibliography, 229; kinds of wave in, 213, 219, 220, 305, 306; mechanism,
    302-304; pacemakers, 312-314; recipes, 301, 302; resetting phase, 170;
    rotors in, 236, 240, 246, 248, 251, 253-256; wave singularity, 68-70,
    124, 125, 272, 273, 308-312
Map: genetic, 6; topological, defined, 6
Mariner's convention, 9, 31
Mathematical metaphors for inner clocks, 200
Meandering of rotor's core, 228, 251, 270-272, 309, 335
Measurement, 29, 75-77
Menstrual cycle: and circadian rhythms, 451, 457, 458; human, 78, 175, 274,
    275, 319, 450-458 (Chapter 23); no type 0 resetting, 70, 110, 455-457
Metric, lack of a, 163, 352
Migraine headache, 336
Minimum size for patterns, 227, 228, 310, 332
Mitochondria, 212, 256, 285, 292
Mitogen, 385, 440-443
Mitotic cycle: 319, 435-449 (Chapter 22); as a relaxation oscillator, 443-
    448; as a simple clock, 78, 91, 92; no type 0 resetting, 70, 110; two
    oscillators, 208, 209; unsmooth, 173, 274
Modulation of angular velocity: 81, 85, 143, 163, 320; inadequacy, 136
Moire, 113
Morphallaxis, defined, 355, 356
Morphogenesis. See Cuticle, Dictyostelium, Nectria, Neurospora, or
    Regeneration
Morphogenetic gradient, 249, 250, 267, 346, 352, 361
Multi-oscillator. See Clockshop
Muscle fibers, helically striated, 269
Mutual entrainment: actively resisted, 116-118, 281, 324, 399, 400; among
    slime mold cells, 119, 343; among simple clocks, 112-121; among
    relaxation oscillators, 210, 277-284 (Chapter 11); among yeast cells,
    60-63, 297-299, 398; among attractor-cycle oscillators, 207-212; among
    circadian clocks, 263, 396-400; among neural pacemakers, 119, 120, 277,
    279, 323-329, 398; critical point.115, 210, 281, 398
Mutual repulsion among clocks. See Mutual entrainment actively resisted
Mycelium, defined, 367
Myxomycetes. See Physarum
Naive flies, 64, 196, 201, 264, 418-421, 434
Native period distribution, 115-120, 210, 281, 325, 326, 398
Nectria cinnabarina: 64, 121-124, 215-230, 266, 344, 370-374; probably mis-
    named, 370
Negative resistance, 277, 278, 320
Neon glow tube, 117, 169, 277-284 (Chapter 11), 304
Nerve impulse, 70, 110, 315, 316
Neural rhythms. See Pacemaker neuron
Neurospora crassa circadian clock, 63, 64, 121-124, 213-230, 266, 368, 369
New phase, defined, 31, 412
Nicotinamide adenine dinucleotide (NADH), 60-62, 265, 285-299 (Chater 12)
Notational change, 180
Observables, 29, 72, 75-77, 176-179
Observation window, defined, 179
Old phase, defined, 31, 412
Organizer, 273
Organizing waves, 66
Oscillation and reaction complexity, 287, 391, 392, 440
Oscillators: chemical. See Chemical oscillator; enumerating, 104, 205, 206;
    excitable, 233; independent, 95-112, 121-125, 205-207, 212-224; long
    period, 378; relaxation. See Relaxation oscillator; verbal
    rationalization of, 289
Ovulation, 78, 175, 319
Pacemaker: in gastro-intestinal tract, 330; in heart muscle (ectopic
    focus), 313; in malonic acid reaction, 310-314; in slime mold, 66, 313;
    potential, 319, 325
Pacemaker neuron: models, 153-156; mutual entrainment, 119, 120, 208, 209,
    277, 279, 323-329, 398; phase resetting, 24, 110, 111, 317-325;
    unsmooth, 70, 170-172, 274
Parameter, defined, 76
Parameters, 79-82, 132, 228
Parametric pumping, 121
Pasteur effect, 60, 161, 288, 289, 292, 294
Pattern formation, minimum size for, 227, 228, 310, 332
Penicillium diversum, 64, 222, 369, 370
Periodism. See Rhythmicity

Peristalsis. See Intestine
Phase compatibility, 116-119, 208, 281
Phase compromise, 24-28, 61-63, 123, 162, 163, 208, 298, 299, 444-448
Phase dispersion. See Incoherence
Phase distribution, two-horned, 419, 420
Phase gradients: 48, 109, 235. 294; in cuticle, 364; in fungi, 121-125,
    211-224, 370; in limbs, 250, 346, 347; in malonic acid reagent, 121-125,
    303-306, 312
Phase jump, real, 98, 101, 102, 163, 172, 173, 318, 383-385, 438, 439 445-
    448, 456, 457
Phase jump, unreal, 107, 173, 322, 346, 383-385, 427-430
Phase map , defined, 13
Phase reference event, 33, 35, 321, 412
Phase resetting: by acetaldehyde, 60, 111, 118, 265, 288, 295-298; by ADP-AMP
    212, 291; by Cyclic AMP, 172, 274, 290, 342, 343; by divalent ions, 265,
    295, 296; by electrical impulses, 24, 109, 111, 274, 323-325; by female
    hormones, 455-457; by fusion with another oscillator. See Phase
    compromise; by light. See Avena, Circadian rhythm and species; by
    mechanical force, 134-137, 188, 385; by osmotic potential, 109, 156,
    265, 275; by oxygen, 60-63, 161, 265, 287-289, 292-296; by pyruvate,
    296, 298; by repeated stimuli, 164. 176-204 (Chapter 7), 318. See also
    Entrainment, Phase 'compromise; by temperature change, 60, 408, 418, 433;
    curve (Drosophila), 108, 176-204 (Chapter 7), 383, 416; curve
    (diagrams). See Phase resetting curve; curve, negative slope, 83, 85,
    89, 132, 133, 163, 439, 440; data formats, 32-39, 110, 180, 182, 185,
    186, 317, 320-323; in glycolytic oscillator, 60-63, 265, 290-296; in
    malonic acid reaction, 170, 312; only types 1 and 0 found, 38, 39, 112,
    164-168; surface, defined, 89, 414-416; types defined, 14, 37, 38, 107,
    322, 323; vs. phase shifts, 383-387. See also Phase resetting data
    formats
Phase resetting curves (diagrams): Dictyostelium, 343; Drosophila, 108,
    180-196, 383, 416, 419; female, 456; Kalanchoe, 428, 430; models, 32-34,
    82-89, 142, 143, 408; Physarum, 438; reference list, 110, 111
Phase response curve (PRC). See Phase shift
Phase scatter. See Incoherence
Phase shift: curves (diagrams), 87, 88, 142, 143, 383, 417, 438; defined,
    87, 143, 423; format, 107, 322, 323, 438
Phase singularities, paired, 25-28. 58, 61-63, 162, 298, 299
Phase singularity: at developmental origin of oscillator , 64, 196, 201,
    222, 264, 266, 417-420; at zero amplitude, 103-106, 133, 260; defined,
    40; due to incoherence, 84, 104, 112, 203, 263, 420, 433, 434; existence
    theorem, 25, 26; geographical, 13, 260; impossible in ring device media,
    234; in Avena, 109, 198, 265; in Dictyostelium, 66, 269, 270; in
    Kalanchoe, 59, 60, 109, 264, 422, 431-434; in adjustable compass, 24-27,
    260; in chemical liquid, 68-70, 124, 125, 272, 273; in developmental
    fields, 65, 266, 267, 347, 349, 354; in excitable media, 130, 234, 235,
    238-252; in glycolysis, 60-63, 265, 291-299; in heart muscle, 67, 68,
    240, 270-272, 318, 333-335; in microtomed cuticle, 65, 269, 363-366; in
    mycelia, 215-230, 266; in pacemaker neurons, 109, 318; not observed in
    mitotic cycle, 448; of local time, 43, 260; of maps, 25-30, 258; of
    navigator, 42, 43, 260; of pendulum-like models, 143, 144; of simple
    clock population, 103, 106; of the tides, 46-48, 260, 261; one-
    dimensional, 67-70, 162, 172, 230, 252-256, 272, 310-312, 446; split,
    206, 207; unpaired, 52; vs. state singularity, 221, 222, 246, 267
Phase transition: physics, 115. See also Phase resetting
Phase, defined, 30, 122, 139, 412
Phase, latent. See Latent phase
Phaseless set, 146, 153-155, 216, 222, 265, 274
Phaselessness vs. timelessness, 157, 260, 270
Phosphofructokinase (PFK), 233, 288-292, 320
Photoperiodism, 377
Phycomyces, 82
Physarum polycephalum, 63, 114, 173, 208, 211, 229, 274, 319, 439, 440,
    443, 444-449
Piecewise linear kinetics, 208, 210, 233, 234, 240-242, 273, 341, 342
Pilobolus, 369
Pinwheel experiment: attractor-cycle oscillator, 217-220; defined, 48;
    fluorescent, 294, 295; fly's circadian clock, 48-64, 178, 194-198, 414-
    417; simple clock, 90
Plateaus of frequency, 326-330
Poincare return map, 189
Polarity, 250
Poles, geographical, 12, 43, 45, 63, 260
Polymorphism in conidiation patterns, 216, 220-230, 370-374

' of attractor cycle oscillators, 205-230 (Chapter 8); of
ı oscillators, 210, 277-284 (Chapter 11); of ring devices, 95-
ıpter 4)
.ve plane, 25
.owaves (near-uniform period), 124-127, 212-215, 220, 295, 304-307;
  Winfree errors, 305
Pulsars, 242
Pyruvate, 287, 296-298
QUERIES, 16, 38, 39, 64, 110, 112, 117, 129, 153, 166, 201, 203, 204, 223, 224,
  230, 264, 265, 266, 272, 273, 275, 295, 314, 329, 332, 340, 355-358,
  380, 382, 390, 393, 394, 398, 399, 406, 421, 440, 441, 446, 447, 457
Recipes for malonic acid reagent, 301, 302
Reflex ovulator, 78, 450, 451
Regeneration: 58, 64, 65, 238, 250, 266-268, 273, 345-360 (Chapter 16); vs.
  reduplication, 356
Regulatory kinetics of enzymes, 287-290, 381
Relaxation oscillator: 71, 163, 169-175, 273, 274, 298; circadian clock as
  a, 169-175, 277-284 (Chapter 11), 385, 389; female cycle as a, 457;
  glycolysis as a, 286; mitotic cycle as a, 443, 446-448; mutual
  entrainment, 210, 277-284 (Chapter 11); generalized, defined, 93, 94,
  147
Repelling cycle, 156
Rephasing. See Phase resetting
Rescheduling. See Phase resetting
Resetting curve. See Phase resetting curve
Resetting map. See Phase resetting
Resetting surface, algebraic description, 414-416
Retina, electrical waves in, 240, 336
Reverberator. See Rotor
Rhythmicity: imprecision of, 185, 373, 398, 415, 443, 452-455; precision
  enhanced, 120; selective value of, 391, 392; systematic departures from.
  See Transients
Ring devices, 74-94 (Chapter 3), 171, 247-250, 261, 436-440; defined, 7
Ring, defined, 4
Rotating waves: discontinuous in discontinuous models, 171, 338-340; not
  rotors, 50-52, 60, 109, 130, 218, 294, 295, 416
Rotor; anatomy, 235-240; chemical, 66-70, 73, 124, 125, 272, 273, 308-312;
  defined, 235; impossible using ring devices, 247-250; in Dictyostelium,
  295, 338, 342; in brain tissue, 335, 336; in heart muscle, 67, 270, 330-
  335; multi-armed, 340; spontaneous generation , 270-272, 331, 332, 340;
  temperature coefficients, 246; three-dimensional, 250-257
Rotor core: 66, 231, 235-252, 309, 310; defined, 235, 273; meandering, 228, 251,
  270-272, 309, 335
Rowboat, 18, 42, 52, 260
Rubber sheet geometry, 163
R1, R2, R3, defined, 5
Saccharomyces carlsbergensis. See Yeast
Salisbury cathedral, 244
Sarcophaga circadian rhythm, 58, 59, 264, 402, 409, 421, 422, 425, 426
Scatter. See Incoherence
Screw-like surface: and type 0 resetting, 53, 109; in cuticle, 66, 365,
  366; in Drosophila clock, 51, 57, 414-417, 425; in excitable media, 254;
  in glycolysis, 61, 291, 295; in Kalanchoe, 59, 431-433; in model
  oscillators, 141, 161, 162; in neural pacemakers, 318; in Sarcophaga,
  58, 426
Scroll ring wave: 251-257, 310-312; knotted, 254, 272; twisted, 253-256
Self-sustaining vs. self-exciting, 156
Sensitivity rhythm, 117, 147, 399; defined, 117
Separatrix, 154, 172
Sequential transcription. See Chronon
Simple clock: 74-130 (Chapters 3 and 4); and homeostasis of period, 86,
  382; circadian clock as, 391; defined, 78, 147; Drosophila clock not,
  421-423; incoherence, 104, 112; mitotic oscillator as, 78, 91, 92, 436-
  440; mutual entrainment, 112-121; Neurospora clock not, 224; phase
  resetting curves, 82-90, 143; population (phase singularity), 103, 106;
  relaxation oscillator, 169; three, 102-105; two, 97-101
Single-factor description, 267, 268
Single-factor dynamics, 248-250. See also Ring device, Relaxation
  oscillator
Singularity: defined, 71-73, 261, 262; critical point, 115, 210, 281, 398;
  in a magnetic bubble, 72, 365; mere coordinate, 72, 260; of a map, 14,
  77, 261, 365; trap, 54, 55, 275, 414;
Slime mold: cellular. See Dictyostelium; true. See Physarum
Slow waves, 319, 325

Smooth, defined, 7
Smoothness, 108, 164, 168–175, 273–275, 349, 457
Sn, defined, 18, 138
Social amoeba. See Dictyostelium
Solar wind, 242
Space defined: composition, 8; source, 6; target, 7; topological, 4
Spike referenced, 317, 322, 323
Spiral locus: of conidiation, 63, 64, 123, 215–230, 247–250, 266, 369–374;
    of electron density, 65, 66, 363–366
Spiral wave, 66, 236, 242–250, 308, 330–336, 338, 342; multi-armed, 250,
    370
Splitting of rhythms, 105, 112, 113, 117, 202, 211, 281, 282, 323, 324,
    396, 397, 458
Spores. See Conidiation
Spreading depression, 68, 171, 246, 335, 336
Star-formation, 242
State space, 29, 132
State, defined, 74
Stationary state. See Steady-state
Steady-state, 73, 153, 157, 162, 215, 225, 233, 235, 292, 343, 344, 391;
    attracting, 156, 172, 174, 198, 263, 265, 274, 275, 296, 455; of
    circadian clock, 261–265. See also Diffusive, Homeostasis, Phase
    singularity at developmental and at zero amplitude
Stereo viewing, instructions, 293
Stimulus: as a mapping, 100, 104–106, 135–140, 160, 161, 166–168, 191–195,
    212–223; latency, 34, 35, 141, 322; plane (diagrams), 50, 51, 53–55, 62,
    90, 101, 105, 142, 165, 195, 206, 219; referenced, 317, 322, 323, 431
Stomach wall, 329, 330
Stopping the clock, first anticipated, 422
Stopwatch, 81
Strange attractors. See Chaos
Subthreshold rhythmicity. See Pacemaker potential
Supernovae, 242
Suprachiasmatic nucleus, 451, 458
Symmetry, mirror: in circadian clocks, 202, 208, 396, 397; in development,
    266–268, 351, 357–359. See also Phase compromise
Synchronization. See Entrainment
Synchrony, defined, 114
System, defined, 74
S1, S2, defined, 5
Tachycardia, 67, 240, 270–272, 330–335
Television, 17
Temperature: dependence of reaction-diffusion systems, 246; gradients, 125,
    305, 327–329; Winfree, defined, 72; -compensated reaction rates, 380–
    382; -sensitivity of circadian period, 173, 377–382, 424; -sensitivity
    of circadian phase, 60, 403, 418, 433
The Geometry of Biological Time, 259
Theorem, only: applications, 28, 53, 57, 72, 102, 103, 124, 125, 129, 157,
    161, 218–220, 249, 255, 269, 291, 298, 310, 334, 336
Three simple clocks, 102–105
Threshold (not in relaxation oscillator): 139, 140, 148, 310, 319, 326,
    404–406; for mutual synchronization, 115, 210, 281, 398; of
    excitability, 233, 234, 308, 319, 326. See also Relaxation oscillator
Thyroid follicles, 118
Tides, 44–48, 260, 344
Time crystal: Avena, 109; Drosophila melanogaster, 56, 57, 109; Drosophila
    pseudoobscura, 51, 109, 414–416; glycolysis, 109, 292–297; Kalanchoe,
    59, 60, 109, 431–433; non-simple-clock models, 141–144; pacemaker
    neurons, 109; Sarcophaga, 58, 426; simple clocks, 89, 90
Time delay, 158, 262
Time machine, 177, 178, 413
Time machine experiment: defined, 178–180; Drosophila, 176–204 (Chapter 7);
    Kalanchoe, 198; yeast, 198
Time, one-dimensional sense of, 131
Timelessness vs. phaselessness, 157
Tissue specificity: 346; space, 350–360
Tn, defined, 18, 138
Tornado, 72, 260, 272
Toroidal coordinates for maps: from ring to ring, 6, 15–28, e.g. phase
    resetting, 33, 37, 38, 88, 107; from a product of rings, e.g. phase
    compromise, 62, 446–448, e.g. two simple clocks, 97–101, 114
Transients: in Drosophila (central), 185–189, 201; in Drosophila
    (peripheral), 94, 97, 384, 406–408; in female cycle, 454; in Kalanchoe,
    428–432; in mitotic cycle, 438, 440; only in multicellulars, 394

ation. See Avena

r waves (near-uniform speed), 126, 127, 220, 234, 305-307

ulence, 228, 271, 272, 335. See also Chaos

o attractor-cycle oscillators, 206-209

Two simple clocks, 97-102, 112-116

Two-oscillator model: circadian clocks (A-B), 94, 209, 406; circadian
  clocks (left-right), 117, 202, 208, 209, 282, 396, 397; glycolysis, 290;
  mitotic cycle, 209, 440, 441

Two-pulse resetting, 164, 176-204 (Chapter 7), 318

Type 0 resetting: aggregate rhythm, 105; bibliography, 110-112; defined,
  37, 38; Dictyostelium, 172, 274, 343; Drosophila, 52, 108, 182-196;
  glycolysis, 60, 291, 292, 296; impossible in models, 171, 391;
  Kalanchoe, 427-430, 433; Neurospora, 368; not found, 70, 110, 448;
  reality, 107-109, 385, 427-430, 457; Sarcophaga, 59; sheet of
  oscillators, 217, 266

Type 1 resetting, defined, 37, 38, 107

Ultraviolet light, 170, 213, 303, 312

Van der Pol oscillator, 114, 116, 208, 211, 326, 330

Variability. See Rhythmicity, Imprecision

Variables of state, 74-77, 132

Vortex atoms, 256, 257

Wave: broom, 121, 314; chemical, in Dictyostelium, 66, 240, 246, 337-339,
  343, 344; chemical, in malonic acid reagent, 66-70, 73, 124, 219, 220,
  272, 273, 295, 304-312; colliding, 311-313, 344; creation of endpoint,
  236-238; eclosion, 49, 178, 416; fertilization, 246; front, 218, 219,
  249, 333; on rings, 22, 126-129, 227, 330; organizing, 66; rotating
  about a hole, 130, 236, 237, 242, 330, 335, 336, 339, 340; rotating (not
  rotor), 50-52, 60, 109, 130, 218, 294, 295, 416; shearing, 236-239, 313;
  temperature dependence of speed, 246

Winding number: conservation of, 127, 215; defined, 14, 15; direction maps,
  18, 20; fungus polymorphism, 63, 64, 121-124, 215-230; hue maps, 17, 42;
  limb handedness, 266, 347, 348, 354; multiple resetting, 164; non-zero.
  See Theorem; optical maps, 22; spiral waves, 68-70, 247-250; three-
  oscillator singularity, 103

Yeast mitotic cycle, 436, 437, 440. See also Glycolysis

Zeitgeber. See Entrainment, not mutual

Zhabotinsky reagent. See Malonic acid reagent